Y0-BTA-473

COMBINATORIAL SET THEORY: PARTITION RELATIONS FOR CARDINALS

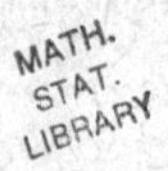

STUDIES IN LOGIC

AND

THE FOUNDATIONS OF MATHEMATICS

VOLUME 106

NORTH-HOLLAND PUBLISHING COMPANY
AMSTERDAM • NEW YORK • OXFORD

STUDIES IN LOGIC AND THE FOUNDATIONS OF MATHEMATICS

COMBINATORIAL SET THEORY: PARTITION RELATIONS FOR CARDINALS

PAUL ERDŐS,
ANDRÁS HAJNAL,
ATTILA MÁTÉ,
RICHARD RADO

1984

NORTH-HOLLAND PUBLISHING COMPANY
AMSTERDAM • NEW YORK • OXFORD

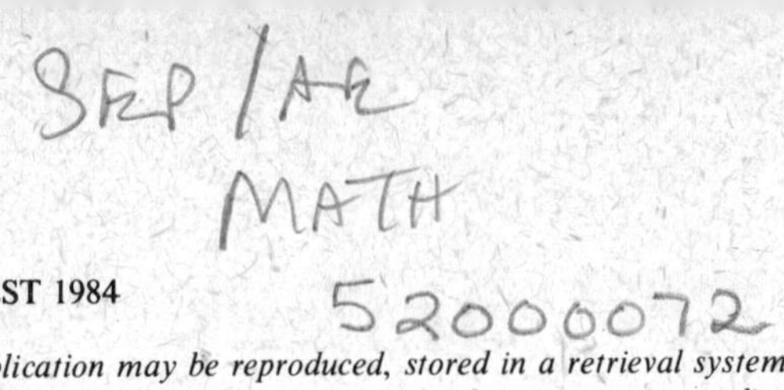

This monograph was published
as vol. 13 in the series Disquisitiones Mathematicae Hungaricae
ISBN 963 05 2877 0

Publishers

NORTH-HOLLAND PUBLISHING COMPANY
AMSTERDAM · NEW YORK · OXFORD

and

AKADÉMIAI KIADÓ
BUDAPEST

Sole distributors for the U.S.A. and Canada

ELSEVIER NORTH-HOLLAND, INC.
52 VANDERBILT AVENUE
NEW YORK, N. Y. 10017

for the East European countries, Democratic People's Republic of Korea, People's Republic of China, People's Republic of Mongolia, Republic of Cuba and Socialist Republic of Vietnam

AKADÉMIAI KIADÓ, BUDAPEST

Library of Congress Cataloging in Publication Data
Main entry under title:

Combinatorial set theory.

(Studies in logic and the foundations of mathematics; v. 106)
Bibliography: p.
1. Combinatorial set theory. I. Erdős, Paul, 1913-
II. Series.
QA248.C616 1984 511.3'22 83-4121
ISBN 0-444-86157-2

PRINTED IN HUNGARY

PREFACE

Ramsey's classical theorem in its simplest form, published in 1930, says that if we put the edges of an infinite complete graph into two classes, then there will be an infinite complete subgraph all edges of which belong to the same class. The partition calculus developed as a collection of generalizations of this theorem. The first important generalization was the Erdős–Dushnik–Miller theorem which says that for an arbitrary infinite cardinal κ, if we put the edges of a complete graph of cardinality κ into two classes then either the first class contains a complete graph of cardinality κ or the second one contains an infinite complete graph. An earlier result of Sierpiński says that in case $\kappa = 2^{\aleph_0}$ we cannot expect that either of the classes contains an uncountable complete graph. The first work of a major scope, which sets out to give a 'calculus' of partitions as its aim, was the paper "A partition calculus in set theory" written by Erdős and Rado in 1956. In 1965, Erdős, Hajnal, and Rado gave an almost complete discussion of the ordinary partition relation for cardinals under the assumption of the generalized continuum hypothesis. At that time there were hardly any general results for ordinals, though there were some results of Specker, and the paper of Erdős and Rado quoted above also contains some results for them. The situation has now changed considerably. The advent of Cohen's forcing method, and later Jensen's theory of the constructible universe gave new spurs to the development of the partition calculus. The main contributors in the next ten years were J. E. Baumgartner, C. C. Chang, F. Galvin, J. Larson, R. A. Laver, E. C. Milner, K. Prikry, and S. Shelah, to mention but a few. Independence results are beyond the scope of this book, though it will occasionally be useful to quote some of them in order to put theorems of set theory into their real perspective. An attempt to give a survey that deals also with independence results was made by Erdős and Hajnal in their paper "Solved and unsolved problems in set theory" which appeared in the Tarski symposium volume in 1974. The progress here is, however, so rapid that this survey was obsolete to a certain extent already when it appeared in print.

Our aim in writing this book is to present what we consider to be the most important combinatorial ideas in the partition calculus, and we also want to give a discussion of the ordinary partition relation for cardinals without the assumption of the generalized continuum hypothesis; we tried to make this latter as complete as possible. A separate section describes the main partition symbols scattered in the literature. A chapter on the applications of the combinatorial methods in partition calculus includes a section on topology with Arhangel'skii's famous result that a first countable compact Hausdorff space has cardinality at most the continuum, several sections on set mappings, and an account of recent inequalities for cardinal powers that were obtained in the wake of Silver's breakthrough result saying that the continuum hypothesis cannot first fail at a singular cardinal of uncountable cofinality. Large cardinals are discussed up to measurability, in slightly more detail than would be necessary strictly from the viewpoint of the partition calculus.

We assume some acquaintance with set theory on the part of the reader, though we tried to keep this to a minimum by the inclusion of an introductory chapter. The nature of the subject matter made it inevitable that we make some demands on the reader in the way of mathematical maturity.

And we make another important assumption: *the axiom of choice*, that is the axiomatic framework in this book is Zermelo–Fraenkel set theory always with the axiom of choice. There are interesting results in the partition calculus which do not need the axiom of choice, but we have never made an attempt to avoid using it. There are many interesting assertions that are consistent with set theory without the axiom of choice but contradict this latter, and there are many important theorems of set theory plus some interesting additional assumption, e.g. the axiom of determinacy, that is known to contradict the axiom of choice. We did not include any of these; unfortunate though this may be, we had to compromise; we attempted to discuss infinity, but had to accomplish our task in finite time.

CONTENTS

CHAPTER I

INTRODUCTION

The notations generally used in this book are given in the first section, then there follows a section listing the axioms of Zermelo–Fraenkel set theory. The most important notions involving ordinals and cardinals are presented, and finally such important tools in set theory as transfinite recursion, Mostowski's Collapsing Lemma, the Wellordering Theorem, Hausdorff's Maximal Chain Theorem, Zorns's Lemma, and Hausdorff's Cofinality Lemma are discussed.

1. NOTATION AND BASIC CONCEPTS

Although we shall rarely ever directly refer to any axiom of set theory below, a natural, perhaps the most natural framework for the considerations in this book is Zermelo–Fraenkel set theory (denoted by ZF) with the Axiom of Choice (denoted by AC; the axiom system ZF + AC is shortly written as ZFC). A list of the axioms of ZFC is given in the next section. We shall make no effort to avoid the use of AC even if this could be done, as it is beyond our aims to clarify the exact role of this axiom in the topics discussed. We assume that the reader is familiar with the basics of ZFC. Consistency results are occasionally mentioned, but their proofs are never given, so the reader need not be acquainted with the Axiom of Constructibility (written as $V=L$) and with forcing or the theory of Boolean-valued models.

Following the convention introduced by J. von Neumann, we identified ordinals with the set of their predecessors, and cardinals with their initial ordinals. Ordinals are usually denoted by lower case Greek letters (excepting ε, ι, o, υ, φ, χ, and ψ) with or without subscripts or superscripts. Among these, ω has a specific meaning (the least infinite ordinal), and the letters κ, λ, ρ, σ, and τ *always* denote cardinals, even if we do not mention this explicitly. Cardinals and ordinals may be denoted also by other letters if specifically mentioned. Integers, which are defined as finite ordinals, are usually denoted by the lower case Roman letters i, j, k, l, m, n, r, and s.

When writing logical formulas, we denote the logical connectives 'not', 'and', 'or', 'implies', 'if', 'if and only if' (this last one is occasionally abbreviated as '*iff*') by $\neg$, &, $\vee$, $\Rightarrow$, $\Leftarrow$, $\Leftrightarrow$ in turn; the existential and universal quantifiers are written as $\exists x$ ('there is an x such that') and $\forall x$ ('for all x'), respectively. The *language of* ZF (which is of course, the same as the language of ZFC), often called the *primitive language of* ZF, contains, aside from these symbols, also the signs $\in$ (membership or elementhood) and $=$ (equality). Constants are not admitted in the primitive language of ZF; variables (usually denoted below by $x, y, z, u, \ldots$ with or without subscripts or superscripts) will occasionally be called set variables, since in the intuitive interpretation they are meant to denote sets. As is usual, we extend the primitive language of ZF by introducing certain constant symbols and abbreviations commonly used in set theory. In this way, we shall use the quantifier $\exists! x$, meaning there is exactly one x, and the *bounded* (or *restricted*) quantifiers $\forall x \in y$, $\exists x \in y$, and $\exists! x \in y$. If $\varphi(x)$ is a formula of ZF, then the *classifier* $\{x: \varphi(x)\}$ is intuitively interpreted as the *class* (or collection) of those sets x for which $\varphi(x)$ holds. The precise formal interpretation is a collection of rules about how to eliminate a classifier from a formula whenever it occurs in it. According to these rules, the formula $y \in \{x: \varphi(x)\}$ has to be replaced by the formula $\varphi(y)$. If A and B are two *classes* (or, rather, classifiers), then $A = B$ is defined as $\forall x[x \in A \Leftrightarrow x \in B]$, where x is a variable which does not occur (or, at least, is not free) in A or B; this is an implicit reference to the Axiom of Extensionality, as generalized for classes. The case when x is free in A or B (so that one must choose another variable instead of x) is called a *clash* (*of variables*). As any set x can be considered the same as the class $\{y: y \in x\}$, this also explains the meanings of the formulas $x = A$ and $A = x$, where A is a class and x is a set. Finally, the formula $A \in B$, where A and B are classes, is interpreted as $\exists x[x = A \;\&\; x \in B]$, where x is a variable not occurring in A or B. A class A is a *real class* if $\forall x[\neg\, x = A]$ (assume x does not occur in A); in other words, A is a real class if and only if it is not (equal to) a set. Note that if A is a real class then the formula $A \in B$ is false for any class B. The classifier $\{x: x \in y \;\&\; \varphi(x)\}$ always defines a set in virtue of the Axiom of Replacement (see the next section); an often used shorter notation for this set is $\{x \in y: \varphi(x)\}$. In Sections 1–4, classes will be denoted by upper case letters and sets by lower case ones. From Section 5 on, classes will hardly ever be used, and upper case letters will also denote sets unless the contrary is explicitly mentioned.

Further abbreviations added to the language of set theory are negations of equality and membership ($\neq$ and $\notin$, respectively; other binary predicates are also often negated in this way, by crossing off), *inclusion in the wider sense* ($\subseteq$; $A \subseteq B$ allows the equality of A and B), *strict* (or *proper*) *inclusion* ($\subset$; equality is not allowed), *union of two classes* ($\cup$), *union* of the elements of a class ($\bigcup$), *difference* of two classes ($\setminus$), *intersection* of two classes ($\cap$), and *intersection* of the elements of a class ($\bigcap$). We have $\bigcap A = \{x: \forall y[y \in A \Rightarrow x \in y]\}$, if clashes of variables are avoided; hence, if A equals the *empty set* $\{x: x \neq x\}$ (denoted by the constant symbol 0) then $\bigcap A = \{x: x = x\}$, this latter being a shorthand for the *universal* class, or the class of all sets. x and y are called *disjoint* if $x \cap y = 0$. If x is a set, then $\mathscr{P}(x)$ denotes the power set of x, i.e., $\mathscr{P}(x) = \{y: y \subseteq x\}$.

Singleton x is the one element set $\{x\}$, which can be defined as $\{y: y = x\}$. The unordered pair $\{xy\}$ is the set $\{z: z = x \vee z = y\}$. In cases when we think it practical we may define larger sets as well by enumerating its elements: $\{ab \ldots c\}$ or $\{a, b, \ldots, c\}$ denotes the set $\{x: x = a \vee x = b \vee \ldots \vee x = c\}$, where the dots ... are to be replaced by symbols according to a rule that should be straightforward (occasionally infinite sets are also written in this way; the last classifier mentioned above is then, strictly speaking, meaningless, since it contains a formula with infinitely many disjunctions, and there is no such formula in ZF; but it will always be possible to replace it with a formula of ZF).

An *ordered pair* $\langle xy \rangle$ is defined as the set $\{\{x\}\{xy\}\}$; a set z defined in some other way from x and y so that one could, just by looking at z, tell what x and y were would be equally good. A *relation* is a class of ordered pairs; for a relation R, $\langle xy \rangle \in R$ is often written as xRy. If $\langle xy \rangle \in R$ implies $x, y \in C$ for some class C, then we say that R is a relation *on* the class C. The *restriction* $R|A$ of a relation R to a class A is defined as the relation $\{\langle xy \rangle \in R: x, y \in A\}$. (Given two classes A and B, the symbol $A|B$ will occasionally be used in other senses as well; it will always denote a kind of restriction. This ambiguity of notation will always be cleared up by the context.) An *operation* F is a relation such that, by looking at the first element of a pair in F, one can always tell which the second element is, i.e., $\forall xyz[[\langle xy \rangle \in F \,\&\, \langle xz \rangle \in F] \Rightarrow y = z]$ (avoid clashes!). An operation F is *one-to-one* or 1-1 if by looking at the second element of a pair in F one can tell which the first element is, i.e., $\forall xyz[[\langle yx \rangle \in F \,\&\, \langle zx \rangle \in F] \Rightarrow y = z]$. If F is an operation that is a set then F is called a *function*. For a class F, $\mathrm{dom}(F)$ denotes its *domain*, i.e., the class $\{y: \exists z[\langle yz \rangle \in F]\}$ and $\mathrm{ra}(F)$ denotes its *range*, i.e., the class $\{y: \exists z[\langle zy \rangle \in F]\}$. These concepts have intuitive motivation only in case F is an operation, but it will be useful in formal arguments that we have defined them for an arbitrary class F (cf. e.g. the proof of the Wellordering Theorem in Subsection 4.6 below).

If F is an operation and $x \in \mathrm{dom}(F)$, then the value of F at x, written as $F(x)$, is defined as the unique set y with $\langle xy \rangle \in F$. If $x \notin \mathrm{dom}(F)$ then $F(x)$ is defined as the empty set 0 in the strict formal sense; intuitively, however, it is better to think that $F(x)$ is not defined at all in this case. If F is an operation and X is a class then the restriction of F to X is denoted by $F \restriction X$, i.e., we put

$$F \restriction X = \{\langle xy \rangle \in F: x \in X\}.$$

We shall also occasionally use Gödel's notation

$$F``X = \mathrm{ra}(F \restriction X) \; (= \{F(x): x \in X \cap \mathrm{dom}(F)\}).$$

Finally, if x and y are sets, then ${}^x y$ (say: y pre x) denotes the set of all functions *from x into y*, i.e.,

$$ {}^x y = \{f\colon f \text{ is a function, dom } (f) = x \text{, and ra } (f) \subseteq y\} . $$

${}^x y$ can be defined in exactly the same way in the case when y is a real class; we shall, however, hardly ever be concerned with real classes; therefore here, and in what follows, we shall be content with explaining the basic concepts for sets. The notation $f\colon x \to y$ means that f is a function from x into y, i.e., that $f \in {}^x y$. $f\colon x \mapsto y$ means, on the other hand, that f is a function, $x \in \operatorname{dom}(f)$, and $f(x) = y$. Given two functions f and g, their composition $f \circ g$ is defined as the function h such that $h(x) = f(g(x))$ whenever $x \in \operatorname{dom}(g)$ and $g(x) \in \operatorname{dom}(f)$; in other words, we put

$$ f \circ g = \{\langle xy \rangle\colon \exists z[\langle xz \rangle \in g \,\&\, \langle zy \rangle \in f]\} . $$

A function is occasionally called an *indexed set;* in this case, its domain is called the *index set*.

If F is an operation with domain D and $F = \{\langle xz_x \rangle\colon x \in D\}$, where z_x is a set somehow determined by x, then we may write F as $\langle z_x\colon x \in D \rangle$ (this latter is the 'classifier' for operations; note that its definition lacks mathematical precision, and so it is only a visual device to inform the reader rather than a part of the extended language of ZF).

A *sequence* is a function defined on an ordinal. A (usually finite) sequence $\langle x_i\colon x_0 = a, \ldots, x_{k-1} = d \rangle$ may also be written as $\langle a, \ldots, d \rangle$ (the commas may be omitted). In case $k = 2$ the sequence $\langle ad \rangle$ of length two can be confused with the ordered pair $\langle ad \rangle$ in this notation. In the strict formal sense, this confusion is unpardonable, since these two are different sets. Intuitively, however, there is usually no need to distinguish between them; in formal definitions, it is best to think about $\langle ad \rangle$ as an ordered pair unless it is clear from the context that it should be a sequence of length two.

If f is a function, then a *choice function for f* is a function g such that $\operatorname{dom}(g) = \operatorname{dom}(f)$ and $g(x) \in f(x)$ for every $x \in \operatorname{dom}(f)$; the *Cartesian product* $\mathsf{X} f$ is the set of all choice functions for f (in case f assumes the empty set 0 as a value [at some element of its domain] then, obviously, $\mathsf{X} f$ is also empty). The *Cartesian product $X \times Y$ of two classes* X and Y is usually defined by

$$ X \times Y = \{\langle uv \rangle\colon u \in X \,\&\, v \in Y\} . $$

Here $\langle uv \rangle$ is an ordered pair. Note that if f is the function $\{\langle 0x \rangle, \langle 1y \rangle\}$, where x and y are arbitrary sets, then the Cartesian product $\mathsf{X} f$ is not the same as $x \times y$, though there is no harm in identifying the two in informal considerations. In a similar way, the two (formally different) Cartesian products $(x \times y) \times z$ and

$x \times (y \times z)$ are also identified usually, and simply written as $x \times y \times z$. A *matrix* (or *double sequence*) is a function defined on the Cartesian product of two ordinals.

A choice function is also called a *transversal*, but by transversal one frequently means a 1-1 choice function. Given a set x of pairwise disjoint nonempty sets, a set $y \subseteq \bigcup x$ such that $u \cap y$ consists of exactly one element for any $u \in x$ will also be called a *transversal* (*of* x); the logic behind this terminology is that y can be identified with a 1-1 choice function for the identity function on x in a straightforward manner. The reader will have no difficulty in finding out from the context in which sense the word transversal is used.

The chapters in this book are numbered by Roman numerals; each chapter is divided into sections, which are numbered by Arabic numerals independently of chapter numbers. Theorems, lemmas, corollaries, definitions, and problems are indexed by two numerals, the first of which is the section number and the second is consecutively increasing within each chapter, and they are referred to by these numbers. Some of the formulas in each section are indexed by a single number in parentheses. In the same section these formulas are referred to by their numbers in parentheses; when referring to a formula in a different section, we placed the section number in front of the formula number. Occasionally we found it clearer to divide a section into subsections. In this case, the subsections are numbered by two numerals, and theorems, lemmas, corollaries, etc. are not numbered at all. Each subsection contains at most one theorem, lemma, corollary, etc. (but it may contain one of each); so it will not lead to misunderstanding if we refer to e.g. the theorem in Subsection 4.1 as Theorem 4.1.

2. THE AXIOMS OF ZERMELO–FRAENKEL SET THEORY

Here we present a list of the axioms of Zermelo–Fraenkel set theory; the axioms are written in the language of ZF, extended as described in the preceding section.

2.1. The Axiom of the Empty Set:

$$\exists x \forall y \neg\, y \in x \,.$$

This axiom simply says that there is a set with no elements (called the *empty set*, and denoted by 0).

2.2. The Axiom of Extensionality:

$$\forall xy[\forall z[z \in x \Leftrightarrow z \in y] \Leftrightarrow x = y] \,.$$

Instead of the second equivalence sign $\Leftrightarrow$, one can write $\Rightarrow$ only, since the implication $\Leftarrow$ follows from an axiom of logic involving equality saying that equal objects can replace each other in a formula.

2.3. The Axiom of Pairing:

$$\forall xy \exists z \forall u [u \in z \Leftrightarrow [u = x \vee u = y]] .$$

This axiom together with the preceding one confirms the existence of an unordered pair; namely, $\{xy\}$ is the unique z such that the part after the quantifier $\exists z$ of the above formula holds. Singleton x can here be considered as the unordered pair $\{xx\}$.

2.4. The Axiom Scheme of Replacement: This axiom says that for every operation F and for any set x, $F``x$ is a set. As the quantifier 'for every operation F' is inadmissible in the (extended) language of ZF (namely, quantification can be made only over sets), this is actually not a single axiom, but *infinitely many;* that is why it is called an *axiom scheme*. Formally, this scheme can be described as follows: for every formula $\varphi(x, y, \vec{z})$ of ZF (here $\vec{z}$ stands for a finite sequence $z_0, \ldots, z_{n-1}$ of variables; $\forall \vec{z}$ stands for $\forall z_0 \ldots z_{n-1}$, etc.) we have

$$\forall \vec{z} \, [\forall xyy' \, [(\varphi(x, y, \vec{z}) \,\&\, \varphi(x, y', \vec{z})) \Rightarrow y = y'] \Rightarrow$$

$$\Rightarrow \forall u \exists v \forall y [y \in v \Leftrightarrow \exists x \in u \varphi(x, y, \vec{z})]] .$$

It is easy to derive from this axiom scheme and from Axiom 2.1 the *Axiom Scheme of Comprehension*, which says that the classifier $\{x \in u : \varphi(x, \vec{z})\}$ always defines a set (we leave the formal description of this axiom scheme to the reader).

2.5. The Axiom of Union:

$$\forall x \exists y \forall u [u \in y \Leftrightarrow \exists z [u \in z \,\&\, z \in x]] .$$

The unique set y satisfying this axiom is $\bigcup x$. Here, of course, one needs only the implication $\Rightarrow$, since the other implication can be made to hold with the aid of (an instance of) the Axiom (Scheme) of Replacement (or, rather, Comprehension). This economy is, however, usually only worth making if one wants to prove that the axioms of set theory hold in a model.

2.6. The Axiom of the Power Set:

$$\forall x \exists y \forall u [u \subseteq x \Leftrightarrow u \in y] .$$

The unique set y satisfying this axiom is the power set of x, denoted by $\mathscr{P}(x)$.

2.7. The Axiom of Infinity:

$$\exists x[0 \in x \,\&\, \forall y[y \in x \Rightarrow y \cup \{y\} \in x]] .$$

This axiom confirms the existence of a set containing the sets 0, $1=\{0\}$, $2=\{0, \{0\}\}$, etc; the smallest such set is the ordinal ω.

2.8. The Axiom of Regularity:

$$\forall x[x \neq 0 \Rightarrow \exists y \in x[x \cap y = 0]] .$$

The Axiom of Regularity is convenient, but not necessary, for the development of set theory. With the aid of the other axioms of ZFC (the Axiom of Choice (AC) is needed for this), it can be shown to be equivalent to the following statement: there is no infinite sequence $\langle x_i : i < \omega \rangle$ of sets such that $\ldots \in x_2 \in x_1 \in x_0$ holds. It is important to see the difference between the Axiom of Regularity and well-foundedness; this is discussed in Subsection 4.5 below.

The axioms mentioned so far constitute ZF set theory. If the next axiom is also added, then we obtain the theory ZFC.

2.9. The Axiom of Choice (AC): Given a set x such that $0 \notin x$, *call a function* $f: x \to \bigcup x$ a *choice function on* x if $f(y) \in y$ holds for every $y \in x$ (a choice function *for* another function was defined above). The Axiom of Choice says that there is a choice function on every set not containing the empty set.

2.10. This completes our list of the axioms of ZFC set theory. We shall assume that the reader is familiar with the way to develop set theory from them, or from other similar axioms (see e.g. Cohen [1966], Devlin [1973], Gödel [1940], Jech [1971], Kelley [1955], Takeuti–Zaring [1971], etc.) although usually a familiarity with intuitive set theory also suffices. Yet to give some help to the reader we shall prove some of the basic results of set theory in Section 4. Our discussion cannot, however, be regarded as complete, since we shall accept some simple facts of set theory as well known; in particular, after giving the definition of ordinals in Section 3 we shall assume as known such standard results as e.g. the fact that the ordinals are wellordered by the membership relation.

2.11. There are two more axioms that are occasionally mentioned in this book when discussing consistency results. They are Gödel's *Axiom of Constructibility*, usually abbreviated as $V=L$, and *Martin's Axiom*; the latter has a cardinal parameter κ and is abbreviated as MA_κ. The axiom $V=L$ says that all sets are constructible, but we cannot go into details as to what this means (see e.g. Gödel [1940], Cohen [1966], Devlin [1973], etc.); this axiom is well known to be consistent with the axioms of ZFC provided ZF is consistent. The axiom MA_κ is the following proposition: Given any notion of forcing c satisfying the countable chain condition, for any set f of cardinality $\leq\kappa$ of dense open subsets of c there is an f-generic filter on c. A *notion of forcing* c is a partially ordered set with a largest element, usually denoted as 1. Two elements of c are compatible if they have a common lower bound. c satisfies the *countable chain condition* if any uncountable subset of c contains two distinct compatible elements. Denoting the partial order on c by $\leq$, a subset x of c is *dense open* if $\forall p\in c\exists q\in x\ q\leq p$ and $\forall p\in c\forall q\in x[p\leq q\Rightarrow p\in x]$. If f is a set of dense open subsets of c, then a set $d\subseteq c$ is called an *f-generic filter* if (i) $1\in d$, (ii) $\forall p\in d\forall q\in c[p\leq q\Rightarrow q\in d]$, (iii) $\forall p,q\in d\exists r\in d\,[r\leq p,q]$, and (iv) $\forall x\in f[x\cap d\neq 0]$.) As is well known, ZFC + MA_κ always implies that $2^{\aleph_0}>\kappa$. The phrase "$2^{\aleph_0}=\lambda$ and Martin's Axiom holds" means that $2^{\aleph_0}=\lambda$ and MA_κ holds for all $\kappa<\lambda$. It is well known that this assertion is consistent relatively to ZFC whenever $\lambda>\omega$ is regular and is defined in a 'simple' way (e.g. $\lambda=\aleph_1, \aleph_2, \aleph_3, \ldots$) (see e.g. Jech [1971], Martin–Solovay [1972], and Solovay–Tennenbaum [1971]).

3. ORDINALS, CARDINALS, AND ORDER TYPES

A set x is *transitive* if $x\subseteq\mathscr{P}(x)$; it is an *ordinal* if it is transitive and all its elements are also transitive; this definition is applicable only if the Axiom of Regularity (see 2.8) is adopted, and in its absence another, more complicated definition must be given. The above definition implies that, as mentioned in Section 1, we accept J. von Neumann's convention that an ordinal is the set of all smaller ordinals. The class of all ordinals is often denoted by *On*. Greek lower case letters other than $\varepsilon, \iota, o, \upsilon, \varphi, \chi$ and ψ (as for ω, see below) will always denote ordinals, unless the contrary is explicitly mentioned. Ordinals may, of course, be denoted also by other letters. If $\alpha\in\beta$, then we may also write $\alpha<\beta$ (α is less than, or precedes, etc., β); $\alpha\leq\beta$ means that α is less than or equal to β, i.e., that $\alpha<\beta\vee\alpha=\beta$. For a set or class X of ordinals, $\bigcup X$ and $\bigcap X$ will usually be denoted by sup X and min X, respectively; if X has a largest element (e.g. if it is finite), then we may write max X instead of sup X; if X is empty, then min X is the universal class according to this definition, but it is better to think intuitively that min X is not defined at all. The (*ordinal*) *sum* of the ordinals α and β is denoted by $\alpha\dot{+}\beta$, and their (*ordinal*) *product*, by $\alpha\cdot\beta$ (the latter notation is ambiguous, see below). We do not define ordinal multiplication here, as the reader is probably familiar with it (see e.g. Devlin [1973], Takeuti–Zaring [1971], etc.); we only point out, for the reader's orientation, that according to (the somewhat arbitrary) tradition we have e.g. $2\cdot\omega=2\dot{+}2\dot{+}2\dot{+}\ldots=\omega$, and $\omega\cdot 2=$

$=\omega\dot{+}\omega(\neq\omega)$. Given a sequence $\langle\alpha_\nu: \nu<\eta\rangle$ of ordinals, their sum is written as

$$\dot{\sum}\langle\alpha_\nu: \nu<\eta\rangle=\dot{\sum_{\nu<\eta}}\alpha_\nu.$$

For an ordinal α, $\alpha\dot{-}1$ is the ordinal immediately preceding α if there is such an ordinal; otherwise $\alpha\dot{-}1=\alpha$. Given an *integer* (i.e., a finite ordinal) r, we put $\alpha\dot{-}(r\dot{+}1)=(\alpha\dot{-}r)\dot{-}1$ and $\alpha\dot{-}0=\alpha$. This defines $\alpha\dot{-}r$ for all integers r. Let $\alpha>0$; if $\alpha\dot{-}1=\alpha$, then α is called a limit ordinal, and otherwise a *successor ordinal*. We shall occasionally be concerned with intervals of ordinals: $[\alpha, \beta]=$ $=\{\xi: \alpha\leq\xi\leq\beta\}$, $[\alpha, \beta)=\{\xi: \alpha\leq\xi<\beta\}$; $(\alpha, \beta]$ and (α, β) are defined analogously.

Two sets x and y are *equipollent* if there is a 1-1 function f from x onto y (i.e., such that f is 1-1, dom $(f)=x$, and ra $(f)=y$). One often says *equivalent* instead of equipollent, but the former term is clearly overused, and so we prefer to avoid it. A *cardinal* is an ordinal that is not equipollent with any smaller ordinal; in this way, as mentioned in Section 1, we make no distinction between cardinals and their *initial ordinals*. The lower case Greek letters κ, λ, ρ, σ and τ (with or without subscripts or superscripts) always denote cardinals; of course, cardinals may also be denoted by other letters. The cardinality $|x|$ of the set x is, by definition, the (only) cardinal equipollent to x (the existence of such a cardinal follows from the Wellordering Theorem — see Subsection 4.6). The (*cardinal*) *sum* of κ and λ is denoted by $\kappa+\lambda$, and their (*cardinal*) *product*, by $\kappa\cdot\lambda$. It might lead to confusion that two very different concepts, ordinal product and cardinal product, are denoted in the same way. In practice, if α and β are both cardinals, then $\alpha\cdot\beta$ will always denote the cardinal product when not mentioned otherwise explicitly; and if α or β is not a cardinal, then $\alpha\cdot\beta$ cannot but denote the ordinal product. Infinite cardinal sums will be denoted by the symbol Σ; e.g. if κ_x, $x\in I$ are cardinals, then their sum is denoted by $\sum_{x\in I}\kappa_x$ or $\Sigma\langle\kappa_x: x\in I\rangle$. This sum is defined as the cardinality of the set $\bigcup\{z_x: x\in I\}$, where the sets z_x, $x\in I$, are pairwise disjoint and such that $|z_x|=\kappa_x$. One might be tempted to use also the notation $\Sigma\{\kappa_x: x\in I\}$, but this is incorrect, since e.g. if $\kappa_x=\kappa$ for all $x\in I$, then $\{\kappa_x: x\in I\}=$ $=\{\kappa\}$, and so we should have $\Sigma\{\kappa_x: x\in I\}=\Sigma\{\kappa\}=\kappa$, and this is clearly not what one wants. A similar remark applies to cardinal multplication, denoted by Π. The cardinal immediately succeeding κ is denoted by κ^+. (Thus, e.g. for finite κ we have $\kappa^+=\kappa+1=\kappa\dot{+}1$.) κ^- denotes the cardinal immediately preceding κ if there is such a cardinal; otherwise we put $\kappa^-=\kappa$. Let $\kappa>0$; if $\kappa^-=\kappa$ then κ is called a limit cardinal, and otherwise, a successor cardinal. *Cardinal subtraction* can also be defined: if κ and λ are cardinals, then $\kappa-\lambda$ denotes the cardinality of the set $\kappa\setminus\lambda$. If κ is infinite, then clearly $\kappa-\lambda=\kappa$ or 0 according as $\lambda<\kappa$ or $\lambda\geq\kappa$.

Cardinal *exponentiation* is defined as follows: we put $\kappa^\lambda = |{}^\lambda\kappa|$, $\underset{\smile}{\kappa}^\lambda = \sum_{\rho<\kappa} \rho^\lambda$, $\kappa^{\underset{\smile}{\lambda}} = \sum_{\rho<\lambda} \kappa^\rho$, and $\underset{\smile}{\kappa}^{\underset{\smile}{\lambda}} = \sum_{\rho<\kappa,\sigma<\lambda} \rho^\sigma$. One way of reading these symbols is, in turn: "κ to the power λ", "weak κ to the power λ", "κ to the weak power λ", and "weak κ to the weak power λ". For a set x and a cardinal κ we write $[x]^\kappa = \{y: y \subseteq x \,\&\, |y| = \kappa\}$ and $[x]^{<\kappa} = \{y: y \subseteq x \,\&\, |y| < \kappa\}$. The symbols $[x]^{\leq\kappa}$, $[x]^{>\kappa}$, and $[x]^{\geq\kappa}$ may be defined analogously. It is well known that $|\mathscr{P}(x)| = 2^{|x|}$ holds for any set x; and, moreover, for infinite x we also have $|[x]^\kappa| = |x|^\kappa$ provided $\kappa \leq |x|$ (if $\kappa > |x|$, then $|[x]^\kappa| = 0$), and $|[x]^{<\kappa}| = |x|^{\underset{\smile}{\kappa}}$ provided $\kappa < |x|^+$. We shall often use the symbol exp for *iterated exponentation*:

$$\exp_0(\kappa) = \kappa \qquad \text{and} \qquad \exp_{i+1}(\kappa) = \exp_i(2^\kappa), \tag{1}$$

where i is an integer; this defines $\exp_i(\kappa)$ for every integer i recursively. The *logarithm* operation, discussed in detail in Section 7, will be defined as follows:

$$L_\lambda(\kappa) = \min\{\rho: \exists\lambda_0 < \lambda[\lambda_0^\rho \geq \kappa]\}. \tag{2}$$

Here, in a natural way, λ is called the *base* of the logarithm. The *iterated logarithm* (of base 3) is defined by recursion as follows:

$$L^0(\kappa) = \kappa, \qquad \text{and} \qquad L^{i+1}(\kappa) = L_3(L^i(\kappa)), \tag{3}$$

where i is an integer; this defines $L^i(\kappa)$ for every integer i.

For each ordinal α, $\aleph_\alpha$ denotes the $(\alpha+1)$-st infinite cardinal; ω_α means the same thing. The existence of these two different notations is due to the tradition according to which $\aleph_\alpha$ denoted the cardinal and ω_α its initial ordinal; present day axiomatic set theory, however, usually identifies the two, and, as mentioned above, we also follow this practice. Nonetheless, occasionally this tradition may be helpful, as we shall see this when discussing partition symbols in Subsection 8.2 (cf. the remarks made after (8.5)). One usually writes ω instead of ω_0, but never $\aleph$ instead of $\aleph_0$. Let $\kappa = \aleph_\alpha$; then $\kappa^+ = \aleph_{\alpha \dot{+} 1}$ and $\kappa^- = \aleph_{\alpha \dot{-} 1}$ according to the notations introduced above; $\aleph_{\alpha \dot{+} \beta}$ will also be denoted as $\kappa^{(+\beta)}$, and, for an integer r, $\aleph_{\alpha \dot{-} r}$ as $\kappa^{(-r)}$.

The *Continuum Hypothesis*, denoted as CH, is the assertion that $2^{\aleph_0} = \aleph_1$ holds. The *Generalized Continuum Hypothesis*, denoted as GCH, says that $2^\kappa = \kappa^+$ holds for any infinite cardinal κ. A well-known result of Gödel [1940] (see also Cohen [1966], Devlin [1973], Jech [1971], etc.) says that GCH cannot be disproved in ZFC, and an also well-known result of Cohen [1963] and [1964] says that even CH cannot be proved in ZFC (see also e.g. Shoenfield [1971], Jech [1971]).

A *partial order* $\langle x, \prec \rangle$ is an ordered pair that consists of a set x and of a relation $\prec$ on x such that the relation $\prec$ is irreflexive, antisymmetric and transitive. A partial order $\langle x, \prec \rangle$ is called an *order* (occasionally: a *total order* or a *linear order*) if it is trichotomous, i.e., if $\forall u, v \in x\ [u \prec v \vee u = v \vee v \prec u]$. The relation $\prec$ itself is called a (partial or total, according to which is the case) *ordering*. A *wellorder* $\langle x, \prec \rangle$ is a total order such that each nonempty set $y \subseteq x$ has a $\prec$*-least* element (i.e., $\exists u \in y \forall v \in y [u \preceq v]$, where, as usual, $u \preceq v$ stands for $u \prec v \vee u = v$. A partial order is often called a *partially ordered set*, and the analogous terms are also used for linear orders and wellorders.

Two partial orders $\langle x, \prec \rangle$ and $\langle x', \prec' \rangle$ are called *similar* if they are isomorphic as structures, i.e., if there is a 1-1 function f from x onto x' such that $\forall u, v \in x [u \prec v \Leftrightarrow f(u) \prec' f(v)]$. It is well known that each wellorder is similar to an ordinal α (more precisely to the wellorder $\langle \alpha, \in | \alpha \rangle$, where $\in | \alpha$ is the elementhood relation restricted to α; this latter wellorder is usually written imprecisely as $\langle \alpha, \in \rangle$ or $\langle \alpha, < \rangle$). The *order type* of a wellorder $\langle x, \prec \rangle$, denoted as $\mathrm{tp} \langle x, \prec \rangle$, is defined as this unique ordinal. The *order type* $\mathrm{tp} \langle x, \prec \rangle$ of an arbitrary partial order $\langle x, \prec \rangle$ can be defined in a more complicated way as follows: If $\langle x, \prec \rangle$ is a partial order, then put

$$\mathrm{tp} \langle x, \prec \rangle = \{\langle |x|, \prec' \rangle \text{: the partial order } \langle |x|, \prec' \rangle \text{ is similar to } \langle x, \prec \rangle\};$$

so as not to run into contradiction with the definition of order type for wellorders, the latter definition must not be applied if $\langle x, \prec \rangle$ is a wellorder. Since the definition of the cardinality $|x|$ of x as given above depends on the Wellordering Theorem, i.e. on the Axiom of Choice, the definition of order type given here also depends on this axiom. By formula (4.14) we shall give a possible definition of cardinality that is independent of the Axiom of Choice. In an entirely analogous way, the order type of a partially ordered set could also be defined independently of this axiom.

Let d and d' be two order types. We write $d \leq d'$ if there are two partial orders $\langle x, \prec \rangle$ and $\langle x', \prec' \rangle$ such that $\mathrm{tp} \langle x, \prec \rangle = d$, $\mathrm{tp} \langle x, \prec \rangle = d'$, $x \subseteq x'$, and the restriction $\prec' | x$ of $\prec'$ onto x is $\prec$. If $d \leq d'$ and $d' \not\leq d$, then one writes $d < d'$. We must emphasize that the relation $\leq$ is not an ordering even for *linear order types* (i.e., for order types of linearly ordered sets). To give a simple example, denote by η the order type of the set of rational numbers. Then $\eta \leq \eta \dotplus 1 \leq \eta$, where $\dotplus$ denotes addition of order types. If d is an order type, then d^* denotes the reverse of d; i.e., if $d = \mathrm{tp} \langle x, \prec \rangle$, then $d^* = \mathrm{tp} \langle x, \succ \rangle$, where $u \succ v$ means, as usual, the same as $v \prec u$ for any $u, v \in x$.

Let $\langle x, \prec\rangle$ be a partial order and d an order type. We put

$$[x, \prec]^d = \{y : y \subseteq x \;\&\; \mathrm{tp}\,\langle y, \prec|y\rangle = d\},$$

and

$$[x, \prec]^{<d} = \{y : y \subseteq x \;\&\; \mathrm{tp}\,\langle y, \prec|y\rangle < d\};$$

the sets $[x, \prec]^{\leq d}$, $[x, \prec]^{\geq d}$, and $[x, \prec]^{>d}$ can be defined analogously. When there is no danger of misunderstanding, we may write $[x]^d$, $[x]^{<d}$, etc. instead of the above symbols. (There might be a real danger of misunderstanding, especially in case d is a cardinal, since then d is an order type as well according to our terminology.)

If $\langle x, \prec\rangle$ and $\langle x', \prec'\rangle$ are total orders such that $x' \subseteq x$, $\prec' = \prec|x'$, and $\forall y \in x' \forall z \in x\; [z \prec y \Rightarrow z \in x']$ then we say that $\langle x', \prec'\rangle$ is an *initial segment of* $\langle x, \prec\rangle$; if $\forall y \in x' \forall z \in x\; [y \prec z \Rightarrow z \in x']$ holds instead, then we say that $\langle x', \prec'\rangle$ is a *final segment of* $\langle x, \prec\rangle$. Alternatively, we may say that x' is an *initial* or *final segment*, respectively, *in* (or, occasionally, *of*) $\langle x, \prec\rangle$.

For a total order $\langle x, \prec\rangle$ and a set $y \subseteq x$, y is said to be *cofinal in* $\langle x, \prec\rangle$ if $\forall u \in x$ $\exists v \in y [u \preceq v]$. If there is no danger of misunderstanding, then we may say that y is cofinal in x. For an ordinal α, its *cofinality*, $\mathrm{cf}\,(\alpha)$, is defined as

$$\mathrm{cf}\,(\alpha) = \min\{\mathrm{tp}\,\langle x, <\rangle : x \subseteq \alpha \text{ and } x \text{ is cofinal in } \langle \alpha, <\rangle\}. \tag{4}$$

Note here that $\mathrm{tp}\,\langle x, <\rangle$ on the right-hand side is an ordinal, and so the minimum there is meaningful. If α and β are two ordinals, then the phrase 'α is *cofinal to* β' means that $\mathrm{cf}\,(\alpha) = \mathrm{cf}\,(\beta)$. Clearly $\mathrm{cf}\,(\alpha) = 0$ implies $\alpha = 0$, $\mathrm{cf}\,(\alpha) = 1$ holds exactly if α is a successor ordinal, and $\mathrm{cf}\,(\alpha) \geq \omega$ holds otherwise. The basic facts about the operation cf will be proved in Subsection 4.10, and in Subsection 4.11 this operation will be extended to order types.

Let $\kappa \geq \omega$ be a cardinal. κ is called *regular* if $\mathrm{cf}\,(\kappa) = \kappa$, *singular* if $\mathrm{cf}\,(\kappa) < \kappa$, *strong limit* if $2^\lambda < \kappa$ for any $\lambda < \kappa$ (in this case $\kappa^- = \kappa$, i.e., κ is a limit cardinal), and *inaccessible* if it is an uncountable regular strong limit cardinal. An ordinal is called *accessible* if it is not an inaccessible cardinal. A cardinal is said to be a *regular limit cardinal* if it is a regular cardinal that is a limit cardinal; an uncountable regular limit cardinal is occasionally called *weakly inaccessible*. According to earlier terminology, a cardinal that we called inaccessible here was called *strongly inaccessible*; we shall, however, never use the terms strongly and weakly inaccessible.

4. BASIC TOOLS OF SET THEORY

We do not intend to present a full development of set theory from the axioms; for example, we entirely skip the proofs of fundamental facts concerning ordinals and cardinals. Our aim here is to give 'axiomatic' proofs of such basic results as the Transfinite Recursion Theorem, the Wellordering Theorem, and Zorn's Lemma, etc, and to prove some fundamental facts about the cofinality operation. The reader need not be acquainted with such axiomatic proofs, since an acquaintance with the development of intuitive set theory usually suffices for the understanding of the material below; we felt, however, that our introductory chapter would be incomplete without the inclusion of this section. E.g. the proof of the Generalized Transfinite Recursion Theorem, given in Subsection 4.2, is hard to find in textbooks on set theory.

4.1. TRANSFINITE RECURSION THEOREM. *For every operation G there is a unique operation F on the class of all ordinals such that $F(\alpha)=G(F \upharpoonright \alpha)$ holds for all ordinals α.*

This formulation exploits the fact that $G(x)$ was defined (as the empty set) even in case the set x does not belong to the domain of G. Observe also that the above theorem cannot be formalized in the language of ZF, as we cannot quantify over classes because, by the definition of classes above, this would mean quantification over formulas. Hence, this theorem is not a theorem *of* the formal system of ZF (or ZFC); rather it is a theorem *about* that formal system. Such a theorem is called *metatheorem*. Metatheorems in this section are Theorems 4.1, 4.2, Lemmas 4.1, 4.3, 4.4, and the R-Induction Principle in Subsection 4.4.

PROOF. Define F by the classifier

$$F=\{\langle\alpha z\rangle:\exists f[f \text{ is a function } \& \\ \& \operatorname{dom}(f)=\alpha\dot{+}1 \,\&\, \forall\beta\leq\alpha[f(\beta)=G(f\upharpoonright\beta)] \,\&\, z=f(\alpha)]\}, \tag{1}$$

where α runs over ordinals. For every α there is at most one z such that $\langle\alpha z\rangle\in F$; in fact if α is the least ordinal for which there are z and z' with $\langle\alpha z\rangle, \langle\alpha z'\rangle\in F$ and $z\neq z'$, then there are functions f and f' such that $\operatorname{dom}(f)=\operatorname{dom}(f')=\alpha\dot{+}1$, $f\upharpoonright\alpha=f'\upharpoonright\alpha$, $z=f(\alpha)=G(f\upharpoonright\alpha)$ and $z'=f'(\alpha)=G(f'\upharpoonright\alpha)\neq z$, which is absurd. For every ordinal α there is a z such that $\langle\alpha z\rangle\in F$. In fact, if α is the least ordinal for which there is no such z, then $f'=\{\langle\beta y\rangle\in F:\beta<\alpha\}$ is a set by the Axiom of Replacement, and so it is obviously a function on α. Putting $z=G(f')$, it follows from (1) with $f=f'\cup\{\langle\alpha z\rangle\}$ that $\langle\alpha z\rangle\in F$. Hence F is indeed an operation on all ordinals. The uniqueness of F also easily follows: If F and F' are two operations such that $F(\alpha)=G(F\upharpoonright\alpha)$ and $F'(\alpha)=G(F'\upharpoonright\alpha)$ for all ordinals α, then we get a contradiction for the first α for which $F(\alpha)\neq F'(\alpha)$. The proof is complete.

4.2. Recursion on well-founded relations. A relation R on a class X is well-founded if $\{y\in X: yRx\}$ is a set for every $x\in X$ and

$$\forall u\subseteq X[u\neq 0\Rightarrow\exists x\in u\forall y\in u\neg yRx]. \quad (2)$$

The element x whose existence is claimed in this formula is called *R-minimal in u*. From the Axiom of Choice it easily follows (e.g. with the aid of the preceding theorem) that the second requirement (in the presence of the first one) is equivalent to saying that there is no infinite sequence $\langle x_i: i<\omega\rangle$ of elements of X such that $x_{i+1}Rx_i$ holds for any $i<\omega$.

Generalized Transfinite Recursion Theorem. *If R is a well-founded relation on a class X then for every operation G there is a unique operation F on X such that*

$$F(x)=G(F\upharpoonright\{y\in X: yRx\}) \quad (3)$$

holds for every $x\in X$.

Usually, this result is proved as a corollary of the preceding theorem. Such a proof is intuitively quite clear and proceeds as follows. First one defines an R-rank, that is, an operation rk_R on X into the ordinals with the following property: if xRy, then $\mathrm{rk}_R(x)<\mathrm{rk}_R(y)$ holds; having done this, the definition of F can be given with the aid of *common transfinite recursion* (i.e., with the aid of the preceding theorem). Of course, one has to define the operation rk_R also by common transfinite recursion, which is a little complicated; its definition by *generalized transfinite recursion* (i.e., with the aid of the present theorem) is rather simple: we put

$$\mathrm{rk}_R(x)=\bigcup[\mathrm{rk}_R(y)\dot{+}1: yRx\} \quad (4)$$

for any $x\in R$.

As the proof outlined above is formally quite complicated, we shall rather give a direct proof analogous to the proof of the preceding theorem. The advantage of this proof is that it immediately gives the definition of the operation F. To this end, for a relation R on a class X, call a class $D\subseteq X$ *R-transitive* if

$$\forall x\in D\forall y[yRx\Rightarrow y\in D].$$

We need the following lemma (this lemma is needed also in the alternative proof outlined above):

Transitive Closure Lemma. *If a relation R on a class X is such that the class* $\{y\in X: yRx\}$ *is a set for every* $x\in X$, *then for every set* $u\subseteq X$ *there is an R-transitive set d including u.*

We shall use this lemma only in case R is a well-founded relation; such a relation satisfies the requirements of the lemma by the definition of well-foundedness.

PROOF OF THE LEMMA. Given R, X and $u \subseteq X$, define the operation F by transfinite recursion (i.e., by Theorem 4.1) as follows:

$$F(\alpha) = u \cup \{y: \exists \beta < \alpha \exists z \in F(\beta) [y = z \vee yRz]\}.$$

It is very important here that $F(\alpha)$ is always a set, and never a real class; the reason why this is so important can be seen from the formal definition of F given according to Theorem 4.1: For any set $v \subseteq On \times X$ put

$$G(v) = \{y \in X: y \in u \vee \exists \alpha z [\langle \alpha z \rangle \in v \,\&\, (y = z \vee yRz)]\}.$$

$G(v)$ *is a set* for every v by the assumptions of the lemma to be proved; *hence* we can define an operation G as

$$G = \{\langle vG(v) \rangle : v \subseteq On \times X\}.$$

We can now define the operation F by the stipulation that $F(\alpha) = G(F \upharpoonright \alpha)$ for every α, as is done in Theorem 4.1.

Write now

$$d = \bigcup_{n<\omega} F(n).$$

It is easy to see that d is R-transitive, since if $x \in d$ then $x \in F(n)$ for some $n < \omega$, and then $y \in F(n+1)$ for any $y \in X$ with yRx. On the other hand, $u \subseteq F(\alpha)$ clearly holds for every ordinal α, and so $u \subseteq d$; hence d is an R-transitive set including u (actually, as is easily seen, it is the smallest such set; hence it might be called the *R-transitive closure* of u). The proof is complete.

PROOF OF THE THEOREM. Given R, X and G, write analogously to (1) that

$$\begin{aligned} F = \{\langle xz \rangle : x \in X \,\&\, \exists fd[&f \text{ is a function } \,\&\\ &\,\&\, d = \text{dom}(f) \,\&\, d \subseteq X \text{ is } R\text{-transitive } \,\&\, x \in d \,\&\\ &\,\&\, \forall v \in d[f(v) = G(f \upharpoonright \{y \in X: yRv\})] \,\&\, z = f(x)]\}. \end{aligned} \tag{5}$$

We repeat the proof of Theorem 4.1 with minor changes.

First we claim that, for every x, there is at most one z such that $\langle xz \rangle \in F$. In fact, assume that $\langle xz_0 \rangle, \langle xz_1 \rangle \in F$ and $z_0 \neq z_1$; then there are f_0, d_0 and f_1, d_1 such that the formula after the existential quantifier in (5) holds with z_0, f_0, d_0 and z_1, f_1, d_1 replacing z, f, d, respectively. Write $d' = d_0 \cap d_1$; then d' is obviously R-transitive. Let x' be an *R-minimal* element in the set $\{y \in d': f_0(y) \neq f_1(y)\}$ (see the definition right after formula (2)). Note that this set is not empty since x belongs to it, and so there is such an x' by well-foundedness. Then $f_0(x') \neq f_1(x')$, and yet $f_0(y) = f_1(y)$ for every y in the set $\{y \in d' : yRx'\} = \{y \in X : yRx'\}$ (this latter equality holds by the R-transitivity of d'). This is a contradiction, since we have

$$f_0(x') = G(f_0 \upharpoonright \{y \in X : yRx'\}) = G(f_1 \upharpoonright \{y \in X : yRx'\}) = f_1(x').$$

Hence our first claim is proved.

Secondly, we claim that for every $x \in X$ there is a z such that $\langle xz \rangle \in F$. In fact, assume $x_0 \in X$ is such that $\langle x_0 z \rangle \in F$ holds for no set z, and let c be an R-transitive set containing x_0; there is such a c by the lemma just proved.

Put

$$d' = \{x \in c : \exists z \langle xz \rangle \in F\}.$$

We assert that d' is an R-transitive set. In fact, if $x \in d'$, then $\langle xz \rangle \in F$ for some z, and so the formula after the existential quantifier in (5) is satisfied with some f and d. If now yRx holds then $y \in d$ by the R-transitivity of d, and so $\langle y f(y) \rangle \in F$ holds by (5). As $y \in c$ by the R-transitivity of c, we can conclude that $y \in d'$. This proves our assertion that d' is R-transitive.

Write

$$f' = \{\langle yz \rangle \in F : y \in d'\}.$$

Since for every $y \in X$ there is at most one z such that $\langle yz \rangle \in F$ according to our first claim, f' is a set (and so a function) by the Axiom of Replacement. dom $(f') = d'$ holds by the definition of d'. Let now x be an R-minimal element in $c \setminus d'$; as $x_0 \in c \setminus d'$ by the definitions of c, x_0, and d', $c \setminus d'$ is not empty, and so there is such an x by (2). Noting that

$$\{y \in X : yRx\} \subseteq d' \tag{6}$$

holds by the R-minimality of x and by the R-transitivity of d', put

$$z = G(f \upharpoonright \{y \in X : yRx\}),$$

$$f = f' \cup \{\langle xz \rangle\},$$

and

$$d = d' \cup \{x\}.$$

Noting that d is R-transitive by (6) since d' is so, we can see that x, z, f and d satisfy the formula after the existential quantifier in (5); hence $\langle xz \rangle \in F$. This is a contradiction, proving our second claim that $\forall x \in X \exists z \langle xz \rangle \in F$.

Finally, we claim that there is only one operation F satisfying (3). Assume, on the contrary, that F and F' are two such operations with $F(x_0) \neq F'(x_0)$ for some $x_0 \in X$. Let $d \subseteq X$ be an R-transitive set containing x_0 (there is such a d by the lemma just proved), and let x be an R-minimal element of the set $\{y \in d: F(y) \neq F'(y)\}$; this set is not empty since x_0 belongs to it, and so, by (2), there is an x as required. We now have $\{y \in X: yRx\} = \{y \in d: yRx\}$ by the R-transitivity of d, and so $F(y) = F'(y)$ holds for every element y of this set by the R-minimality of x. Hence, by (3), we have

$$F(x) = G(F \upharpoonright \{y \in X: yRx\}) = G(F' \upharpoonright \{y \in X: yRx\}) = F'(x),$$

which is a contradiction, since $F(x) \neq F'(x)$ by the choice of x. This contradiction establishes the uniqueness of F. The proof of the theorem is complete.

4.3. Here we formulate a principle that was several times used in the proof of the preceding theorem:

R-INDUCTION PRINCIPLE. *Let R be a well-founded relation on a class X, and let $\varphi(x, \vec{z})$ be a formula of* ZF. *Assume that*

$$\forall \vec{z} \forall x \in X[\forall y \in X[yRx \Rightarrow \varphi(y, \vec{z})] \Rightarrow \varphi(x, \vec{z})]. \tag{7}$$

Then

$$\forall \vec{z} \forall x \in X \; \varphi(x, \vec{z})$$

holds.

The proof of this principle is based upon the following simple lemma saying that real classes have R-minimal elements.

LEMMA. *Let R be a well-founded relation on a class X, and let $Y \subseteq X$ be a nonempty class. Then there is an R-minimal element x in Y, i.e., an $x \in Y$ such that* $\forall y \in Y \; \neg yRx$.

PROOF OF THE LEMMA. Let $y_0 \in Y$ be arbitrary, and choose an R-transitive set $d \subseteq X$ that contains y_0; there is such a d by the lemma in the preceding subsection. Let x be an R-minimal element in the set $d \cap Y$; such an x exist by (2). As $\{y \in X: yRx\} \subseteq d$ by the R-transitivity of d, it follows that x is R-minimal also in the class Y. The proof is complete.

PROOF OF THE R-INDUCTION PRINCIPLE. Assume, on the contrary, that $\neg \varphi(x_0, \vec{z}_0)$ holds for some $x_0 \in X$ and $\vec{z}_0$. Put $Y = \{y \in X: \neg \varphi(y, \vec{z}_0)\}$ and, using the lemma just proved, pick an R-minimal element x in Y. Then $\forall y \in X[yRx \Rightarrow \varphi(y, \vec{z}_0)]$, and so $\varphi(x, z_0)$ holds according to (7). This contradicts the choice of x according to which $x \in Y$. The proof is complete.

The formulation of the R-induction principle is too long in comparison to its proof from the above lemma. Hence one often prefers to use the lemma and repeat the proof of the principle every time. In fact this is what one always does when using *transfinite induction* (i.e. using $<$-induction, where $<$ is the 'less than' relation on ordinals): one always uses phrases like 'let α be the smallest ordinal for which the assertion fails'.

4.4. The following lemma plays a key role in studying models of set theory. In the present book it will only be used in Section 30, which is not an integral part of this book. So the reader wishing to skip Section 30 can also skip this subsection.

MOSTOWSKI'S COLLAPSING LEMMA (Mostowski [1949]). *If R is a well-founded relation on a class X and*

$$\forall x, y \in X[\forall z \in X[zRx \Leftrightarrow zRy] \Rightarrow x = y], \tag{8}$$

then there is a transitive (i.e., $\in$-transitive) class Y and a 1-1 operation F from X onto Y such that

$$\forall x, y \in X[xRy \Leftrightarrow F(x) \in F(y)]. \tag{9}$$

Note that (8) means that the structure $\langle X, =, R \rangle$ satisfies the Axiom of Extensionality, and the existence of a 1-1 operation F satisfying (9) means that the structure $\langle X, =, R \rangle$ is isomorphic to $\langle Y, =, \in \rangle$. Hence a shorter formulation of the above lemma is the following (this is the formulation usually given):

If R is a well-founded relation on a class X such that the structure $\langle X, =, R \rangle$ satisfies the Axiom of Extensionality, then there is a transitive class Y such that $\langle X, =, R \rangle$ and $\langle Y, =, \in \rangle$ are isomorphic.

PROOF. Using the Generalized Transfinite Recursion Theorem, for any $x \in X$ put

$$F(x) = \{F(y) \colon yRx\}$$

and write $Y = F``X$. Y is clearly transitive.

We claim that F is 1-1. Assume, on the contrary, that this is not the case and, using the lemma in the preceding subsection, let x_0 be an R-minimal element in the class $\{x \in X \colon \exists y \in X[x \neq y \,\&\, F(x) = F(y)]\}$, and let $y_0 \in X$ be such that $x_0 \neq y_0$ and $F(x_0) = F(y_0)$. We claim that

$$\forall z \in X[zRx_0 \Leftrightarrow zRy_0] \tag{10}$$

holds. This will of course mean that $x_0 = y_0$ in view of (8), which contradicts the choice of x_0 and y_0, proving our claim. To see (10), let $z \in X$ be such that zRx_0 holds. Then

$$F(z) \in F(x_0) = F(y_0) = \{F(z') \colon z'Ry_0\},$$

i.e., there is a $z' \in X$ such that $z'Ry_0$ and $F(z) = F(z')$ hold. It follows by $z'Rx_0$ and the R-minimality of x_0 from this latter relation that $z = z'$; hence zRy_0 holds in view of $z'Ry_0$. Let now $z \in X$ be such that zRy_0 holds. Then we have

$$F(z) \in F(y_0) = F(x_0) = \{F(z') \colon z'Rx_0\},$$

i.e., there is a $z' \in X$ such that $z'Rx_0$ and $F(z) = F(z')$ hold. It follows now from this latter relation by $z'Rx_0$ and the R-minimality of x_0 that $z = z'$; hence zRx_0 holds in view of $z'Rx_0$. Thus (10), and along with it also our claim, is established.

We have yet to prove that F is an isomorphism. If xRy holds then, obviously, $F(x) \in F(y)$ also holds by the definition of F. To prove the other implication in (9), assume that $F(x) \in F(y)$ holds for some $x, y \in X$. Then we have

$$F(x) \in F(y) = \{F(z): zRy\},$$

i.e., $F(x) = F(z)$ for some $z \in X$ with zRy. But then $x = z$ as F is 1-1, i.e., xRy holds, which we wanted to show. The proof is complete.

4.5. Epsilon Recursion and Epsilon Induction. The relation $\in = \{\langle xy \rangle: x \in y\}$, which is often called the epsilon relation, is well-founded according to the Axiom of Regularity; hence, by 4.2 and 4.3, we can perform recursion and induction on it; these are called $\in$-recursion and $\in$-induction, respectively. (Note here that if $\mathfrak{M} = \langle M, =, E \rangle$ is a model of set theory then the fact that $\mathfrak{M}$ satisfies the Axiom of Regularity does not mean that E is well-founded. In fact, the Axiom of Regularity for $\mathfrak{M}$ means that

$$\forall u \in M[\exists z\, zEu \Rightarrow \exists x[xEu \,\&\, \forall y[yEu \Rightarrow \neg\, yEx],$$

while the well-foundedness of E means that

$$\forall u \subseteq M[u \neq 0 \Rightarrow \exists x \in u \forall y \in u \neg\, yEx];$$

an enormous difference is made by the fact that in the first formula we have $\forall u \in M$, and in the second, $\forall u \subseteq M$. The remaining parts of the two formulas are entirely analogous. There are various well-known examples of *nonstandard* models of set theory, i.e., models $\mathfrak{M} = \langle M, =, E \rangle$ of set theory such that E is not well-founded. Conversely, it is of course true that if E is well-founded, then $\mathfrak{M}$ satisfies the Axiom of Regularity.)

An important example for $\in$-recursion is the definition of the rank operation. The rank of a set x, rk (x), is defined recursively as

$$\text{rk}\,(x) = \bigcup \{\text{rk}\,(y) \dotplus 1 : y \in x\}; \tag{11}$$

a more general rank operation was defined above, in (4). Define the operation V as follows: for every ordinal α put

$$V(\alpha) = \bigcup_{\beta < \alpha} \mathscr{P}(V(\beta)). \tag{12}$$

Note that $V(\alpha)$ is always a set (formally, this can be seen by observing that an operation always has sets as values; but this argument is a vicious circle, since one has to check that the operation used in the recursive definition of V is well defined, i.e. that it always has sets as values — cf. the proof of the lemma in

Subsection 4.2). The reader can easily show that

$$V(\alpha)=\{x\colon \mathrm{rk}\,(x)<\alpha\} \tag{13}$$

holds. Hence the sets having rank below a certain ordinal always form a set; this is an important fact which we shall exploit shortly, but first it seems appropriate to make another remark about (13). (12) and (13) give a method to define the rank of a set by common transfinite recursion, thus avoiding $\in$-recursion. (This can also be done for the general rank operation defined under (4); cf. also the remark in parentheses at the end of this paragraph.) The snag of this approach is that $\in$-induction is still needed to show that the rank operation so defined is defined for every set; this is partly the reason why we did not derive the Generalized Transfinite Recursion Theorem from the Transfinite Recursion Theorem above. (In fact, the case of an arbitrary well-founded relation R on a class X is even more complicated, because (12) and (13) cannot be directly generalized to define the operation rk_R in (4); in fact it may even happen that the analogue of $V(0)$, *i.e.*, $\{x\in X\colon \neg\exists y\; yRx\}$, is a real class.)

The formula in (13) (or, rather, the class $V(\alpha)$ being a set) enables us to define the cardinality of a set even in the absence of the Axiom of Choice. We can call the cardinality of x the set

$$\{y\colon y \text{ is equipollent with } x\ \&\ \forall z[\mathrm{rk}\,(z)< \\ <\mathrm{rk}\,(y)\Rightarrow z \text{ is not equipollent with } x]\}\,. \tag{14}$$

As we shall always accept the Axiom of Choice, we shall never use this definition of cardinality. As mentioned in the preceding section, an analogous definition can be given for the order type of a (partial) order in the absence of the Axiom of Choice.

4.6. The Wellordering Theorem (Zermelo). *For every set x there is an ordinal* α *and a* 1-1 *function f from* α *onto x.*

The name of this theorem is justified by the obvious fact that f and α can be used to define a wellordering $\prec$ of x by stipulating that, for any $u, v\in x$, we put $u\prec v$ if and only if $u=f(\xi)$ and $v=f(\eta)$ for some ξ, $\eta<\alpha$ with $\xi<\eta$.

Proof. Let $g\colon (\mathscr{P}(x)\setminus\{0\})\to x$ be a choice function on the set $\mathscr{P}(x)\setminus\{0\}$; i.e., let g be such that $g(u)\in u$ holds for every nonempty set $u\subseteq x$. Such a g exists by the Axiom of Choice. We shall use transfinite recursion to define an operation F, which will in fact turn out to be a function f as required by the theorem to be proved. To this end, put

$$G(u)=g(x\setminus \mathrm{ra}\,(u))$$

for every set u; here we exploit our stipulation that ra (u) was defined even in case

u is not a function; note also that even though $0 \notin \mathrm{dom}\,(g)$, the value $g(0)$ is (formally) defined as 0 according to a remark made in Section 1. Define F on the class of all ordinals by the recursion formula $F(\xi)=G(F \restriction \xi)$ (cf. Theorem 4.1) and, remembering that $\subset$ denotes strict inclusion, put

$$A=\{\xi : \mathrm{ra}\,(F \restriction \xi) \subset x\} \qquad \text{and} \qquad F'=F \restriction A. \tag{15}$$

Note that $\mathrm{ra}\,(F') \subseteq x$. In fact, if $\xi \in A$, then

$$F'(\xi)=F(\xi)=g(x \setminus \mathrm{ra}\,(F \restriction \xi)) \in x \setminus \mathrm{ra}\,(F \restriction \xi),$$

since $x \setminus \mathrm{ra}\,(F \restriction \xi)$ is not empty. Next, observe that F' is 1-1. In fact, pick $\xi, \eta \in A$ with $\eta < \xi$. Then, as we have just seen, $F'(\xi) \in x \setminus \mathrm{ra}\,(F \restriction \xi)$; on the other hand as $\eta \in \xi$, we have $F'(\eta)=F(\eta) \in \mathrm{ra}\,(F \restriction \xi)$. Hence $F'(\xi) \neq F'(\eta)$. Since we saw right after (15) that $\mathrm{ra}\,(F') \subseteq x$, we have

$$A=\mathrm{dom}\,(F')=\{\xi \in \mathrm{dom}\,(F') : \exists y \in x[F'(\xi)=y]\}$$

and

$$F'=\{\langle \xi y \rangle : \xi \in \mathrm{dom}\,(F') \;\&\; y \in x \;\&\; F'(\xi)=y\};$$

as F' is 1-1, it follows from the Axiom of Replacement that these classes are sets. It is clear from the definition of A in (15) that it is transitive; all its elements being ordinals (and so transitive), we can see that A is an ordinal. Write $\alpha=A$ and $f=F'$. The only thing that we have yet to prove is that $\mathrm{ra}\,(f)=\mathrm{ra}\,(F')=x$. Assume, on the contrary, that $\mathrm{ra}\,(F') \neq x$. Since we saw above, right after (15), that $\mathrm{ra}\,(F') \subseteq x$, we must have $\mathrm{ra}\,(F') \subset x$. As $F'=F \restriction A$ and A is an ordinal, this means that $A \in A$ by the definition in (15) of A. This, however, cannot hold for any ordinal. (In fact, $u \in u$ cannot hold for any set in view of the Axiom of Regularity; for, if $u \in u$, then the set $\{u\}$ fails to satisfy this axiom.) This contradiction proves the theorem.

It is easy to derive the Axiom of Choice from the Wellordering Theorem in the presence of the axioms of ZF: If x is a set such that $0 \notin x$, then let $\prec$ be a wellordering of $\bigcup x$. For any $y \in x$ define $g(y)$ as the smallest element of y in the wellordering $\prec$; then g is a choice function on x.

The Wellordering Theorem also enables one to define the cardinality $|x|$ of a set x as

$$|x|=\min\{\alpha : \exists f[f \text{ is a 1-1 function from } \alpha \text{ onto } x]\}. \tag{16}$$

This is the definition of cardinality that we adopted in Section 3 above and which we shall always use in this book.

4.7. Let $\langle x, \prec \rangle$ be a partial order. Call $y \subseteq x$ a *chain* (*in* $\langle x, \prec \rangle$) if $\langle y, \prec | y \rangle$ is a total order. Call a chain y *maximal* (*in* $\langle x, \prec \rangle$) if there is no chain z in $\langle x, \prec \rangle$ such that $y \subset z$. We shall now prove the

MAXIMAL CHAIN THEOREM (Hausdorff). *In every partial order there is a maximal chain.*

An insignificant modification of this result is called Zorn's lemma (see below, in the next subsection). Though Zorn did not know Hausdorff's result, the priority obviously belongs to Hausdorff, who found his theorem more than 30 years earlier. This should not, however, diminish Zorn's merit since, along with Kuratowski, it was he who recognized the importance of this result, and it was through him that the mathematical community became aware of it.

PROOF. Let $\langle x, \prec\rangle$ be a partial order, and let $g: (\mathscr{P}(x)\setminus\{0\})\to x$ be a choice function on $\mathscr{P}(x)\setminus\{0\}$. Put

$$G(u)=g(\{y\in x\setminus \text{ra}\,(u)\colon \text{ra}\,(u)\cup\{y\} \text{ is a chain in } \langle x, \prec\rangle\}).$$

Define F on the class of all ordinals by the recursion formula $F(\alpha)=G(F\upharpoonright\alpha)$ (cf. Theorem 4.1), and put

$$A=\{\xi\colon \text{ra}\,(F\upharpoonright\xi)\subseteq x\;\&\;\text{ra}\,(F\upharpoonright\xi) \text{ is a chain in } \langle x, \prec\rangle \text{ that is not maximal}\} \tag{17}$$

and

$$F'=F\upharpoonright A.$$

Similarly as in the proof of the preceding theorem, we can show that ra $(F')\subseteq x$ and that the operation F' is 1-1. We have

$$A=\text{dom}\,(F')=\{\xi\in\text{dom}\,(F')\colon \exists y\in x[F'(\xi)=y]\}$$

and

$$F'=\{\langle\xi y\rangle\colon \xi\in\text{dom}\,(F')\;\&\;y\in x\;\&\;F'(\xi)=y]\};$$

again, it follows from the Axiom of Replacement that these classes are sets. A is clearly transitive, and so it is an ordinal.

We claim that ra (F') is a chain in $\langle x, \prec\rangle$. In fact, in the contrary case there is a $\xi\in\text{dom}\,(F')=A$ such that ra $(F'\upharpoonright\xi)\cup\{F'(\xi)\}=$ ra $(F\upharpoonright\xi)\cup\{F(\xi)\}$ is not a chain. By the definition of A in (17), ra $(F\upharpoonright\xi)$ is a chain that is not maximal, and so the set

$$\{y\in x\setminus\text{ra}\,(F\upharpoonright\xi)\colon \text{ra}\,(F\upharpoonright\xi)\cup\{y\} \text{ is a chain in } \langle x, \prec\rangle\}$$

is not empty. By the definition of G, we can see that $F(\xi)=G(F\upharpoonright\xi)$ belongs to this set, i.e., that ra $(F\upharpoonright\xi)\cup\{F(\xi)\}$ is a chain, which is a contradiction. Our claim is established.

Finally, we show that ra $(F')=$ ra $(F\upharpoonright A)$ is a maximal chain in $\langle x, \prec\rangle$. In fact, in the contrary case we could conclude from the definition of A in (17) that $A\in A$, since A is an ordinal. This is impossible, and so the proof of the theorem is complete.

The Axiom of Choice, the Wellordering Theorem, and the Maximal Chain Theorem are equivalent relatively to the axioms of ZF; a part of this equivalence was seen in the preceding subsection and, moreover, it is easy to derive the Wellordering Theorem from the above theorem in the presence of the axioms of ZF. Hint: given a set x, consider the following ordering of wellorders on subsets of x: say that one such wellorder is less than or equal to another if the former is an initial segment of the latter. Taking a maximal chain y in this ordering, it is easily seen that $x = \bigcup \operatorname{dom}(y)$, and one can put together the wellorderings on the elements of $\operatorname{dom}(y)$ to obtain a wellordering of x. The drawback of such a proof is that if one wants to obtain the form of the Wellordering Theorem that we announced, then one has to define the ordinal α and the 1-1 function f by a transfinite recursion argument that is almost as complicated as the proof of this theorem itself.

4.8. Let $\langle x, \prec \rangle$ be a partial order; for a set $u \subseteq x$ and an element y of x, call y an *upper bound* (*in* $\langle x, \prec \rangle$) of u if $\forall z \in u[z \preceq y]$. Call $y \in x$ *maximal* (*in* $\langle x, \prec \rangle$) if $y \prec z$ holds for no $z \in x$. We have

Zorn's Lemma (general formulation). (Kuratowski [1922], Zorn [1935].) *If $\langle x, \prec \rangle$ is a partial order such that every chain has an upper bound, then x has a maximal element.*

Proof. By the preceding theorem, there is a maximal chain in $\langle x, \prec \rangle$. An upper bound of this chain is clearly a maximal element. The proof is complete.

Conversely, Hausdorff's Maximal Chain Theorem easily follows from Zorn's lemma. In fact, one only has to apply the latter to the set of all chains, as ordered by inclusion in a partial order, so as to obtain a maximal chain.

4.9. We give a slightly modified but equivalent formulation of Zorn's lemma. To this end, call a set y nested if it is totally ordered by inclusion, i.e., if $\forall u,v \in y[u \subseteq v \vee v \subseteq u]$. Call $u \in x$ maximal (in x, *with respect to inclusion*) if $\forall v \in x \neg u \subset v$. Then we have.

Zorn's Lemma (original formulation). *If x is a set such that for every nested set $y \subseteq x$ we have $\bigcup y \in x$, then x has a maximal element.*

The proof is obvious from the former version of Zorn's lemma; it is also easy to see that this version implies the former one.

4.9. Hausdorff's Cofinality Lemma. *For every total order $\langle x, \prec \rangle$ there is a cofinal subset y of x such that $\langle y, \prec | y \rangle$ is a wellorder and* tp $\langle y, \prec | y \rangle$ *is* $\leq |x|$.

PROOF. Let $\prec'$ be a wellordering of x such that $\operatorname{tp}\langle x, \prec'\rangle = |x|$. Put

$$y = \{v \in x : \forall u \in x[u \prec' v \Rightarrow u \prec v]\}. \tag{18}$$

We first claim that

$$u \prec' v \Leftrightarrow u \prec v \tag{19}$$

holds for any two $u, v \in y$. In fact, the implication $\Rightarrow$ here is trivial by the definition of y. Moreover, if $\neg u \prec' v$ holds, then we have either $v \prec' u$ or $v = u$, i.e., either $v \prec u$ or $v = u$ by the implication just established; hence $\neg u \prec v$ holds in this case. Thus (19) is proved. Note that (19) means that the orderings $\prec'$ and $\prec$ coincide on y, and so $\prec$ wellorders y and

$$\operatorname{tp}\langle y, \prec\rangle = \operatorname{tp}\langle y, \prec'\rangle \leq \operatorname{tp}\langle x, \prec'\rangle = |x|.$$

We have yet to prove that y is cofinal in $\langle x, \prec\rangle$. Assume, on the contrary, that there is a $v \in x$ such that $u \prec v$ holds for all $u \in y$, and let v_0 be the minimal such element in the wellordering $\prec'$. We shall get a contradiction by showing that $v_0 \in y$ holds according to (18). In fact, if $u \in x$ is such that $u \prec' v_0$, then the minimality of v_0 implies that we have $u \preceq u_0$ for some $u_0 \in y$. But then $u_0 \prec v_0$ holds (since this holds for any element u' of y replacing u_0); thus we can conclude that $u \prec v_0$. Hence (18) confirms that indeed $v_0 \in y$, which is a contradiction completing the proof.

4.10. The Lemma just proved enables one to prove the main facts about the cofinality operation cf defined in (3.4).

THEOREM. a) *For each ordinal* α, $\operatorname{cf}(\alpha)$ *is a cardinal*; b) *if* κ *is an infinite cardinal then* $\operatorname{cf}(\kappa)$ *is the least cardinal* λ *such that* κ *can be represented as the sum of* λ *cardinals* $<\kappa$.

As we trivially have $\operatorname{cf}(\operatorname{cf}(\alpha)) = \operatorname{cf}(\alpha)$, a) in fact means that $\operatorname{cf}(\alpha)$ is a regular cardinal for any limit ordinal α.

PROOF. *Ad* a). If α is not a limit ordinal then $\operatorname{cf}(\alpha) = 0$ or 1, which are cardinals. Assume therefore that α is a limit ordinal, and let $x \subseteq \alpha$ be a set cofinal in α such that $\operatorname{cf}(\alpha) = \operatorname{tp}\langle x, <\rangle$; the preceding lemma implies that there is an $y \subseteq x$ such that y is cofinal in $\langle x, <\rangle$ and $\operatorname{tp}\langle y, <\rangle \leq |x| \leq \operatorname{tp}\langle x, <\rangle = \operatorname{cf}(\alpha)$, where the second inequality holds by the definition of cardinality. We must actually have equality in place of both inequalities since y is obviously also cofinal in α, and so the contrary case would contradict the definition of $\operatorname{cf}(\alpha)$ in (3.4). Hence $\operatorname{cf}(\alpha) = |x|$, i.e., $\operatorname{cf}(\alpha)$ is indeed a cardinal.

Ad b). Let λ be the least cardinal λ' such that κ can be represented as the sum of λ' cardinals $<\kappa$. We clearly have $\lambda \leq \operatorname{cf}(\kappa)$ since if x is cofinal in κ and

$\operatorname{tp}\langle x, <\rangle=\operatorname{cf}(\kappa)$, then

$$\Sigma\langle|\xi|;\ \xi\in x\rangle$$

is such a representation of κ for $\lambda'=\operatorname{cf}(\kappa)$. To prove $\operatorname{cf}(\kappa)\leq\lambda$, note that we obviously have $\operatorname{cf}(\kappa)\leq\kappa$, and so this is trivial in case $\lambda=\kappa$. Assume therefore that $\lambda<\kappa$. We have

$$\kappa=\Sigma\langle\kappa_\alpha:\alpha<\lambda\rangle$$

with some cardinals $\kappa_\alpha<\kappa$ $(\alpha<\lambda)$ by the definition of λ, and it is easy to conclude from the assumption $\lambda<\kappa$ that the set $x=\{\kappa_\alpha:\ \alpha<\lambda\}$ is cofinal in κ. By the preceding lemma, there is then a set $y\subseteq x$ with $\operatorname{tp}\langle y, <\rangle\leq|x|\leq\lambda$ that is cofinal in x. y is cofinal also in κ, which shows that $\operatorname{cf}(\kappa)\leq\operatorname{tp}\langle y, <\rangle\leq\lambda$, completing the proof.

4.11. Hausdorff's Cofinality Lemma also allows one to extend the definition of cofinality in a meaningful way to an arbitrary linear order type. For an arbitrary linear order type d, let $\langle x, \prec\rangle$ be a linear order of order type d, and put

$$\operatorname{cf}(d)=\min\{\operatorname{tp}\langle y, \prec|y\rangle: y \text{ is cofinal in } x \text{ and } \langle y, \prec|y\rangle \text{ is a wellorder}\}. \tag{20}$$

It is clear that $\operatorname{cf}(d)$ is an ordinal; it is 0 if x is empty, it is 1 if $\langle x, \prec\rangle$ has a maximal element, and it is infinite according to Lemma 4.9 if neither of these is the case. The following result about this extended notion of cofinality is important:

THEOREM. *Let $\langle x, \prec\rangle$ be a linear order, and assume that y is cofinal in x. Then*

$$\operatorname{cf}(\operatorname{tp}\langle x, \prec\rangle)=\operatorname{cf}(\operatorname{tp}\langle y, \prec\rangle). \tag{21}$$

PROOF. It is obvious that the inequality $\leq$ holds in place of equality in (21). To prove also $\geq$, we shall use Lemma 4.9: Let z be a set cofinal in x that is wellordered by $\prec$ and is such that $\operatorname{tp}\langle z, \prec\rangle$ is the least possible (i.e., is equal to $\operatorname{cf}(\operatorname{tp}\langle x; \prec\rangle)$), and let z' be a set cofinal in y that is wellordered by $\prec$. For each $u\in z$, let $f(u)$ be the $\prec$-least element of z' such that $u\preceq f(u)$. It is easy to see that

$$z''=\{f(u): u\in z\}\subseteq z'\subseteq y$$

is also cofinal in z and that

$$\operatorname{tp}\langle z'', \prec\rangle\leq\operatorname{tp}\langle z, \prec\rangle$$

(in fact, one needs transfinite recursion to show this latter). As the right-hand side here equals $\operatorname{cf}(\operatorname{tp}\langle x, \prec\rangle)$, this shows that the inequality $\geq$ also holds in (21). The proof is complete.

Noting that $\operatorname{cf}(\operatorname{cf}(d))=\operatorname{cf}(d)$ holds for any order type d by the (trivial part) of the theorem just proved, Theorem 4.10.a implies that $\operatorname{cf}(d)$ is always a cardinal, in fact, 0, 1, or a regular cardinal.

CHAPTER II

PRELIMINARIES

In the three sections of this chapter we discuss important results concerning the theory of stationary sets, which is largely a theory of certain functions sending ordinals to ordinals; classical results on cardinal exponentiation (recent important results on powers of singular cardinals will be presented in Sections 47 and 48); finally, a theory of the logarithm operation is developed.

5. STATIONARY SETS

Throughout this section, η denotes a fixed limit ordinal with $\mathrm{cf}(\eta)>\omega$. A subset X of η is called *closed* (in η) if it is closed in the topology of η generated by its natural ordering; in other words, if $\sup(X\cap\alpha)\in X$ holds for any $\alpha<\eta$ with $X\cap\alpha\neq 0$. $X\subseteq\eta$ is called *unbounded* (*in* η) if it is cofinal in η, i.e., if $\sup X=\eta$. A set $X\subseteq\eta$ which is closed and unbounded is called a *closed unbounded set*, but we shall usually use the shorter acronym *club* (*in* η). A set $X\subseteq\eta$ is called *stationary* (*in* η) if it intersecs every club in η. Obviously, a stationary set must be cofinal to η and, clearly, η itself is a stationary set.

Stationary sets are mainly of interest if η is an uncountable regular cardinal (though we shall see in the proof of Theorem 5.7 that there is some advantage in considering the general case of any ordinal η with $\mathrm{cf}(\eta)>\omega$). First we need a simple lemma:

LEMMA 5.1. *The intersection of two clubs in η is a club in η. Consequently in η, if C is a club and S is stationary then $S\cap C$ is also stationary.*

PROOF. It is easy to see that the first assertion holds. In fact, if D and E are clubs, then their intersection, the intersection of two closed sets, is closed. It is also cofinal to η. In fact, we have to show that $(D\cap E)\setminus\alpha$ is nonempty for any $\alpha<\eta$. Let $\langle\alpha_k\colon k<\omega\rangle$ be an increasing sequence such that $\alpha_0>\alpha$ and $\alpha_k\in D$ or $\alpha_k\in E$ according as k is even or odd. Then clearly $\alpha_\omega\in D\cap E$ holds with $\alpha_\omega=\sup\{\alpha_k\colon k<\omega\}$.

We are now going to establish the second assertion. To this end, let D be an arbitrary club. The set $(S\cap C)\cap D=S\cap(C\cap D)$ is not empty, since S, being stationary, meets the club $C \cap D$. So $S \cap C$ meets every club, which we wanted to show.

A consequence of this lemma is the following theorem, confirming our claim made above that stationary sets are of real interest only in regular uncountable cardinals:

THEOREM 5.2. *Assume* $\langle \alpha_\xi : \xi < \mathrm{cf}(\eta) \rangle$ *is an increasing sequence of ordinals tending to* η *that is continuous, i.e.,* $\alpha_\xi = \sup \{\alpha_\gamma : \gamma < \xi\}$ *holds for every limit ordinal* $\xi < \mathrm{cf}(\eta)$. *Then* $S \subseteq \eta$ *is stationary in* η *if and only if the set*

$$\{\xi < \mathrm{cf}(\eta) : \alpha_\xi \in S\}$$

is stationary in $\mathrm{cf}(\eta)$.

PROOF. The set $C = \{\alpha_\xi : \xi < \mathrm{cf}(\eta)\}$ is obviously a club in η. Hence it easily follows from the first assertion of the above lemma that a set is stationary in η if and only if it meets every club $D \subseteq C$. The assertion of the theorem can now be established by observing that a set $D \subseteq C$ is a club in η if and only if

$$\{\xi < \mathrm{cf}(\eta) : \alpha_\xi \in D\}$$

is a club in $\mathrm{cf}(\eta)$. The proof is complete.

We are going to give a very useful characterization of stationary sets, but we shall need a few more definitions for this purpose.

A function f sending ordinals to ordinals, i.e., a function f such that $\mathrm{dom}(f)$, $\mathrm{ra}(f) \subseteq On$ is called *regressive* if $f(\xi) < \xi$ whenever ξ is a nonzero element of its domain. Occasionally, when this does not lead to confusion, we may call f regressive even if $f(\xi) < \xi$ holds only for 'large' ξ's in $\mathrm{dom}(f)$ (when studying stationary subsets of η, large means $\xi > \alpha$ for some $\alpha < \eta$ depending on f); the reader can easily see that this modification of the definition does not affect the validity of the theorems below, but may occasionally render technical details simpler. Given a function f with $\mathrm{dom}(f)$, $\mathrm{ra}(f) \subseteq \eta$, we call f *divergent* (in η) if the inverse image under f of bounded sets is bounded. In other words, f is divergent in η if

$$\forall \alpha < \eta \exists \beta < \eta \forall \gamma \in \mathrm{dom}(f) [f(\gamma) < \alpha \Rightarrow \gamma < \beta]. \tag{1}$$

Note that this definition is meaningful even if we allow $\mathrm{cf}(\eta) = \omega$, and this fact will be useful to us in a technical sense later. The fundamental connection between stationary sets and regressive functions is expressed by the following theorem, where the assumption $\mathrm{cf}(\eta) > \omega$ made above must of course be added.

THEOREM 5.3 (Neumer [1951]). *A set* $S \subseteq \eta$ *is stationary if and only if there is no divergent regressive function* f *with* S *as its domain.*

PROOF. "If". Assume S is not stationary, and let $C \subseteq \eta \setminus S$ be a club. Then

$$f(\xi) = \sup(C \cap \xi)$$

is obviously a divergent regressive function on S.

"Only if". Let f be a divergent regressive function with S as its domain. Put

$$C = \{\alpha < \eta : \alpha \neq 0 \;\&\; \forall \gamma \in S[f(\gamma) < \alpha \Rightarrow \gamma < \alpha]\}.$$

We claim that C is a club. C is clearly closed; we have to show that it is also unbounded, i.e., that for any $\zeta < \eta$ the set $C \setminus \zeta$ is not empty. To this end, let $\alpha_0 = \zeta$, and if $\alpha_n < \eta$ has already been defined for some $n < \omega$, let α_{n+1} be an ordinal with $\alpha_n \leq \alpha_{n+1} < \eta$ such that formula (1) above holds with $\alpha = \alpha_n$ and $\beta = \alpha_{n+1}$. Then $\alpha = \sup_{n<\omega} \alpha_n < \eta$ as $\operatorname{cf}(\eta) > \omega$, and, clearly, $\alpha \in C$. This establishes our claim that C is a club. It is easy to see that $C \cap S$ is empty. In fact, if $\alpha \in S$ and $\alpha > 0$ then $f(\alpha) < \alpha$, so the substitution $\gamma = \alpha$ in the definition of C shows that $\alpha \notin C$. Hence S does not meet the club C, i.e., S is not stationary. The proof is complete.

THEOREM 5.4 (Fodor [1956]). *Assume $S \subseteq \eta$ is a stationary set and f is a regressive function on S. Then f is bounded on some stationary set, i.e., there is a stationary set $X \subseteq S$ and an ordinal $\xi < \eta$ such that $f``X \subseteq \xi$.*

PROOF. Write $\kappa = \operatorname{cf}(\eta)$ and $\xi_0 = 0$. Let $\langle \xi_\alpha : \alpha < \kappa \rangle$ be an increasing sequence tending to η that is continuous (i.e., $\xi_\alpha = \sup_{\beta<\alpha} \xi_\beta$ for all limit $\alpha < \kappa$), and put

$$S_\alpha = \{\gamma \in S : \xi_\alpha \leq f(\gamma) < \xi_{\alpha+1}\}$$

for any $\alpha < \kappa$. As the intervals $[\xi_\alpha, \xi_{\alpha+1}) = \{\xi : \xi_\alpha \leq \xi < \xi_{\alpha+1}\}$ cover η, we clearly have

$$S = \bigcup_{\alpha<\kappa} S_\alpha .$$

We are going to prove that at least one S_α is stationary, and this will clearly suffice. For this purpose, assume on the contrary, that S_α is not stationary for any $\alpha < \kappa$, and let f_α be a divergent regressive function defined on S_α; such an f_α exists in view of the preceding theorem. We are going to define a divergent regressive function g the domain of which is S; this will contradict the assumption that S is stationary. To this end, put

$$g(\gamma) = \max\{\xi_\alpha, f_\alpha(\gamma)\}$$

if $\alpha < \kappa$ and $\gamma \in S_\alpha$. This formula defines $g(\gamma)$ for any $\gamma \in \bigcup_{\alpha<\kappa} S_\alpha = S$. g is clearly regressive, as both f_α and f are, and $\xi_\alpha \leq f(\gamma)$ for $\gamma \in S_\alpha$. We are going to show that g is divergent. In fact, if $\xi<\eta$ then $\xi\le\xi_\alpha$ for some $\alpha<\kappa$, and so

$$\{\gamma: g(\gamma)<\xi\}\subseteq \bigcup_{\beta<\kappa} \{\gamma\in S_\beta: g(\gamma)<\xi_\alpha\}\subseteq$$

$$\subseteq \bigcup_{\beta<\alpha} \{\gamma\in S_\beta: f_\beta(\gamma)<\xi_\alpha\}.$$

None of the sets on the right-hand side is cofinal in η as f_β is divergent. As there are less than $\kappa=\mathrm{cf}(\eta)$ of these sets, their union is not cofinal in η either, showing that g is indeed divergent. By the preceding theorem, this contradicts the assumption that S is stationary, completing the proof.

From here we immediately obtain:

THEOREM 5.5 (Fodor [1956]). *Assume $\kappa>\omega$ is regular, and let $S\subseteq\kappa$ be a set stationary in κ. If f is a regressive function on S, then f is constant on a set stationary in κ.*

COROLLARY 5.6 (Fodor [1956]). *Assume $\kappa>\omega$ is regular, and let N_ξ, $\xi<\kappa$, be sets nonstationary in κ. Then $N=\bigcup_{\xi<\kappa}(N_\xi\setminus(\xi\dot{+}1))$ is nonstationary in κ. (In particular: the union of less than κ nonstationary subsets of κ is also nonstationary.)*

PROOF. Put $M_\xi=N_\xi\setminus\bigcup_{\alpha<\xi}N_\alpha$, and for $\gamma\in M_\xi\setminus(\xi\dot{+}1)$ put $f(\gamma)=\xi$. Then f is a regressive function on the set $\bigcup_{\xi<\kappa}(M_\xi\setminus(\xi\dot{+}1))=N$. We have $\{\gamma\in N: f(\gamma)=\xi\}\subseteq$ $\subseteq M_\xi\subseteq N_\xi$ for any $\xi<\kappa$, i.e., f is not constant on any stationary set; hence N is not stationary. The last sentence of the corollary, i.e., the remark in parentheses, is a simple consequence of the main assertion. The proof is complete.

Our next aim is to prove a deep result of Solovay saying that every stationary set splits up into many disjoint stationary sets.

THEOREM 5.7 (Solovay [1971]). *Let $\kappa>\omega$ be a regular cardinal. Then every stationary subset of κ can be represented as the union of κ pairwise disjoint stationary sets.*

To prove this theorem, we need the following concept: For a set X of ordinals, define the class $\mathrm{nst}(X)$ of ordinals *nonstationary* with respect to X by stipulating that, for an arbitrary ordinal α, $\alpha\in\mathrm{nst}(X)$ if either α is cofinal to ω or $\mathrm{cf}(\alpha)>\omega$ and $X\cap\alpha$ is nonstationary in α. $\mathrm{nst}(X)$ is obviously a real class, as it contains all limit ordinals above $\sup X$, but we shall only be interested in the set $X\cap\mathrm{nst}(X)$. The proof proceeds via two lemmas, which are also due to Solovay. In these lemmas, κ denotes an uncountable regular cardinal.

LEMMA 5.8 *If X is a stationary subset of κ, then so is $A = X \cap \text{nst}(X)$.*

PROOF. Assuming, on the contrary, that A is nonstationary, let $C \subseteq \kappa \setminus A$ be a club in κ. Let C' be the set of all limit points of C, i.e., put

$$C' = \{\xi : 0 < \xi < \kappa \;\&\; \xi = \sup(C \cap \xi)\}.$$

It is clear that C' is also a club in κ. X being stationary, the set $X \cap C'$ is nonempty; let α be its first element. α is a limit ordinal, as C' contains only limit ordinals. If $\text{cf}(\alpha) = \omega$, then $\alpha \in \text{nst}(X)$ by the definition of the right-hand side, and so $\alpha \in X \cap \text{nst}(X) \cap C' \subseteq A \cap C$, which contradicts the choice of C. Hence we must have $\text{cf}(\alpha) > \omega$. In this case $C' \cap \alpha$ is a club in α, and so the set $X \cap \alpha$, being disjoint from $C' \cap \alpha$ in view of the choice of α, is nonstationary in α; hence $\alpha \in \text{nst}(X)$. Therefore we again have $\alpha \in X \cap \text{nst}(X) \cap C' \subseteq A \cap C$, which contradicts the choice of C. This contradiction completes the proof of the lemma.

The next lemma depends on the assumption that Theorem 5.7 fails, so it is of no interest in itself.

LEMMA 5.9. *Let X be a stationary subset of κ, and assume that X cannot be decomposed into κ pairwise disjoint stationary sets. Then, for every regressive function f defined on a stationary subset of X, there is a nonstationary set $N \subseteq \text{dom}(f)$ such that f is bounded on $\text{dom}(f) \setminus N$, i.e., there is an $\alpha < \kappa$ such that $f(\xi) < \alpha$ whenever $\xi \in \text{dom}(f) \setminus N$. (In other words: f is essentially bounded.)*

PROOF. Let Z be the set of all ordinals $\zeta < \kappa$ such that the set

$$X_\zeta = \{\xi \in \text{dom}(f) : f(\xi) = \zeta\}$$

is stationary. Then $|Z| < \kappa$, since otherwise one could easily define a decomposition of X into κ pairwise disjoint stationary sets with the aid of the sets X_ζ, $\zeta \in Z$. On the other hand, the set $N = \text{dom}(f) \setminus \bigcup_{\zeta \in Z} X_\zeta$ is nonstationary in view of Fodor's Theorem 5.5. For any $\xi \in \text{dom}(f) \setminus N$ we have $f(\xi) < \alpha$ with $\alpha = \sup Z$. Here $\alpha < \kappa$, as $|Z| < \kappa$, which completes the proof of the lemma.

We are now in a position to accomplish the

POOF OF THEOREM 5.7. Let X be a stationary subset of κ and assume, on the contrary, that X cannot be decomposed into κ pairwise disjoint stationary sets. On account of this assumption we shall be free to use Lemma 5.9.

Consider the set $A = X \cap \text{nst}(X)$. For any $\alpha \in A$, there is a regressive function f_α defined on $X \cap \alpha$ that is divergent in α. In fact, if $\text{cf}(\alpha) > \omega$, then this follows by Theorem 5.3 as $X \cap \alpha$ is nonstationary in α, and if $\text{cf}(\alpha) = \omega$, then this is trivially true since then one can easily define a divergent regressive function on the whole of α, and one can take the restriction of this function to the set $X \cap \alpha$ as f_α. For any $\xi \in X$ put

$$f(\xi) = \min \{\beta < \xi: A_\beta^\xi \text{ is stationary in } \kappa\}, \tag{2}$$

where

$$A_\beta^\xi = \{\alpha \in A: \xi < \alpha \;\&\; f_\alpha(\xi) = \beta\}.$$

Note that the set after the minimum sign in (2) is not empty. In fact, A is stationary in view of Lemma 5.8, and we clearly have

$$A \setminus (\xi \dot{+} 1) = \bigcup_{\beta < \xi} A_\beta^\xi;$$

so there must be a $\beta < \xi$ for which A_β^ξ is stationary (cf. the sentence in parentheses in Corollary 5.6). Hence (2) defines a regressive function on X. The preceding lemma now implies that f is essentially bounded, i.e., that there is a $\zeta < \kappa$ and a nonstationary subset M of X such that

$$f(\xi) < \zeta \qquad \text{whenever} \qquad \xi \in X \setminus M. \tag{3}$$

So far, we have made use of the fact that the functions f_α are regressive. We are now going to exploit their being divergent. For any $\alpha \in A$ with $\alpha > \zeta$ put

$$g(\alpha) = \sup \{\xi \in X \cap \alpha: \quad f_\alpha(\xi) < \zeta\}.$$

Note that the fact that f_α is a function defined on $X \cap \alpha$ that is divergent in α implies that $g(\alpha) < \alpha$ holds, i.e., that g is a regressive function defined on $A \setminus (\zeta \dot{+} 1)$. Observing that A is a stationary subset of X (cf. Lemma 5.8), the preceding lemma again implies that g is essentially bounded; i.e., there is a $\gamma < \kappa$ and a nonstationary set $N \subseteq A$ such that $g(\alpha) < \gamma$ whenever $\alpha \in A \setminus N$ (assume $\zeta \dot{+} 1 \subseteq N$, so that g is defined everywhere on $A \setminus N$). In view of the definition of g, this means that $f_\alpha(\xi) \geq \zeta$ whenever $\alpha \in A \setminus N$, $\xi \in X$ and $\gamma \leq \xi < \alpha$. Now turning back to the definition of f in (2), this implies that $f(\xi) \geq \zeta$ for any ξ with $\gamma \leq \xi \in X$, as $A_\beta^\xi \subseteq N$ holds for any $\beta < \zeta$ in this case. This, however, contradicts (3), completing the proof of Theorem 5.7.

An alternative, model-theoretic proof of the above theorem is given in Baumgartner–Hajnal–Máté [1975].

6. EQUALITIES AND INEQUALITIES FOR CARDINALS

The present section considers classical equalities and inequalities for cardinal exponentiation. The recent advance in the study of the powers of singular cardinals, initiated by Silver and further developed by Galvin and Hajnal, needs too complicated methods to be a feature of this section: we shall, however, devote Sections 47 and 48 to the discussion of these and other new results on cardinal exponentiation.

We start with two inequalities; we shall refer to either of them as König's theorem (cf. J. König [1905]). The second one of these was the first known nontrivial inequality for infinite cardinals apart from Cantor's result saying that $2^\kappa > \kappa$ for all $\kappa \geq \omega$.

THEOREM 6.1. a) *Let* $\langle \kappa_\xi : \xi < \alpha \rangle$ *be a sequence of cardinals such that* $\kappa_\xi \geq 2$ *for every* $\xi < \alpha$. *Then*

$$\sum_{\xi<\alpha} \kappa_\xi \leq \prod_{\xi<\alpha} \kappa_\xi .$$

b) *Let* $\langle \kappa_\xi : \xi < \alpha \rangle$ *and* $\langle \lambda_\xi : \xi < \alpha \rangle$ *be two sequences of cardinals such that* $\kappa_\xi < \lambda_\xi$ *for every* $\xi < \alpha$. *Then*

$$\sum_{\xi<\alpha} \kappa_\xi < \prod_{\xi<\alpha} \lambda_\xi .$$

PROOF. *Ad* a). The case $\alpha \leq 2$ is rather easy, but requires a different approach; hence we assume $\alpha > 2$. Let A_ξ, $\xi < \alpha$, be pairwise disjoint sets such that $|A_\xi| = \kappa_\xi$. In order to prove the above inequality, it is enough to define a 1-1 function f from $\bigcup_{\xi<\alpha} A_\xi$ into the set of *transversals* of the set $\{A_\xi : \xi < \alpha\}$; here a transversal of a set of *pairwise disjoint* (nonempty) sets is defined as a set intersecting each element of the set in exactly one element. Let $\{a^i_\xi : \xi < \alpha\}$, $i = 0, 1$, be two disjoint transversals of $\{A_\xi : \xi < \alpha\}$ ($a^i_\xi \in A_\xi$; note that the assumption $\kappa_\xi \geq 2$ is used here); if $x \in A_\eta$, $\eta < \alpha$, then put

$$f(x) = \{x\} \cup \{a^i_\xi : \xi < \alpha \;\&\; \xi \neq \eta\},$$

where $i = 0$ if $x \neq a^0_\eta$, and $i = 1$ otherwise. It is clear that f is 1-1; note that the assumption $\alpha > 2$ is used here.

Ad b). We may assume that $\kappa_\xi > 0$ always holds. Then $\lambda_\xi \geq 2$ holds for every $\xi < \alpha$, and so a) implies that $\leq$ holds. We are going to show that equality cannot hold. To this end, let A_ξ and B_ξ ($\xi < \alpha$) be sets any two of which are disjoint such that $|A_\xi| = \kappa_\xi$ and $|B_\xi| = \lambda_\xi$ hold for every $\xi < \alpha$. Let f be a mapping of $\bigcup A_\xi$ into the set of *transversals* of the set $\{B_\xi : \xi < \alpha\}$ (see the definition in the first part of the proof); it suffices to show that f is not onto. Let $C_\xi = B_\xi \cap (\bigcup f``A_\xi)$; $|C_\xi| \leq |A_\xi| = \kappa_\xi < \lambda_\xi = |B_\xi|$, so let $b_\xi \in B_\xi \setminus C_\xi$. Clearly, $\{b_\xi : \xi < \alpha\} \notin \operatorname{ra}(f)$, which completes the proof.

COROLLARY 6.2. *Let* $\langle \kappa_\xi : \xi < \alpha \rangle$ *be a strictly increasing sequence of cardinals, where* α *is a limit ordinal. Then*

$$\sum_{\xi<\alpha} \kappa_\xi < \prod_{\xi<\alpha} \kappa_\xi .$$

PROOF. The result follows from the above theorem with $\lambda_\xi = \kappa_{\xi+1}$.

THEOREM 6.3 (Bernstein–Hausdorff–Tarski; cf. Hausdorff [1904], Tarski [1925]). *Assume κ is infinite and $0<\lambda<\mathrm{cf}(\kappa)$. Then $\underset{\sim}{\kappa}^\lambda \cdot \kappa = \kappa^\lambda$.*

PROOF. We obviously have $\underset{\sim}{\kappa}^\lambda \cdot \kappa \le \kappa^\lambda$. To prove the converse inequality, note that $\kappa^\lambda = |[\kappa]^\lambda|$; clearly for any $x \in [\kappa]^\lambda$ there is a $\xi<\kappa$ such that $x \subseteq \xi$, as $\lambda < \mathrm{cf}(\kappa)$. Hence

$$[\kappa]^\lambda = \bigcup_{\xi<\kappa} [\xi]^\lambda,$$

and so

$$\kappa^\lambda \le \sum_{\xi<\kappa} |\xi|^\lambda \le \Sigma\langle \xi^+ \cdot \xi^\lambda : \xi<\kappa \text{ is a cardinal}\rangle = \kappa \cdot \underset{\sim}{\kappa}^\lambda.$$

The proof is complete.

COROLLARY 6.4. a) *If κ is infinite and $0<\lambda\le\kappa$ then $(\kappa^+)^\lambda = \kappa^\lambda \cdot \kappa^+$.*

b) *If α is a limit ordinal, $\langle \kappa_\xi : \xi<\alpha\rangle$ is a strictly increasing sequence of cardinals such that $\sum_{\xi<\alpha} \kappa_\xi = \kappa$, and λ is a cardinal with $0<\lambda<\mathrm{cf}(\alpha)$, then $\sum_{\xi<\alpha} \kappa_\xi^\lambda = \kappa^\lambda$.*

PROOF. a) follows from the equality $(\underset{\sim}{\kappa}^+)^\lambda = \kappa^\lambda$. To prove b) observe that $\kappa = \sup_{\xi<\alpha} \kappa_\xi$ and $\mathrm{cf}(\kappa) = \mathrm{cf}(\alpha)$. Hence

$$\kappa^\lambda = \kappa \cdot \underset{\sim}{\kappa}^\lambda \le \kappa \cdot \sum_{\eta<\kappa} |\eta|^\lambda \le \kappa \cdot \sum_{\xi<\alpha} \kappa_\xi^\lambda \cdot \kappa_{\xi+1} \le \sum_{\xi<\alpha} \kappa_\xi^\lambda \le \kappa^\lambda,$$

which completes the proof.

THEOREM 6.5 (Tarski [1925]). *Assume κ is an infinite cardinal and $\kappa = \sum_{\xi<\mathrm{cf}(\kappa)} \kappa_\xi$, where $2 \le \kappa_\xi < \kappa$ holds for every $\xi < \mathrm{cf}(\kappa)$. Then*

$$\prod_{\xi<\mathrm{cf}(\kappa)} \kappa_\xi = \kappa^{\mathrm{cf}(\kappa)}.$$

PROOF. The inequality $\prod_{\xi<\mathrm{cf}(\kappa)} \kappa_\xi \le \kappa^{\mathrm{cf}(\kappa)}$ is obvious. To prove the converse inequality, we may assume that $\mathrm{cf}(\kappa) < \kappa$, as $2^\kappa = \kappa^\kappa$. In this case we may further assume that the sequence $\langle \kappa_\xi : \xi < \mathrm{cf}(\kappa)\rangle$ is strictly increasing; indeed we can bring this about by rearranging and thinning the original sequence. Write $\lambda = \mathrm{cf}(\kappa)$. Then, using a simple distributive law, we obtain

$$\kappa^\lambda = \prod_{\zeta<\lambda} \sum_{\xi<\lambda} \kappa_\xi = \sum_{f \in {}^\lambda\lambda} \prod_{\zeta<\lambda} \kappa_{f(\zeta)} \le 2^\lambda \cdot \prod_{\xi<\lambda} \kappa_\xi = \prod_{\xi<\lambda} \kappa_\xi;$$

the inequality here holds since for each $f \in {}^\lambda\lambda$ there is a strictly increasing $g \in {}^\lambda\lambda$ such that $f(\xi) \le g(\xi)$ for every $\xi < \lambda$. The theorem is proved.

Corollary 6.6. *Under the assumptions of the preceding theorem we have*

$$\kappa^{\lambda} = \prod_{\xi < \mathrm{cf}(\kappa)} \kappa_{\xi}^{\lambda}$$

whenever $\lambda \geq \mathrm{cf}(\kappa)$.

The proof is trivial from the preceding theorem. From Theorem 6.3 and Corollary 6.6 we obtain:

Corollary 6.7. *Let* κ *be infinite,* $\lambda > 0$, *and assume that* $\rho^{\lambda} \leq \kappa$ *whenever* $\rho < \kappa$. *Then* $\kappa^{\lambda} = \kappa$ *if* $\lambda < \mathrm{cf}(\kappa)$, *and* $\kappa^{\lambda} = \kappa^{\mathrm{cf}(\kappa)}$ *if* $\lambda \geq \mathrm{cf}(\kappa)$.

Corollary 6.8. *For any infinite cardinal* κ *we have* $\kappa^{\mathrm{cf}(\kappa)} > \kappa$.

Proof. If κ is regular, then we have $\kappa^{\mathrm{cf}(\kappa)} = \kappa^{\kappa} = 2^{\kappa} > \kappa$. If κ is singular, then represent it as the sum of a strictly increasing sequence $\langle \kappa_{\xi} : \xi < \mathrm{cf}(\kappa) \rangle$ of cardinals, and use Theorem 6.5 and Corollary 6.2.

Theorem 6.9 (J. König [1905]). *If* κ *and* λ *are cardinals,* $\kappa \geq 2$, *and* λ *is infinite, then* $\mathrm{cf}(\kappa^{\lambda}) > \lambda$.

Proof. Put $\rho = \kappa^{\lambda}$. By the previous result we have $\rho^{\mathrm{cf}(\rho)} > \rho$, whereas $\rho^{\lambda} = \rho$ holds, establishing $\mathrm{cf}(\kappa^{\lambda}) = \mathrm{cf}(\rho) > \lambda$. A direct proof using König's theorem is equally simple. In fact, assume, on the contrary, that $\kappa^{\lambda} = \sum_{\xi < \lambda} \kappa_{\xi}$, where $\kappa_{\xi} < \kappa^{\lambda}$. Then $\sum_{\xi < \lambda} \kappa_{\xi} < \prod_{\xi < \lambda} \kappa^{\lambda}$ holds by König's theorem, which is a contradiction as both sides here are equal to κ^{λ}.

We need a number of equalities and inequalities for the weak power $\kappa^{\underline{\lambda}}$.

Theorem 6.10. *Assume* $\kappa \geq 2$ *and* $\lambda \geq \omega$. *Then*

a) $\kappa^{\underline{\lambda}} \geq \lambda$.

b) $\kappa^{\underline{\lambda}} = \sum_{\rho_0 \leq \rho < \lambda} \kappa^{\rho}$ *for every* $\rho_0 < \lambda$.

c) *Either* (i) *there is a* $\rho_0 < \lambda$ *such that* $\kappa^{\underline{\lambda}} = \kappa^{\rho}$ *holds for every* ρ *with* $\rho_0 \leq \rho < \lambda$, *or* (ii) *there is an increasing sequence* $\langle \lambda_{\xi} : \xi < \mathrm{cf}(\lambda) \rangle$ *of cardinals tending to* λ *such that the sequence* $\langle \kappa^{\lambda_{\xi}} : \xi < \mathrm{cf}(\lambda) \rangle$ *is strictly increasing and* $\kappa^{\underline{\lambda}} = \sum_{\xi < \mathrm{cf}(\lambda)} \kappa^{\lambda_{\xi}}$ *holds.*

d) $\mathrm{cf}(\kappa^{\underline{\lambda}}) \geq \lambda$ *if* c(i) *holds, and* $\mathrm{cf}(\kappa^{\underline{\lambda}}) = \lambda$ *if* c(ii) *holds.*

e) *We have*

$$(\kappa^{\underline{\lambda}})^{\rho} = \begin{cases} \kappa^{\underline{\lambda}} & \text{if } 0 < \rho < \mathrm{cf}(\lambda), \\ \kappa^{\lambda} & \text{if } \mathrm{cf}(\lambda) \leq \rho < \lambda, \\ \kappa^{\rho} & \text{if } \lambda \leq \rho. \end{cases}$$

f) *We have*

$$(\kappa^{\underset{\cdot}{\lambda}})^{\varrho}=\begin{cases}\kappa^{\underset{\cdot}{\lambda}} & \text{if} \quad 1<\rho\leq \operatorname{cf}(\lambda),\\ \kappa^{\lambda} & \text{if} \quad \operatorname{cf}(\lambda)<\rho\leq\lambda,\\ \kappa^{\varrho} & \text{if} \quad \lambda<\rho.\end{cases}$$

Proof. *Ad* a). For every $\rho<\lambda$ we have $\kappa^{\underset{\cdot}{\lambda}}\geq\kappa^{\rho}\geq 2^{\rho}>\rho$, and so we must have $\kappa^{\underset{\cdot}{\lambda}}\geq\lambda$.

Ad b). This is rather simple and will be left to the reader.

Ad c). If $\lambda=\rho^{+}$ for some ρ, then $\kappa^{\rho}=\kappa^{\underset{\cdot}{\lambda}}$, and so (i) holds. If λ is a limit cardinal, then either the sequence $\langle\kappa^{\rho}\colon \rho_0\leq\rho<\lambda\rangle$ is constant for some $\rho_0<\lambda$, in which case (i) holds, or $\langle\kappa^{\rho}\colon\rho<\lambda\rangle$ has a strictly increasing subsequence of type $\operatorname{cf}(\lambda)$, in which case (ii) holds.

Ad d). If c(i) holds with some $\rho_0<\lambda$, then $\operatorname{cf}(\kappa^{\underset{\cdot}{\lambda}})=\operatorname{cf}(\kappa^{\rho})>\rho$ holds for all ρ with $\rho_0\leq\rho<\lambda$ in view of Theorem 6.9, and so we indeed have $\operatorname{cf}(\kappa^{\underset{\cdot}{\lambda}})\geq\lambda$. If c(ii) holds, then $\kappa^{\underset{\cdot}{\lambda}}$ is the limit of a strictly increasing sequence of type $\operatorname{cf}(\lambda)$ of cardinals, and so we have $\operatorname{cf}(\kappa^{\underset{\cdot}{\lambda}})=\operatorname{cf}(\lambda)$.

Ad e). Assume first that c(i) holds. The first and the third case is obvious then, and to see that the equality is valid also if the second alternative holds, i.e., when $\operatorname{cf}(\lambda)\leq\rho<\lambda$, one only has to observe that $\kappa^{\underset{\cdot}{\lambda}}=\kappa^{\lambda}$ holds. To see this latter equality, observe that λ must be singular in this case. Represent λ in the form $\lambda=\sum_{\xi<\operatorname{cf}(\lambda)}\lambda_{\xi}$, where $\lambda_{\xi}<\lambda$, and we may also assume that $\lambda_{\xi}\geq\rho_0\geq\operatorname{cf}(\lambda)$, where ρ_0 is as described in c(i). Then

$$\kappa^{\lambda}=\kappa^{\sum_{\xi<\operatorname{cf}(\lambda)}\lambda_{\xi}}=\prod_{\xi<\operatorname{cf}(\lambda)}\kappa^{\lambda_{\xi}}=\prod_{\xi<\operatorname{cf}(\lambda)}\kappa^{\rho_0}=\kappa^{\rho_0}=\kappa^{\underset{\cdot}{\lambda}},$$

which we wanted to show.

Next assume that c(ii) holds; in this case we have $\kappa^{\underset{\cdot}{\lambda}}=\sum_{\xi<\operatorname{cf}(\lambda)}\kappa^{\lambda_{\xi}}$. If $\rho<\operatorname{cf}(\lambda)$, then, using Corollary 6.4.b and making the harmless assumption $\lambda_{\xi}\geq\rho$ for all $\xi<\operatorname{cf}(\lambda)$, we can see that

$$(\kappa^{\underset{\cdot}{\lambda}})^{\rho}=\sum_{\xi<\operatorname{cf}(\lambda)}(\kappa^{\lambda_{\xi}})^{\rho}=\sum_{\xi<\operatorname{cf}(\lambda)}\kappa^{\lambda_{\xi}}=\kappa^{\underset{\cdot}{\lambda}},$$

which proves the first equality. If $\rho\geq\operatorname{cf}(\lambda)$, then, using Corollary 6.6 while observing that $\operatorname{cf}(\kappa^{\underset{\cdot}{\lambda}})=\operatorname{cf}(\lambda)$ holds by virtue of d) of the theorem being proved, we can conclude that

$$(\kappa^{\underset{\cdot}{\lambda}})^{\rho}=\prod_{\xi<\operatorname{cf}(\lambda)}(\kappa^{\lambda_{\xi}})^{\rho}=\kappa^{\sum_{\xi<\operatorname{cf}(\lambda)}\lambda_{\xi}\cdot\rho}=\kappa^{\lambda\cdot\rho}$$

which completes the proof of the second and third equality as well.

Ad f). This is a direct consequence of the results obtained in the preceding point e).

COROLLARY 6.11. (Bukovsky [1965], Hechler [1973]). *Let $\kappa \geq 2$ and $\lambda \geq \omega$ be cardinals, and assume that λ is singular. If there is a $\rho_0 < \lambda$ such that $\kappa^\rho = \kappa^{\rho_0}$ whenever $\rho_0 \leq \rho < \lambda$, then $\kappa^{\underline{\lambda}} = \kappa^\lambda = \kappa^{\rho_0}$.*

PROOF. According to e) of the theorem just proved, we have $(\kappa^{\underline{\lambda}})^{\mathrm{cf}(\lambda)} = \kappa^\lambda$, and by the assumptions we can conclude that

$$(\kappa^{\underline{\lambda}})^{\mathrm{cf}(\lambda)} = \kappa^{\rho_0 \cdot \mathrm{cf}(\lambda)} = \kappa^{\rho_0},$$

as $\rho_0 \leq \rho_0 \cdot \mathrm{cf}(\lambda) < \lambda$, completing the proof. A direct proof is equally simple. In fact, if $\lambda = \sum_{\xi < \mathrm{cf}(\lambda)} \lambda_\xi$, where $\rho_0 \leq \lambda_\xi < \lambda$, then

$$\kappa^\lambda = \prod_{\xi < \mathrm{cf}(\lambda)} \kappa^{\lambda_\xi} = \prod_{\xi < \mathrm{cf}(\lambda)} \kappa^{\rho_0} = \kappa^{\rho_0 \cdot \mathrm{cf}(\lambda)} = \kappa^{\rho_0}.$$

THEOREM 6.12. *Assume* GCH, *and let $\kappa \geq 2$ and $\lambda \geq \omega$ be cardinals. Then*

$$\kappa^\lambda = \begin{cases} \kappa & \text{if } \lambda < \mathrm{cf}(\kappa), \\ \kappa^+ & \text{if } \mathrm{cf}(\kappa) \leq \lambda \leq \kappa, \\ \lambda^+ & \text{if } \kappa < \lambda, \end{cases}$$

and

$$\kappa^{\underline{\lambda}} = \begin{cases} \kappa & \text{if } \lambda \leq \mathrm{cf}(^\kappa), \\ \kappa^+ & \text{if } \mathrm{cf}(\kappa) < \lambda \leq \kappa^+, \\ \lambda & \text{if } \kappa^+ < \lambda. \end{cases}$$

PROOF. The case $\kappa < \omega$ is partly vacuous and partly obvious, so assume $\kappa \geq \omega$. Consider first the results on κ^λ. We use GCH. In case $\lambda < \mathrm{cf}(\kappa)$ we have $\kappa \leq \kappa^\lambda = \kappa^{\underline{\lambda}} \cdot \kappa \leq \kappa$ by Theorem 6.3. In case $\mathrm{cf}(\kappa) \leq \lambda \leq \kappa$ we have $\kappa < \kappa^\lambda \leq \kappa^\kappa = 2^\kappa = \kappa^+$ by Corollary 6.8, and, finally, in case $\lambda > \kappa$ we have $2^\lambda \leq \kappa^\lambda \leq \lambda^\lambda = 2^\lambda = \lambda^+$. The results on $\kappa^{\underline{\lambda}}$ are easy consequences of these results.

THEOREM 6.13. *If $\kappa \geq 3$ and $\lambda > 0$, then $\kappa^{\underline{\lambda}} \geq \lambda$.*

PROOF. $\kappa^{\underline{\lambda}} \geq 2^\rho > \rho$ holds whenever $\rho < \lambda$.

THEOREM 6.14. *Let $\kappa \geq 3$, $\lambda \geq \omega$ and assume $\rho < \kappa^{\underline{\lambda}}$. Then there are $\kappa_0 < \kappa$ and $\lambda_0 < \lambda$ such that $\rho < \kappa_0^{\lambda_0}$.*

PROOF. The case $\rho<\omega$ is trivial, so assume $\rho\geq\omega$. Assuming that the assertion is false, we can conclude that

$$\rho<\kappa^{\lambda}=\Sigma\langle\kappa_0^{\lambda_0}:\kappa_0<\kappa\ \&\ \lambda_0<\lambda\rangle\leq\Sigma\langle\rho:\kappa_0<\kappa\ \&\ \lambda_0<\lambda\rangle=$$

$$=\rho\cdot|\{\sigma:\sigma<\kappa\}|\cdot|\{\sigma:\sigma<\lambda\}|,$$

i.e., either $\rho<|\{\sigma:\sigma<\kappa\}|\leq\kappa^-$ or $\rho<|\{\sigma:\sigma<\lambda\}|\leq\lambda^-$, where σ, as usual, runs over cardinals. So we have accordingly either $\rho^+<\kappa$, in which case the result follows with $\kappa_0=\rho^+$ and $\lambda_0=1$, or $\rho^+<\lambda$, in which case the result follows with $\kappa_0=2$ and $\lambda_0=\rho$, completing the proof.

7. THE LOGARITHM OPERATION

Assuming that $\lambda\geq 3$, we define the *logarithm* $L_\lambda(\kappa)$ of κ with *base* λ as follows:

$$L_\lambda(\kappa)=\min\{\rho:\exists\lambda_0<\lambda[\lambda_0^\rho\geq\kappa]\}.$$

Note that the set on the right-hand side is obviously not empty as $2^\kappa>\kappa$.

There would be several alternative ways to define the logarithm operation for infinite cardinals. $L_\lambda(\kappa)$ is not an extension of the ordinary logarithm function since the definition of $L_\lambda(\kappa)$ does not give ${}^l\log k=L_l(k)$ for positive integers k and l, and even for infinite κ we do not necessarily have $\lambda^{L_\lambda(\kappa)}=\kappa$. The operation $L_\lambda(\kappa)$ nevertheless retains many important properties of the ordinary logarithm function, and it will be very convenient to use it in formulating our main theorems concerning ordinary partition relations without assuming GCH. The aim of the present section is to prove several basic properties of the logarithm operation. An important corollary of our results will be Theorem 7.12 which says in particular that, for any cardinal κ, κ^ρ assumes only finitely many values as ρ runs over all cardinals satisfying $2^\rho<\kappa$. We start out with rather simple observations.

THEOREM 7.1. a) *We have* $L_{\kappa^+}(\kappa)=1$ *for every infinite* κ.

b) *The logarithm operation has the following monotonicity properties*:

$$L_\lambda(\kappa)\geq L_{\lambda'}(\kappa)\qquad \textit{for}\qquad 3\leq\lambda\leq\lambda',$$

and

$$L_\lambda(\kappa)\leq L_\lambda(\kappa')\qquad \textit{for}\qquad \kappa\leq\kappa'.$$

The proof is obvious. In view of a) of this theorem, it will be natural to assume $\lambda\leq\kappa$ when considering the function $L_\lambda(\kappa)$.

THEOREM 7.2. *Assume* $3\leq\lambda\leq\kappa$ *and* $\kappa\geq\omega$. *Then* $\omega\leq L_\lambda(\kappa)\leq\kappa$.

The proof is obvious. The following property extends a well-known property of the ordinary logarithm.

THEOREM 7.3. *Assume $3 \le \lambda \le \kappa \le \rho$ and $\kappa \ge \omega$. Then $L_\lambda(\kappa) \cdot L_{\kappa^+}(\rho) = L_\lambda(\rho)$ holds and, consequently, we also have $L_\lambda(\kappa) \cdot L_\kappa(\rho) = L_\lambda(\rho)$.*

PROOF. We clearly have $L_\lambda(\kappa) \le L_\lambda(\rho)$ and $L_{\kappa^+}(\rho) \le L_\lambda(\rho)$ by the monotonicity properties described in Theorem 7.1.b, and as all these logarithms are infinite cardinals, the inequality $L_\lambda(\kappa) \cdot L_{\kappa^+}(\rho) \le L_\lambda(\rho)$ follows. To prove that we also have $\ge$ here, we only have to show that $\lambda_0^{L_\lambda(\kappa) \cdot L_{\kappa^+}(\rho)} \ge \rho$ holds for some $\lambda_0 < \lambda$. But the definition of the logarithm shows that we have $\lambda_0^{L_\lambda(\kappa)} \ge \kappa$ for some $\lambda_0 < \lambda$ and we have $\kappa^{L_{\kappa^+}(\rho)} \ge \rho$. We get the desired result by combining these two inequalities. The second equality in the theorem easily follows from the first one by the observation that $L_{\kappa^+}(\rho) \le L_\kappa(\rho) \le L_\lambda(\rho)$ holds.

THEOREM 7.4. *For every $\kappa \ge \omega$ the following three propositions are equivalent:* (i) *κ is a strong limit cardinal,* (ii) *$L_3(\kappa) = \kappa$, and* (iii) *$L_\lambda(\kappa) = \kappa$ holds for every λ with $3 \le \lambda \le \kappa$.*

The proof is straightforward.

THEOREM 7.5. *Given cardinals λ, κ with $3 \le \lambda \le \kappa$ and $\kappa \ge \omega$, the cardinal $\bar{\lambda}$ defined by*

$$\bar{\lambda} = \min\{\lambda' : 3 \le \lambda' \le \lambda \;\&\; L_{\lambda'}(\kappa) = L_\lambda(\kappa)\}$$

either equals 3 or is an infinite successor cardinal.

PROOF. Put $\rho = L_\lambda(\kappa)$. As $\rho = L_{\bar{\lambda}}(\kappa)$, by the definition of the logarithm there is a minimal $\sigma < \bar{\lambda}$ such that $\sigma^\rho \ge \kappa$. If σ is finite then, noting that ρ is infinite according to Theorem 7.2 we can see that $2^\rho = \sigma^\rho \ge \kappa$. Hence $L_3(\kappa) \le \rho$, i.e., $L_3(\kappa) = \rho$ (as $L_3(\kappa) \ge L_\lambda(\kappa)$ holds by one of the monotonicity properties in Theorem 7.1.b). Therefore we have $\bar{\lambda} = 3$ in this case. If σ is infinite then $L_{\sigma^+}(\kappa) \le \rho$, i.e., $L_{\sigma^+}(\kappa) = \rho$. Hence $\bar{\lambda} \le \sigma^+$; as we have $\bar{\lambda} \ge \sigma^+$ in view of the choice of σ, we have $\bar{\lambda} = \sigma^+$, and so $\bar{\lambda}$ is indeed a successor cardinal.

THEOREM 7.6. *For each cardinal $\kappa \ge \omega$ there are an integer $n \ge 1$ and cardinals $2 = \lambda_0 < \lambda_1 < \ldots < \lambda_n = \kappa$ and $\rho_0 > \ldots > \rho_{n-1}$ such that*

$$L_\lambda(\kappa) = L_{\lambda_i^+}(\kappa) = \rho_i \text{ whenever } i < n \text{ and } \lambda_i < \lambda \le \lambda_{i+1}, \tag{1}$$

$$\lambda_i^{\rho_i} \ge \kappa \quad \text{if} \quad i < n, \quad \text{and} \tag{2}$$

$$\lambda^{\rho_i} < \kappa \quad \text{if} \quad i < n \quad \text{and} \quad \lambda < \lambda_i. \tag{3}$$

PROOF. $L_\lambda(\kappa)$ is a nonincreasing function of λ ($3 \le \lambda \le \kappa$) in view of Theorem 7.1.b, so it assumes only finitely many values, since otherwise we would have an infinite decreasing sequence of ordinals. Let these values be $\rho_0, \ldots, \rho_{n-1}$ in decreasing order. For any $i<n$ take the least cardinal $\lambda \geq 3$ such that $L_\lambda(\kappa)=\rho_i$, and define λ_i by $\lambda=\lambda_i^+$ (note that λ is always a successor cardinal in view of the preceding theorem). We clearly have $\lambda_0=2$, as $\rho_0=L_3(\kappa)$ by the definition of ρ_0. Putting $\lambda_n=\kappa$, the sequence of λ's has been defined. (1), (2), and (3) are obviously fulfilled.

THEOREM 7.7. *If $\rho=L_\lambda(\kappa)$ and $\rho_0=\underset{\smile}{\lambda}^\rho$, where $\lambda \geq 3$ and $\kappa \geq \omega$, then $L_{\rho_0}(\kappa)\geq \rho$. If, in addition, λ is a limit cardinal, then $L_{\rho_0}(\kappa)=\rho$.*

PROOF. Assume $\rho_0'<\rho_0$. Then $\rho_0'<\lambda'^{\rho'}$ for some $\lambda'<\lambda$ and $\rho'<\rho$ by Theorem 6.14 (observe that $\rho \geq \omega$ by Theorem 7.2), and so, for any $\rho''<\rho$ we have $\rho_0'^{\rho''} \leq \leq \lambda'^{\rho' \cdot \rho''}<\kappa$, as $\lambda'<\lambda$ and $\rho' \cdot \rho''<\rho=L_\lambda(\kappa)$. Hence we indeed have $L_{\rho_0}(\kappa)\geq \lambda$. If λ is a limit cardinal, then $\rho_0=\underset{\smile}{\lambda}^\rho \geq \lambda$, and so we also have $L_{\rho_0}(\kappa)\leq L_\lambda(\kappa)=\rho$ by Theorem 7.1.b, i.e., $L_{\rho_0}(\kappa)=\rho$, which completes the proof.

THEOREM 7.8. *If $\kappa \geq \omega$ and $\lambda \geq 3$ is the largest cardinal $\lambda' \leq \kappa$ with $L_{\lambda'}(\kappa)=\rho$, then $\underset{\smile}{\lambda}^\rho \leq \lambda$. Hence, using the notations of Theorem 7.6, we have $\underset{\smile}{\lambda_i}^{\rho_i} \leq \lambda_i$, whenever $1 \leq i \leq n$.*

Note that, by Theorem 7.6, if there is a $\lambda' \leq \kappa$ satisfying $L_{\lambda'}(\kappa)=\rho$, then there is a largest λ' with this property.

PROOF. The second sentence simply restates the assertion given in the first sentence, so we have to prove only the first assertion. Put $\rho_0=\underset{\smile}{\lambda}^\rho$. We have $L_{\rho_0}(\kappa)\geq L_\lambda(\kappa)=\rho$ by the preceding theorem; note that this implies $\rho_0 \leq \kappa$. If we have $\rho_0>\lambda$, then $L_{\rho_0}(\kappa)\leq L_\lambda(\kappa)=\rho$ also holds by Theorem 7.1.b, i.e., $L_{\rho_0}(\kappa)=\rho$ holds, which contradicts the maximality of λ.

Our next result shows that $L_\lambda(\kappa)$ has properties similar to the ordinary logarithm function.

THEOREM 7.9. *We have $\underset{\smile}{\lambda}^{L_\lambda(\kappa)} \leq \kappa \leq \lambda^{L_\lambda(\kappa)}$ for every λ, κ with $3 \leq \lambda \leq \kappa$ and $\kappa \geq \omega$.*

PROOF. If $\kappa<\underset{\smile}{\lambda}^{L_\lambda(\kappa)}$, then by Theorem 6.14 we have $\kappa<\lambda_0^{\rho_0}$ for some $\lambda_0<\lambda$ and $\rho_0<L_\lambda(\kappa)$ which contradicts the definition of the logarithm operation. The second inequality is obvious.

THEOREM 7.10. *If $L_\lambda(\kappa)<L_3(\kappa)$, where $\lambda \geq 3$ and $\kappa \geq \omega$, then $\lambda>L_3(\kappa)$.*

PROOF. A direct proof is quite easy, but we can make a shortcut by using the above results. Using Theorem 7.5 or 7.6, let λ_1 be the largest λ' such that $L_{\lambda'}(\kappa)=L_3(\kappa)$. By Theorem 7.8 we obtain:

$$\lambda>\lambda_1 \geq \underset{\smile}{\lambda_1}^{L_3(\kappa)} \geq L_3(\kappa).$$

THEOREM 7.11. *Using the notations of Theorem 7.6, we have $\operatorname{cf}(\lambda_i)=\rho_i$ for $1 \leq i<n$, and λ_i is a singular cardinal with $\lambda_i=\underset{\smile}{\lambda_i}^{\rho_i}$ for $1<i<n$. As a corollary, $L_\lambda(\kappa)$ is a regular cardinal whenever $L_\lambda(\kappa)<L_3(\kappa)$.*

Proof. Assume $1 \le i < n$. Then we have $\lambda^{\rho_i} \le \lambda_i$ for every $\lambda < \lambda_i$, since $\lambda_i^{\rho_{i-1}} \le \lambda_i$ holds by the last sentence of Theorem 7.8. So Corollary 6.7 tells us that we have $\lambda_i^{\rho_i} = \lambda_i$ ($< \kappa$, as $i < n$) if $\rho_i < \mathrm{cf}(\lambda_i)$, and $\lambda_i^{\rho_i} = \lambda_i^{\mathrm{cf}(\lambda_i)}$ otherwise. Here we must have the second alternative, as $\lambda_i^{\rho_i} \ge \kappa$ according to (2) in Theorem 7.6. Hence $\rho_i \ge \mathrm{cf}(\lambda_i)$ and $\lambda_i^{\rho_i} = \lambda_i^{\mathrm{cf}(\lambda_i)} \ge \kappa$. The latter relation means that $\mathrm{cf}(\lambda_i) \ge L_{\lambda_i^+}(\kappa) = \rho_i$, i.e., we indeed have $\mathrm{cf}(\lambda_i) = \rho_i$. If we assume that $1 < i < n$, then $\lambda_i > \lambda_1 \ge \mathrm{cf}(\lambda_1) = \rho_1 > \rho_i = \mathrm{cf}(\lambda_i)$, i.e., λ_i is a singular cardinal. Hence $\lambda_i^{\rho_{i-1}} \ge \lambda_i$. We have $\lambda_i^{\rho_{i-1}} \le \lambda_i$ according to the last sentence of Theorem 7.8, so in fact $\lambda_i^{\rho_{i-1}} = \lambda_i$ holds. As for the last sentence of the theorem to be proved, $L_\lambda(\kappa) = \rho_i = \mathrm{cf}(\lambda_i)$ for some i with $1 \le i \le n$ under the assumptions, i.e., $L_\lambda(\kappa)$ is indeed regular. The proof is complete.

Theorem 7.12. *Assume $\kappa \ge \omega$. Then, using the notations of Theorem 7.6, we have*

$$\kappa^\rho = \begin{cases} \kappa & \text{if } 1 \le \rho < \rho_{n-1} \text{ and } \rho < \mathrm{cf}(\kappa), \\ \kappa^{\mathrm{cf}(\kappa)} & \text{if } \mathrm{cf}(\kappa) \le \rho < \rho_{n-1}, \\ \kappa^{\rho_i} & \text{if } \rho_i \le \rho \le \rho_{i-1} \text{ for } 1 \le i < n, \\ 2^\rho & \text{if } \rho_0 \le \rho. \end{cases}$$

Hence κ^ρ assumes only finitely many values for $\rho < \rho_0 = L_3(\kappa)$ (i.e., for ρ with $2^\rho < \kappa$).

Proof. Assume first that $1 \le \rho < \rho_{n-1}$. Then we have $\kappa^\rho \le \kappa$ by Theorem 7.8, and so indeed $\kappa^\rho = \kappa$ if $\rho < \mathrm{cf}(\kappa)$ and $\kappa^\rho = \kappa^{\mathrm{cf}(\kappa)}$ if $\rho \ge \mathrm{cf}(\kappa)$ according to Corollary 6.7. Assume now $\rho_i \le \rho < \rho_{i-1}$, where $1 \le i < n$. We have $\lambda_i^\rho \ge \lambda_i^{\rho_i} \ge \kappa$ by (2) in Theorem 7.6; as $\lambda_i < \kappa$, we obtain $\kappa^\rho = \lambda_i^\rho$. Noting that $\rho \ge \rho_i = \mathrm{cf}(\lambda_i)$ and $\lambda_i^\rho \le \lambda_i^{\rho_{i-1}} = \lambda_i$ by the preceding theorem, Corollary 6.7 implies that $\lambda_i^\rho = \lambda_i^{\rho_i}$ holds, so we have $\kappa^\rho = \lambda_i^{\rho_i}$. Observe now that $\lambda_i^{\rho_i} = \kappa^{\rho_i}$ (in fact, we obtained the equality $\lambda_i^\rho = \kappa^\rho$ for any ρ with $\rho_i \le \rho < \rho_{i-1}$ just before). Hence we indeed have $\kappa^\rho = \kappa^{\rho_i}$. Finally, assume $\rho \ge \rho_0$. We have $2^{\rho_0} \ge \kappa$ according to (2) in Theorem 7.6 (note that $\lambda_0 = 2$), and so $2^\rho \ge \kappa$. Thus $2^\rho = \kappa^\rho$ indeed holds. The proof is complete.

Theorem 7.13. *Let κ be infinite, and assume that $2^{L_3(\kappa)} < \lambda' < \lambda \le \kappa$. Then $\lambda'^{L_\lambda(\kappa)} < \kappa$.*

Proof. As we clearly have $L_3(\lambda') = L_3(\lambda) = L_3(\kappa)$, the preceding theorem implies that λ'^ρ assumes only finitely many values for $\rho < L_\lambda(\kappa) (\le L_3(\lambda'))$; thus there is a $\rho_0 < L_\lambda(\kappa)$ such that $\lambda'^\rho \le \lambda'^{\rho_0}$ for any $\rho < L_\lambda(\kappa)$. Then $\lambda'^{L_\lambda(\kappa)} \le \lambda'^{\rho_0} \cdot L_\lambda(\kappa) < \kappa$, since $\lambda'^{\rho_0} < \kappa$ and $L_\lambda(\kappa) \le L_3(\kappa) \le 2^{L_3(\kappa)} < \kappa$.

Theorem 7.14. *If $L_\lambda(\kappa) < L_{\lambda'}(\kappa)$, then $L_\lambda(\kappa) = \mathrm{cf}(L_\lambda(\kappa)) \le \mathrm{cf}(L_{\lambda'}(\kappa))$ ($\lambda' \ge 3$; $\lambda, \omega \le \kappa$). A more precise result is the following: using the notations of Theorem 7.6, if $L_3(\kappa)$ is singular, then $2^{L_3(\kappa)} = \lambda_1$ and $\mathrm{cf}(L_3(\kappa)) = \mathrm{cf}(\lambda_1)$ hold.*

PROOF. The first assertion is an easy consequence of the main part of the theorem. In fact, first note that $L_\lambda(\kappa) < L_{\lambda'}(\kappa) \le L_3(\kappa)$, and so $L_\lambda(\kappa) = \mathrm{cf}\,(L_\lambda(\kappa))$ according to the last sentence of Theorem 7.11. We have to show that $L_\lambda(\kappa) < \mathrm{cf}\,(L_{\lambda'}(\kappa))$. This is obvious in case $L_{\lambda'}(\kappa)$ is regular, which is the case if $L_{\lambda'}(\kappa) < L_3(\kappa)$ according to the last sentence of Theorem 7.11. Hence we may assume that $L_{\lambda'}(\kappa) = L_3(\kappa) = \rho_0$ is singular (the notations of Theorem 7.6 are used). Then $L_3(\kappa) > L_\lambda(\kappa) = \rho_i$ for some i with $1 \le i < n$, and so $\mathrm{cf}\,(\rho_i) = \rho_i \le \rho_1 = \mathrm{cf}\,(\lambda_1) = \mathrm{cf}\,(L_3(\kappa))$ according to the main part of the present theorem and to Theorem 7.11.

We now turn to the proof of the second assertion. Put $\rho_0 = L_3(\kappa)$. By Theorem 7.8 we have $2^{\underline{\rho_0}} \le \lambda_1^{\underline{\rho_0}} \le \lambda_1$. We have $(2^{\underline{\rho_0}})^{\mathrm{cf}\,(\rho_0)} = 2^{\rho_0} > \kappa$ in view of Theorem 6.10.e and (2) above, and so the assumption $2^{\underline{\rho_0}} < \lambda_1$ would imply $L_{\lambda_1}(\kappa) \le \mathrm{cf}\,(\rho_0) < \rho_0$ (note that $\rho_0 = L_3(\kappa)$ was assumed to be singular), which contradicts Theorem 7.6. Hence $2^{\underline{\rho_0}} = \lambda_1$. According to Theorem 6.10.d, either $\mathrm{cf}\,(2^{\underline{\rho_0}}) = \mathrm{cf}\,(\rho_0)$ holds, or alternative c(i) holds in that theorem, in which case Theorem 6.11 implies that $2^{\underline{\rho_0}} = 2^{\rho_0}$ holds. In the latter case $2^{\rho_0} = 2^{\rho'}$ for some $\rho' < \rho$, which contradicts the equality $\rho_0 = L_3(\kappa)$. Hence the former alternative must hold, i.e., $\mathrm{cf}\,(\rho_0) = \mathrm{cf}\,(2^{\underline{\rho_0}})$, and this equals $\mathrm{cf}\,(\lambda_1)$ as we have established $\lambda_1 = 2^{\underline{\rho_0}}$ already, which completes the proof.

THEOREM 7.15. *If κ is regular then $L_\kappa(\kappa)$ is regular. If $L_\kappa(\kappa)$ is singular then $L_3(\kappa) = L_\kappa(\kappa)$, $2^{\underline{L_\kappa(\kappa)}} = \kappa$, and* $\mathrm{cf}\,(\kappa) = \mathrm{cf}\,(L_\kappa(\kappa))$.

PROOF. Assume first that κ is regular. If $L_\kappa(\kappa) < L_3(\kappa)$ then $L_\kappa(\kappa)$ is regular by Theorem 7.11; so assume $\rho = L_3(\kappa) = L_\kappa(\kappa)$, and assume that ρ is singular, i.e., $\mathrm{cf}\,(\rho) < \rho$. Let $\rho = \sum_{\xi < \mathrm{cf}\,(\rho)} \rho_\xi$, where $\rho_\xi < \rho$. Then

$$\kappa \le 2^\rho = \prod_{\xi < \mathrm{cf}\,(\rho)} 2^{\rho_\xi},$$

and $2^{\rho_\xi} < \kappa$ here. As κ is regular and $\mathrm{cf}\,(\rho) < \rho = L_3(\kappa) \le \kappa$, writing

$$\lambda = \sum_{\xi < \mathrm{cf}\,(\rho)} 2^{\rho_\xi},$$

we have $\lambda < \kappa$; on the other hand,

$$\kappa \le \prod_{\xi < \mathrm{cf}\,(\rho)} 2^{\rho_\xi} \le \lambda^{\mathrm{cf}\,(\rho)},$$

which implies that $L_\kappa(\kappa) \le \mathrm{cf}(\rho)$, contradicting $L_\kappa(\kappa) = \rho > \mathrm{cf}\,(\rho)$.

To prove the second part, assume now that $L_\kappa(\kappa)$ is singular. Then we must have $L_\kappa(\kappa) = L_3(\kappa)$ in view of the last sentence of Theorem 7.11. So $n = 1$ and

$\lambda_1=\kappa$ hold in the notations of Theorem 7.6. Hence the second part of Theorem 7.14 implies that $2^{L_\kappa(\kappa)}=2^{L_3(\kappa)}=\kappa$ and $\operatorname{cf}(L_\kappa(\kappa))=\operatorname{cf}(L_3(\kappa))=\operatorname{cf}(\kappa)$ hold, completing the proof.

THEOREM 7.16. *If κ is a limit cardinal, $3\leq\lambda<\kappa$, then either $2^{L_3(\kappa)}=\kappa$ and $L_3(\kappa)=L_\lambda(\kappa)=L_\kappa(\kappa)$ hold, or there is a regular cardinal σ such that $\lambda<\sigma<\kappa$, $2^{L_3(\kappa)}<\sigma$ and $L_\lambda(\sigma)=L_\lambda(\kappa)$ hold.*

PROOF. Assume first that $2^{L_3(\kappa)}=\kappa$ holds. Then, using Theorem 7.8, we obtain that

$$\kappa=2^{L_3(\kappa)}=2^{\rho_0}\leq\lambda_1^{\rho_0}\leq\lambda_1\leq\kappa,$$

i.e., $\lambda_1=\kappa$ and $n=1$ hold in Theorem 7.6. Therefore we indeed have $L_3(\kappa)=L_\lambda(\kappa)=L_\kappa(\kappa)$ in this case.

Assume now $2^{L_3(\kappa)}<\kappa$. We distinguish two cases: a) $L_\lambda(\kappa)>L_\kappa(\kappa)$ and b) $L_\lambda(\kappa)=L_\kappa(\kappa)$. Consider case a) first. The least cardinal $\sigma_0\leq\kappa$ for which $L_{\sigma_0}(\kappa)=L_\kappa(\kappa)$ holds is a successor cardinal in view of Theorem 7.5, hence, κ being a limit cardinal, we have $\sigma_0<\kappa$. Put $\sigma=(\sigma_0\cdot\lambda\cdot 2^{L_3(\kappa)})^+$. Then σ is a regular cardinal less than κ, and $L_\sigma(\kappa)=L_\kappa(\kappa)<L_\lambda(\kappa)$ holds. As we have $L_\lambda(\sigma)\cdot L_\sigma(\kappa)=L_\lambda(\kappa)$ according to Theorem 7.3, this implies that $L_\lambda(\sigma)=L_\lambda(\kappa)$ does indeed hold.

Consider case b) now, i.e., assume $L_\lambda(\kappa)=L_\kappa(\kappa)$. Write $\sigma_0=(\lambda\cdot 2^{L_\kappa(\kappa)})^+$. Then Theorem 7.13 entails that $\sigma_0^{L_\kappa(\kappa)}=\sigma_0^{L_{\sigma_0^+}(\kappa)}<\kappa$ holds. Put $\sigma=(\sigma_0^{L_\kappa(\kappa)})^+$. Then σ is $<\kappa$ and regular. As $\lambda^{L_\lambda(\kappa)}\leq\sigma_0^{L_\kappa(\kappa)}<\sigma$, we have $L_\lambda(\sigma)\geq L_\lambda(\kappa)$. As $L_\lambda(\sigma)\leq L_\lambda(\kappa)$ is obvious by $\sigma\leq\kappa$, we obtain $L_\lambda(\sigma)=L_\lambda(\kappa)$, which completes the proof.

We shall need the following inequalities:

THEOREM 7.17. a) *If $\lambda\geq 3$, $\rho\geq\omega$, and $\kappa=\underset{\sim}{\lambda}^\varrho$, then $L_\kappa(\kappa^+)\geq\rho$.*

b) *If $\lambda\geq 2$, $\rho\geq\omega$ and $\kappa=\lambda^\varrho$ then $L_{\kappa^+}(\kappa^+)\geq\operatorname{cf}(\rho)$.*

c) *If $\lambda\geq 2$, $\rho\geq\omega$ and $\kappa=\lambda^\rho$ then $L_{\kappa^+}(\kappa^+)\geq\rho^+$.*

PROOF. *Ad* a). We clearly have $L_\lambda(\kappa^+)\geq\rho$; hence, writing $\sigma=L_\lambda(\kappa^+)$, Theorem 7.7 implies the second inequality below (the others are obvious):

$$L_\kappa(\kappa^+)\geq L_{\underset{\sim}{\lambda}^\varrho}(\kappa^+)\geq\sigma=L_\lambda(\kappa^+)\geq\rho.$$

Ad b). We have $\kappa^{\rho'}=\kappa$ if $0<\rho'<\operatorname{cf}(\rho)$ in view of Theorem 6.10.e, hence $L_{\kappa^+}(\kappa^+)\geq\operatorname{cf}(\rho)$ indeed holds.

Ad c). This follows from b) with ρ^+ replacing ρ.

We shall also consider Tarski's function $p(\kappa)$, defined as follows:

DEFINITION 7.18. *For any infinite cardinal κ put*

$$p(\kappa)=L_{\kappa^+}(\kappa^+)=\min\{\rho: \kappa^\rho>\kappa\}.$$

As a corollary of Theorem 7.15, $p(\kappa)$ is always regular; this is a theorem of Tarski [1925]. We have $\omega\le p(\kappa)\le \mathrm{cf}(\kappa)$; indeed, the first inequality is given in Theorem 7.2, and the second one follows from the relation $\kappa^{\mathrm{cf}(\kappa)}>\kappa$ (see Corollary 6.8).

THEOREM 7.19. *Assume* GCH, *and let $\kappa\ge\omega$. Then we have*

$$\kappa=L_3(\kappa)=L_\kappa(\kappa) \quad \textit{if } \kappa \textit{ is a limit cardinal,} \tag{4}$$

$$L_\lambda(\kappa^+)=\kappa \quad \textit{for every } \lambda \textit{ with } 3\le\lambda\le\kappa, \tag{5}$$

and, moreover,

$$L_{\kappa^+}(\kappa^+)=p(\kappa)=\mathrm{cf}(\kappa). \tag{6}$$

The proof is obvious. As a corollary, we have $L_\lambda(\kappa)=\kappa^-$ for any λ with $3\le\lambda\le\kappa$ and for every infinite κ that is not the successor of a singular cardinal, provided that GCH holds. If $\kappa=\rho^+$, where ρ is singular, then $L_\lambda(\kappa)=\kappa^-=\rho$ for $\lambda<\kappa$ and $L_\lambda(\kappa)=\mathrm{cf}(\kappa^-)=\mathrm{cf}(\rho)$ for $\lambda=\kappa$ (GCH is again assumed).

4*

CHAPTER III

FUNDAMENTALS ABOUT PARTITION RELATIONS

In the present chapter we first give a comprehensive guide to partition symbols; some of the symbols introduced are not used in this book, and they are defined here only for the sake of completeness. After that, we present such basic results involving ordinary partition relations as Ramsey's theorem and the Erdős–Dushnik–Miller theorem. The chapter is concluded with a result saying that strong negative square-bracket relations hold with infinite superscripts.

8. A GUIDE TO PARTITION SYMBOLS

8.1. Partition symbols are very powerful in giving a unified expression to various combinatorial assertions, and by their displaying the variables graphically, separated according to the different roles they play, they afford an efficient classification of different types of problems. The aim of the present section is to define the most important partition symbols occurring in the literature, regardless of the role they play in this book. Hence the reader may choose, *and*, indeed, *is strongly advised to skip this section* upon first reading (except its two subsections, 8.2 and 8.3 where the ordinary partition symbol is defined, which is the main topic of the next and later sections), and only return to it to learn the definition of a partition symbol as the need arises.

A *partition* (*on* H) of a set x is an indexed set (i.e., a function) $\langle x_h : h \in H \rangle$ of subsets of x such that $x = \bigcup_{h \in H} x_h$. We shall often say that this is a partitioning of x into $|H|$ parts. A partition of $[x]^r$, where r is a cardinal (or, possibly, an order type if x is ordered) is also called an *r-partition* of x. If the sets x_h are pairwise disjoint, then the above partition is called a *disjoint partition*. We could restrict our attention to disjoint partitions if we wanted to, but there is a slight advantage in not doing this, as many of the counter-examples that naturally arise involve nondisjoint partitions, and a further technical step is needed to render these disjoint. A *coloring* of x is simply a function f whose domain is x. If $f: x \to H$ is a coloring then we may call it a *coloring* of x *with* H (or with $|H|$ *colors*). With a coloring $f: x \to H$ of x we can naturally associate a disjoint partition $\langle x_h : h \in H \rangle$

of x in a 1-1 way by putting $x_h = \{z \in x: f(z) = h\}$ for every $h \in H$. Similarly to the terminology used for partitions, a coloring of $[x]^r$, where r is a cardinal (or, possibly, an order type if x is ordered) is called an r-coloring of x. Whenever it is possible, the domain of a partition and the range of a coloring will be a cardinal or, at least, an ordinal; thus a partition will usually be a sequence.

8.2. *The ordinary partition symbol.* The most frequently used partition symbol, the ordinary partition symbol

$$a \to (b_h)^r_{h \in H}, \tag{1}$$

read as "a arrows b_h (super) r (sub) $h \in H$", is meaningful if H is an arbitrary set, and each of the other letters (except, of course, h) denotes either a cardinal or an order type (we shall define this symbol under more general circumstances below; see Subsection 8.8). Provided that these conditions are satisfied, the above partition relation is said to hold if, given any partition $I = \langle I_h : h \in H \rangle$ of the set $[A]^r$, where A is an arbitrary set of *kind* a (i.e., of cardinality or order type a according as a denotes a cardinal or an order type), there is an $h \in H$ and a set B_h of kind b_h such that $[B_h]^r \subseteq I_h$. Note that the symbol $[\cdot]^r$ has different meanings in the above definition according as r is a cardinal or an order type. The negation of the relation in (1) is written as

$$a \nrightarrow (b_h)^r_{h \in H}$$

and is read as "a does not arrow (or 'nonarrows') b_h, r, $h \in H$". In general, the negation of any partition relation is expressed by crossing the arrow.

The set B_h in the above definition is often called a homogeneous set: given a partition $I = \langle I_h : h \in H \rangle$ of $[A]^r$, the set $B \subseteq A$ is called *homogeneous of class h with respect to* the partition I if $[B]^r \subseteq I_h$ $(h \in H)$. Similarly, if $f:[A]^r \to H$ is a coloring then a set $B \subseteq A$ is called *homogeneous of color h with respect to* the coloring f if $f``[B]^r \subseteq \{h\}$. If B is homogeneous of class (or color) h for some $h \in H$ (with respect to the above partition or coloring), then we may simply say that B is homogeneous. The term homogeneous is often helpful in discussing partition relations. E.g. the definition of the partition relation in (1) can be restated by using this term as follows: (1) holds if, given any partition on H of the set $[A]^r$, where A is of kind a, there is an $h \in H$ and a set B_h of kind b_h such that B is homogeneous of class h.

If we want to show that (1) fails, then we have to find a coloring or a partition for which there are no homogeneous sets of the required size. If such a coloring (or partition) is found, it is convenient to say that it *verifies* or *establishes* the negation of (1).

The display of variables in (1) may seem somewhat arbitrary; the rationale behind separating a from the b's is a difference in their monotonicity properties: if we increase a, then a valid partition relation in (1) will continue to hold, and the same is true if we decrease any of the b_h's.

One usually takes an ordinal (or even a cardinal) ϑ as the set H in the partition symbol in (1) unless it is inconvenient to do so; hence one usually considers a partition relation like

$$a \to (b_\xi)^r_{\xi<\vartheta}, \tag{2}$$

where $\xi<\vartheta$, of course, means the same thing as $\xi \in \vartheta$. This does not involve any restriction of generality provided the Axiom of Choice is accepted, since any set H can then be mapped onto a cardinal by a 1-1 function.

If $b_\xi = b$ for all $\xi<\vartheta$ in (2), then we may write

$$a \to (b)^r_\vartheta$$

instead of (2).

The partition relation

$$a \to (b_{\xi,\eta})^r_{\langle\xi\eta\rangle\in H}$$

will often be written as

$$a \to ((b_{\xi,0})_{\xi<\xi_0}, \ldots, (b_{\xi,\eta})_{\xi<\xi_\eta}, \ldots)^r_{\eta<\vartheta}, \tag{3}$$

where the ordinals ξ_η and ϑ are such that $H=\{\langle\xi\eta\rangle : \eta<\vartheta \,\&\, \xi<\xi_\eta\}$; this notation will be especially convenient when ϑ is a small finite ordinal, e.g. 2 or 3. When the latter notation is used, then the indexed set $\langle b_{\xi,\eta} : \langle\xi\eta\rangle \in H\rangle$ will usually be identified with the sequence

$$\langle (b_{\xi,0})_{\xi<\xi_0}, \ldots, (b_{\xi,\eta})_{\xi<\xi_\eta}, \ldots \rangle_{\eta<\vartheta}, \tag{4}$$

defined as

$$\langle c_\alpha : \alpha < \dot{\sum_{\eta<\vartheta}} \xi_\eta \rangle,$$

where $c_\alpha = b_{\xi,\eta}$ if α is of form $\left(\dot{\sum_{\eta'<\eta}} \xi_{\eta'}\right) \dot{+} \xi (\xi<\xi_\eta)$. If $b_{\xi,\eta}=b_\eta$ for some $\eta<\vartheta$ and for all $\xi<\xi_\eta$, then we may write $(b_\eta)_{\xi_\eta}$ instead of $(b_{\xi,\eta})_{\xi<\xi_\eta}$ $(=(b_\eta)_{\xi<\xi_\eta})$ in (3) or (4) (in this case ξ_η is usually a cardinal), and if $\xi_\eta=1$, then we may simply write b_η instead of $(b_\eta)_1$. The conventions introduced so far explain the meanings e.g. of the following partition relations

$$a \to (b)^r_\vartheta, \quad a \to ((b_0)_{\tau_0}, (b_1)_{\tau_1})^r \quad \text{and} \quad a \not\to (b_0, b_1, b_2)^r.$$

Similar conventions will be used in connection with the partition symbols introduced below.

Ramsey's classical theorem, proved below in Section 10, may be expressed in partition symbols by saying that $\omega \to (\omega)^m_n$ holds for any integers m and n, and it is irrelevant whether ω denotes a cardinal or an ordinal (i.e., order type) here. One is not always so fortunate: the partition relation

$$\eta \to (\eta, \aleph_0)^2 \tag{5}$$

where η denotes the order type of the rational numbers, is true if $\aleph_0$ denotes a cardinal and it is false if $\aleph_0$ denotes an ordinal. Hence the true meaning of a partition relation can only be discerned if we tell which of the symbols denote cardinals and which of them denote order types. This will usually be clear from the context; note also that there is no difficulty if a denotes a cardinal or a wellordered order type (i.e., an ordinal) in (1), since then mixing up cardinals and order types obviously does not then affect the meaning of the partition relation in (1). As an extra precaution, $\omega, \omega_1, \ldots$ will always denote ordinals in partition relations, while $\aleph_0, \aleph_1, \ldots$ will always denote cardinals. Apart from this slightly artificial distinction, ω_α and $\aleph_\alpha$ will always mean the same thing. (One may reflect that the proper way of avoiding the possibility of misunderstanding here would be to introduce additional variables which identify an argument as a cardinal or as an order type; this would, however, lead to a real loss of clarity and simplicity in the notation. It would be even worse to abandon our identification of cardinals with their initial ordinals since this identification has many advantages elsewhere.)

8.3. *Disjunction in partition symbols.* In the ordinary partition symbol in (1), and in other partition symbols to be defined, we can give a wider interpretation of the letters b_h. So far, b_h denoted either a cardinal or an order type; it is often advantageous to allow b_h to be a symbol of the form $\bigvee_{h' \in H_h} b_{h,h'}$, where $b_{h,h'}$ denotes the sorts of objects that were allowed to feature as b_h previously, say cardinals or order types. To say that a set is of *kind* $b_h = \bigvee_{h' \in H} b_{h,h'}$ means that there is an $h' \in H_h$ such that b_h is of kind $b_{h,h'}$. The meaning of the partition symbol in (1), as well as of other partition symbols to be introduced later, is defined by substituting this extended interpretation of the phrase "of kind b_h".

The partition relation $\eta \to (\eta, \aleph_0)^2$ considered above in (5) can clearly be rewritten with the aid of a disjunction as

$$\eta \to (\eta, \omega \vee \omega^*)^2 ,$$

where $b_{00} \vee b_{01}$ stands for $\bigvee_{\alpha<2} b_{0\alpha}$. In words, this relation can be expressed by saying that if we color the (unordered) pairs of the rationals with two colors, 0 and 1, then there is either a homogeneous set of order type η (the type of the rationals) that has color 0 or a homogeneous set of kind $\omega \vee \omega^*$ (i.e., either of order type ω or of order type ω^*) that is of color 1.

8.4. *The square bracket symbol.* The square bracket symbol

$$a \to [b_\xi]^r_{\xi < \vartheta}, \tag{6}$$

read as "a arrows square bracket b_ξ, r, ξ less than ϑ", is meaningful exactly if the analogous ordinary partition symbol (i.e., that in (2)) is so, i.e., if ϑ denotes an ordinal, and each of the other symbols denotes either a cardinal or an order type. Provided that these conditions are satisfied, the partition relation in (6) is said to hold if, given any *disjoint* partition $I = \langle I_\xi : \xi < \vartheta \rangle$ of the set $[A]^r$, where A is an arbitrary set of kind a, there is a $\xi < \vartheta$ and a set B_ξ of kind b_ξ such that $[B_\xi]^r \subseteq [A]^r \setminus I_\xi$. In this case we say that B_ξ *omits* class ξ of the partition, or that it omits color ξ of the associated coloring.

The partition relation

$$a \to [b_h]^r_{h \in H}$$

can be defined analogously to the relation in (6). We do not bother to give the definition, since, given the Axiom of Choice, this more general notation is hardly needed (cf. our remarks above made in the paragraph containing formula (2), when we compared relations (1) and (2)). In what follows, we shall define all partition symbols only in cases that correspond to the situation in formula (2) rather than (1).

Given a partition I of $[A]^r$, a subset B of the set A is called *completely inhomogeneous* with respect to I if $[B]^r$ intersects each class of I. If all the b_ξ's are equal to b, then the negation of the square bracket relation in (6) can be expressed as follows: there is a disjoint partition I of $[A]^r$ into $|\vartheta|$ parts, where A is a set of kind a, such that any subset of kind b of A is completely inhomogeneous with respect to I.

Analogously to the case of the ordinary partition symbol, the negation of the partition relation in (6) is written as

$$a \not\to [b_\xi]^r_{\xi < \vartheta},$$

and we shall also use the notation corresponding to the one mentioned in (3), as well as the notational conventions introduced in the paragraph containing formulas (3) and (4). Thus the meaning e.g. of the partition relation

$$a \to [(b_\xi)_{\xi < \vartheta}, (d)_\tau, c]^r$$

should be clear to the reader.

The partition relation $2^{\aleph_0} \nrightarrow (\aleph_1)^2_2$ is a theorem of set theory. If we assume the continuum hypothesis then this just says that $\aleph_1 \nrightarrow (\aleph_1)^2_2$. A much stronger statement is true under this assumption; in fact, one can prove from CH that $\aleph_1 \nrightarrow [\aleph_1]^2_{\aleph_1}$. The moral is that if one knows an ordinary negative partition relation then it is often worth asking whether a stronger assertion, a negative square bracket relation, is also true.

8.5. *The weak square bracket symbol.* The weak square bracket symbol

$$a \to [b]^r_{\tau, <\sigma} \tag{7}$$

read as "a arrows square bracket b super r sub τ, less than σ", is meaningful if τ and σ denote cardinals and each of the other letters denote either a cardinal or an order type. Provided that these conditions are satisfied, the relation in (7) is said to hold if, given any disjoint partition $I = \langle I_\xi : \xi < \tau \rangle$ of the set $[A]^r$, where A is a set of kind a, there is a set $B \subseteq A$ of kind b such that the set $[B]^r$ meets less than σ classes of the partition I (i.e., if $|\{\xi < \tau : I_\xi \cap [B]^r \neq 0\}| < \sigma$). The set B that confirms the validity of the relation in (7) may be called *relatively* $<\sigma$-*homogeneous*.

The partition relation

$$a \to [b]^r_{\tau, \sigma}$$

is defined as being equivalent to $a \to [b]^r_{\tau, <\sigma^+}$. A relatively $<\sigma^+$ homogeneous set confirming this relation may more simply be called a *relatively* σ-*homogeneous* set.

The weak square bracket relation can be used e.g. to express Chang's conjecture in a concise way. The original, model-theoretic formulation of this conjecture says that any structure of type $\langle \omega_2, \omega_1 \rangle$ has an elementary submodel of type $\langle \omega_1, \omega \rangle$. (Here a structure $\mathfrak{A}$ can only have countably many finitary relation symbols, and, by definition, $\mathfrak{A}$ is said to be of *type* $\langle \kappa, \lambda \rangle$ if the underlying set A of $\mathfrak{A}$ has cardinality κ, and the language of $\mathfrak{A}$ has a distinguished unary relation symbol U such that the set $\{x \in A : U^{\mathfrak{A}}(x)\}$ has cardinality λ, where $U^{\mathfrak{A}}$ denotes the interpretation of U in $\mathfrak{A}$.) As it can be shown, this is equivalent to saying that

$$\aleph_2 \to [\aleph_1]^2_{\aleph_1, \aleph_0}.$$

8.6. *The relation* $\alpha \to (\beta)^{<\omega}_\tau$. Given ordinals α, β, and a cardinal τ, the partition relation

$$\alpha \to (\beta)^{<\omega}_\tau, \tag{8}$$

read as "α arrows β, less than ω, τ", is said to hold if, given partitions $I_n = \langle I_{n\xi} : \xi < \tau \rangle$ of $[\alpha]^n$ for each $n < \omega$, there is a set $B \subseteq \alpha$ of order type β such that for any $n<\omega$ there is a $\xi<\tau$ for which $[B]^n \subseteq I_{n\xi}$. The set B here is called *homogeneous* with respect to the partitions I_n or the associated coloring f: $[\alpha]^{<\omega} \to \tau$, defined as $f(x)=\xi$ if and only if $x \in I_{n\xi}$ for some $n<\omega$.

It will be proved in Section 12 that the partition relation

$$\alpha \to (\omega)_2^\omega$$

does not hold for any ordinal α; that is, the existence of an ordinal α for which this holds contradicts the Axiom of Choice. The symbol in (8) was invented in order to express something stronger than the relation $\alpha \to (\beta)_\tau^n$ $(n<\omega)$ without contradicting the Axiom of Choice. The validity of the relation in (8) is usually a very strong condition on α. The relation

$$\alpha \to (\omega)_2^{<\omega}$$

implies that α is an inaccessible cardinal (in fact, a "very large" inaccessible cardinal), and the existence of an α such that

$$\alpha \to (\aleph_1)_2^{<\omega}$$

contradicts the Axiom of Constructibility; this is a result of Rowbottom [1971]. The subscript 2 in these relations is often omitted (i.e., one often writes $\alpha \to (\aleph_1)^{<\omega}$); we shall not follow this practice here.

Analogously to the relation in (8), one can define the square-bracket relation

$$\alpha \to [\beta]_\tau^{<\omega}, \tag{9}$$

where α, β are ordinals and τ is a cardinal, which is said to hold if, given disjoint partitions $I_n = \langle I_{n\xi}: \xi<\tau \rangle$ of $[\alpha]^n$ for any $n<\omega$, there is a set $B \subseteq \alpha$ of order type β such that for each n there is a $\xi<\tau$ such that $[B]^n \cap I_{n\xi}=0$.

8.7. *Polarized partitions.* The polarized partition symbol

$$\begin{pmatrix} a_0 \\ \vdots \\ a_{s-1} \end{pmatrix} \to \begin{pmatrix} b_{\xi 0} \\ \vdots \\ b_{\xi, s-1} \end{pmatrix}_{\xi<\vartheta}^{r_0, \dots, r_{s-1}} \tag{10}$$

is meaningful if s is an integer, ϑ is an ordinal, and each of the other symbols denotes either a cardinal or an order type. Provided that these conditions are satisfied, this symbol is said to hold if, given a partition $I=\langle I_\xi: \xi<\vartheta \rangle$ of the Cartesian product $[A_0]^{r_0} \times \dots \times [A_{s-1}]^{r_{s-1}}$, where A_i is a set of kind a_i $(i<s)$, there is a $\xi<\delta$ and there are sets $B_{\xi i} \subseteq A_i$, $i<s$ of kind $b_{\xi i}$ such that $[B_{\xi 0}]^{r_0} \times \dots \times [B_{\xi, s-1}]^{r_{s-1}} \subseteq I_\xi$.

It is clear that we may assume in this definition that the sets A_i, $i<s$, are pairwise disjoint. This allows us to deal with simpler objects than Cartesian

products in our considerations. For pairwise disjoint sets A_i, $i<s$, denote by

$$[A_0, \ldots, A_{s-1}]^{r_0, \ldots, r_{s-1}} \tag{11}$$

the set of all sets $X \subseteq \bigcup_{i<s} A_i$ such that $X \bigcap A_i \in [A_i]^{r_i}$ holds for all $i<s$. If all the r_i's are equal to 1, then we may omit them, i.e., $[A_0, \ldots, A_{s-1}]$ stands for $[A_0, \ldots, A_{s-1}]^{1, \ldots, 1}$. It is clear that we can restate the definition of the relation in (10) as follows: given any partition $I = \langle I_\xi \colon \xi < \vartheta \rangle$ of the set $[A_i, \ldots, A_{s-1}]^{r_0, \ldots, r_{s-1}}$, where A_i, $i<s$, is a set of kind a_i and the A_i's are pairwise disjoint, there is a $\xi < \vartheta$ and there are sets $B_{\xi i} \subseteq A_i$, $i<s$ of kind $b_{\xi i}$ such that $[B_{\xi 0}, \ldots, B_{\xi s-1}]^{r_0, \ldots, r_{s-1}} \subseteq I_\xi$. Here the set $\bigcup_{i<s} B_{\xi i}$ is called *homogeneous* of class (or color) ξ with respect to the partition I (or with respect to the associated coloring, provided that I is a disjoint partition).

Using disjunctions in polarized partitions (cf. Subsection 8.3 above) is especially useful. It was quite useless with ordinary partition symbols involving only cardinals, as cardinals are totally ordered by inclusion, it had some use with partition relations involving order types, and it occurs frequently in polarized partitions. The meaning of the symbol

$$\begin{pmatrix} a_0 \\ \vdots \\ a_{s-1} \end{pmatrix} \to \left(\bigvee_{\alpha < \gamma_\xi} \begin{matrix} b_{\xi 0 \alpha} \\ \vdots \\ b_{\xi, s-1, \alpha} \end{matrix} \right)^{r_0, \ldots, r_{s-1}}_{\xi < \vartheta} \tag{12}$$

can be reconstructed with the aid of Subsection 8.3, but we give the definition here for the sake of completeness: this relation is said to hold if, given any partition $I = \langle I_\xi \colon \xi < \vartheta \rangle$ of the set $[A_0, \ldots, A_{s-1}]^{r_0, \ldots, r_{s-1}}$, where A_i, $i<s$, is a set of kind a_i and the A_i's are pairwise disjoint, there is a $\xi < \vartheta$ and there is a set $X \subseteq \bigcup_{i<s} A_i$ with $[X \cap A_0, \ldots, X \cap A_{s-1}]^{r_0, \ldots, r_{s-1}} \subseteq I_\xi$ such that there is an $\alpha < \gamma_\xi$ for which $X \cap A_i$, $i<s$, is of kind $b_{\xi i \alpha}$.

It should be clear how to define the polarized square-bracket and weak square-bracket symbols, possibly with disjunctions. The details are left to the reader.

Lemmas involving polarized partition relations are often useful in proving theorems on ordinary partition relations. One of the oldest results involving polarized partition relations is

$$\begin{pmatrix} \aleph_1 \\ \aleph_0 \end{pmatrix} \to \begin{pmatrix} \aleph_1 & \aleph_0 \\ \aleph_0 & \aleph_0 \end{pmatrix}^{1,1},$$

due to Erdős and Rado [1956]. An example showing the usefulness of disjunction in polarized partition relations is the following: one can prove under the assumption of CH that

$$\binom{\aleph_1}{\aleph_1} \nrightarrow \binom{\aleph_1}{\aleph_0} \vee \binom{\aleph_0}{\aleph_1}_2^{1,1} \tag{13}$$

holds (Erdős–Rado–Hajnal [1965]), which is a strengthening of the result that CH implies $\aleph_1 \nrightarrow (\aleph_1)_2^2$.

8.8. *Classes of hypergraphs in partition relations.* Assume first that r is a cardinal. Then an *r-hypergraph* is an ordered pair $\langle x, y\rangle$ such that $y \subseteq [x]^r$. A 2-hypergraph is simply called a *graph.* Let $\langle x, y\rangle$ and $\langle x', y'\rangle$ be two r-hypergraphs. $\langle x, y\rangle$ is called a *subgraph* of $\langle x', y'\rangle$ if $x \subseteq x'$ and $y \subseteq y'$, and $\langle x, y\rangle$ is called *isomorphic* to $\langle x', y'\rangle$ if there is a 1-1 function f from x onto x' such that $\forall z \subseteq x[z \in y \Leftrightarrow f``z \in y']$. Let now r be either a cardinal or an order type. A *partially ordered r-hypergraph* is now a triple $\langle x, \prec, y\rangle$ such that $\langle x, \prec\rangle$ is a partial order and $y \subseteq [x]^r$ or $y \subseteq [x, \prec]^r$ according as r is a cardinal or an order type. Let $\langle x, \prec, y\rangle$ and $\langle x', \prec', y'\rangle$ be two partially ordered hypergraphs. $\langle x, \prec, y\rangle$ is called a *subgraph* of $\langle x', \prec', y'\rangle$ if $x \subseteq x'$, $y \subseteq y'$, and $\prec = \prec' | x$. An *isomorphism* between $\langle x, \prec, y\rangle$ and $\langle x', \prec', y'\rangle$ now has to be also an order isomorphism between $\langle x, \prec\rangle$ and $\langle x', \prec'\rangle$; in other respects, it is defined in an obvious way, as above. If $\langle x, y\rangle$ is a hypergraph (or $\langle x, \prec, y\rangle$ is an ordered hypergraph) then, according to usual graph-theoretic terminology, the elements of x are called *vertices* (or *points*), and the elements of y *edges* of this hypergraph.

A partition relation of the types described above can be given an obvious meaning if some of the letters a, b in them denote a class of (possibly partially ordered) r-hypergraphs instead of denoting a cardinal or an order type. Indeed, if d denotes a class of (partially ordered) hypergraphs then the phrase "*D is of kind d*" has to be interpreted as follows: D is an r-hypergraph, and there are r-hypergraphs D_1 and D_2 such that D_1 is a subgraph of D, D_1 is isomorphic to D_2, and $D_2 \in d$. Making this adjustment, all the definitions of partition relations given above (except the one with superscript $<\omega$ described in Subsection 8.6) can be interpreted in this general situation in a more or less obvious way.

For example, if a and b_ξ, $\xi < \vartheta$, denote classes of r-hypergraphs and r is a cardinal then the relation

$$a \rightarrow (b_\xi)_{\xi<\vartheta}^r$$

means the following: if $\langle A, A'\rangle$ is (an r-hypergraph having a subgraph isomorphic to) an element of a and $I = \langle I_\xi : \xi < \vartheta\rangle$ is a partition of A' then there is a $\xi < \vartheta$ and an r-hypergraph $\langle B_\xi, B'_\xi\rangle$ with $B'_\xi \subseteq I_\xi$ such that $\langle B_\xi, B'_\xi\rangle$ has a subgraph that is isomorphic to an element of b_ξ. For this ξ, the hypergraph $\langle B_\xi, B'_\xi \rangle$ is called *homogeneous* (of class ξ with respect to the partition I).

We are going to give a few examples of partition relations featuring graphs (i.e., 2-hypergraphs). To this end, for any two cardinals κ and λ denote by $(\kappa\colon \lambda)$ the class of *complete bipartite graphs* with κ plus λ vertices, i.e., the class of all graphs

$$\langle X \cup Y, \quad \{\{xy\} : x \in X \,\&\, y \in Y\} \rangle,$$

where $|X| = \kappa$, $|Y| = \lambda$, and X and Y are disjoint.

The relation

$$\aleph_1 \nrightarrow [(\aleph_0 : \aleph_1)]^2_{\aleph_1},$$

provable from CH (see (49.1) below), is a stronger statement than the one $\aleph_1 \nrightarrow [\aleph_1]^2_{\aleph_1}$, the proof of which also needs CH. An interesting unsolved problem is whether the relation

$$\aleph_{\omega+1} \nrightarrow ((\aleph_\omega : \aleph_{\omega+1}), \aleph_1)^2$$

holds if we assume GCH. What one knows is that GCH implies

$$\aleph_{\omega+1} \nrightarrow (\aleph_{\omega+1}, \aleph_1)^2$$

and

$$\aleph_{\omega+1} \nrightarrow ((\aleph_\omega : \aleph_{\omega+1}), (3)_{\aleph_0})^2;$$

a corollary of the second relation here is

$$\aleph_{\omega+1} \nrightarrow ((\aleph_\omega : \aleph_{\omega+1}), \aleph_2)^2$$

(see Erdős–Hajnal–Rado [1965]). On the other hand, the relation

$$\aleph_{\omega+1} \rightarrow (\aleph_{\omega+1}, (\aleph_\omega : \aleph_\omega))^2$$

also follows from GCH (loc. cit.). It is often customary to write $\binom{\kappa}{\lambda}$ instead of $(\kappa\colon \lambda)$ in these partition relations, especially if printing space permits this; it is irrelevant which letter is written above. One may write e.g. the last centered line equivalently as

$$\aleph_{\omega+1} \rightarrow \left(\aleph_{\omega+1}, \binom{\aleph_\omega}{\aleph_\omega}\right)^2.$$

Such partition relations should not be confused with polarized partition relations. For example, the relation

$$(\aleph_1 : \aleph_1) \nrightarrow ((\aleph_0 : \aleph_1))^2_2,$$

also written as

$$\binom{\aleph_1}{\aleph_1} \nrightarrow \left(\binom{\aleph_0}{\aleph_1}\right)^2_2,$$

is to be distinguished from the polarized relation

$$\binom{\aleph_1}{\aleph_1} \nrightarrow \binom{\aleph_0}{\aleph_1}^{1,1}_2$$

Indeed, the former relation expresses the same thing as the polarized relation in (13), while the latter formulates an apparently weaker assertion.

To give a further example, for two ordinals α, β, let $(\alpha:\beta)_<$ denote the set containing (exactly) the following ordered graph:

$$\langle \alpha \dot{+} \beta, <, \{\{\xi\eta\}: \xi<\alpha \,\&\, \alpha \le \eta < \alpha \dot{+} \beta\}\rangle \tag{14}$$

A theorem of Hajnal says that

$$\kappa^+ \nrightarrow (\kappa^+, (\kappa:2)_<)^2$$

holds for any regular cardinal κ satisfying $2^\kappa = \kappa^+$; this sets a limit to possible generalizations of the Erdős–Dushnik–Miller theorem (see Section 11).

Finally, we give an example reflecting a more general situation. To this end, call the *coloring number* of a graph G the least cardinal κ such that G is isomorphic to a graph $\langle \alpha, E\rangle$ that has the following property: for every $\xi<\alpha$, the set $\{\eta<\xi: \{\eta\xi\} \in E\}$ has cardinality $<\kappa$ (i.e., less than κ edges go back from ξ). A theorem of Shelah says the following. If $V=L$ and G is a graph with $\aleph_1$ vertices that has coloring number $>\omega$ then

$$\aleph_1 \nrightarrow (\{G\})^2_2$$

holds (see Shelah [1975]). With an abuse of notation, this relation is usually written as

$$\aleph_1 \nrightarrow (G)^2_2 .$$

Other generalizations of partition relations could also be considered since our discussion here does not cover e.g. the case of relation (18.11) below.

9. ELEMENTARY PROPERTIES OF THE ORDINARY PARTITION SYMBOL

9.1. In Subsection 8.2 of the preceding section we defined the ordinary partition relation

$$a \to (b_x)^r_{x\in I} \qquad \text{or} \qquad a \to (b_\xi)^r_{\xi<\vartheta} \tag{1}$$

(here and below in this section, each of the letters a, b, c, d, and r denotes either an order type or a cardinal; I and J denote index sets). Our aim here is to give an account of simple properties of this relation.

9.2. Invariance Under Permutations of Arguments. *If $f: J \to I$ is a 1-1 function onto I and $d_x = b_{f(x)}$ for each $x \in J$, then*

$$a \to (b_x)^r_{x \in I} \quad \text{iff} \quad a \to (d_x)^r_{x \in J}.$$

The trivial proof is left to the reader. This rule shows that it is enough to study the second relation in (1) (provided we assume the Axiom of Choice).

9.3. Omission Rule. *If J is a subset of I, r and b_x, $x \in I \setminus J$, denote the same kind of objects (i.e., either all denote cardinals or all denote order types), and $b_x \geq r$ for all $x \in I \setminus J$, then*

$$a \to (b_x)^r_{x \in I} \quad \text{implies} \quad a \to (b_x)^r_{x \in J}.$$

The proof is trivial. Note that the assumption $b_x \geq r$ for all $x \in I \setminus J$ is necessary, as $3 \to (4, 0)^1$ but $3 \not\to (4)^1$.

9.4. Reduction Rule. *For an arbitrary cardinal τ, the relation*

$$a \to (b_\xi)^r_{\xi < \vartheta} \tag{2}$$

is equivalent to

$$a \to ((b_\xi)_{\xi < \vartheta}, (r)_\tau)^r \tag{3}$$

provided $\vartheta \neq 0$.

Proof. (3) implies (2) by the omission rule. We prove that $\neg$ (3) implies $\neg$ (2) as well. Indeed, a set of cardinality (or order type) r is always homogeneous for an r-coloring; hence, if we want to make (3) fail, then no color η with $\vartheta \leq \eta < \vartheta \dotplus \tau$ must be allowed; i.e., a coloring establishing $\neg$ (3) also establishes $\neg$ (2). (This argument does not apply to the anomalous case when $\vartheta = 0$, since there is no coloring with range 0; this case is, however, of absolutely no interest. To illustrate the situation, one can easily see that a *strictly formal* application of the definition of the ordinary partition relation gives that $1 \to (b_\xi)^2_{\xi < 0}$ holds but $1 \to ((b_\xi)_{\xi < 0}, 2)^2$ fails.)

9.5. Range of Parameters. *If a, $b_\xi (\xi < \vartheta)$, and r denote the same kind of objects (i.e., either all denote cardinals or all denote order types), then the relation*

$$a \to (b_\xi)^r_{\xi < \vartheta} \tag{4}$$

is of interest only if

$$\vartheta \geq 2 \quad \text{and} \quad 1 \leq r < b_\xi \leq a \quad \text{for each} \quad \xi < \vartheta. \tag{5}$$

In fact if $\vartheta \leq 1$, $r = 0$, or $b_\xi < r$ for some $\xi < \vartheta$ then the truth value of (4) can be computed trivially (e.g., assuming $\vartheta \neq 0$, in the first two cases (4) holds if and only if $b_\xi \le a$ for all $\xi < \vartheta$, and in the third case it can fail only in case $b_\xi > a$ holds for the ξ in question). All those b_ξ's for which $b_\xi = r$ can be omitted by Subsection 9.4 unless $\vartheta = 0$; finally, if the first three inequalities are satisfied but $b_\xi > a$ for some $\xi < \vartheta$, then (4) is trivially false.

(5) could be formulated also for the case when some of the letters a, $b_\xi (\xi < \vartheta)$, and r denote cardinals and some denote order types, but it is hardly worth doing so.

9.6. MONOTONICITY PROPERTIES. a) *If $a \le a'$, then $a \to (b_\xi)^r_{\xi<\vartheta}$ implies $a' \to (b_\xi)^r_{\xi<\vartheta}$.*
b) *If $b'_\xi \le b_\xi$ for all $\xi < \vartheta$ then $a \to (b_\xi)^r_{\xi<\vartheta}$ implies $a \to (b'_\xi)^r_{\xi<\vartheta}$.*

Both of these properties are trivial. The interpretation of the inequality $a \le a'$ here also requires that a and a' are the same type of objects (i.e., either both are cardinals or both are order types). A similar remark applies to other inequalities; e.g., in b), b_ξ and b'_ξ must be the same type of objects (if ξ and η are different then, of course, b_ξ and b'_η can be objects of different type).

9.7. MONOTONICITY IN THE SUPERSCRIPT. a) *Let κ, λ_ξ $(\xi < \vartheta)$, and ρ be cardinals, r an integer, and assume that*

$$\kappa \nrightarrow (\lambda_\xi)^r_{\xi<\vartheta}. \tag{6}$$

Then

$$\kappa \nrightarrow (\lambda_\xi + \rho)^{r+\rho}_{\xi<\vartheta} \tag{7}$$

holds.

PROOF. Let $f: [\kappa]^r \to \vartheta$ be a coloring establishing (6). For each $X \in [\kappa]^r$ and $Y \in [\kappa]^\rho$ with $\max X < \min Y$ put $g(X \cup Y) = f(X)$. Then g: $[\kappa]^{r+\rho} \to \vartheta$ is a coloring establishing (7).

An analogous result holds with order types; we formulate it only for ordinals in a special case:

b) *Let r and s be integers, and α, β_ξ $\Phi\xi < \vartheta)$ ordinals. Let $\beta'_\xi = \beta_\xi$ if β_ξ is a limit ordinal and $\beta'_\xi = \beta_\xi \dot{+} s$ otherwise. Then $\alpha \nrightarrow (\beta_\xi)^r_{\xi<\vartheta}$ implies $\alpha \nrightarrow (\beta'_\xi)^{r \dot{+} s}_{\xi<\vartheta}$.*

$r \dot{+} s$ is of course the same as $r + s$. The proof is (almost) identical. From b) we get the following corollary:

c) *Let r and s be integers, α an ordinal, and β_ξ $(\xi < \vartheta)$ limit ordinals. Then $\alpha \nrightarrow (\beta_\xi)^r_{\xi<\vartheta}$ implies $\alpha \nrightarrow (\beta_\xi)^{r+s}_{\xi<\vartheta}$.*

In particular, this result is applicable if the β_ξ's are infinite cardinals; this special case also follows from a) (note that, as we remarked in the preceding section, while one has to make a distinction between cardinals and order types in a partition relation, no such distinction need be made if all the occurring order types are ordinals).

9.8. TRANSITIVITY RULE. *Assume that we have*

$$a \to (b_\xi)^r_{\xi<\vartheta}, \tag{8}$$

and

$$b_\xi \to (b_{\xi\eta})^r_{\eta<\nu_\xi} \tag{9}$$

holds for all $\xi<\vartheta$. *Then*

$$a \to (b_{\xi\eta})^r_{\xi<\vartheta,\,\eta<\nu_\xi} \tag{10}$$

also holds.

PROOF. Put $I=\bigcup_{\xi<\vartheta}(\{\xi\}\times\nu_\xi)$, let A be a set of order type or cardinality a (shortly: *kind* a), and let $f:[A]^r\to I$ be a coloring. We have to show that there is a homogeneous set of kind $b_{\xi\eta}$ and of color $\langle\xi\eta\rangle$ for some $\langle\xi\eta\rangle\in I$. If $f(X)=\langle\xi\eta\rangle$ for some $X\in[A]^r$ then put $\xi=f_1(X)$ and $\eta=f_2(X)$. Then $f_1:[A]^r\to\vartheta$ is a coloring, so there is a homogeneous set B_ξ of kind b_ξ and of color ξ for some $\xi<\vartheta$ by (8). Now, by (9), there is a homogeneous set $B_{\xi\eta}$ of kind $b_{\xi\eta}$ and of color η for some $\eta<\nu_\xi$ with respect to the coloring $f_2 \restriction [B_\xi]^r:[B_\xi]^r\to\nu_\xi$. It is obvious that $B_{\xi\eta}$ is a homogeneous set of kind $b_{\xi\eta}$ and of color $\langle\xi\eta\rangle$ with respect to f. The proof is complete.

9.9. THE CASE OF EXPONENT 1. a) *Let* κ *and* λ_ξ $(\xi<\vartheta)$ *be cardinals. Then*

$$\kappa\to(\lambda_\xi)^1_{\xi<\vartheta} \tag{13}$$

holds if and only if

$$\sum_{\xi<\vartheta}\lambda'_\xi<\kappa \tag{14}$$

holds for every sequence of cardinals $\lambda'_\xi<\lambda_\xi$ $(\xi<\vartheta)$.

The proof is obvious. The following corollary is worth mentioning:

b) *If* $\kappa\geq\omega$, $\rho<\mathrm{cf}(\kappa)$ *and* λ, $\sigma<\kappa$, *then*

$$\kappa\to((\kappa)_\rho,(\lambda)_\sigma)^1 \tag{16}$$

holds.

10. RAMSEY'S THEOREM

The aim of this section is to prove Ramsey's theorem, which says that $\omega\to(\omega)^r_k$ holds for any integers r and k. We shall give two proofs, one using ultrafilters and the other using trees. The reason we give both proofs is that they employ important ideas, and it is perhaps useful to illustrate these ideas in a relatively simple case.

In order to be able to present the first proof, we are going to consider ultrafilters. A *filter* on a set X is a nonempty subset F of $\mathscr{P}(X)$ such that (i) $0 \notin F$, (ii) if $x \in F$, then $y \in F$ for every y with $x \subseteq y \subseteq X$, and (iii) if $x, y \in F$, then $x \cap y \in F$. An *ultrafilter* F on X is a filter on X satisfying in addition the following property: (iv) if $x \cup y = X$, then either $x \in F$ or $y \in F$. A filter on X that is not properly included in any other filter on X is called a *maximal filter*. It is obvious by Zorn's lemma that any filter is included in a maximal filter, as the union of a set of filters totally ordered by inclusion is clearly a filter. The importance of maximal filters is underlined by

LEMMA 10.1. *A filter is maximal if and only if it is an ultrafilter.*

PROOF. "If part". Assume F is an ultrafilter on X that is not maximal, i.e., one that is properly included in another filter, say G. Choose an element x of $G \setminus F$. Then we have $X \setminus x \in F \subset G$ by property (iv) of ultrafilters above, as $x \notin F$. So $x, X \setminus x \in G$, which implies $0 = x \cap (X \setminus x) \in G$ by (iii); but this contradicts (i).

"Only if" part. Assume F is a maximal filter on X that is not an ultrafilter, i.e., we have $x, y \notin F$ and $x \cup y = X$ for some subsets x and y of X. Then the set

$$G = \{u \colon \exists v \in F\, [u \supseteq x \cap v]\}$$

is easily seen to be a filter on X. In fact, we only have to observe that $x \cap v \neq 0$ for any $v \in F$, since the contrary would mean that $v \subseteq y$, i.e., we would have $y \in F$ according to (ii) above, and this would contradict our assumptions on y. As G properly contains F ($x \in G \setminus F$), this contradicts the maximality of F, completing the proof.

This lemma shows us that there are many ultrafilters. If $p \in X$, then the set $\{x \subseteq X \colon p \in x\}$ is obviously an ultrafilter; an ultrafilter of this form is called *trivial.* There are *nontrivial* ultrafilters as well; in fact, any maximal filter extending (i.e., including) the *co-finite* filter

$$\{x \subseteq X \colon |X \setminus x| < \omega\}$$

on X is clearly a nontrivial ultrafilter. We are now in a position to prove Ramsey's theorem:

THEOREM 10.2 (Ramsey [1930]). *For any positive integers r and k we have*

$$\omega \rightarrow (\omega)^r_k, \tag{1}$$

i.e., if we partition the r-tuples of an infinite set X into finitely many parts, then there will be an infinite subset of X all r-tuples of which belong to the same class of the partition.

It is easy to see that it is enough to prove this theorem only in case $k=2$. In fact, the case $k=1$ is trivial, and if the case $k=2$ is known (i.e., if we know that $\omega \to (\omega)^r_2$), then the general case follows by induction on k with the aid of the Transitivity Rule (see Subsection 9.8). We note that the assumption $k=2$, which we are now allowed to make, will not make the proof below much easier, but the ideas will perhaps be clearer if their presentation is not burdened by a superfluous parameter.

PROOF. Assume $k=2$. We use induction on r. The case $r=1$ is obvious, so assume $r \geq 2$, and assume that the relation

$$\omega \to (\omega)^{r-1}_2 \tag{2}$$

has already been established. Let $I = \langle I_0, I_1 \rangle$ be a partition of $[\omega]^r$; we then have to construct an infinite homogeneous set. To this end, let U be a nontrivial ultrafilter on ω. First, we define a set $X = \{x_n : n < \omega\}$ that will include the homogeneous set. Choose $0, \ldots, r-2$ as $x_0, \ldots, x_{r-2}$. Assume that $X_n = \{x_0, \ldots, x_{n-1}\}$ has been defined for some integer $n \geq r-1$. For any $u \in [X_n]^{r-1}$ the set

$$Y_{u,j} = \{y \in \omega \setminus u : u \cup \{y\} \in I_j\}$$

belongs to the ultrafilter U for $j = f(u) = 0$ or 1; in fact, $Y_{u,0} \cup Y_{u,1} = \omega \setminus u$, and u is finite, and so $u \notin U$. Put

$$Y^n = \bigcap \{Y_{u,f(u)} : u \in [X_n]^{r-1}\}.$$

Obviously, we have $Y^n \in U$, and so $Y^n \setminus \{l : l \leq x_{n-1}\}$ is not empty. Let x_n be the first element of the latter set. This completes the definition of the set $X = \{x_n : n < \omega\}$. Observe also that the function f is unambiguously defined for any $u \in [X]^{r-1}$ and assumes 0 or 1 as values.

By the induction hypothesis (2), there is an infinite homogeneous set $Z \subseteq X$ with respect to the coloring $f : [X]^{r-1} \to 2$. We may assume, without loss of generality, that $f(u) = 0$ for any $u \in [Z]^{r-1}$. We are then going to show that

$$[Z]^r \subseteq I_0, \tag{3}$$

i.e., that Z is a homogeneous set with respect to the partition I. To this end, let $v \in [Z]^r$ be arbitrary, let x_n be the largest element of v, and put $u = v \setminus \{x_n\}$. Then $x_n \in Y_{u,f(u)} = Y_{u,0}$, which is the same as saying that $v = u \cup \{x_n\} \in I_0$, establishing (3). The proof is complete.

An alternative proof of Ramsey's theorem uses trees. A *tree* is a partially ordered set (i.e., a partial order) $\langle T, \prec \rangle$ such that the set of predecessors $\mathrm{pr}(x) = \mathrm{pr}(x, \langle T, \prec \rangle) = \{y \in T : y \prec x\}$ of any $x \in T$ is wellordered by $\prec$ (we assume that $\prec$ is irreflexive, i.e., $x \prec x$ does not hold for any $x \in T$). A subset of T

linearly ordered by $\prec$ is said to be a *chain* (*of* or *in* T). A maximal chain is called a *branch*. For any element x of T, $\mathrm{o}(x)=\mathrm{o}(x, \langle T, \prec\rangle)$ denotes the order type of $\mathrm{pr}(x)$; $\mathrm{o}(x)$ is called the *order* of x. For any ordinal α, the set $\{x \in T : \mathrm{o}(x)=\alpha\}$ is called the αth *level* of T. The *length* of T is defined as $\mathrm{lth}\,(T)=\mathrm{lth}\,(\langle T, \prec\rangle)=$ $=\sup\{\alpha\dot{+}1$: the αth level of T is not empty$\}$. For an element x of T, $\mathrm{ims}\,(x)=\mathrm{ims}\,(x, \langle T, \prec\rangle)$ will denote the set of *immediate successors* of x; i.e., $\mathrm{ims}\,(x)=\{y \in T : x \prec y \;\&\; \mathrm{o}(y)=\mathrm{o}(x) \dot{+} 1\}$. In this section we shall only need trees of length $\leq\omega$. The key result that we need for the proof of Ramsey's theorem is König's lemma on trees, which describes a property of ω that we shall call tree property later in a more general setting (see Definition 29.2).

LEMMA 10.3 (D. König [1927]). *If $\langle T, \prec\rangle$ is a tree of length ω each level of which is finite then T has an infinite branch.*

PROOF. We are going to construct a sequence of subtrees $\langle T_n, \prec \mid T_n\rangle$, $n<\omega$, such that $T=T_0 \supseteq T_1 \supseteq T_2, \ldots$ and $\bigcap_{n<\omega} T_n$ is an infinite branch. Assume that T_n has already been constructed in such a way that its length is ω, and, for any $k<n$, its kth level contains only one element x_k. For any x in the nth level of T_n, put

$$T_{n,x}=\{x_k : k<n\} \cup \{y \in T_n : x \preceq y\}\,.$$

As $T_n=\bigcup_x T_{n,x}$, where x runs over all elements of the nth level in T_n, we have

$$\omega=\mathrm{lth}\,(T_n)=\sup_x \mathrm{lth}\,(T_{n,x})\,.$$

As the nth level in T_n is finite by our assumptions, we have $\mathrm{lth}\,(T_{n,x})=\omega$ for some x in the nth level of T. Put $x_n=x$ and $T_{n+1}=T_{n,x}$ for such an x. It is easy to see that $\{x_n : n<\omega\}=\bigcap_{n<\omega} T_n$ is an infinite branch of T.

We now give an alternative proof of Ramsey's Theorem 10.2:

We may assume $r\geq 2$. We proceed by induction on r, so assume that the relation

$$\omega \rightarrow (\omega)^{r-1}_k \tag{4}$$

has already been established.

Given a coloring f: $[\omega]^r \rightarrow k$, we have to construct an infinite homogeneous set. In order to do this, we first build a tree $\langle T, \prec\rangle$ of length ω level by level such that $T \subseteq \omega$ and each level of T is finite. Denote by $T[n]$ the nth level of the tree to be defined. As an auxiliary to the definition of T, for each element x of T we shall define a set $S(x)$ which will include the set $\{y \in T: x \prec y\}$. Intuitively, $S(x)$ will be the set of potential successors of x in the homogeneous set to be defined. For each integer n, the sets $S(x)$, $x \in T[n]$, will be pairwise disjoint, but the really important thing is that we shall have

$$|\omega \setminus \bigcup_{x \in T[n]} S(x)| < \omega.$$

We now set out to define T. To this end, for each $n \leq r-2$ put $T[n] = \{n\}$ and $S(n) = \omega \setminus \{m: m \leq n\}$ $(= \omega \setminus (n+1))$. Suppose next that $T[i]$ and $S(x)$ have already been defined for any $i \leq n$ and $x \in \bigcup_{j \leq n} T[j]$, where $n \geq r-2$. Let $x \in T[n]$ be arbitrary. We are going to define the set of immediate successors of x in the tree, i.e., the set $T(x) = \{y \in T[n+1]: x \prec y\}$. If $S(x) = 0$, then put $T(x) = 0$. If not, then consider the following equivalence relation $\equiv_x$ on $S(x)$: for any $y, z \in S(x)$, $y \equiv_x z$ if $f(u \cup \{y\}) = f(u \cup \{z\})$ holds for any $u \in [\{x': x' \preceq x\}]^{r-1}$, where we recall that f is the given coloring of $[\omega]^r$. As there are only finitely many possibilities for u (in fact, x has n predecessors), and f assumes only $k < \omega$ values, there are only finitely many equivalence classes. For each nonempty equivalence class C, choose a $y \in C$ as the immediate successor of x and put $S(y) = C \setminus \{y\}$. This completes the definition of the tree up to and including the $(n+1)$-st level. It is easy to see that all the requirements on $S(y)$ $(y \in T[n+1])$, are satisfied, and so the definition of $\langle T, \prec \rangle$ is complete.

As each level of T is finite, T has an infinite branch X by the preceding lemma. Define the coloring g: $[X]^{r-1} \to k$ as follows. For any $u \in [X]^{r-1}$, put $g(u) = f(u \cup \{x\})$, where x is an arbitrary element of X exceeding every element of u in the ordering $\prec$. Observe that this is a sound definition, since if $x, y \in X$ exceed every element of u in the ordering $\prec$, then we have $x \equiv_{\max_\prec u} y$, and so $f(u \cup \{x\}) = f(u \cup \{y\})$ holds. The induction hypothesis (4) implies that there is an infinite homogeneous set $Y \subseteq X$ with respect to the coloring g, i.e., there is an $i < k$ such that $g(u) = i$ holds for any $u \in [Y]^{r-1}$. The definition of g clearly entails then that $f(v) = i$ holds for any $v \in [Y]^r$, i.e., that Y is a homogeneous set with respect to the coloring f. The proof is complete.

Ramsey's theorem has the following consequence, often called the finite Ramsey theorem:

THEOREM 10.4. *For any three positive integers k, r, and m there is an integer n such that*

$$n \to (m)_k^r.$$

We shall study later, in Section 26, how n depends on k, r, and m. There are several proofs of this theorem; in particular, there is one using ultrafilters and one using König's lemma on trees above. We shall present the latter.

PROOF. Assume that the conclusion of the theorem does not hold, and for each integer $n > r-1$ let f_n: $[n]^r \to k$ be a coloring with respect to which there is no homogeneous set of m elements. The set

$$T = \{f_n \restriction [j]^r : j \le n \,\&\, n < \omega\}$$

with inclusion $\subset$ as the ordering $\prec$ is obviously a tree of length $\le \omega$. As each level of T is finite and T is infinite, the length of T must be ω, and so T must have an infinite branch F by Lemma 10.3. Put $f = \bigcup F$; then f is a coloring $f: [\omega]^r \to k$, and so, by Theorem 10.2, there is an infinite set $X \subseteq \omega$ homogeneous with respect to f. Let Y be an arbitrary subset of m elements of X, and let $g \in F$ be such that $Y \subseteq \operatorname{dom}(g)$. Let n and $j \le n$ be integers such that $g = f_n \restriction [j]^r$. Then Y is a set of m elements that is homogeneous with respect to f_n. This contradicts the definition of f_n, completing the proof.

Finally, we want to point out that Ramsey's theorem does not seem to have any proper generalization; indeed, it might be interesting to observe that none of the general results in Chapter IV imply Ramsey's theorem. The particular case $r = 2$ does, however, have an important generalization; this is the topic of the next section.

11. THE ERDŐS–DUSHNIK–MILLER THEOREM

The theorem mentioned in the title is a generalization of Ramsey's theorem for $r = 2$ (cf. Dushnik–Miller [1941]; these authors proved (1) in case κ is regular, and the singular case is due to Erdős).

THEOREM 11.1 (Erdős–Dushnik–Miller). *For any infinite cardinal κ we have*

$$\kappa \to (\kappa, \omega)^2. \tag{1}$$

Making use of the Transitivity Rule (see Subsection 9.8), it follows from here by induction on k that we have

$$\kappa \to (\kappa, (\omega)_k)^2 \tag{2}$$

for any integer k and any infinite cardinal κ.

In the proof of the above theorem, we fix a coloring $f: [\kappa]^2 \to 2$ and, for the sake of lucidity, we call the color 0 red, and the color 1 blue. A homogeneous set e.g. of color 0 will be called a red homogeneous set. We write

$$B(x) = \{y \in \kappa \setminus \{x\} : f(\{xy\}) = 1\} \tag{3}$$

for any $x \in \kappa$, i.e., $B(x)$ is the set of points that are connected to x with a blue edge (edge is a synonym for pair). We need the following lemma:

LEMMA 11.2. *If κ does not have an infinite blue homogeneous subset, then there is a set $X \subseteq \kappa$ of cardinality κ such that $|B(x) \cap X| < \kappa$ holds for any $x \in X$.*

PROOF. Assume the contrary, and define x_n and X_n, $n<\omega$, by recursion. Let $X_0=\kappa$. Assume that $X_n\subseteq\kappa$ has already been defined for some integer n and X_n has cardinality κ. Let $x_n\in X_n$ be such that $|B(x_n)\cap X_n|=\kappa$, and put $X_{n+1}=$ $=B(x_n)\cap X_n$. Then the set $\{x_n: n<\omega\}$ is an infinite blue homogeneous set; in fact, if $n<m<\omega$, then $x_m\in X_{n+1}\subseteq B(x_n)$, and so the edge $\{x_n x_m\}$ is blue. The proof is complete.

PROOF OF THEOREM 11.1. We distinguish two cases: a) κ is regular, and b) κ is singular.

Ad a). Suppose there is no red homogeneous set of cardinality κ. We are going to prove that then there is an infinite blue homogeneous set by showing that the conclusion of the preceding lemma fails. In fact, let $X\subseteq\kappa$ be a set of cardinality κ, and let $Y\subseteq X$ be a red homogeneous set that is *maximal* in X; i.e., let Y be a red homogeneous subset of X such that $Y\cup\{x\}$ is not a red homogeneous set for any $x\in X\setminus Y$. Then $|Y|<\kappa$, and for any $x\in X\setminus Y$ there is an $y\in Y$ such that the edge $\{xy\}$ is blue, that is

$$X\setminus Y\subseteq\bigcup_{y\in Y}(B(y)\cap X).$$

As κ is regular and $|Y|<|X|=\kappa$, we must have $|B(y)\cap X|=\kappa$ for some $y\in Y(\subseteq X)$. This shows that the conclusion of the preceding lemma does indeed fail, completing the proof in case a).

Ad b). Assume there is no infinite blue homogeneous subset. Then there is a set $X\subseteq\kappa$ of cardinality κ such that

$$|B(x)\cap X|<\kappa \tag{4}$$

holds for any $x\in X$. Let $\langle\kappa_\xi: \xi<\mathrm{cf}(\kappa)\rangle$ be an increasing sequence of regular cardinals $>\mathrm{cf}(\kappa)$ tending to κ, and let X_ξ, $\xi<\mathrm{cf}(\kappa)$, be pairwise disjoint subsets of X such that $|X_\xi|=\kappa_\xi$. Using the partition relation $\kappa_\xi\to(\kappa_\xi,\omega)^2$, which we proved just before, as κ_ξ is regular, we can see that X_ξ has a red homogeneous subset Y_ξ of cardinality κ_ξ. We have

$$Y_\xi=\bigcup_{\alpha<\mathrm{cf}(\kappa)}\{x\in Y_\xi: |B(x)\cap X|<\kappa_\alpha\}$$

in view of (4). As κ_ξ is regular and $>\mathrm{cf}(\kappa)$, there is an $\alpha<\mathrm{cf}(\kappa)$ for which the set after the union sign has cardinality κ_ξ. Denote such an α by $h(\xi)$ and write

$$Z_\xi=\{x\in Y_\xi: |B(x)\cap X|<\kappa_{h(\xi)}\}.$$

Then $|Z_\xi|=\kappa_\xi$.

Let $\langle \alpha_\xi : \xi < \mathrm{cf}(\kappa) \rangle$ be a strictly increasing sequence of ordinals such that $h(\alpha_\eta) < \alpha_\xi < \mathrm{cf}(\kappa)$ holds for any η, ξ with $\eta < \xi < \mathrm{cf}(\kappa)$, and put

$$S_\xi = Z_{\alpha_\xi} \setminus \bigcup \{B(x) : x \in \bigcup_{\eta < \xi} Z_{\alpha_\eta}\}$$

for any $\xi < \mathrm{cf}(\kappa)$. Then $|S_\xi| = \kappa_{\alpha_\xi}$ and $S = \bigcup_{\xi < \mathrm{cf}(\kappa)} S_\xi$ is a red homogeneous set of cardinality κ. The proof is complete.

Next we prove a slightly stronger result for regular $\kappa > \omega$ than Theorem 11.1 above. We shall do this by using a tree argument in a way similar to the one used in the second proof of Ramsey's theorem (Theorem 10.2).

THEOREM 11.3. *If κ is a regular cardinal $> \omega$, then*

$$\kappa \to (\kappa, \omega \dot{+} 1)^2 .$$

PROOF. Let $f : [\kappa]^2 \to 2$ be a coloring, and, similarly as before, call the color 0 red, and the color 1 blue, and assume there is no red homogeneous set of cardinality κ. We are going to define a tree $\langle T, \prec \rangle$ of length ω such that each element of T is a red homogeneous set and, simultaneously, for each element X of T we shall define a set $S(X) \subseteq \kappa$ which will include its successors in $\prec$. The definition uses recursion. T will have a single least element; this can be chosen as an arbitrary maximal red homogeneous set X_0; put

$$S(X_0) = \kappa .$$

Let n be an integer. Assume that, for each $i \leq n$, the ith level $T[i]$ of T has already been defined, and for each $X \in T[i]$ assume that $S(X)$ has also been defined, and suppose that X is a maximal red homogeneous subset of $S(X)$. Choose an arbitrary $X \in T[n]$; we are going to define the set ims(X), i.e., the set of immediate successors of X in $\langle T, \prec \rangle$, and also, for every $Y \in$ ims(X), we are going to define $S(Y)$. To this end, for each $\xi \in X$ put

$$S_\xi(X) = B(\xi) \cap (S(X) \setminus (\sup X \dot{+} 1)), \tag{5}$$

where $B(\xi)$ is the set of points that are connected with a blue edge to ξ (see (3) above), and let X_ξ be a maximal red homogeneous subset of the set

$$S'_\xi(X) = S_\xi(X) \setminus \bigcup_{\eta < \xi, \eta \in X} S_\eta(X). \tag{6}$$

The set ims(X) of the *immediate successors* of X in $\langle T, \prec \rangle$ will be defined as

$$\mathrm{ims}(X) = \{X_\xi : \xi \in X \quad \text{and} \quad S'_\xi(X) \neq 0\}; \tag{7}$$

note here that $S'_\xi(X) \neq 0$ implies $X_\xi \neq 0$. For each X_ξ in this set put

$$S(X_\xi) = S'_\xi(X). \tag{8}$$

As X was an arbitrary element of the nth level $T[n]$ of T, this completes the definition of the $(n+1)$st level $T[n+1]$ of T. Noting that $T = \bigcup_{n<\omega} T[n]$, the definition of the tree $\langle T, \prec \rangle$ is complete. (A remark on the sets $S'_\xi(X)$ defined in (6): The pairwise disjointness of these sets ensures that the sets X_ξ, $\xi \in X$, are also pairwise disjoint, i.e., the nonempty ones among them are pairwise distinct. This, together with (5), excludes the possibility that we add the same sets as elements to T on different 'occasions'. This is a technical advantage in a sense, since otherwise the same sets added on different occasions would have to be distinguished in some other way. In later proofs, when we shall have at our disposal a general framework, presented in Section 14, of tree arguments, we shall not need such a 'technical advantage'.)

We claim that T has an infinite branch H such that

$$\bigcap \{S(X) : X \in H\} \neq 0. \tag{9}$$

To this end, observe that, for an arbitrary element X of T, we have

$$S(X) \setminus (\sup X \dot{+} 1) = \bigcup_{\xi \in X} S_\xi(X)$$

with the sets defined in (5), since X is a maximal red homogeneous subset of $S(X)$ by the definition of T; note also that

$$|X| < \kappa \tag{10}$$

also follows, since κ has no red homogeneous subset of cardinality κ by our assumption. Hence it is clear by the regularity of κ from (6), (7), and (8) that

$$|S(X) \setminus \bigcup \{S(Y) : Y \in \text{ims}\,(X)\}| < \kappa. \tag{11}$$

(10) and (7) imply that $|\text{ims}\,(X)| < \kappa$. It follows by induction on n from this that for each integer n we have

$$|T[n]| < \kappa, \tag{12}$$

and so it also easily follows by induction from (11) that we have

$$|\kappa \setminus \bigcup \{S(X) : X \in T[n]\}| < \kappa \tag{13}$$

for each integer n, since κ was assumed to be regular. Now, for a finite or infinite branch H of T, put

$$\bar{S}(H) = \bigcap \{S(X) : X \in H\}.$$

We claim that we have

$$\bigcap_{n<\omega} \bigcup \{S(X): X \in T[n]\} \subseteq \bigcup_H \bar{S}(H), \tag{14}$$

where H runs over all branches of T (actually, equality holds here by a simple distributive law). In fact, if ξ belongs to the set on the left-hand side, then $H=\{X \in T: \xi \in S(X)\}$ is easily seen to be a branch such that $\xi \in \bar{S}(H)$. Now either there is a branch as required in (9), or $\bar{S}(H)$ is empty for all infinite branches; assume the second alternative holds. In this case, with H' running over all finite branches of T, we have

$$|\bigcup_{H'} \bar{S}(H')| = \kappa$$

according to (14), since the left-hand side of (14) has cardinality κ by (13). As $|T| = |\bigcup_{n<\omega} T[n]| < \kappa$ in view of (12), T has fewer than κ finite branches. Hence $|\bar{S}(H')| = \kappa$ holds for a finite branch H'. Let X be the largest element in $\prec$ of H'. Then $S(X) = \bar{S}(H')$ has cardinality κ, and so X has an immediate successor Y (e.g. by (11)). This is impossible, since a branch is a maximal chain in a tree. Hence (9) must be valid.

Let now $H=\{X^n: n<\omega\}$ be an infinite branch of T satisfying (9), where $X^n \in T[n]$. Define $\xi_n \in X^n$ by the formula

$$X^{n+1} = X^n_{\xi_n}$$

(cf. (7)), let ξ_ω be an arbitrary element of $\bar{S}(H) = \bigcap_{n<\omega} S(X^n)$, and put $X^\omega = \{\xi_\omega\}$. It is easy to see that $\{\xi_\alpha: \alpha \leq \omega\}$ is a blue homogeneous set. In fact, if $n < \alpha \leq \omega$, then

$$\xi_\alpha \in X^\alpha \subseteq S(X^{n+1}) \subseteq S_{\xi_n}(X^n),$$

and so the edge $\{\xi_n \xi_\alpha\}$ is blue by (5). As $\xi_n \in X^n$, it also follows from here by (5) that $\xi_n < \xi_\alpha$. Hence $\{\xi_\alpha: \alpha \leq \omega\}$ has order type $\omega \dot{+} 1$. The proof is complete.

R. Laver and F. Galvin asked the following

PROBLEM 11.4. Does

$$\kappa \to (\kappa, \omega \dot{+} 1)^2$$

hold in ZFC if κ is an infinite cardinal with $\operatorname{cf}(\kappa) > \omega$?

As a corollary of Lemma 28.1 and the previous result the answer is affirmative if GCH is assumed. Without GCH, the answer is not known, even for $\kappa = \aleph_{\omega_1}$. As for possible strengthenings of the above theorem, it is known that

$$\aleph_1 \to (\aleph_1, \omega \dot{+} 2)^2 \tag{15}$$

does not necessarily hold. In fact, Hajnal [1960] proved the following:

THEOREM 11.5. *Let λ be a regular cardinal, and assume that $\kappa=2^\lambda=\lambda^+$. Then*

$$\kappa\nrightarrow((\kappa:\kappa),(\lambda:2)_<)^2 \tag{16}$$

holds.

Here $(\kappa:\kappa)$ denotes a bipartite graph with κ plus κ vertices, and $(\lambda:2)_<$ a bipartite graph with λ vertices below and 2 vertices above (see Subsection 8.8, and especially (8.14)). (16) obviously implies that $\kappa\nrightarrow(\kappa,\lambda\dot{+}2)^2$ holds under the given assumptions. The case $\lambda=\omega$ is relevant for the relation in (15). As for the case $\lambda>\omega$, we mention only that it will be clear from our results below that the relation

$$\lambda^+\rightarrow(\lambda^+,\mathrm{cf}(\lambda)\dot{+}1)^2 \tag{17}$$

holds if we assume GCH. (Cf. (17.25) below. Only the relation $\lambda^+\rightarrow(\lambda^+,\mathrm{cf}(\lambda))^2$ is given there, but it is clear that Theorem 17.1 implies this stronger result.) It is not known whether Theorem 11.5 can be extended to the case of singular λ; in view of (17), the simplest question that can be asked is the following:

PROBLEM 11.6. Assume GCH. Does then

$$\aleph_{\omega+1}\nrightarrow(\aleph_{\omega+1},\omega\dot{+}2)^2 \tag{18}$$

hold?

PROOF OF THEOREM 11.5. Using the assumption $\kappa=2^\lambda=\lambda^+$, write

$$[\kappa]^\lambda=\{A_\alpha:\alpha<\kappa\}.$$

We are going to define a coloring $f:[\kappa]^2\rightarrow 2$ establishing (16); to this end, we shall first define the sets

$$F_\xi=\{\eta<\xi: f(\{\eta\xi\})=1\} \tag{19}$$

by transfinite recursion on $\xi<\kappa$. If $\xi<\lambda$ then put $F_\xi=0$, and if $\lambda\le\xi<\kappa$ and F_η has already been defined for each $\eta<\xi$, then we shall describe F_ξ in the form

$$F_\xi=\{\delta_{\xi\eta}:\eta<\lambda\},$$

where the ordinals $\delta_{\xi\eta}<\xi$ are again defined by transfinite recursion. To this end, let $g_\xi:\lambda\rightarrow\xi$ be a 1-1 function onto ξ and, given an $\eta<\lambda$, assume that $\delta_{\xi\zeta}$ has already been defined for all $\zeta<\eta$. Choose $\delta_{\xi\eta}$ as the least element of the set

$$D_{\xi\eta}=((A_{g_\xi(\eta)}\cap\xi)\setminus\{\delta_{\xi\zeta}:\zeta<\eta\})\setminus\bigcup_{\zeta<\eta}F_{g_\xi(\zeta)}, \tag{20}$$

provided this set is not empty; otherwise put $\delta_{\xi\eta}=0$. This completes the definition of the sets F_ξ, $\xi<\kappa$, and so (19) defines the coloring f.

Our first claim is that there is no homogeneous graph of color 1 that is of kind (i.e., that has a subgraph isomorphic to the single element of) $(\lambda:2)_<$ (cf. (8.14)).

Assume the contrary. This means that there is a set $X \subseteq \kappa$ of order type λ and there are two ordinals α and ξ with $\sup X \leq \alpha < \xi < \kappa$ such that

$$X \subseteq F_\alpha \cap F_\xi .$$

This is, however, impossible since, clearly,

$$|F_\alpha \cap F_\xi| < \lambda \tag{21}$$

holds for any α, ξ with $\alpha < \xi < \kappa$. In fact, if $\alpha = g_\xi(\eta)$ for some $\eta < \lambda$, then we have

$$F_\alpha \cap F_\xi \subseteq \{0\} \cup \{\delta_{\xi\zeta} : \zeta < \eta\} .$$

Hence our first claim is established.

Our second claim is that there is no homogeneous set of kind $(\kappa : \kappa)$ (cf. Subsection 8.8) and of color 0. Assume the contrary. Then there are disjoint sets $X, Y \subseteq \kappa$ of cardinality κ such that X and Y form a bipartite graph of color 0, i.e.,

$$f(\{\delta\xi\}) = 0 \tag{22}$$

holds for any $\delta \in X$ and $\xi \in Y$. Write

$$Z = \{\gamma < \kappa : |F_\gamma \cap X| = \lambda\} .$$

We distinguish two cases: a) $|Z| < \lambda$, and b) $|Z| \geq \lambda$.

Ad a). Let A be an arbitrary subset of cardinality λ of $X \setminus (\sup Z \dotplus 1)$, and let $\alpha < \kappa$ be such that $A = A_\alpha$. Choose $\xi \in Y \setminus (\sup A \dotplus 1)$ with $\xi > \alpha$ arbitrarily, and define $\eta < \lambda$ by the equality $g_\xi(\eta) = \alpha$. Then it is easy to see by the regularity of λ from (20) and the definition of Z that $D_{\xi\eta}$ has cardinality λ, and so it is not empty. Hence

$$\delta_{\xi\eta} \in D_{\xi\eta} \subseteq A_\alpha \subseteq X .$$

Now, the pair $\{\delta_{\xi\eta}\xi\}$ has color 1 in view of (19), since $\delta_{\xi\eta} \in F_\xi$ by the definition of this latter set. This contradicts (22).

Ad b) Let γ_ν, $\nu < \lambda$, be the first λ elements of Z, and put

$$A = \bigcup_{\nu < \lambda} (X \cap F_{\gamma_\nu}) .$$

Each set on the right-hand side has cardinality λ; hence so has A. Let $\alpha < \kappa$ be such that $A = A_\alpha$, and choose a $\xi \in Y \setminus (\sup A \dotplus 1)$ with $\xi > \alpha$ arbitrarily. Define $\eta < \lambda$ by the equality $g_\xi(\eta) = \alpha$. We claim that $D_{\xi\eta}$ is not empty. To show this, pick a $\nu < \lambda$ such that $\gamma_\nu \notin \{g_\xi(\zeta) : \zeta < \eta\}$. It follows from (21) and from the definition of Z by the regularity of λ that the set

$$(X \cap F_{\gamma_\nu}) \setminus \bigcup_{\zeta < \eta} F_{g_\xi(\zeta)}$$

has cardinality λ. As

$$X \cap F_{\gamma_\xi} \subseteq A = A_\alpha = A_{g_\xi(\eta)} \subseteq \xi ,$$

it follows from (20) that $D_{\xi\eta}$ has cardinality λ, and so it is indeed not empty, as we claimed above. Hence

$$\delta_{\xi\eta} \in D_{\xi\eta} \subseteq A_\alpha \subseteq X .$$

Now, the pair $\{\delta_{\xi\eta}, \xi\}$ has color 1 in view of (19), since $\delta_{\xi\eta} \in F_\xi$. This again contradicts (22). The proof is complete.

A further result of the type given in the above theorem was established by R. Laver [1975]. He proved that

$$\aleph_2 \nrightarrow (\aleph_2, (\omega\colon 2)_<)^2$$

if $2^{\aleph_0} = \aleph_2$ and Martin's Axiom holds. He also showed that if κ is a Mahlo cardinal, then there is a Boolean extension of the universe in which $2^{\aleph_0} = \kappa$ is weakly Mahlo, and

$$2^{\aleph_0} \nrightarrow (2^{\aleph_0}, (\omega\colon 2)_<)^2 \tag{23}$$

holds (a regular cardinal κ is *Mahlo* if the inaccessible cardinals $< \kappa$ form a stationary set in κ; κ is *weakly Mahlo* if the regular limit cardinals $< \kappa$ form a stationary set in κ). J. E. Baumgartner showed that if GCH holds and κ is a regular cardinal, then there is a cofinality-preserving Boolean extension in which $2^{\aleph_0} = \kappa^+$ and

$$2^{\aleph_0} \nrightarrow (2^{\aleph_0}, (\omega\colon 2)_<)^2 \tag{24}$$

holds.

In the other direction, a theorem of Erdős and Rado, to be discussed later, says that we have

$$(2^{\aleph_0})^+ \to ((2^{\aleph_0})^+, \aleph_1)^2 \tag{25}$$

(see Corollary 17.5 below), and Kunen showed that if κ is real-valued measurable then

$$\kappa \to (\kappa, \alpha)^2$$

holds for any $\alpha < \omega_1$ (measurable and real-valued measurable cardinals will be defined in Section 34; see Kunen [1971]). To compare this with (16), (23), (24), and (25), observe that it is consistent relatively to ZFC+ "there exists a measurable cardinal" that $2^{\aleph_0}$ is real-valued measurable, according to a theorem of Solovay [1971].

Finally we remark that, as in the case of Ramsey's theorem, we shall not obtain the Erdős–Dushnik–Miller theorem as a special case of a general theorem in Chapter IV.

12. NEGATIVE RELATIONS WITH INFINITE SUPERSCRIPTS

Strong negative partition relations can be proved with infinite superscripts. Namely, we have

THEOREM 12.1. *Given any two cardinals κ and λ with $\omega \leq \kappa \leq \lambda$, the relation*

$$\lambda \nrightarrow [\kappa]^{\kappa}_{2^{\kappa}} \tag{1}$$

holds.

The meaning of the relation in (1) can be discerned by consulting the definition of the square bracket symbol, given in Subsection 2.4, but, for the sake of the reader's convenience, here is a translation of the above result into words: for any κ, λ with $\omega \leq \kappa \leq \lambda$, there is a partition of $[\lambda]^{\kappa}$ into 2^{κ} pairwise disjoint classes such that, for any set $X \subseteq \lambda$ of cardinality κ, the set $[X]^{\kappa}$ intersects each class of the partition. This result has the following corollary (Erdős–Rado [1952]): given κ, λ with $\omega \leq \kappa \leq \lambda$, we have

$$\lambda \nrightarrow (\kappa)^{\kappa}_{2}. \tag{2}$$

In fact, let $\langle I_{\xi}: \xi < 2^{\kappa} \rangle$ be a partition verifying (1); then $\langle I_0, \bigcup_{1 \leq \xi < 2^{\kappa}} I_{\xi} \rangle$ verifies (2).

PROOF. First we show that

$$\kappa \nrightarrow [\kappa]^{\kappa}_{2^{\kappa}} \tag{3}$$

holds for any infinite cardinal κ by defining a coloring $f: [\kappa]^{\kappa} \to 2^{\kappa}$ that verifies this relation. To this end, let $\langle X_{\xi}: \xi < 2^{\kappa} \rangle$ be an enumeration of $[\kappa]^{\kappa}$ such that each set is mentioned 2^{κ} times. For each $\xi < 2^{\kappa}$, choose a set $Y_{\xi} \subseteq X_{\xi}$ of cardinality κ in such a way that $Y_{\xi} \neq Y_{\eta}$ holds for any ξ, η with $\xi < \eta < 2^{\kappa}$. For any $\xi < 2^{\kappa}$ put

$$f(Y_{\xi}) = \mathrm{tp}\, \langle \{\alpha: \alpha < \xi \,\&\, X_{\alpha} = X_{\xi}\}, < \rangle,$$

and for any $Y \in [\kappa]^{\kappa}$ that differs from all the Y_{ξ}'s, define $f(Y)$ arbitrarily. It is easy to see that, for any $X \in [\kappa]^{\kappa}$ and any $\eta < 2^{\kappa}$, there is a set $Y \subseteq X$ of cardinality κ such that $f(Y) = \eta$. In fact, let X_{γ} be the $(\eta \dotplus 1)$-st occurrence of X in the sequence $\langle X_{\xi}: \xi < 2^{\kappa} \rangle$. Then the set Y_{γ} will have color η. Thus relation (3) is established.

We are now going to prove (1). Let $M \subseteq [\lambda]^{\kappa}$ be a maximal system of almost disjoint sets; i.e., let M be such that $|X \cap Y| < \kappa$ holds for any two distinct sets $X, Y \in M$, and if $X \in [\lambda]^{\kappa} \setminus M$, then there is an $Y \in M$ such that $|X \cap Y| = \kappa$. It is easily seen by Zorn's lemma that there is such an M. For each $X \in M$, let f_X: $[X]^{\kappa} \to 2^{\kappa}$ be a coloring verifying the relation in (3) (note that $|X| = \kappa$). Define the coloring f: $[\lambda]^{\kappa} \to 2^{\kappa}$ as follows. If $Z \in [\lambda]^{\kappa}$ and there is an $X \in M$ with $Z \subseteq X$, then observing that this X is uniquely determined, put $f(Z) = f_X(Z)$; if there is no such X then define $f(Z)$ arbitrarily.

We claim that the coloring f verifies relation (1). To this end, it is sufficient to show that, given a set $S \in [\lambda]^\kappa$, we have

$$f``[S]^\kappa = 2^\kappa. \tag{4}$$

In view of the maximality of M, there is an $X \in M$ such that $|S \cap X| = \kappa$. Then f coincides with f_X on $[S \cap X]^\kappa$, and so we have

$$f``[S]^\kappa \supseteq f``[S \cap X]^\kappa = f_X``[S \cap X]^\kappa = 2^\kappa,$$

where the last equality holds because f_X was assumed to verify relation (3); so (4) does indeed hold. The proof is complete.

A generalization of the above result by F. Galvin and K. Kunen, proved along the same lines, says the following: for any two cardinals κ, λ with $\omega \le \kappa \le \lambda$, there is a disjoint partition of $[\lambda]^\kappa$ into 2^κ classes in such a way that, for any set of κ pairwise disjoint pairs of elements of λ, the set of all transversals of these pairs intersects each class of the partition.

Kunen showed that if $V = L$, then

$$\lambda \nrightarrow [\tau]^\kappa_{\tau^\kappa}$$

holds for all cardinals κ, λ and τ with $\omega \le \kappa \le \tau \le \lambda$. Without $V = L$ this has not been shown, and the simplest problem is whether

$$\aleph_{\omega+1} \nrightarrow [\aleph_2]^{\aleph_0}_{\aleph_2}$$

holds if we assume GCH. Without assuming GCH, the simplest open problem is whether

$$\lambda \nrightarrow [(2^{\aleph_0})^+]^{\aleph_0}_{(2^{\aleph_0})^+}$$

holds for any cardinal $\lambda > (2^{\aleph_0})^+$.

A 'simultaneous' nonarrow relation is expressed by a result of K. Kunen: if $V = L$ then the following assertion (A) holds: for any cardinal $\lambda \ge \omega$, there is a coloring $f: [\lambda]^{\aleph_0} \to \lambda^{\aleph_0}$ such that, for any cardinal τ with $\omega \le \tau < \lambda$, each set $X \subseteq \lambda$ of cardinality τ carries the first $\tau^{\aleph_0}$ colors, i.e. $\tau^{\aleph_0} \subseteq f``[X]^{\aleph_0}$. He also showed that if (A) holds, then there is no supercompact cardinal (for the definition of supercompact cardinal, see e.g. Jech [1973], Reinhardt–Solovay–Kanamori [1978]).

A theorem of Erdős and Hajnal saying that the relation

$$\kappa \nrightarrow [\kappa]^{\aleph_0}_\kappa$$

is true for any infinite cardinal κ will be proved in Section 55 (Theorem 55.2).

CHAPTER IV

TREES AND POSITIVE ORDINARY PARTITION RELATIONS

A general framework for tree arguments is presented, and this is applied in order to derive the main results of the form $\kappa \to (\lambda_\xi)^r_{\xi<\vartheta}$, where κ is a regular cardinal upon which no "large cardinal assumptions" (see Chapter VII) are imposed.

13. TREES

We defined trees in Section 10, where we needed the concept for the second proof of Ramsey's theorem, but, for the sake of completeness, we recall the definition here:

A *tree* is a partial order $\langle T, \prec \rangle$ such that, for any $x \in T$, the set of its predecessors, $\mathrm{pr}\,(x) = \mathrm{pr}\,(x, \langle T, \prec \rangle) = \{y \in T : y \prec x\}$ is wellordered by $\prec$ (we assume that $\prec$ is irreflexive, i.e., that $\neg\, x \prec x$ holds for any $x \in T$, though this is not an important assumption). We sometimes write T instead of $\langle T, \prec \rangle$. For any $x \in T$, the *order* $\mathrm{o}(x) = \mathrm{o}(x, \langle T, \prec \rangle)$ of the element x is the order type of $\mathrm{pr}\,(x)$. For any ordinal α, the set $T[\alpha] = \{x \in T : \mathrm{o}(x) = \alpha\}$ is called the αth *level* of T, and $T|\alpha = \{x \in T : \mathrm{o}(x) < \alpha\}$ is the restriction of T to α; clearly, $\langle T|\alpha, \prec \rangle$ is also a tree. The *length* of T is defined as $\mathrm{lth}\,(T) = \mathrm{lth}\,(\langle T, \prec \rangle) = \min\,\{\alpha : T[\alpha] = 0\}$. A subset of T linearly ordered by $\prec$ is called a *chain* (*of* or *in* T), a maximal chain is a *branch*, and an initial segment of a branch is a *path*. The order type of a path (or branch) will often be referred to as its *length*. For any path p of T denote by $\mathrm{ls}\,(p) = \mathrm{ls}\,(p, \langle T, \prec \rangle)$ the set of *least successors* in $\prec$ of p, i.e.,

$$\mathrm{ls}\,(p) = \{x \in T : \forall y \in p\ y \prec x \;\&\; \forall t \in T\ \exists z \in p\ \neg\,(z \prec t \;\&\; t \prec x)\}.$$

For any $x \in T$, $\mathrm{ims}\,(x) = \mathrm{ims}\,(x, \langle T, \prec \rangle)$ is the set of *immediate successors* in $\prec$ of x, i.e., $\mathrm{ims}\,(x) = \mathrm{ls}(\{y : y \preceq x\})$.

As we saw in Sections 10 and 11, trees can be useful in constructing large homogeneous sets with respect to a partition. In order to prove Ramsey's theorem, we had to find an infinite branch of a certain tree. In order to prove theorems stating the existence of larger homogeneous sets, we shall usually have to look for longer branches. An important tool in finding long branches will be the following theorem.

THEOREM 13.1. *Let $\langle T, \prec \rangle$ be a tree, and assume that $\kappa \le |T|$ is regular. Let λ be a cardinal with $\omega \le \lambda < \kappa$. Then T has a path of length $\lambda \dot{+} 1$ provided one of the following two conditions is satisfied.*

a) $|\mathrm{ls}\,(p)| < \kappa$ *holds for every path p of T, and $\kappa_0^{\lambda_0} < \kappa$ for every $\kappa_0 < \kappa$ and $\lambda_0 < \lambda$.*

b) *There is a cardinal σ with $\sigma^{\lambda} < \kappa$ such that $|\mathrm{ls}\,(p)| \le \sigma^{|[p]^{<\omega}|}$ holds for every path p of T. (If $\lambda > \omega$, then one can write $\sigma^{|p|^{\omega}}$ instead of $\sigma^{|[p]^{<\omega}|}$ here.)*

PROOF. We have to prove that T has length $\ge \lambda \dot{+} 1$, as this is clearly equivalent to the existence of a path of length $\lambda \dot{+} 1$; i.e., we have to show that $T|\lambda \ne T$. As T was assumed to have cardinality $\ge \kappa$, it is enough to show to this end that $|T|\lambda| < \kappa$. λ being an infinite cardinal, it is a limit ordinal, and so $T|\lambda = \bigcup_{\alpha < \lambda} T|\alpha$. As $\lambda < \kappa$ and κ is regular, it will be enough to show that $|T|\alpha| < \kappa$ holds for any $\alpha < \lambda$. We are going to show that, for any $\alpha < \lambda$ there are less than κ paths in $T|\alpha$. This will be enough, as $x \mapsto \{y \in T|\alpha : y \preceq x\}$ is a 1-1 mapping of $T|\alpha$ into the set of its paths, i.e., $T|\alpha$ has at least as many paths as elements. Assume that the assertion to be proved fails, and let $\eta < \lambda$ be the least ordinal such that $T|\eta$ has at least κ paths. We shall now deal separately with conditions a) and b) in the theorem to be proved.

Ad a). Each level $T[\alpha]$, $\alpha < \eta$, of T has cardinality $< \kappa$; in fact, any element of $T[\alpha]$ belongs to the set $\mathrm{ls}\,(p)$ of least successors of a path p in $T|\alpha$; $T|\alpha$ has less than κ paths and each path has less than κ successors, and so the regularity of κ implies $|T[\alpha]| < \kappa$. Hence there is a $\kappa_0 < \kappa$ such that $|T[\alpha]| < \kappa_0$ holds for any $\alpha < \eta$. Each path of $T|\eta$ is a transversal of the set $\{T[\alpha] : \alpha < \eta\}$. This set has cardinality $\le \kappa_0^{|\eta|}$, and since $|\eta| < \lambda$, this number is less than κ according to our assumptions, which contradicts the choice of η.

Ad b). For every path p of $T|\eta$, let $\{x_{p,\alpha} : \alpha < |\mathrm{ls}\,(p)|\}$ be an enumeration of the least successors of p. For any $\xi < \mathrm{tp}\,\langle p, \prec \rangle$, put $f_p(\xi) = \alpha$ if the ξth element of p is $x_{q,\alpha}$ with $q = p \cap T|\xi$; this defines the function f_p with $\mathrm{tp}\,\langle p, \prec \rangle \le \eta$ as its domain, and it is obvious that if p and q are different paths of $T|\eta$, then $f_p \ne f_q$. Since we have $|\mathrm{ls}\,(p)|) \le \sigma^{|[p]^{<\omega}|} \le \sigma^{|[\eta]^{<\omega}|} = \sigma^{\lambda_0}$ for any path p in $T|\eta$, where $\lambda_0 = |[\eta]^{<\omega}| < \lambda$ (note that $|\eta| < \lambda$ and $\lambda \ge \omega$), the range of each f_p is included in σ^{λ_0}. Hence the number of the f_p's is $\le \sum_{\alpha \le \eta} (\sigma^{\lambda_0})^{|\alpha|} \le \sigma^{\lambda} < \kappa$ so $T|\eta$ has less than κ paths, which contradicts our choice of η, establishing b). (The remark in parentheses can be proved similarly; or, alternatively, it is a direct consequence of the main result if we add ω new levels at the bottom of T.) The proof is complete.

Another result establishing the existence of long branches is

THEOREM 13.2. *Let κ, λ, ρ, and τ be cardinals, and suppose that κ is regular, $\omega \le \lambda$, $(0 <) \tau < \rho < \lambda$, and $\lambda^{\rho} < \kappa$. Let ϑ be an ordinal, and let T be a set of functions $f \in \bigcup_{\alpha \le \vartheta} {}^{\alpha}\tau$, and assume that if $f \in T$, then $f \restriction \alpha \in T$ for any $\alpha \le \vartheta$. T ordered by* *inclusion is a tree. Assume that* $|T| \geq \kappa$ *and that*

$$|\{\xi \in \mathrm{dom}(f) : f(\xi) > 0\}| < \rho \tag{1}$$

holds for any $f \in T$. *Then* T *has a path of length* $\lambda \dotplus 1$.

PROOF. It is obvious that T is a tree. As before, it will again be sufficient to prove that $|T|\lambda| < \kappa$ holds. Noting that κ is regular and $\lambda (\leq \lambda^{\varrho}) < \kappa$, this will follow if we show that $|T|\eta| < \kappa$ holds for any $\eta < \lambda$. Fix $\eta < \lambda$, and observe that

$$T|\eta \subseteq \bigcup_{\alpha<\eta} {}^{\alpha}\tau .$$

The number of functions $f \in \bigcup_{\alpha<\eta} {}^{\alpha}\tau$ satisfying (1) is $\leq$

$$\sum_{\alpha<\eta} |\alpha|^{\varrho} \cdot \varrho^{\tau} \leq |\eta| \cdot (\rho \cdot |\eta|)^{\varrho} \leq \lambda^{\varrho} < \kappa ,$$

i.e., $|T|\eta| < \kappa$, which we wanted to prove.

14. TREE ARGUMENTS

One is often interested in constructing a 'long' sequence $\langle x_\alpha : \alpha < \vartheta \rangle$ of elements of a set S that has 'nice' properties. For example, in the proof of Ramsey's theorem we were given a coloring $f : [S]^r \to k$, and we constructed a sequence $\langle x_i : i < \omega \rangle$ in such a way that we had $f(u \cup \{x_m\}) = f(u \cup \{x_n\})$ whenever $m \leq n < \omega$ and $u \subseteq [\{x_i : i < m\}]^{r-1}$. The procedure we followed in the second proof of Ramsey's theorem was that, when we had already constructed a finite sequence $\{x_i : i < l\}$ that had the above property (with $m \leq n < l$), then we continued this sequence in several ways that, in a sense, represented all possible continuations, thus ensuring that at least one of the sequences had an infinite continuation. In constructing these sequences, it was helpful to define the set $S(\langle x_i : i < l \rangle)$ of the potential successors of the sequence $\langle x_i : i < l \rangle$. The situation in Ramsey's theorem was simplified by the fact that we did not have to consider sequences of transfinite length. The proof of Theorem 11.3 above was in many ways analogous, though an additional complication was caused by the fact that the elements of the 'long sequence' to be constructed were subsets, rather than elements, of the underlying set. We are now going to consider a general framework for arguments of this type, called tree arguments, and we shall prove a few results concerning the sizes of 'nice' sequences that can be obtained in this way.

The underlying concept of tree arguments is that of a partition tree, also called *ramification system*:

DEFINITION 14.1. A *partition tree* is a pair $\langle E, S\rangle$, where E is a set (called the underlying set), and S is a function with the following properties:

(i) ra $(S)\subseteq\mathscr{P}(E)$; (ii) dom $(S)=\bigcup_{\alpha<\vartheta}{}^{\alpha}\mu$ for some ordinals ϑ and μ; (iii) $S(0)=E$ (note that $0\in{}^{0}\mu=\{0\}$); (iv) if $f, g\in$ dom (S) and $f\subseteq g$, then $S(g)\subseteq S(f)$; and (v) if $f\in$ dom (S), and $\eta=$ dom (f) is a limit ordinal, then $S(f)=\bigcap_{\alpha<\eta} S(f\upharpoonright\alpha)$.

S will usually be defined by transfinite recursion with the aid of the recursion formula

$$S(f)=G(f, S|\text{dom } f)), \tag{1}$$

where we put

$$S|\xi=S\upharpoonright\bigcup_{\alpha<\xi}{}^{\alpha}\mu \tag{2}$$

for any ordinal ξ. Here G is a given function whose range is a subset of $\mathscr{P}(E)$ and whose domain is what it ought to be according to (1). In order for (1) to define a partition tree, G has to fulfil the following conditions with $X=S|\text{dom }(f)$:

$$G(0, 0)=E\,, \tag{iii$'$}$$

$$G(f, X)\subseteq S(f\upharpoonright\alpha) \tag{iv$'$}$$

for any $\alpha<$ dom (f), and

$$G(f, X)=\bigcap_{\alpha<\eta} S(f\upharpoonright\alpha) \tag{v$'$}$$

if $\eta=$ dom (f) is a limit ordinal, so that (iii), (iv), and (v) of the above definition be satisfied. If these conditions are satisfied, then (1) obviously defines a partition tree. (iii$'$) and (v$'$) mean that we have no freedom at all in specifying $G(f, X)$ unless dom (f) is an ordinal of form $\alpha\dot{+}1$ for some α; that is, in order to define a partition tree, we have to specify $G(f, X)$ only in case dom (f) is a successor ordinal.

We are going to define some auxiliary concepts concerning a partition tree $\langle E, S\rangle$. Noting that $\subset$ always denotes strict inclusion here, for any $f\in$ dom (S) write

$$R(f)=S(f)\setminus\bigcup\{S(g): f\subset g\in\text{dom }(S)\} \tag{3}$$

and, moreover, put

$$T=\{f\in\text{dom }(S): S(f)\neq 0\}. \tag{4}$$

In view of (iv) of the above definition, T ordered by inclusion $\subset$ is a tree (this justifies the word tree in the term partition tree), and we have

$$R(f)=S(f)\setminus\bigcup\{S(g): f\subseteq g\in T\ \&\ \mathrm{dom}(g)=\mathrm{dom}(f)\dot{+}1\}.$$

Our main interest below will be how long the branches of T are, as a branch of T will define a 'nice' sequence of the same length. We shall occasionally refer to a partition tree as a quadruple $\langle E, S, R, T\rangle$, where R and T are defined from S in the way described above.

We now give some explanation about the intuitive meanings of the letters E, S, R, and T. In the simplest case, we want to construct a long 'nice' sequence $\langle x_\alpha: \alpha<\eta\rangle$ of pairwise distinct elements of E. We construct this sequence step by step, making alternative choices at each step, the number of alternatives being limited by $|\mu|$, where μ is the ordinal in (ii) of the above definition. Having constructed a sequence $\langle x_\alpha: \alpha<\xi\rangle$ of length ξ, a function $f: \xi\to\mu$ keeps track of the alternative choices made, and the sequence $\langle x_\alpha: \alpha<\xi\rangle$ is obtained in the form $\langle s(f\upharpoonright\alpha): \alpha<\xi\rangle$, where s is a function from T into E defined by transfinite recursion (simultaneously, or subordinated to, the recursive definition of S as described in (1)) in such a way that $s(f)\in S(f)$ holds for any $f\in T$. Usually, one can pick an arbitrary element of $S(f)$ as $s(f)$, that is, $S(f)$ is the set of potential successors of the sequence $\langle s(f\upharpoonright\alpha): \alpha<\mathrm{dom}(f)\rangle$. The set $R(f)$ includes the element $s(f)$, which is chosen just after having constructed $S(f)$ and will not be chosen again, and, possibly, some other elements of $S(f)$ that have to be thrown away as they cannot be potential successors of the sequence $\langle s(f\upharpoonright\alpha): \alpha\leq\mathrm{dom}(f)\rangle$.

There are variations to the above pattern. In the above description we can make no alternative choices at limit steps. Indeed, if $\xi=\mathrm{dom}(f)$ is a limit ordinal then f is uniquely determined by $\langle f\upharpoonright\alpha: \alpha<\xi\rangle$, and so the sequence $\langle s(f\upharpoonright\alpha): \alpha<\xi\rangle$ has only one continuation, namely $s(f)$. If we want to have alternative choices after having constructed a sequence $\langle x_\alpha: \alpha<\xi\rangle$ of limit length, then what we have to do is to define $s(f)$ only for those $f\in T$ for which $\mathrm{dom}(f)$ is not a limit ordinal. In that case, having constructed a sequence $\langle s(f\upharpoonright\alpha): \alpha<\xi$ and α is not limit$\rangle$ where ξ is a limit ordinal and $\mathrm{dom}(f)=\xi$, this sequence can be continued in different ways according to how we define f at ξ, i.e., the least successor of the above sequence can be any of the elements $s(f\cup\{\langle\xi\gamma\rangle\})$, where $\gamma<\mu$.

Using the results of the preceding section, we are going to prove a theorem on the length of branches in a partition tree which will be one of our main tools in constructing large homogeneous sets. We first need a lemma, which simply says that each element of E is thrown away (or used up) at some stage:

LEMMA 14.2. *For any partition tree* $\langle E, S, R, T\rangle$ *we have*

$$E=\bigcup\{R(f): f\in T\}.$$

PROOF. For an arbitrary $x \in E$ put

$$T_x = \{g \in T : x \in S(g)\}.$$

This set is not empty in view of (iii) in Definition 14.1, and $\langle T_x, \subset \rangle$ is obviously a tree by (iv) there. By Hausdorff's Maximal Chain Theorem (see Subsection 4.7), there is a branch b of this tree; put $f = \bigcup b$. We claim that $f \in b$. This is obvious if b has a last element in the ordering $\subset$, as this last element can only be f. Hence, assuming that $f \notin b$, $\xi = \mathrm{tp}\langle b, \subset \rangle = \mathrm{dom}(f)$ must be a limit ordinal (we cannot have $\xi = 0$, since b is not empty, as $T_x \neq 0$). Then

$$x \in \bigcap \{S(g) : g \in b\} = \bigcap \{S(f \upharpoonright \alpha) : \alpha < \xi\} = S(f),$$

where the last equality holds in view of (v) in Definition 14.1; observe that (ii) there is also used to show $f \in \mathrm{dom}(S)$ (this is why it is an important technicality to take $\mathrm{dom}(S) = \bigcup_{\alpha \leq \vartheta} {}^{\alpha}\mu$ instead of $\mathrm{dom}(S) = \bigcup_{\alpha < \vartheta} {}^{\alpha}\mu$). So $f \in T_x$; therefore $f \notin b$ means that $b \cup \{f\}$ is a chain of T_x properly including b which contradicts the assumption that b is a branch. This establishes the claim $f \in b \subseteq T_x$. Hence $x \in S(f)$. If g is an element of $\mathrm{dom}(S)$ that properly includes f then $x \notin S(g)$, since otherwise $b \cup \{g\}$ would be a chain of T_x that properly includes b which would be a contradiction. Hence $x \in R(f)$ holds by the definition of R in (3) above. As $x \in E$ was arbitrary, this completes the proof.

The main theorem of this section is simply a reformulation of the results of the preceding section in the light of the lemma just proved. We recall that $\mathrm{ims}(f, T)$ denotes the set of immediate successors in T of the element f. In the present case we have

$$\mathrm{ims}(f, T) = \{g \in T : f \subset g \ \&\ \mathrm{dom}(g) = \mathrm{dom}(f) \dot{+} 1\}.$$

THEOREM 14.3. *Let $\langle E, S, R, T \rangle$ be a partition tree; let κ be a regular cardinal with $\kappa \leq |E|$, and assume that $|R(f)| < \kappa$ holds for any $f \in T$. Let λ be a cardinal with $\omega \leq \lambda < \kappa$. Then T has a path of length $\lambda \dot{+} 1$ provided one of the following three conditions is satisfied:*

a) *$|\mathrm{ims}(f)| < \kappa$ holds for every $f \in T$, and $\kappa_0^{\dot{\lambda}_0} < \kappa$ for every $\kappa_0 < \kappa$ and $\lambda_0 < \lambda$.*

b) *There is a cardinal σ with $\sigma^{\lambda} < \kappa$ such that $|\mathrm{ims}(f)| \leq \sigma^{|[|f|+1]^{<\omega}|}$ holds for every $f \in T$. (If $\lambda > \omega$, then one can write $\sigma^{(|f|+1)^{\omega}}$ instead of $\sigma^{|[|f|+1]^{<\omega}|}$ here.)*

c) *There are cardinals ρ and τ with $(0<)\tau < \rho < \lambda$ and $\lambda^{\rho} < \kappa$, we have $T \subseteq \bigcup_{\alpha \leq \vartheta} {}^{\alpha}\tau$ for some ordinal ϑ, and $|\{\xi \in \mathrm{dom}(f) : f(\xi) > 0\}| < \rho$ holds for any $f \in T$.*

PROOF. $|R(f)| < \kappa$ implies $|T| \geq \kappa$ by virtue of the preceding lemma. Hence the result is a simple restatement of Theorems 13.1 and 13.2. Note that each path of limit length has at most one successor in case of a partition tree $\langle E, S, R, T \rangle$.

Hence, in cases a) and b), the inequalities for $|\mathrm{ls}(p)|$ required in Theorem 13.1 follow from the analogous inequalities for $|\mathrm{ims}(f)|$ here.

There is a possibility of generalizing the concept of partition trees to cases when $S(f)$ assumes values in any partially ordered set instead of the power set of a given set E. A case in point is e.g. the following simple result observed by K. Kunen: given a cardinal κ, any $(\kappa^+, 3)$-distributive complete Boolean algebra B that satisfies the κ-chain condition is atomic. To prove this result, one has to use partition trees where $S(f) \in B$. (For the definitions of the concepts here, see e.g. Sikorski [1960]; $(\kappa^+, 3)$-distributivity here is to be understood in the sense of $(<\kappa^+, <3)$-distributivity, and not of $(\leq\kappa^+, \leq 3)$-distributivity.)

15. END-HOMOGENEOUS SETS

Let α be an ordinal and τ a cardinal, and let $f: [\alpha]^{<\omega} \to \tau$ be a coloring. Call a set $X \subseteq \alpha$ *end-homogeneous* with respect to f if for every finite set $u \subseteq X$ and for every μ, $\nu \in X$ with $\max u < \mu, \nu$, we have

$$f(u \cup \{\mu\}) = f(u \cup \{\nu\}). \tag{1}$$

If f is only a *partial coloring* of $[\alpha]^{<\omega}$ i.e., if f is a function from a subset of $[\alpha]^{<\omega}$, then call $X \subseteq \alpha$ end-homogeneous with respect to f if (1) holds whenever $u \in [X]^{<\omega}$, $\max u < \mu, \nu$, and $u \cup \{\mu\}$, $u \cup \{\nu\} \in \mathrm{dom}(f)$. The partition relation

$$\alpha \to \langle\beta\rangle_\tau \tag{2}$$

means the following: for every coloring $f: [\alpha]^{<\omega} \to \tau$ there is an end-homogeneous set of order type β.

End-homogeneous sets are very important in establishing positive partition relations; in fact, they were already used in both proofs of Ramsey's theorem in Section 10. The following simple lemma throws some light upon the situation.

LEMMA 15.1. *Let $r \geq 1$ be an integer, τ a cardinal, and let α, β, ν_ξ $(\xi < \tau)$ be ordinals. Assume*

$$\beta \to (\nu_\xi)^{r-1}_{\xi<\tau}. \tag{3}$$

Then $\alpha \to \langle\beta\rangle_\tau$ implies

$$\alpha \to (\nu_\xi)^r_{\xi<\tau}, \tag{4}$$

and $\alpha \to \langle\beta \dot{+} 1\rangle_\tau$ implies

$$\alpha \to (\nu_\xi \dot{+} 1)^r_{\xi<\tau}. \tag{5}$$

An instance of relation (4) was used above in the proof of Ramsey's theorem, and relation (5) will be used below in the proof of the Stepping-up Lemma (Lemma 16.1).

PROOF. Let $f\colon [\alpha]^r \to \tau$ be a coloring. First we verify (5). Assume to this end that $\alpha \to \langle \beta \dot{+} 1 \rangle_\tau$ holds; then there is a set X of order type $\beta \dot{+} 1$ that is end-homogeneous with respect to f. Let ζ be the maximal element of X, write $X' = X \setminus \{\zeta\}$, and define the coloring $g\colon [X']^{r-1} \to \tau$ as follows: put $g(u) = f(u \cup \{\zeta\})$ for every $u \in [X']^{r-1}$. By (1), there is a $\xi < \tau$ and a set $Y \subseteq X'$ of order type ν_ξ such that Y is homogeneous of color ξ with respect to g, i.e., $g``[Y]^{r-1} = \{\xi\}$. But then we have

$$f(v) = f((v \setminus \{\max v\}) \cup \{\zeta\})$$

for any $v \in [Y \cup \{\zeta\}]^r$ by the end-homogeneity of X, and the right-hand side here equals $g(v \setminus \{\max v\}) = \xi$ by the definition of g. Hence $Y \cup \{\zeta\}$ is a homogeneous set of order type $\nu_\xi \dot{+} 1$ and of color ξ with respect to f, which proves (5).

In the proof of (4), assume first that β is a limit ordinal. The relation $\alpha \to \langle \beta \rangle_\tau$ implies the existence of an end-homogeneous set $X \subseteq \alpha$ of order type β with respect to the coloring $f\colon [\alpha]^r \to \tau$. Given an arbitrary $u \in [X]^{r-1}$, put $g(u) = f(u \cup \{v\})$ for a $v \in X$ with $v > \max(u)$; there is such a v, since $\operatorname{tp} X = \beta$ was assumed to be a limit ordinal, and $g(u)$ does not depend on the choice of v in view of the end-homogeneity of X. By (3), there is a $\xi < \tau$ and a set $Y \subseteq X$ of order type ν_ξ that is homogeneous of color ξ with respect to the coloring $g\colon [X]^{r-1} \to \tau$; clearly, Y is then also homogeneous of color ξ with respect to f. This establishes (4) in case β is a limit ordinal.

If $\beta = \beta' \dot{+} 1$ for some β', then we have to work with the last element of X as we did in the proof of (5). It is, however, simpler to observe that (3) implies

$$\beta' \to (\nu_\xi \dot{-} 1)^{r-1}_{\xi < \tau}$$

in this case (indeed, by taking away the last element of β, we can take away only the last element of a homogeneous set), and so we have

$$\alpha \to ((\nu_\xi \dot{-} 1) \dot{+} 1)^r_{\xi < \tau}$$

in view of (5) with β' replacing β. The proof is complete.

The following lemma, due to Erdős and Rado, confirms the existence of large end-homogeneous sets. It will usually be applied together with the preceding lemma to prove the existence of large homogeneous sets. It is, however, also important in itself: in Section 45 we shall give some applications of it concerning set mappings in cases where the corresponding partition relations would lead to weaker results.

LEMMA 15.2. *For any cardinals $\kappa \geq \omega$ and $\lambda \geq 2$ we have*

$$(\lambda^\kappa)^+ \to \langle \kappa \dot{+} 1 \rangle_\lambda. \tag{6}$$

Proof. Writing $E=(\lambda^{\kappa})^{+}$, let $f\colon [E]^{<\omega}\to\lambda$ be a coloring. We are going to define a partition tree $\langle E, S, R, T\rangle$, called the *canonical partition tree* associated with the coloring f. To this end, write

$$\operatorname{dom}(S)=\bigcup_{\alpha\leq\vartheta}{}^{\alpha}\mu\,, \tag{7}$$

where μ and ϑ will be specified later. As we mentioned in the preceding section, we can define S by the recursion formula

$$S(g)=G(g, S|\operatorname{dom}(g)),$$

where G has to be specified only in case dom (g) is a successor ordinal and $S(g\frown\xi)\neq 0$ for any $\xi<\operatorname{dom}(g)$ (cf. (14.1) and the remarks made afterwards). Write dom $(g)=\alpha\dot{+}1$, and assume that $S|(\alpha\dot{+}1)$ has already been defined, i.e., that $S(h)\subseteq E$ has already been defined for any h with $h\in\operatorname{dom}(S)$ and dom $(h)\subseteq\alpha$. For any such h, define $s(h)$ as the least element in $S(h)$ provided this latter is not empty (note that E is a cardinal, and so $S(h)\subseteq E$ consists of ordinals); i.e., if $S(h)\neq 0$ then put

$$s(h)=\min S(h). \tag{8}$$

In order to define $S(g)$, write $g'=g\frown\alpha$ and put

$$E_{g'}=\{s(g'\frown\beta)\colon \beta\leq\alpha\} \tag{9}$$

and, for each $\xi\in S(g')\setminus\{s(g')\}$, define the coloring $f_{g',\xi}\colon [E_{g'}]^{<\omega}\to\lambda$ by putting

$$f_{g',\xi}(u)=f(u\cup\{\xi\}) \tag{10}$$

for any $u\in[E_{g'}]^{<\omega}$. For any two $\xi,\eta\in S(g')\setminus\{s(g')\}$, put

$$\xi\equiv_{g'}\eta \qquad \text{iff} \qquad f_{g',\xi}=f_{g',\eta}. \tag{11}$$

$\equiv_{g'}$ is obviously an equivalence relation. The number of equivalence classes under $\equiv_{g'}$ is $\leq$ the number of possibilities for the coloring $f_{g',\xi}$, i.e., is at most

$$\lambda^{|[\alpha\dot{+}1]^{<\omega}|}=\lambda^{|[|g'|+1]^{<\omega}|}, \tag{12}$$

where we recall that $\alpha=\operatorname{dom}(g')$. Let $S_{g',\nu}$, $\nu<\mu$, be an enumeration of the equivalence classes under $\equiv_{g'}$. Here μ (cf. (7)) has to be large enough so that we may enumerate all the equivalence classes (as these classes are pairwise disjoint, $\mu=E=|E|$ clearly suffices); if $|\mu|$ is larger than the number of these equivalence classes, then put $S_{g,\nu}=0$ as many times as necessary. Write

$$S(g)=S_{g',g(\alpha)}\,,$$

and, for further reference, if $S(g)\neq 0$, then write

$$f_g=f_{g',\xi}\,, \tag{13}$$

where $\xi \in S(g)$; clearly, f_g does not depend on the particular choice of ξ here. By (10), we have

$$f_g(u) = f(u \cup \{\xi\}) \tag{14}$$

for any finite set $u \subseteq \{s(g \wedge \beta): \beta < \mathrm{dom}(g)\}$ and any $\xi \in S(g)$. This finishes the definition of S. Choose ϑ so large that $S(g) = 0$ for any $g \in {}^{\vartheta}\mu$; clearly, $\vartheta = E^+$ suffices since, for any g with $S(g) \neq 0$, the set

$$\{s(g \wedge \alpha): \alpha \leq \mathrm{dom}(g)\}$$

consists of pairwise distinct elements of the cardinal E. Putting, as usual,

$$T = \{h \in \mathrm{dom}(S): S(h) \neq 0\},$$

we clearly have $R(h) = \{s(h)\}$ for any $h \in T$, i.e., we have

$$|R(h)| = 1 \tag{15}$$

for any $h \in T$. Taking $h = g'$ in (12), we can conclude that

$$|\mathrm{ims}(h)| \leq \lambda^{|[|h|+1]^{<\omega}|} \tag{16}$$

holds for any $h \in T$, where ims (h) denotes the set of immediate successors of h in T; in fact, exactly one immediate successor of h in T corresponds to each (nonempty) equivalence class under $\equiv_h$.

Noting that $|E| = E = (\lambda^{\kappa})^+$, we can conclude from Theorem 14.3.b with $(\lambda^{\kappa})^+$, κ, and λ replacing κ, λ, and σ, respectively, that T has a path of length $\kappa \dot{+} 1$, i.e., that there is a $g \in T$ with dom $(g) = \kappa$. Write

$$\xi_\alpha = s(g \wedge \alpha)$$

for each $\alpha \leq \kappa$. Observe first that $\xi_\alpha < \xi_\beta$ holds whenever $\alpha < \beta \leq \kappa$. In fact, using (8), we obtain that

$$\xi_\alpha = \min S(g \wedge \alpha) < \xi_\beta$$

provided $\alpha < \beta \leq \kappa$, since

$$\xi_\beta \in S(g \wedge \beta) \subseteq S(g \wedge \alpha) \setminus \{\xi_\alpha\}$$

holds by our construction above. Hence the set $X = \{\xi_\alpha: 1 \leq \alpha \leq \kappa\}$ has order type $\kappa \dot{+} 1$. We claim that X is an end-homogeneous set with respect to f. In fact, if $u \in [X]^{<\omega}$ and max $u = \xi_\alpha < \xi_\beta, \xi_\gamma$ (or $\alpha = 0$ and $\xi_\beta, \xi_\gamma \in X$ in case u is empty), then we have $\alpha \dot{+} 1 \leq \beta, \gamma$, and so $\xi_\beta, \xi_\gamma \in S(g \wedge (\alpha \dot{+} 1))$. Hence (14) implies that

$$f(u \cup \{\xi_\beta\}) = f_{g \wedge (\alpha \dot{+} 1)}(u) = f(u \cup \{\xi_\gamma\})$$

holds; this shows that X is indeed end-homogeneous. The proof is complete.

COROLLARY 15.3. *Let κ and τ be cardinals with $\tau<\kappa$ and $\kappa\geq\omega$. Then*

$$(2^\kappa)^+\to\langle\kappa^+\dot{+}1\rangle_{2^\kappa} \tag{17}$$

and

$$(2^\kappa)^+\to\langle\kappa\dot{+}1\rangle_\tau \tag{18}$$

hold.

PROOF. For (17), replace λ and κ with 2^κ and κ^+, and, for (18), with τ and κ, respectively.

(18) will be used for the proof of the Stepping-up Lemmas in the next section; (19) will be used in connection with set mappings (see Section 45), and not in order to establish a partition relation.

16. THE STEPPING-UP LEMMA

The following lemma, one of our most important tools in establishing positive ordinary partition relations, is an easy consequence of the results of the preceding section:

LEMMA 16.1 (Stepping-up Lemma). *Let $\kappa\geq\omega$ and τ be cardinals, $r\geq 2$ an integer, and ν_ξ $(\xi<\tau)$ be ordinals. If*

$$\kappa\to(\nu_\xi)^{r-1}_{\xi<\tau} \tag{1}$$

holds, then we have

$$(2^\kappa)^+\to(\nu_\xi\dot{+}1)^r_{\xi<\tau}. \tag{2}$$

PROOF. We may assume that $\nu_\xi\geq r$ holds for all $\xi<\tau$ (cf. Subsection 9.5). In this case, noting that κ is infinite and $r\geq 2$, (1) implies that $\tau<\kappa$; hence, we have

$$(2^\kappa)^+\to\langle\kappa\dot{+}1\rangle_\tau$$

according to (15.18) in Corollary 15.3. Hence (2) follows from (15.5) in Lemma 15.1 with $(2^\kappa)^+$ and κ replacing α and β, respectively. The proof is complete.

This result alone has nontrivial corollaries. As an illustration, observe that the relation

$$(2^{\aleph_0})^+\to(\aleph_1)^2_\omega,$$

which is a theorem of Erdős and Rado, follows from the trivial relation $\aleph_1\to(\aleph_1)^1_\omega$. This is, however, not the appropriate place to go into discussing the corollaries of the above lemma, since as we shall soon see, stronger results can be obtained in the case $r=2$ by more refined methods. This will be done in the next section.

The aim of the rest of this section is to derive a result saying that certain partition relations hold simultaneously; this result will have an important application in the proof of the General Canonization Lemma, given in Section 28. We start with the following

DEFINITION 16.2. Let α and β be ordinals, N a set of integers ≥ 1, and τ_i $(i \in N)$ cardinals. Then the partition symbol

$$\alpha \overset{\text{sim}}{\to} (\beta)_{\tau_i}^{i \in N},$$

called the *simultaneous ordinary partition symbol* and read as "α arrows β simultaneously for $i \in N$ with τ_i colors", is said to hold if we have the following: Given an arbitrary coloring $f_i: [\alpha]^i \to \tau_i$ for each $i \in N$, there is a set $X \subseteq \alpha$ of order type β that is homogeneous with respect to each $f_i, i \in N$, i.e., for any $i \in N$ there is a $\xi < \tau_i$ with $f_i``[X]^i = \{\xi\}$. The negation of the above symbol will, as usual, be written as

$$\alpha \overset{\text{sim}}{\nrightarrow} (\beta)_{\tau_i}^{i \in N}.$$

We mention a few natural variations of the above symbol. Where k and l are integers, the relation

$$\alpha \overset{\text{sim}}{\to} (\beta)_{\tau_i}^{k \leq i < l}$$

means the same thing as

$$\alpha \overset{\text{sim}}{\to} (\beta)_{\tau_i}^{i \in \{j: k \leq j < l\}}.$$

The meaning of the symbol

$$\alpha \overset{\text{sim}}{\to} (\beta)_{\tau_{k_0}, \ldots, \tau_{k_{n-1}}}^{k_0, \ldots, k_{n-1}}$$

will be defined as

$$\alpha \overset{\text{sim}}{\to} (\beta)_{\tau_i}^{i \in \{k_0, \ldots, k_{n-1}\}}.$$

Note also that $\alpha \overset{\text{sim}}{\to} (\beta)_{\tau}^{i \in \omega}$ means the same thing as $\alpha \to (\beta)_{\tau}^{<\omega}$ (cf. Subsection 8.6).

Using the ideas of the proof of Lemma 15.1, it is easy to establish the following:

LEMMA 16.3. *Let $r(\geq 1)$ be an integer, α, β, and ν ordinals, and τ_i $(2 \leq i \leq r)$ cardinals. Assume that*

$$\alpha \to \langle \beta \dotplus 1 \rangle_{\tau} \tag{3}$$

holds with $\tau = \max\{\tau_i: 2 \leq i \leq r\}$. Then

$$\beta \overset{\text{sim}}{\to} (\nu)_{\tau_{i+1}}^{1 \leq i \leq r-1} \tag{4}$$

implies

$$\alpha \overset{\text{sim}}{\rightarrow} (\nu \dot{+} 1)_{\tau_i}^{2 \le i \le r}. \tag{5}$$

PROOF. For each i with $2 \le i \le r$, let $f_i\colon [\alpha]^i \rightarrow \tau_i$ be a coloring. Put $f = \bigcup_{2 \le i < r} f_i$; then f maps a subset of $[\alpha]^{<\omega}$ into $\tau = \max_{2 \le i < r} \tau_i$. By (3), there is a set $X \subseteq \alpha$ of order type $\beta \dot{+} 1$ that is end-homogeneous with respect to f. Write $X = X' \cup \{\zeta\}$, where ζ is the maximal element of X. For each integer i with $2 \le i \le r$ define the coloring $g_i\colon [X']^{i-1} \rightarrow \tau$ by putting $g_i(u) = f_i(u \cup \{\zeta\})$ for any $u \in [X']^{i-1}$. Then, by (4), there is a set $Y \subseteq X$ of order type ν that is homogeneous with respect to each g_i, $2 \le i \le r$. We can now conclude in exactly the same way as we did in the proof of (15.5) in Lemma 15.1 that $Y \cup \{\zeta\}$ is a homogeneous set of order type $\nu \dot{+} 1$ with respect to each coloring f_i, $2 \le i \le r$. The proof is complete.

We can now prove the following:

LEMMA 16.4 (Simultaneous Stepping-up Lemma). *Let r be an integer, $\kappa \ge \omega$ and $\tau_i \ge 1$ $(2 \le i \le r)$ cardinals, and ν an ordinal. Assume that*

$$\kappa \overset{\text{sim}}{\rightarrow} (\nu)_{\tau_{i+1}}^{1 \le i \le r-1} \tag{6}$$

holds. Then we have

$$(2^\kappa)^+ \overset{\text{sim}}{\rightarrow} (\nu \dot{+} 1)_{\tau_i}^{2 \le i \le r}. \tag{7}$$

PROOF. We may assume $r \ge 2$ and $\nu \ge r$ here (in fact, if e.g. $2 \le \nu < r$, then we can replace r with $r' = \nu$ in (7)). Then (6) implies that $\tau_i < \kappa$ for each τ with $2 \le i \le r$; hence we have

$$(2^\kappa)^+ \rightarrow \langle \kappa \dot{+} 1 \rangle_\tau$$

with $\tau = \max_{2 \le i \le r} \tau_i$ according to (15.18). Using now the preceding lemma with $\alpha = (2^\kappa)^+$ and $\beta = \kappa$, we can therefore see that (6) indeed implies (7). The proof is complete.

Define the operation exp by induction as follows:

$$\exp_0(\kappa) = \kappa \quad \text{and} \quad \exp_{n+1}(\kappa) = \exp_n(2^\kappa),$$

where κ is a cardinal and n runs over integers. The result we are aiming at (the one we shall need in the proof of the General Canonization Lemma in Section 27) is the following:

COROLLARY 16.5. *Let $r \ge 1$ be an integer and $\kappa \ge \omega$ a cardinal. Then we have*

$$(exp_{r-1}(\kappa))^+ \overset{\text{sim}}{\rightarrow} (\kappa^+)_{\exp_{r-i}(\kappa)}^{1 \le i \le r}. \tag{8}$$

PROOF. For $r=1$ the assertion says $\kappa^+\to(\kappa^+)^1_\kappa$, which is obviously true. If (8) holds with some $r\geq 1$, then the preceding lemma implies that

$$(\exp_r(\kappa))^+ \overset{\text{sim}}{\to} (\kappa^+)^{2\leq i\leq r+1}_{\exp_{r+1-i}(\kappa)}. \tag{9}$$

In order to complete the induction step we also have to cover the case $i=1$, i.e., we have to prove

$$(\exp_r(\kappa))^+ \overset{\text{sim}}{\to} (\kappa^+)^{1\leq i\leq r+1}_{\exp_{r+1-i}(\kappa)}; \tag{10}$$

that is, we have to show that if we are given colorings

$$f_i\colon[(\exp_r(\kappa))^+]^i\to\exp_{r+1-i}(\kappa),$$

where $1\leq i\leq r+1$, then there is a set of cardinality κ^+ that is homogeneous with respect to each f_i. Obviously, there is a homogeneous set X of cardinality $(\exp_r(\kappa))^+$ with respect to the coloring f_1. Then, using (9), we obtain that there is a set $Y\subseteq X$ of cardinality κ^+ that is homogeneous with respect to each f_i, $2\leq i<r+1$. Then Y is homogeneous also with respect to f_1. This verifies (10). The proof is complete.

17. THE MAIN RESULTS IN CASE $r=2$ AND κ IS REGULAR; AND SOME COROLLARIES FOR $r\geq 3$

The primary aim of this section is to study the partition relation

$$\kappa\to(\nu_\xi)^2_{\xi<\tau}$$

in case κ is regular. We shall again use tree arguments to establish our results. As we shall see later, more sophisticated arguments involving the canonical partition tree alone, which was defined in the proof of Lemma 15.2, would be sufficient. Our aim in defining also a different kind of tree is to give a wider illustration of tree arguments. The main theorem we are going to prove is as follows (we recall that $L_\lambda(\kappa)$ is the logarithm operation defined in Section 7):

THEOREM 17.1. *Let κ, λ, ρ and τ be cardinals, and ν_ξ $(\xi<\tau)$ ordinals. Suppose that κ is regular, $\kappa\geq\omega$, $3\leq\lambda\leq\kappa$, $2\leq\rho\leq L_\lambda(\kappa)$, $\rho<\kappa$ and, moreover, $\nu_\xi\geq 2$ for any $\xi<\tau$. If*

$$\rho\to(\nu_\xi)^1_{\xi<\tau} \tag{1}$$

then

$$\kappa\to(\lambda,(\nu_\xi\dotplus 1)_{\xi<\tau})^2 \tag{2}$$

(the assumptions here allow the case $\tau=0$ for the sake of convenience).

PROOF. We may assume that

$$\lambda \geq \aleph_0, \rho^+ . \tag{3}$$

In fact, if the above assumptions are satisfied with some λ, then they are also satisfied with $\lambda' = \lambda \cdot \rho^+ \cdot \aleph_0$. Assume we are given a coloring $f: [\kappa]^2 \to 1 \dot{+} \tau$. We have to prove that there is either a homogeneous set of color 0 that has order type (or cardinality) λ or a homogeneous set of color $1 \dot{+} \xi$ for some $\xi < \tau$ that has order type $\nu_\xi \dot{+} 1$. We distinguish two cases: a) $\lambda < \kappa$ and b) $\lambda = \kappa$.

Ad a). We use the canonical partition tree defined in the preceding section, but we impose additional stipulations in its definition. Namely, in the preceding section, we took an arbitrary enumeration $S_{g',\xi}$ of the equivalence classes under $\equiv_{g'}$. Here we have a better choice. We shall have

$$\mathrm{dom}\,(S) = \bigcup_{\alpha \leq \kappa^+} {}^{\alpha}(1 \dot{+} \tau),$$

and we are going to define the partition tree $\langle \kappa, S, R, T \rangle$ by transfinite recursion. Assume that $S|(\alpha \dot{+} 1)$ has already been defined for some $\alpha < \kappa^+$, and assume $g \in {}^{(\alpha \dot{+} 1)}(1 \dot{+} \tau)$ and $S(g \restriction \alpha) \neq 0$. Write $g' = g \restriction \alpha$, and put

$$s(g') = \min S(g'),$$

and, moreover, for any $\xi < 1 \dot{+} \tau$, set

$$S_{g',\xi} = \{x \in S(g') \setminus \{s(g')\} : f(\{s(g'), x\}) = \xi\}, \tag{4}$$

and

$$S(g) = S_{g', g(\alpha)} . \tag{5}$$

This completes the definition of the partition tree $\langle \kappa, S, R, T \rangle$. For any $h \in T$ we have

$$s(h) = \min (S(h)) \tag{6}$$

and $R(h) = \{s(h)\}$, i.e.,

$$|R(h)| = 1 . \tag{7}$$

Observe that for any $h \in T$, the sequence $\langle s(h \restriction \alpha) : \alpha \leq \mathrm{dom}\,(h) \rangle$ is a strictly increasing sequence of ordinals less than κ; this easily follows from (6).

Fix now $h \in T$, and note that for any $\alpha < \beta \leq \mathrm{dom}\,(h)$ we have

$$f(\{s(h \restriction \alpha), s(h \restriction \beta)\}) = h(\alpha) ; \tag{8}$$

in fact, this follows from (4) and (5) with $g = h \restriction (\alpha \dot{+} 1)$, since

$$s(h \restriction \beta) \in S(h \restriction \beta) \subseteq S(h \restriction (\alpha \dot{+} 1)) = S(g)$$

holds. Put

$$N_h = \{\xi \in \mathrm{dom}\,(h) : h(\xi) > 0\} .$$

We may assume here that

$$|N_h| < \rho \,. \tag{9}$$

In fact, the contrary easily implies that (2), which we want to prove. To see this, assume that (9) fails, and define the coloring $\tilde{h}: [N_h]^1 \to \tau$ as follows: for any $\alpha \in N_h$ write

$$\tilde{h}(\alpha) = \xi \quad \text{if and only if} \quad h(\alpha) = 1 \dot{+} \xi \,.$$

Then (1) entails that there is an ordinal $\xi < \tau$ and a set $X \subseteq N_h$ of order type ν_ξ such that X is a homogeneous set of color ξ with respect to the coloring $\tilde{h}$. The set

$$Y = \{s(h \frown \alpha): \alpha \in X \cup \{\text{dom}\,(h)\}\}$$

has order type $\nu_\xi \dot{+} 1$, since, as we pointed out above, after formula (7), the sequence $\langle s(h \frown \alpha): \alpha < \text{dom}\,(h) \rangle$ is strictly increasing. (8) implies that Y is a homogeneous set of color $1 \dot{+} \xi$ with respect to the coloring f, showing that (2) indeed holds.

Hence we have to deal only with the case when the assumption in (9) holds. In this case we can use Theorem 14.3.c with $\tau + 1 (= 1 \dot{+} \tau)$ replacing τ. The assumptions there hold in view of (3), (7), and (9); note also that we assumed $\lambda < \kappa$ in the case we are discussing now; $\lambda^\varrho < \kappa$ follows from $\rho \leq L_\lambda(\kappa)$ and $\lambda < \kappa$, and $\tau + 1 < \rho$ follows from (1) in view of the assumptions $\rho \geq 2$ and $\nu_\xi \geq 2$ for any $\xi < \tau$. So T has a path of length $\lambda \dot{+} 1$. Pick an $h \in T$ with $\text{dom}\,(h) = \lambda$, and put

$$X = \lambda \setminus N_h = \{\xi < \lambda : h(\xi) = 0\} \,.$$

X has order type λ in view of (3) and (9), and

$$Y = \{s(g \frown \xi): \xi \in X \cup \{\lambda\}\}$$

is a homogeneous set of color 0 with respect to f according to (8). As Y has order type $\lambda \dot{+} 1$, this completes the proof in case a), i.e., when $\lambda < \kappa$. We proved the sligthly stronger relation

$$\kappa \to (\lambda \dot{+} 1, (\nu_\xi \dot{+} 1)_{\xi < \tau})^2 \tag{10}$$

in case $\lambda < \kappa$.

Ad b). We have $\lambda = \kappa$ in this case. We proceed similarly as in the proof of Theorem 11.3. Assume that there is no homogeneous set of cardinality κ and of color 0 with respect to the coloring $f: [\kappa]^2 \to 1 \dot{+} \tau$. For every set $X \subseteq \kappa$ let $H(X)$ be a maximal homogeneous subset of color 0 of X; the maximality of $H(X)$ means that

$$\forall x \in X \setminus H(X) \, \exists y \in H(X) [f(\{xy\}) \neq 0] \,, \tag{11}$$

and the assumption made just before implies

$$|H(X)|<\kappa. \tag{12}$$

We are going to define a partition tree $\langle\kappa, S, R, T\rangle$. To this end, given $S|(\alpha\dot{+}1)$ for some $\alpha<\vartheta=\kappa^+$, we have to define $S(g)$ for any $g\in{}^{(\alpha\dot{+}1)}\mu$ provided $S(g\restriction\alpha)\neq 0$ (μ will be specified later). Assume we are given such a g. Write $g'=g\restriction\alpha$, and put

$$\tilde{H}(g')=H(S(g')),$$

and, noting that $\tilde{H}(g')$ is a set of ordinals, write

$$R(g')=S(g')\cap(\sup\tilde{H}(g')\dot{+}1); \tag{13}$$

the point in choosing $R(g')$ in this way is that all elements of $S(g')\setminus R(g')$ will then exceed any element of $\tilde{H}(g')$. Note that

$$|R(g')|<\kappa \tag{14}$$

holds by the regularity of κ in view of (12). For any $x\in\tilde{H}(g')$ and any color $1\dot{+}\xi$ with $\xi<\tau$, put

$$S_{g',\kappa\cdot x\dot{+}\xi}=\{y\in S(g')\setminus R(g'): f(\{xy\})=1\dot{+}\xi\}. \tag{15}$$

To unravel the meaning of this definition, one should notice that x is an ordinal, and, as $\xi<\tau<\kappa$ ($\tau<\rho[<\kappa]$ follows from (1)), the second subscript $\kappa\cdot x\dot{+}\xi$ unambiguously identifies the ordinals x and ξ (the multiplication here means ordinal multiplication). If η is an ordinal which cannot be represented in form $\kappa\cdot x\dot{+}\xi$ for any $x\in\tilde{H}(g')$ and $\xi<\tau$, then put $S_{g',\eta}=0$. Writing $\mu=\kappa\cdot\kappa$ (ordinal multiplication again), it is clear that $S_{g,\eta}=0$ for any $\eta\geq\mu$. (11) implies that

$$\bigcup_{\eta<\mu} S_{g',\eta}=S(g')\setminus R(g').$$

That is, putting

$$S(g)=S_{g',g(\alpha)}, \tag{16}$$

we have

$$\bigcup\{S(h): g'\subseteq h \,\&\, h\in{}^{\alpha\dot{+}1}\mu\}=S(g')\setminus R(g'),$$

which means that the definitions, given in (13) and (16), respectively, of $R(g')$ and $S(g)$, are compatible (cf. (14.3)). Noting that $\tilde{H}(g')=H(S(g'))$ has cardinality less than κ in view of (12), we can see that there are fewer than κ ordinals of the form $\kappa\cdot x\dot{+}\xi$ with $x\in\tilde{H}(g')$ and $\xi<\tau$; that is, there are fewer than κ nonempty ones among the sets $S_{g',\eta}$, i.e.,

$$|\text{ims}\,(g')|<\kappa \tag{17}$$

where ims (.) denotes the set of all immediate successors of the element replacing the dot. If $S(g)\neq 0$, then define $s(g)\in \tilde{H}(g\wedge\alpha)$ and $\tilde{f}(g)<\tau$ with the aid of the equation

$$g(\alpha)=\kappa\cdot s(g)\dot{+}\tilde{f}(g), \tag{18}$$

where the multiplication here again means ordinal multiplication. Note that we have dom $(g)=\alpha\dot{+}1$ here; the functions s and $\tilde{f}$ will be defined only for those $h\in T$ for which dom (h) is a successor ordinal. The definitions of the partition tree $\langle\kappa, S, R, T\rangle$ and of the auxiliary functions s and $\tilde{f}$ are complete.

Apply Theorem 14.3.a with $1\dot{+}\rho=\rho+1$ replacing λ. The assumptions of this theorem are satisfied in view of (14) and (17), and in view of the fact that we have $\kappa_0^{\rho_0}<\kappa$ for any $\kappa_0<\kappa$ and $\rho_0<\rho+1$; this latter assertion means the same thing as $\rho+1\leq L_\kappa(\kappa)$, which holds by our assumption $\rho\leq L_\lambda(\kappa)$. In fact, we have $\lambda=\kappa$ in the present case, and $L_\kappa(\kappa)$ is infinite according to Theorem 7.2. So Theorem 14.3.a implies that T has a path of length $1\dot{+}\rho\dot{+}1$. Choose a $h\in T$ with dom$(h)=1\dot{+}\rho$. Write $x_\alpha=s(h\wedge(\alpha\dot{+}1))$ if $\alpha<\rho$ (see (18) for the definition of s), and choose an arbitrary element of $S(h)$ as x_ρ. Noting that $x_\alpha\in\tilde{H}(h\wedge\alpha)\subseteq$ $\subseteq R(h\wedge\alpha)$ for any $\alpha<\rho$, we can easily see from the definition of $R(g')$ in (13) that $\langle x_\alpha\colon\ \alpha\leq\rho\rangle$ is a strictly increasing sequence of ordinals. Putting $X=$ $=\{x_\alpha\colon\ \alpha<\rho\}$, define the coloring $f_h\colon [X]^1\to\tau$ as follows:

$$f_h(\{x_\alpha\})=\tilde{f}(h\wedge(\alpha\dot{+}1))$$

for any $\alpha<\rho$, where $\tilde{f}$ was defined in (18). It follows from (1) that there is a $\xi<\tau$ and a set $Y\subseteq X$ of order type ν_ξ such that Y is a homogeneous set of color ξ with respect to f_h. (15) and (18) imply that $Y\cup\{x_\rho\}$ is a homogeneous set of color $1\dot{+}\xi$ with respect to the coloring f. This set has order type $\nu_\xi\dot{+}1$, which completes the proof.

We now turn to discussing the corollaries of the above result.

Corollary 17.2. *Let κ, λ, ρ, σ, and τ be cardinals, and assume $\kappa\geq\omega$ is regular, $3\leq\lambda\leq\kappa$, ρ, $\tau<L_\lambda(\kappa)<\kappa$, and $\sigma<\operatorname{cf}(L_\lambda(\kappa))$. Then*

$$\kappa\to(\lambda, (L_\lambda(\kappa))_\sigma, (\rho)_\tau)^2. \tag{19}$$

Proof. The result follows from the above theorem with the aid of the simple relation

$$L_\lambda(\kappa)\to((L_\lambda(\kappa))_\sigma, (\rho)_\tau)^1$$

(it is harmless to assume $\rho\geq 2$ here so as to satisfy the requirements of the above theorem).

Corollary 17.3. *Assume κ is inaccessible and $\rho, \tau<\kappa$. Then*

$$\kappa\to(\kappa, (\rho)_\tau)^2. \tag{20}$$

Proof. According to Theorem 17.1, one has to observe only that

$$(\rho\cdot\tau)^+\to(\rho)^1_\tau.$$

Note that in the present case we have $L_\kappa(\kappa)=\kappa$, and this is the case that was not covered by the preceding corollary.

Corollary 17.4. *Assume $\lambda\geq 2$, $\rho\geq\omega$, and $\tau<\mathrm{cf}(\rho)$. Then*

$$(\lambda^\rho)^+\to((\lambda^\rho)^+,(\mathrm{cf}(\rho))_\tau)^2. \tag{21}$$

Proof. Write $\kappa'=\lambda'=(\lambda^\rho)^+$, $\sigma'=\tau$, and $\rho',\tau'=0$. Apply Corollary 17.2 with the parameters having primes. The assumptions there are satisfied. In fact, $L_{\kappa'}(\kappa')$ is regular in view of Theorem 7.15 and so, using Theorem 7.17b, we obtain that

$$\sigma'=\tau<\mathrm{cf}(\rho)\leq L_{\kappa'}(\kappa')=\mathrm{cf}(\mathrm{L}_{\kappa'}(\kappa')). \tag{22}$$

Since $\kappa'=\lambda'$, we have $\sigma'<\mathrm{cf}(L_{\lambda'}(\kappa'))$, as required by the assumptions of Corollary 17.2. The other assumptions are easily checked. Hence we obtain

$$\kappa'\to(\lambda',(L_{\lambda'}(\kappa'))_{\sigma'},(\rho')_\tau)^2.$$

Noting that we have $\mathrm{cf}(\rho)\leq L_{\kappa'}(\kappa')=L_{\lambda'}(\kappa')$ according to (22), relation (21) follows. The proof is complete.

As a special case, we have

Corollary 17.5 (Erdős–Rado [1956]). *Assume $\lambda\geq 2$ and $\rho\geq\omega$. Then*

$$(\lambda^\rho)^+\to((\lambda^\rho)^+,(\rho^+)_\rho)^2. \tag{23}$$

Another interesting consequence of the above theorem is

Corollary 17.6. *Assume $\lambda\geq 3$, $\rho\geq\omega$, $\tau_1<\mathrm{cf}(\rho)$, and $\sigma,\tau_2<\rho$. Then*

$$(\underset{\sim}{\lambda}^\rho)^+\to(\underset{\sim}{\lambda}^\rho,(\rho)_{\tau_1},(\sigma)_{\tau_2})^2. \tag{24}$$

Proof. Noting that $\rho\leq L_{\underset{\sim}{\lambda}^\rho}((\underset{\sim}{\lambda}^\rho)^+)$ according to Theorem 7.17.a, we can apply Theorem 17.1 with $(\underset{\sim}{\lambda}^\rho)^+$, $\underset{\sim}{\lambda}^\rho$, and ρ replacing κ, λ, and ρ; the desired result follows from the simple relation

$$\rho\to((\rho)_{\tau_1},(\sigma)_{\tau_2})^2.$$

Corollary 17.7. *Assume λ is a strong limit cardinal and $\rho,\tau<\lambda$. Then*

$$\lambda\to(\rho)^2_\tau. \tag{25}$$

PROOF. The case $\lambda=\omega$ is covered by Ramsey's theorem (Theorem 10.2); therefore assume $\lambda>\omega$. Let $\sigma=\rho\cdot\tau\cdot\omega$; then

$$(2^\sigma)^+\rightarrow(\rho)^2_\tau$$

in view of (23). The desired result follows from here, since we have $2^\sigma<\lambda$.

Under the assumption of the continuum hypothesis we obtain

COROLLARY 17.8. *Assume* GCH *and let* $\lambda\geq\omega$, $\tau<\mathrm{cf}(\lambda)$. *Then*

$$\lambda^+\rightarrow(\lambda^+,(\mathrm{cf}(\lambda))_\tau)^2 \tag{26}$$

and

$$\lambda^+\rightarrow(\lambda)^2_\tau. \tag{27}$$

PROOF. (26) follows from (21) with $\rho=\lambda$, and (27) follows from (24) with $\rho=\lambda$, $\tau_1=\tau\dot{-}1$, and $\sigma_2=\tau_2=0$.

Using the Stepping-up Lemma (Lemma 16.1), we are now in a position to derive many interesting results from the above corollaries in the case $r\geq 3$. For the convenience of the reader, we recall that the operation exp is defined recursively by putting

$$\exp_0(0)=\lambda \quad\text{and}\quad \exp_{n+1}(\lambda)=\exp_n(2^\lambda)$$

for any cardinal λ and any integer n.

COROLLARY 17.9. *Assume* $\lambda\geq 2$, $\rho\geq\omega$, $\tau<\mathrm{cf}(\rho)$, *and let* $r\geq 2$ *be an integer. Then*

$$(exp_{r-2}(\lambda^\varrho))^+\rightarrow((\lambda^\varrho)^+,\mathrm{cf}(\rho)_\tau)^r, \tag{28}$$

and, in particular,

$$(exp_{r-2}(\lambda^\rho))^+\rightarrow((\lambda^\rho)^+,(\rho^+)_\rho)^r. \tag{29}$$

PROOF. For $r=2$ these results are identical to (21) and (23), respectively, and for $r>2$ they follow by induction, with the aid of the Stepping-up Lemma.

By applying the Stepping-up Lemma to (24), we get

COROLLARY 17.10. *Let* λ, ρ, σ, τ_1, *and* τ_2 *be cardinals satisfying* $\lambda\geq 3$, $\rho\geq\omega$, $\tau_1<\mathrm{cf}(\rho)$, *and* σ, $\tau_2<\rho$, *and let* $r\geq 2$ *be an integer. Then*

$$(exp_{r-2}(\lambda^\varrho))^+\rightarrow(\lambda^\varrho,(\rho)_{\tau_1},(\sigma)_{\tau_2})^r. \tag{30}$$

Taking $\lambda=2$ in (29), we get

COROLLARY 17.11. *Let* $\rho\geq\omega$ *be a cardinal and* $r\geq 2$ *an integer. Then*

$$exp_{r-1}(\rho))^+\rightarrow((2^\rho)^+,(\rho^+)_\rho)^r, \tag{31}$$

and so, a fortiori,

$$(exp_{r-1}(\rho))^{+} \to (\rho^{+})^{r}_{\rho}. \tag{32}$$

Note that (32) can be directly obtained from the trivial relation

$$\rho^{+} \to (\rho^{+})^{1}_{\rho}$$

by induction, using the Stepping-up Lemma. (31) is, however, much stronger, showing that Theorem 17.1 has a special feature not contained in the Stepping-up Lemma. The positive results are in a sense weaker for $r>2$ than for $r=2$, since all results of the form $\kappa \to (\ldots)^{r}$ with $r \geq 3$ and κ regular that we can prove are obtained by repeated applications of the Stepping-up Lemma starting with the case $r=2$, except if κ has some large cardinal property (for this latter remark cf. e.g. Theorem 29.5).

18. A DIRECT CONSTRUCTION OF THE CANONICAL PARTITION TREE

Let κ and τ be cardinals, fix a coloring $f\colon [\kappa]^{<\omega} \to \tau$, and consider the canonical partition tree $\langle \kappa, S, R, T \rangle$ associated with f; see the beginning of the proof of Lemma 15.2 for the definition. Simultaneously with this partition tree we also defined a function $s\colon T \to \kappa$. One can show that s is 1-1 and onto and, moreover, that for any $g, h \in T$ with $g \subseteq h$ we have $s(g) \leq s(h)$. (We in effect showed the latter assertion in the proof of Lemma 15.2; we omit the simple proof of the former here since we need these facts only in order to give some background to what follows.) So we can define an ordering $\prec_f$ of κ which turns κ into a tree isomorphic to $\langle T, \subset \rangle$ by putting

$$\xi \prec_f \eta \Leftrightarrow \exists g, h \in T[g \subset h \;\&\; \xi = s(g) \;\&\; \eta = s(h)], \tag{1}$$

and here $\xi \prec_f \eta$ implies $\xi < \eta$. It is obvious that the tree $\langle \kappa, \prec_f \rangle$ is just as applicable in proving the existence of large homogeneous sets as is the partition tree $\langle \kappa, S, R, T \rangle$. In fact, we shall see that there is some advantage in taking a closer look at the tree $\langle \kappa, \prec_f \rangle$: by so doing we shall be able to prove (a strengthening of) Theorem 17.1 in case b; in the proof we gave in the preceding section we had to define another partition tree. An important feature in considering the tree $\langle \kappa, \prec_f \rangle$ is that there is a simple direct definition of the partial ordering $\prec_f$. We are about to give this definition of $\prec_f$, but we shall not prove the fact that the ordering $\prec_f$ we define is identical to the one defined in (1), as we shall not need this.

As mentioned above, we are given a coloring $f\colon [\kappa]^{<\omega} \to \tau$, where κ and τ are cardinals.

DEFINITION 18.1. (i) Let $X \subseteq \kappa$. Define the equivalence relation $\equiv_{f,X}$ by putting

$$\alpha \equiv_{f,X} \beta \quad \text{iff} \quad \forall u \in [X]^{<\omega} f(u \cup \{\alpha\}) = f(u \cup \{\beta\}) \tag{2}$$

for any $\alpha, \beta < \kappa$.

(ii) For each $\alpha < \kappa$ we define the function $g_\alpha: \alpha \to 2$ by transfinite recursion on $\xi < \alpha$. Assuming that $g_\alpha \restriction \xi$ has already been defined for some $\xi < \alpha$, and, writing

$$X_{\alpha,\xi} = \{\eta < \xi : g_\alpha(\eta) = 1\}, \tag{3}$$

put

$$g_\alpha(\xi) = 1 \quad \text{iff} \quad \xi \equiv_{f, X_{\alpha,\xi}} \alpha. \tag{4}$$

(iii) For any $\alpha, \beta < \kappa$ put

$$\beta \prec_f \alpha \quad \text{iff} \quad \beta < \alpha \;\&\; g_\alpha(\beta) = 1. \tag{5}$$

For any $\alpha < \kappa$, introduce the notation

$$X_\alpha = \{\beta : \beta \prec_f \alpha\} = \{\beta < \alpha : g_\alpha(\beta) = 1\}. \tag{6}$$

Note that we have

$$X_{\alpha,\xi} = X_\alpha \cap \xi, \tag{7}$$

where $X_{\alpha,\xi}$ was defined in (3).

We are going to establish the basic properties of the relation $\prec_f$ so defined.

LEMMA 18.2. (i) *If* $\alpha < \beta$ *and* $g_\beta(\alpha) = 1$, *then* $g_\beta \restriction \alpha = g_\alpha$. (ii) $\langle \kappa, \prec_f \rangle$ *is a tree.* (iii) *For any* $\alpha, \beta < \kappa$ *and* u *with* $\beta \prec_f \alpha$ *and* $u \in [X_\alpha \cap \beta]^{<\omega}$ *we have*

$$f(u \cup \{\beta\}) = f(u \cup \{\alpha\}); \tag{8}$$

that is, for any $v \in [X_\alpha \cup \{\alpha\}]^{<\omega}$, *the value of* $f(v)$ *does not depend on the last element of* v.

PROOF. *Ad* (i). Assume that

$$g_\beta \restriction \xi = g_\alpha \restriction \xi \tag{9}$$

holds for some $\xi < \alpha$. It is then enough to show that

$$g_\beta(\xi) = g_\alpha(\xi) \tag{10}$$

also holds, and the desired result will follow by transfinite induction. We have $\alpha \equiv_{f, X_{\beta,\alpha}} \beta$ according to the assumption $g_\beta(\alpha) = 1$; so we have a fortiori

$$\alpha \equiv_{f, X_{\beta,\xi}} \beta,$$

since $X_{\beta,\xi} \subseteq X_{\beta,\alpha}$. Noting that $X_{\beta,\xi} = X_{\alpha,\xi}$ in view of (9), this implies that $\xi \equiv_{f, X_{\beta,\xi}} \beta$ holds just in case $\xi \equiv_{f, X_{\alpha,\xi}} \alpha$ holds; in other words, $g_\beta(\xi) = 1$ just in case $g_\alpha(\xi) = 1$, which proves (10).

Ad (ii). First we prove that $\prec_f$ is a partial ordering, i.e., that it is an irreflexive, antisymmetric, and transitive relation. $\prec_f$ obviously has the first two of these properties, as $\alpha \prec_f \beta$ implies $\alpha < \beta$ for any $\alpha, \beta < \kappa$; we are going to establish transitivity. Assume to this end that $\alpha \prec_f \beta$ and $\beta \prec_f \gamma$ hold for some $\alpha, \beta, \gamma < \kappa$. Then $\alpha < \beta < \gamma$, $g_\beta(\alpha) = 1$, and $g_\gamma(\beta) = 1$. Hence we have $g_\beta = g_\gamma \upharpoonright \beta$ according to part (i) of the present lemma, and so $g_\gamma(\alpha) = 1$ also holds, which entails that $\alpha \prec_f \gamma$, establishing transitivity.

To prove that $\langle \kappa, \prec_f \rangle$ is a tree it is enough to show that, for any $\alpha < \kappa$, the ordering $\prec_f$ on $X_\alpha = \{\xi : \xi \prec_f \alpha\}$ coincides with the natural ordering $<$ of the ordinals. Assume to this end that $\xi, \eta \in X_\alpha$ and $\xi < \eta$. Then $g_\alpha(\xi) = 1$ and, according to part (i) of the present lemma, $g_\eta = g_\alpha \upharpoonright \eta$; that is, $g_\eta(\xi) = 1$. Hence $\xi \prec_f \eta$, which we wanted to show.

Ad (iii). (8) simply follows from the relation $\alpha \equiv_{f, X_\alpha \cap \beta} \beta$, which holds by virtue of $\beta \prec_f \alpha$ and the equality $X_{\alpha,\beta} = X_\alpha \cap \beta$ mentioned in (7) (cf. (4)).

Next we show that the canonical partition tree suffices to establish also that part of Theorem 17.1 which we proved by constructing another type of partition tree; in fact, it enables one to prove a slightly stronger result:

ALTERNATIVE PROOF OF THEOREM 17.1 IN CASE $\lambda = \kappa > \omega$. We are going to prove a stronger result: the assumptions of Theorem 17.1 with $\lambda = \kappa > \omega$ imply the following: given an arbitrary coloring $f: [\kappa]^2 \to 1 \dotplus \tau$, there is either a homogeneous set of color 0 that is stationary in κ, or there is a $\xi < \tau$ such that there is a homogeneous set of order type $\nu_\xi \dotplus 1$ and of color $1 \dotplus \xi$. Symbolically, this assertion might be expressed as

$$\kappa \to (\text{Stat}(\kappa), (\nu_\xi \dotplus 1)_{\xi < \tau})^2, \tag{11}$$

where Stat (κ) denotes the set of stationary subsets of κ. Here is, however, a note of caution: the notational conventions introduced in Section 8 are inadequate for explaining the meaning of this relation.

For the proof, let $f: [\kappa]^2 \to 1 \dotplus \tau$ be a coloring, and consider the tree $\langle \kappa, \prec_f \rangle$. (Strictly speaking, $\langle \kappa, \prec_f \rangle$ has not been defined, since dom (f) is a proper subset of $[\kappa]^{<\omega}$. So put $f'(u) = f(u)$ whenever $u \in \text{dom}(f)$, and put $f'(u) = 0$ otherwise, i.e., when $u \in [\kappa]^{<\omega} \setminus \text{dom}(u)$, and define $\langle \kappa, \prec_f \rangle$ as being equal to $\langle \kappa, \prec_{f'} \rangle$.) Setting

$$Y_\alpha = \{\xi \in X_\alpha : f(\{\xi\alpha\}) > 0\}, \tag{12}$$

we may assume that

$$|Y_\alpha| < \rho \tag{13}$$

holds for any $\alpha < \kappa$. In fact, assume on the contrary that $|Y_\alpha| \geq \rho$ for some $\alpha < \kappa$. Then (17.1) implies that there is a $\xi < \tau$ and a set $Z \subseteq Y_\alpha$ of order type ν_ξ such that $f(\{\gamma\alpha\}) = 1 \dotplus \xi$ holds for any $\gamma \in Z$. It follows from (8) that $Z \cup \{\alpha\}$ is a homogeneous set of color $1\dot{+}\xi$ with respect to f. As $Z\cup\{\alpha\}$ has order type $\nu_\xi\dot{+}1$, this completes the proof of the theorem in case (13) fails for some $\alpha<\kappa$. Assume therefore that (13) holds for all $\alpha<\kappa$.

Put $\sigma=\omega$ if $\rho<\omega$, $\sigma=\rho$ is ρ is regular, and $\sigma=\rho^+$ if ρ is singular. Note that σ is a regular cardinal and $\sigma<\kappa$. In fact, since we assumed $\kappa>\omega$ and $\rho<\kappa$, the case $\sigma=\kappa$ could occur only if ρ was singular and $\rho^+=\kappa$; but then $L_\kappa(\kappa)\le\mathrm{cf}(\rho)<\rho$, which conflicts with our assumption $\rho\le L_\lambda(\kappa)=L_\kappa(\kappa)$ ($\lambda=\kappa$ in the case considered). Put

$$A=\{\alpha<\kappa:\ \mathrm{cf}(\alpha)=\sigma\}.$$

A is a stationary subset of κ as it contains the σth element of an arbitrary club in κ. The function

$$h(\alpha)=\sup Y_\alpha$$

is a regressive function on A in view of (13). Fodor's theorem (Theorem 5.4) implies that there is an ordinal $\zeta<\kappa$ and a set $B\subseteq A$ stationary in κ such that $h(\alpha)=\zeta$ holds for any $\alpha\in B$. Writing $f_\alpha(\beta)=f(\{\alpha\beta\})$ for any α, β with $\beta<\alpha<\kappa$, consider the functions $f_\alpha\restriction Y_\alpha$ as α runs over elements of B. As $\sup Y_\alpha=h(\alpha)=\zeta$ holds, there are $\le|\zeta|^\varrho$ possibilities for Y_α in view of (13). Given Y_α, there are at most $|Y_\alpha|^\tau\le\varrho^\tau$ possibilities for $f_\alpha\restriction Y_\alpha$; hence the number of possibilities for $f\restriction Y_\alpha$ is $\le|\zeta|^\varrho\cdot\varrho^\tau$. Here $\tau<\rho$ in view of (17.1) since we assumed $\rho\ge2$ and $\nu_\xi\ge2$ for any $\xi<\tau$; so this number is less than κ by virtue of the assumption $\rho\le L_\kappa(\kappa)$. Hence there is a set $C\subseteq B$ stationary in κ and a function $\tilde{f}$ defined on a set $Y\subseteq\zeta$ of cardinality $<\rho$ such that

$$f_\alpha\restriction Y_\alpha=\tilde{f},\quad\text{in particular}\quad Y_\alpha=Y, \tag{14}$$

for any $\alpha\in C$. We claim that C is a chain of the tree $\langle\kappa,\prec_f\rangle$. In fact, let $\alpha,\beta\in C$ with $\alpha<\beta$, and assume, on the contrary, that $\neg\,\alpha\prec_f\beta$. Consider those ordinals γ' for which $\gamma'\preceq_f\alpha$ but $\neg\,\gamma'\preceq_f\beta$ (α is itself such an ordinal by our assumption), and choose the least such γ' as γ. Then we have

$$X_\alpha\cap\gamma=X_\beta\cap\gamma$$

and, for any δ in this set,

$$f(\{\delta\alpha\})=f_\alpha(\delta)=f_\beta(\delta)=f(\{\delta\beta\});$$

the second equality here follows from (14) and (12), since if $\delta\in Y_\alpha=Y_\beta$, then $f_\alpha(\delta)$ and $f_\beta(\delta)$ coincide, and if $\delta\notin Y_\alpha=Y_\beta$ then $f_\alpha(\delta)=f_\beta(\delta)=0$. Using (8), we obtain that $f(\{\delta\gamma\})=f(\{\delta\alpha\})$ since $\delta\prec_f\gamma\preceq_f\alpha$. Comparing this with the last centered line, we obtain that $f(\{\delta\gamma\})=f(\{\delta\beta\})$ holds for any $\delta\in X_\beta\cap\gamma=X_{\beta,\gamma}$; i.e., $\gamma\equiv_{f,X_{\beta,\gamma}}\beta$. In other words, $\gamma\prec_f\beta$, which contradicts the choice of γ. This contradiction establishes that C is a chain of $\langle\kappa,\prec_f\rangle$.

We claim that C is a homogeneous set of color 0. In fact, if $\beta \in C$ then

$$\beta \notin Y_\beta = Y$$

(equality holds here by (14)), since $Y_\beta \subset \beta$ (cf. (12)). Let now $\alpha, \beta \in C$ be such that $\beta < \alpha$. Then $\beta \prec_f \alpha$, and so

$$\beta \in X_\alpha \setminus Y = X_\alpha \setminus Y_\alpha .$$

Hence the definition of Y_α in (12) implies $f(\{\alpha\beta\}) = 0$; this shows that C is in fact a homogeneous set of color 0. As C is stationary, this completes the proof of the theorem.

The reader may easily see that we even proved slightly more than (11). We illustrate the situation only in a special case when $\kappa = \lambda = \aleph_2$, $\rho = \aleph_1$, $\tau = \aleph_1$. The above proof gives e.g. the following: Assume $2^{\aleph_0} = \aleph_1$; let D be a stationary set in $\aleph_2$ with

$$D \subseteq \{\alpha < \aleph_2 : \operatorname{cf}(\alpha) = \aleph_1\}, \tag{15}$$

and let $f: [D]^2 \to 2$ be a coloring. Then there is either a set $C_0 \subseteq D$ stationary in $\aleph_2$ with $f``C_0 = \{0\}$ or a set $C_1$ of cardinality $\aleph_1$ with $f``C_1 = \{1\}$. Using ideas of J. E. Baumgartner [1976], one can show that this assertion is false if we modify our assumptions by requiring that

$$D \subseteq \{\alpha < \aleph_2 : \operatorname{cf}(\alpha) = \omega\}$$

instead of (15).

CHAPTER V

NEGATIVE ORDINARY PARTITION RELATIONS, AND THE DISCUSSION OF THE FINITE CASE

Various methods are developed for establishing negative ordinary partition relations and, as an application of these methods, the chapter ends with a discussion of the relation $\kappa \to (\lambda_\xi)^r_{\xi<\vartheta}$ for finite κ.

19. MULTIPLICATION OF NEGATIVE PARTITION RELATIONS FOR $r=2$

We start with two definitions:

DEFINITION 19.1. Let $A = \mathop{\times}\limits_{\xi<\alpha} A_\xi$. For any two distinct $f, g \in A$ we define the first discrepancy $\delta(f, g) = f \cap g$ of f and g as

$$\delta(f, g) = f \cap g = \min\{\xi < \alpha : f(\xi) \neq g(\xi)\}.$$

DEFINITION 19.2. Let $\langle A_\xi, \prec_\xi \rangle$, $\xi < \alpha$, be ordered sets and let $A = \mathop{\times}\limits_{\xi<\alpha} A_\xi$. We define the *lexicographic ordering* $\prec = \text{lex}\,\langle \prec_\xi : \xi < \alpha \rangle$ of A by putting $f \prec g$ iff $f(f \cap g) \prec_{f \cap g} g(f \cap g)$ for any two distinct $f, g \in A$. The ordering $\prec$ will also be called the *lexicographic product* of the orderings $\prec_\xi$.

The following properties of the first discrepancy and of the lexicographic product are direct consequences of the above definitions:

$$f \cap g = g \cap f \qquad (f, g \in A,\ f \neq g); \tag{1}$$

$$\min\{f \cap h, g \cap h\} \leq f \cap g \tag{2}$$

if $f, g, h \in A$ are pairwise distinct; if $f \cap h \neq g \cap h$, then equality holds here;

$$f(\xi) \preceq_\xi g(\xi) \quad \text{for} \quad \xi \leq f \cap g \quad \text{if} \quad f \prec g. \tag{3}$$

$$\prec \text{ is an ordering of } A. \tag{4}$$

Furthermore, we have

$$\succ = \text{lex}\,\langle \succ_\xi : \xi < \alpha \rangle, \tag{5}$$

where $\prec = \text{lex}\,\langle \prec_\xi : \xi < \alpha \rangle$, and $\succ_\xi$ and $\succ$ denote reverse orderings of $\prec_\xi$ and $\prec$, respectively.

We shall need the following

LEMMA 19.3. *Let $\langle A_\xi, \prec_\xi \rangle$, $\xi < \alpha$, be ordered sets, and let $\prec$ be the lexicographic ordering of the Cartesian product $A = \bigtimes_{\xi<\alpha} A_\xi$. Let κ be a regular cardinal, and $B \subseteq A$ a set of cardinality κ. Assume that B is wellordered by $\prec$. Then there exists a sequence $\langle f_\gamma : \gamma < \kappa \rangle$, strictly increasing in the ordering $\prec$, of elements of B and a nondecreasing sequence $\langle \xi_\gamma : \gamma < \kappa \rangle$ of ordinals less than α such that*

$$f_\mu \cap f_\nu = \xi_\mu \tag{6}$$

holds whenever $\mu < \nu < \kappa$.

An analogous result holds in view of (5) if $\prec$ is replaced with $\succ$.

PROOF. Since B is wellordered by $\prec$, we may assume that its order type is κ. (Indeed, its order type is $\geq \kappa$, and if it is $> \kappa$, then we may omit some elements of B to make its order type κ.) We construct the sequence $\langle f_\gamma : \gamma < \kappa \rangle$ by transfinite recursion; so as to be able to do this, we shall also construct a sequence $\langle B_\gamma : \gamma < \kappa \rangle$ of subsets of B such that B_μ is a nonempty final segment of B and $B_\mu \supseteq B_\nu$ for any ordinals μ and ν with $\mu < \nu < \kappa$. Having constructed B_γ for all $\gamma < \mu$, Where μ is an ordinal less than κ, put $b'_\mu = \bigcap_{\gamma<\mu} B_\gamma$ if $\mu > 0$ and $B'_\mu = B$ if $\mu = 0$. Note that B'_μ is a final segment of B, and it is nonempty in view of the regularity of κ. Let f_μ be the minimal element of B'_μ in the ordering $\prec$, and write

$$\xi_\mu = \min \{ f_\mu \cap f : f \in B'_\mu \setminus \{f_\mu\} \}$$

and

$$B_\mu = \{ f \in B'_\mu \setminus \{f_\mu\} : f_\mu \cap f = \xi_\mu \}.$$

B_μ is clearly not empty, and it is easy to show that B_μ is a final segment of B'_μ, and so of B. This completes the definitions of f_μ, ξ_μ, and B_μ for $\mu < \kappa$. Since we have $f_\nu \in B_\mu$ whenever $\mu < \nu < \kappa$, (6) follows. Let $\mu < \nu < \delta < \kappa$. Then by (6) and (2) we obtain that

$$\xi_\mu = f_\mu \cap f_\nu = f_\mu \cap f_\delta \leq f_\nu \cap f_\delta = \xi_\nu,$$

showing that the sequence $\langle \xi_\gamma : \gamma < \kappa \rangle$ is in fact nondecreasing. The proof is complete.

The simplest result illustrating how orderings can be used to establish negative partition relations is

LEMMA 19.4. *Let $\langle A, \prec \rangle$ be an ordered set, κ a cardinal, and ν_0 and ν_1 ordinals such that $|A| = \kappa$ and $\nu_0, \nu_1^* \nleq \text{tp}\,\langle A, \prec \rangle$. Then $\kappa \nrightarrow (\nu_0, \nu_1)^2$.*

Here ν_1^* is the reverse of ν_1, i.e., $\nu_1^* = \text{tp}\,\langle \nu_1, > \rangle$.

PROOF. Take a wellordering $\prec_0$ of type κ of A, and define the partition $I = \langle I_0, I_1\rangle$ of $[A]^2$ as follows: for any $x, y \in A$ with $x \prec_0 y$ put $\{xy\} \in I_0$ if $x \prec y$ and put $\{xy\} \in I_1$ otherwise. If $X \subseteq A$ is a set of type ν_i in the wellordering $\prec_0$ $(i=0, 1)$, then X is not a homogeneous set of class i; in fact, in the contrary case X would have order type ν_0 or ν_1^* in the ordering $\prec$ according as $i=0$ or 1. The proof is complete.

DEFINITION 19.5. If $\langle A, \prec\rangle$ is an ordered set as in the preceding lemma, then we say that $\langle A, \prec\rangle$ *establishes* the relation $\kappa \nrightarrow (\nu_0, \nu_1)^2$. The partition I defined in the proof is called the ***Sierpiński** partition* of $\langle A, \prec\rangle$ associated with the wellordering $\prec_0$ of A.

The idea of the above proof is due to W. Sierpiński, and he used it to establish the relation

$$2^{\aleph_0} \nrightarrow (\aleph_1, \aleph_1)^2$$

(see (14) below).

The main result in this section is

THEOREM 19.6. *Assume that $\alpha > 0$ and for every $\xi < \alpha$ there is an ordered set $\langle A_\xi, \prec_\xi\rangle$ that establishes the relation $\kappa_\xi \nrightarrow (\mu_\xi, \nu_\xi)^2$, i.e., for which $|A_\xi| = \kappa_\xi$ and $\mu_\xi, \nu_\xi^* \not\leq \operatorname{tp}\langle A_\xi, \prec_\xi\rangle$, where κ_ξ is a cardinal and μ_ξ, ν_ξ are ordinals for every $\xi < \alpha$. Put $\kappa = \prod_{\xi<\alpha} \kappa_\xi$, and let ρ and σ be regular cardinals such that*

$$\rho \to (\mu_\xi)^1_{\xi<\alpha} \qquad \text{and} \qquad \sigma \to (\nu_\xi)^1_{\xi<\alpha}. \tag{7}$$

Then

$$\kappa \nrightarrow (\rho, \sigma)^2 \tag{8}$$

is established by an ordered set, and we also have

$$\kappa \nrightarrow (\rho, (\nu_\xi)_{\xi<\alpha})^2 \tag{9}$$

and

$$\kappa \nrightarrow ((\mu_\xi)_{\xi<\alpha}, (\nu_\xi)_{\xi<\alpha})^2. \tag{10}$$

PROOF. Let $A = \bigtimes_{\xi<\alpha} A_\xi$, and let $\prec$ be the lexicographic ordering of A; clearly A has cardinality κ. We are going to show that $\rho \not\leq \operatorname{tp}\langle A, \prec\rangle$ and $\sigma^* \not\leq \operatorname{tp}\langle A, \prec\rangle$, which mean that the ordered set $\langle A, \prec\rangle$ establishes (8). It suffices to verity the first of these relations; the second one can be proved analogously in view of (5). Assume, on the contrary, that $B \subseteq A$ is a set that has order type ρ in the ordering $\prec$. Apply Lemma 19.3 with ρ replacing κ: we obtain f_γ and $\xi_\gamma (\gamma < \rho)$ such that (6) holds. Define the partition $I = \langle I_\zeta : \zeta < \alpha\rangle$ of ρ by putting $\gamma \in I_\zeta$ iff $\xi_\gamma = \zeta$. The first relation in (7) implies that I_ζ has order type $\geq \mu_\xi$ for some $\zeta < \alpha$. Now for any two distinct $\mu, \nu \in I_\zeta$, we have $f_\mu \cap f_\nu = \zeta$; since the sequence $\langle f_\mu : \mu \in I_\zeta \rangle$ is strictly increasing in the lexicographic ordering, the sequence $\langle f_\mu(\zeta) : \mu \in I_\zeta \rangle$ is strictly increasing in the ordering $\prec_\zeta$. As I_ζ has order type $\geq \mu_\zeta$, this contradicts the assumption $\mu_\zeta \not\leq \mathrm{tp}\, \langle A_\zeta, \prec_\zeta \rangle$. This proves $\rho \not\leq \mathrm{tp}\, \langle A, \prec \rangle$, establishing (8).

Next, instead of proving (9), we are going to prove the completely analogous relation

$$\kappa \nrightarrow ((\mu_\xi)_{\xi < \alpha}, \sigma)^2; \tag{11}$$

by doing so, our notations will be slightly simpler. Define the partition $J = \langle J_0, J_1 \rangle$ as a Sierpiński partition of A: Fix a wellordering $\prec_0$ of type κ of A, and for any $f, g \in A$ with $f \prec_0 g$ put $\{fg\} \in J_0$ if $f \prec g$, and put $\{fg\} \in J_1$ otherwise. Define a partition $I = \langle I_\xi : \xi < \alpha \rangle$ of J_0 by putting $\{fg\} \in I_\zeta$ if $f \cap g = \zeta$ for any f, g with $\{fg\} \in J_0$. Putting $I_\alpha = J_1$, we claim that the partition $I' = \langle I_\xi : \xi \leq \alpha \rangle$ establishes (11). In fact, there is no homogeneous set X such that $\mathrm{tp}\, \langle X, \prec_0 \rangle = \sigma$ and $[X]^2 \subseteq I_\alpha = J_1$, because $\sigma^* \not\leq \mathrm{tp}\, \langle A, \prec \rangle$, as we saw just before. Assume that, for some $X \subseteq A$ and $\xi < \alpha$, we have $\mathrm{tp}\, \langle X, \prec_0 \rangle = \mu_\xi$ and $[X]^2 \subseteq I_\xi$. Then we also have $\mathrm{tp}\, \langle X, \prec \rangle = \mu_\xi$ and so the set $\{f(\xi) : f \in X\}$ has order type μ_ξ in the ordering $\prec_\xi$, which contradicts the assumption $\mu_\xi \not\leq \mathrm{tp}\, \langle A_\xi, \prec_\xi \rangle$.

To prove (10), let I_ξ, $\xi < \alpha$, be the same as just before and, analogously, put

$$I_{\alpha \dot{+} \xi} = \{\{fg\} \in J_1 : f \cap g = \xi\}$$

for any $\xi < \alpha$. We have seen that there is no homogeneous set of order type μ_ξ in $\prec_0$ and of class ξ with respect to the partition $I'' = \{I_\xi : \xi < \alpha \dot{+} \alpha\}$ for any $\xi < \alpha$. One can show in the same way that there is no homogeneous set of order type ν_ξ in $\prec_0$ and of class $\alpha \dot{+} \xi$ for any $\xi < \alpha$. This completes the proof of the theorem.

Corollary 19.7. *Assume $\kappa \geq 2$ and $\lambda \geq \omega$. Then*

$$\kappa^\lambda \nrightarrow ((\kappa \cdot \lambda)^+, \lambda^+)^2 \tag{12}$$

is established by an ordered set, and

$$\kappa^\lambda \nrightarrow ((\kappa \cdot \lambda)^+, (\omega)_\lambda)^2 \tag{13}$$

holds.

Proof. Assume first that $\kappa \geq \omega$. Then $\kappa \nrightarrow (\kappa^+, \omega)^2$ is established by $\langle \kappa, < \rangle$. As $(\kappa \cdot \lambda)^+ \to (\kappa^+)^1_\lambda$ and $\lambda^+ \to (\omega)^1_\lambda$ hold, the assertions follow from (8) and (9), respectively. If $2 \leq \kappa < \omega$ then one has only to note that $\kappa^\lambda = \aleph_0^\lambda$ and $\kappa \cdot \lambda = \aleph_0 \cdot \lambda$.

As a special case of (12), the relation

$$2^\lambda \nrightarrow (\lambda^+, \lambda^+)^2 \tag{14}$$

is established by an ordered set for any $\lambda \geq \omega$; this relation was proved independently by W. Sierpiński [1933] and (later) by D. Kurepa.

Our next result is announced only for singular λ. It remains valid also in case λ is regular, but that case is completely covered by relation (17) below.

COROLLARY 19.8. *Assume $\lambda > \omega$ is singular and $\lambda = \sum_{\xi < \mathrm{cf}(\lambda)} \lambda_\xi$, where $3 \leq \lambda_\xi < \lambda$ for every $\xi < \mathrm{cf}(\lambda)$. Then*

$$2^\lambda \nrightarrow (\lambda_\xi)^2_{\xi < \mathrm{cf}(\lambda)} \tag{15}$$

and, for any cardinal $\rho \geq \lambda$,

$$\rho^\lambda \nrightarrow (\rho^+, (\lambda_\xi)_{\xi < \mathrm{cf}(\lambda)})^2. \tag{16}$$

PROOF. We may assume that $\lambda_\xi \geq \omega$ for each $\xi < \mathrm{cf}(\lambda)$. To prove (15), observe that the relation $2^{\lambda_\xi} \nrightarrow (\lambda_\xi^+, \lambda_\xi^+)^2$ is established by an ordered set according to (14); hence, noting that

$$2^\lambda = 2^{\Sigma_{\xi < \mathrm{cf}(\lambda)} \lambda_\xi} = \prod_{\xi < \mathrm{cf}(\lambda)} 2^{\lambda_\xi},$$

we can conclude by (10) that

$$2^\lambda \nrightarrow ((\lambda_\xi^+)_{\xi < \mathrm{cf}(\lambda)}, (\lambda_\xi^+)_{\xi < \mathrm{cf}(\lambda)})^2;$$

so, by the Omission Rule (see Subsection 9.3) we obtain

$$2^\lambda \nrightarrow (\lambda_\xi^+)^2_{\xi < \mathrm{cf}(\lambda)}.$$

Choose a 1-1 function g from $\mathrm{cf}(\lambda)$ into $\mathrm{cf}(\lambda)$ such that $\lambda_{g(\xi)} \geq \lambda_\xi^+$; then

$$2^\lambda \nrightarrow (\lambda_{g(\xi)})^2{}_{\xi < \mathrm{cf}(\lambda)}$$

by a monotonicity property (cf. Subsection 9.6b). Hence we have

$$2^\lambda \nrightarrow (\lambda_\xi)^2_{\xi < \mathrm{cf}(\lambda)}$$

again by the Omission Rule.

To verify (16), note that the relation $\rho^{\lambda_\xi} \nrightarrow (\rho^+, \lambda_\xi^+)^2$ is established by an ordered set according to (12), and $\rho^+ \rightarrow (\rho^+)^1_{\mathrm{cf}(\lambda)}$, so we have

$$\rho^\lambda = \prod_{\xi < \mathrm{cf}(\lambda)} \rho^{\lambda_\xi} \nrightarrow (\rho^+, (\lambda_\xi^+)_{\xi < \mathrm{cf}(\lambda)})^2$$

by (9); hence, using the function g mentioned in the preceding paragraph, we get (16) by the Omission Rule. The proof is complete.

We next prove the following simple result:

THEOREM 19.9. *Let* $\alpha>0$ *and* $\kappa_\xi\geq 2$ *for each* $\xi<\alpha$. *Then*

$$\prod_{\xi<\alpha}\kappa_\xi\not\to(\kappa_\xi^+)^2_{\xi<\alpha}.$$

PROOF. Let $|A_\xi|=\kappa_\xi$ and $A=\mathop{\times}\limits_{\xi<\alpha}A_\xi$. Define the partition $\langle I_\xi:\xi<\alpha\rangle$ of $[A]^2$ by putting $\{fg\}\in I_\xi$ iff $f\cap g=\xi$ for any two distinct $f,g\in A$. Assuming that $[B]^2\subseteq I_\xi$ holds for some $B\subseteq A$ and $\xi<\alpha$, we have $f(\xi)\neq g(\xi)$ for any two distinct $f,g\in B$; hence $|B|\leq|A_\xi|=\kappa_\xi$. The proof is complete.

As a corollary we have

$$2^\lambda\not\to(3)^2_\lambda \tag{17}$$

for any cardinal $\lambda>0$. This can be expressed alternatively by saying that

$$\kappa\not\to(3)^2_{L_3(\kappa)} \tag{18}$$

holds for any cardinal $\kappa\geq 2$ (see Section 7 for the definition of the logarithm L).

20. A NEGATIVE PARTITION RELATION ESTABLISHED WITH THE AID OF GCH

It is conceivable that relation (19.13) can be improved to

$$\kappa^\lambda\not\to(\kappa^+,(3)_\lambda)^2 \tag{1}$$

for any $\kappa,\lambda\geq\omega$. This is indeed the case if $\kappa^\lambda=2^\lambda$ in view of (19.17) or, trivially, if $\kappa^\lambda\leq\kappa$, but we are unable to prove this in the important case $2^\lambda\leq\kappa<\kappa^\lambda$. The following is the simplest instance of what we cannot settle:

PROBLEM 20.1. Can one prove without assuming GCH that

$$\aleph_\omega^{\aleph_0}\not\to(\aleph_{\omega+1},(3)_{\aleph_0})^2$$

holds?

We can prove (1) under the assumption of GCH. More precisely, using the notation for partition relations with bipartite graphs, introduced in Subsection 8.8, we have

THEOREM 20.2. (Erdős–Hajnal–Rado [1965]). *Assume* κ *is a singular cardinal such that* $2^\kappa=\kappa^+$. *Then*

$$\kappa^+\not\to\left(\binom{\kappa^+}{\kappa},(3)_{\mathrm{cf}(\kappa)}\right)^2. \tag{2}$$

This clearly implies (1) in the nontrivial cases if we assume GCH. For the proof we need a lemma on set mappings. A *set mapping* on a set E is a function $f: E \to \mathscr{P}(E)$ such that $x \notin f(x)$ for any $x \in E$. A set $X \subseteq E$ is free with respect to f if $f(x) \cap X = 0$ holds for any $x \in X$. A more extensive study of set mappings will be given in Sections 44–46; here we need only the following:

Lemma 20.3 (Hajnal–Máté [1975]). *Let λ be a regular cardinal, and let $E = \bigcup_{\xi<\lambda} E_\xi$, where E_ξ is a set of cardinality $>\lambda$ for every $\xi<\lambda$. Let $f: E \to \mathscr{P}(E)$ be a set mapping such that $|f(x) \cap E_\xi| < \lambda$ holds for any $x \in E$ and $\xi < \lambda$. Then there is a set X free with respect to f such that $X \cap E_\xi$ is nonempty for every $\xi < \lambda$.*

Proof. We may assume that the sets E_ξ, $\xi<\lambda$, are pairwise disjoint and have cardinality λ^+. In fact, if this is not already the case, we can always take smaller sets satisfying these requirements. So it means no restriction of generality to assume that $E_\xi = \lambda^+ \times \{\xi\}$. We are going to construct a free set

$$X = \{\langle \alpha_\xi \xi \rangle : \xi < \lambda\} \tag{3}$$

such that $\alpha_\xi < \alpha_\eta (<\lambda^+)$ holds whenever $\xi < \eta < \lambda$. To this end, put

$$Z = \{\alpha < \lambda^+ : \mathrm{cf}(\alpha) = \lambda\},$$

and define the function $g_\xi: Z \to \lambda$ for any $\xi < \lambda^+$ as follows:

$$g_\xi(\alpha) = \sup\{\beta < \alpha : \exists \eta < \xi [\langle \beta \eta \rangle \in f(\langle \alpha \xi \rangle)]\},$$

where $\alpha \in Z$. As λ is regular, the assumptions on the set mapping f imply that the set after the sup sign here has cardinality $<\lambda$; hence it follows from $\mathrm{cf}(\alpha) = \lambda$ that $g_\xi(\alpha) < \alpha$. As Z is stationary in λ^+ (in fact, it contains the λth element of any club in λ^+), Neumer's theorem (Theorem 5.3) implies that there is an ordinal $\vartheta_\xi < \lambda^+$ such that the set

$$S_\xi = \{\alpha \in Z : g_\xi(\alpha) < \vartheta_\xi\}$$

has cardinality λ^+. Put

$$\vartheta = \sup_{\xi<\lambda} \vartheta_\xi .$$

We are going to define the set X in (3) such that $\alpha_\xi > \vartheta$. We do this by transfinite recursion: assuming that $\langle \alpha_\eta : \eta < \xi \rangle$ has already been constructed for some $\xi < \lambda$, choose an $\alpha_\xi \in S_\xi$ such that

$$\alpha_\xi > \max\left\{\vartheta, \sup_{\eta<\xi} \alpha_\eta\right\}$$

and such that

$$\langle\alpha_\xi\xi\rangle\notin\bigcup_{\eta<\xi}f(\langle\alpha_\eta\eta\rangle);\tag{4}$$

this is possible, as $|f(\langle\alpha_\eta\eta\rangle)\cap E_\xi|<\lambda$ holds according to our assumptions. It is easy to see that the set X in (3) so constructed is free with respect to f. In fact, if $\eta<\xi<\lambda$, then $\langle\alpha_\eta\eta\rangle\notin f(\langle\alpha_\xi\xi\rangle)$ holds since $\vartheta<\alpha_\eta<\alpha_\xi\in S_\xi$, and $\langle\alpha_\xi\xi\rangle\notin f(\langle\alpha_\eta\eta\rangle)$ holds in view of (4). The proof is complete.

PROOF OF THEOREM 20.2. Let $\langle\kappa_\nu\colon 1\le\nu<\mathrm{cf}(\kappa)\rangle$ be an increasing sequence of infinite cardinals tending to κ. We shall define a (not necessarily disjoint) partition $I=\langle I_\nu\colon\nu<\mathrm{cf}(\kappa)\rangle$ that verifies relation (2) by first defining set mappings

$$f_\nu\colon\kappa^+\to[\kappa^+]^{\le\kappa_\nu}$$

for each ν with $1\le\nu<\mathrm{cf}(\kappa)$ such that $f_\nu(\xi)\subseteq\xi$ holds for every $\xi<\kappa^+$, and then we shall put

$$I_\nu=\{\{\alpha\beta\}\colon\alpha<\beta<\kappa^+\ \&\ \alpha\in f_\nu(\beta)\}\tag{5}$$

for any ordinal ν with $1\le\nu<\mathrm{cf}(\kappa)$ and

$$I_0=[\kappa]^2\setminus\bigcup_{1\le\nu<\mathrm{cf}(\kappa)}I_\nu.\tag{6}$$

We are going to define the functions f_ν. Let $\xi<\kappa^+$, and assume that $f_\nu(\alpha)$ has already been defined for every $\alpha<\xi$ and every ν with $1\le\nu<\mathrm{cf}(\kappa)$. Using the assumption $2^\kappa=\kappa^+$, let $\langle A_\eta\colon\eta<\kappa^+\rangle$ be an enumeration of the elements of the set $[\kappa^+]^\kappa$, and consider the set

$$S^\xi=\{A_\eta\colon\eta<\xi\ \&\ A_\eta\subseteq\xi\}.$$

S^ξ has cardinality $\le\kappa$; so it can be represented as a disjoint union

$$S^\xi=\bigcup_{1\le\nu<\mathrm{cf}(\kappa)}S^\xi_\nu$$

such that S^ξ_ν has cardinality $\le\kappa_\nu$ for every ν with $1\le\nu<\mathrm{cf}(\kappa)$. Fix ν in this range arbitrarily. In view of the preceding lemma with $\lambda=\kappa_\nu^+$, there is a set X free with respect to $f_\nu\upharpoonright\xi$ such that

$$X\cap A\neq 0$$

for any $A\in S^\xi_\nu$ (note that $f_\nu\upharpoonright\xi$ has already been defined and is such that $|(f_\nu\upharpoonright\xi)(\alpha)|\le\kappa_\nu$ holds for any $\alpha<\xi$). We may assume that $|X|\le|S^\xi_\nu|\le\kappa_\nu$; put

$$f_\nu(\xi)=X.$$

This construction clearly ensures that (i) there are no three distinct ordinals $\alpha, \beta, \xi<\kappa^+$ such that $\{\alpha\beta\}$, $\{\alpha\xi\}$, and $\{\beta\xi\}$ belong to the same I_ν for some ν with $1\leq\nu<\mathrm{cf}(\kappa)$ (the partition I was defined in (5) and (6)). In fact, assuming e.g. $\alpha, \beta<\xi$, we can have $\alpha, \beta\in f_\nu(\xi)$ only if the set $\{\alpha\beta\}$ is free with respect to f_ν. Moreover, (ii) there is no bipartite subgraph of kind $(\kappa:\kappa^+)$ (cf. Subsection 8.8) of κ^+ that is included in I_0. To see this, observe that if $A_\eta\in[\kappa^+]^\kappa$, then $A_\eta\in S^\xi$ for every large enough $\xi<\kappa^+$ (in fact, for any $\xi>\max\{\eta, \sup A_\eta\}$). Given such a ξ, $A_\eta\in S^\xi_\nu$ holds for some ν (depending on ξ) with $1\leq\nu<\mathrm{cf}(\kappa)$; hence $A_\eta\cap f_\nu(\xi)\neq 0$, i.e., $\{\alpha\xi\}\in I_\nu$ for some $\alpha\in A_\eta$. Thus, for every large enough $\xi<\kappa^+$ there is an $\alpha\in A_\eta$ such that $\{\alpha\xi\}\notin I_0$. This verifies (ii). (i) and (ii) establish the desired result, completing the proof.

Noting that we have $(2^{\mathrm{cf}(\kappa)})^+\to((\mathrm{cf}(\kappa))^+)^2_{\mathrm{cf}(\kappa)}$ according to (17.32), and so, in particular, $(2^{\mathrm{cf}(\kappa)})^+\to(3)^2_{\mathrm{cf}(\kappa)}$, and we can conclude from (2) with the aid of the Transitivity Rule (see Subsection 9.8) that

$$\kappa^+\not\to\left(\binom{\kappa^+}{\kappa}, (2^{\mathrm{cf}(\kappa)})^+\right)^2 \tag{7}$$

holds, provided κ is singular and $2^\kappa=\kappa^+$. The strength of this result consists in that it claims the nonexistence of a homogeneous bipartite graph, since if we are not interested in bipartite graphs, then we have

$$\kappa^{\mathrm{cf}(\kappa)}\not\to(\kappa^+, (\mathrm{cf}(\kappa))^+)^2 \tag{8}$$

for any infinite κ according to (19.12), without assuming anything about cardinal exponentiation. If we assume $2^\kappa=\kappa^+$, then one can show that

$$\kappa^+\not\to\left(\binom{\kappa^+}{\kappa^+}, (\mathrm{cf}(\kappa))^+\right)^2, \tag{9}$$

which strengthens (8); an even stronger version of this relation can be expressed with the aid of the square-bracket symbol; that version and its proof will be given below in Section 49 (cf. (49.4)).

21. ADDITION OF NEGATIVE PARTITION RELATIONS FOR $r=2$

If we are given a sequence of pairwise disjoint sets $\langle A_\xi:\xi<\tau\rangle$ and a coloring f_ξ of (unordered) pairs on each of the A_ξ's, as well as a coloring f of pairs of the index set τ, then there is an obvious way to define a coloring f' of pairs of $\bigcup_{\xi<\tau} A_\xi$ by putting $f'(\{xy\})=f_\xi(\{xy\})$ if $x, y\in A_\xi$ and $x\neq y$, and $f'(\{xy\})=f(\{\xi\eta\})$ if $x\in A_\xi$, $y\in A_\eta$, and $\xi\neq\eta$. If f and the f_ξ's each verify a negative partition relation, then f' will also verify one. With the aid of such considerations we can derive the following

THEOREM 21.1. *Assume*

$$\kappa_\xi \nrightarrow (\lambda_{\xi\nu})^2_{\nu<\gamma} \tag{1}$$

for $\xi<\tau$, *and*

$$\tau \nrightarrow (\rho_\nu)^2_{\nu<\gamma}. \tag{2}$$

Let ζ_ξ *run over cardinals and put*

$$\sigma_\nu = \sup \{(\sum_{\xi\in Y} \zeta_\xi)^+ : Y\subseteq\tau \;\&\; |Y|<\rho_\nu \;\&\; \forall\xi\in Y[\zeta_\xi<\lambda_{\xi\nu}]\}$$

for every $\nu<\gamma$. *Then*

$$\sum_{\xi<\alpha} \kappa_\xi \nrightarrow (\sigma_\nu)^2_{\nu<\gamma}. \tag{3}$$

If $\gamma=2$ *and the relations in* (1) *and* (2) *are established by ordered sets, then* (3) *is also established by an ordered set.*

PROOF. Let A_ξ, $\xi<\tau$, be pairwise disjoint sets such that $|A_\xi|=\kappa_\xi$, and assume that the colorings $f_\xi:[A_\xi]^2\to\gamma$ verify the relations in (1), and $f:[\tau]^2\to\gamma$ verifies the one in (2). Write $A=\bigcup_{\xi<\tau} A_\xi$ and, as described above, define $f':[A]^2\to\gamma$ by putting $f'(\{xy\})=f_\xi(\{xy\})$ if $x\neq y$ and $x,y\in A_\xi$ for some $\xi<\tau$, and $f'(\{xy\})= =f(\{\xi\eta\})$ if $x\in A_\xi$ and $y\in A_\eta$ for some $\xi,\eta<\tau$ with $\xi\neq\eta$. If $X\subseteq A$ is a homogeneous set of color ν with respect to f' for some $\nu<\gamma$, then the set

$$Y=\{\xi: X\cap A_\xi\neq 0\}$$

has cardinality $<\rho_\nu$, since Y is a homogeneous set of color ν with respect to f, and f verifies (2). Moreover, for every $\xi\in Y$ the set $X\cap A_\xi$ has cardinality $<\lambda_{\xi\nu}$, since $X\cap A_\xi$ is a homogeneous set of color ν with respect to f_ξ, and f_ξ verifies (1). Hence X has cardinality $<\sigma_\nu$, which completes the proof of (3).

Assume now that $\gamma=2$, the ordered sets $\langle A_\xi, \prec_\xi\rangle$ establish the relations in (1), and $\langle\tau,\prec\rangle$ establishes (2). Assuming again that the A_ξ's are pairwise disjoint, put $A=\bigcup_{\xi<\tau} A_\xi$, and define the ordering $\prec'$ of A by putting $x\prec y$ for any two distinct $x,y\in A$ if either $x,y\in A_\xi$ and $x\prec_\xi y$ for some ξ or $x\in A_\xi$, $y\in A_\eta$ and $\xi\prec\eta$ for some $\xi,\eta<\tau$. We claim that $\sigma_0, \sigma_1^*\not\leq \operatorname{tp}\langle A,\prec'\rangle$. Assuming e.g. that $X\subseteq A$ is a wellordered set in the ordering $\prec'$, we can easily show that $\operatorname{tp}\langle X,\prec'\rangle<\sigma_0$ i.e., that $|X|<\sigma_0$. Indeed, putting

$$Y=\{\xi: X\cap A_\xi\neq 0\}$$

again, we have $|Y|<\rho_0$ and $|X\cap A_\xi|<\lambda_{\xi 0}$ for every $\xi<\tau$, just as before; these inequalities imply $|X|<\sigma_0$. $\sigma_1^*\not\leq \mathrm{tp}\langle A, \prec'\rangle$ can be shown similarly. The proof is complete.

We are now turning to a discussion of the corollaries of the above result.

Corollary 21.2. *Assume $\kappa\geq\omega$ is a cardinal such that*

$$\mathrm{cf}(\kappa)\not\rightarrow(\mathrm{cf}(\kappa), (\lambda_\nu)_{\nu<\gamma})^2 . \tag{4}$$

Then

$$\kappa\not\rightarrow(\kappa, (\lambda_\nu)_{\nu<\gamma})^2 . \tag{5}$$

If $\gamma=1$ and $\mathrm{cf}(\kappa)\not\rightarrow(\mathrm{cf}(\kappa), \lambda_0)^2$ is established by an ordered set, then $\kappa\not\rightarrow(\kappa, \lambda_0)^2$ is also established by an ordered set.

Proof. The case when κ is regular is trivial and uninteresting. So assume that κ is singular, and represent κ as $\kappa=\sum_{\xi<\mathrm{cf}(\kappa)}\kappa_\xi$, where $\kappa_\xi<\kappa$ for every $\xi<\mathrm{cf}(\kappa)$. Observe that we have

$$\kappa_\xi\not\rightarrow(\kappa_\xi^+, (2)_\gamma)^2 \tag{6}$$

trivially. Applying the theorem just proved with $1\dotplus\gamma$ replacing γ, and with $\tau=\mathrm{cf}(\kappa)$, $\lambda_{\xi 0}=\kappa_\xi^+$, $\lambda_{\xi,1\dotplus\nu}=2$, $\rho=\mathrm{cf}(\kappa)$, and $\rho_{1\dotplus\nu}=\lambda_\nu$ $(\xi<\mathrm{cf}(\kappa), \nu<\kappa)$, we obtain (5).

We are now going to establish the assertion given in the last sentence of the corollary to be proved. Note that the assumption that the relation $\mathrm{cf}(\kappa)\not\rightarrow(\mathrm{cf}(\kappa), \lambda_0)^2$ is established by an ordered set means, in particular, that $\lambda_0^*\not\leq\mathrm{tp}\langle\mathrm{cf}(\kappa), \prec\rangle$ for some ordering $\prec$ of $\mathrm{cf}(\kappa)$; hence $\lambda_0\geq\omega$. So, noting that the relation

$$\kappa_\xi\not\rightarrow(\kappa_\xi^+, \omega)^2$$

is established by an ordered set (in fact, any wellordering of κ_ξ will do), we obtain the desired result by using this relation instead of (6). The proof is complete.

Corollary 21.3. *Assume $\kappa\geq\omega$. Then*

$$\kappa\not\rightarrow(\kappa, \mathrm{cf}(\kappa)^+)^2 \tag{7}$$

is established by an ordered set.

Proof. The ordered set $\langle\mathrm{cf}(\kappa), >\rangle$, where $>$ is the reverse of the 'less than' relation on ordinals, establishes $\mathrm{cf}(\kappa)\not\rightarrow(\omega, \mathrm{cf}(\kappa)^+)^2$ and so, a fortiori, $\mathrm{cf}(\kappa)\not\rightarrow(\mathrm{cf}(\kappa), \mathrm{cf}(\kappa)^+)^2$; hence the result follows from the preceding corollary.

Corollary 21.4. *If $\kappa\geq\omega$ and*

$$\mathrm{cf}(\kappa)\not\rightarrow(\mathrm{cf}(\kappa), \mathrm{cf}(\kappa), (\lambda_\nu)_{\nu<\gamma})^2 , \tag{8}$$

then

$$2^\kappa \nrightarrow (\kappa, \kappa, (\lambda_\nu)_{\nu<\gamma})^2 . \tag{9}$$

Furthermore, for any $\kappa \geq \omega$, *we have*

$$2^\kappa \nrightarrow (\kappa, \kappa, (3)_{L_3(\mathrm{cf}(\kappa))})^2 , \tag{10}$$

and the relation

$$2^\kappa \nrightarrow (\kappa^+, \kappa)^2 \tag{11}$$

is established by an ordered set.

Proof. The case $\kappa=\omega$ is trivial, e.g. (10) becomes $\omega \nrightarrow (\omega, \omega, (3)_\omega)^2$ in this case, and we trivially have $\omega \nrightarrow (3)^2_\omega$ (color every pair with a different color). So we may assume $\kappa>\omega$. We distinguish two cases: a) there is a $\rho<\kappa$ such that $2^\kappa=2^\rho$, and b) we have $2^\rho<2^\kappa$ for any $\rho<\kappa$.

Ad a). This case contains nothing new, since according to (19.14), the relation $2^\rho \nrightarrow (\rho^+, \rho^+)^2$ is established by an ordered set.

Ad b). In this case, Theorem 6.10c and d imply that κ is a limit cardinal, $\mathrm{cf}(2^\kappa)=\mathrm{cf}(\kappa)$, and there is a strictly increasing sequence $\langle \kappa_\xi : \xi<\mathrm{cf}(\kappa)\rangle$ of infinite cardinals $<\kappa$ such that $2^\kappa = \sum_{\xi<\mathrm{cf}(\kappa)} 2^{\kappa_\xi}$.

To prove (9), note that

$$2^{\kappa_\xi} \nrightarrow (\kappa_\xi^+, \kappa_\xi^+, (2)_\gamma)^2$$

holds by (19.14) for any $\xi<\mathrm{cf}(\kappa)$, and so the assertion follows from (8) by Theorem 21.1 with 2^{κ_ξ} replacing κ_ξ and $\mathrm{cf}(\kappa)$ replacing τ. To establish (10), observe that $\mathrm{cf}(\kappa) \nrightarrow (3)^2_{L_3(\mathrm{cf}(\kappa))}$ holds according to (19.18); hence we have, a fortiori

$$\mathrm{cf}(\kappa) \nrightarrow (\mathrm{cf}(\kappa), \mathrm{cf}(\kappa), (3)_{L_3(\mathrm{cf}(\kappa))})^2 .$$

Thus (8) holds if we substitute $\lambda_\nu=3$ and $\gamma=L_3(\mathrm{cf}(\kappa))$ there; making the same substitutions in (9), we obtain (10). Finally, to show (11), note that $\mathrm{cf}(\kappa) \nrightarrow (\mathrm{cf}(\kappa)^+, \omega)^2$ is established by the ordered set $\langle \mathrm{cf}(\kappa), <\rangle$, where $<$ is the 'less than' relation between ordinals; as $2^{\kappa_\xi} \nrightarrow (\kappa_\xi^+, \kappa_\xi^+)^2$ is also established by an ordered set (cf. (19.14)), the desired result follows from the last sencence of Theorem 21.1 with 2^{κ_ξ} replacing κ_ξ.

We now take a closer look at the range of validity of the relation $2^\kappa \nrightarrow (\kappa, \kappa)^2$.

Corollary 21.5. *Assume* $\kappa \geq \omega$. *Then any one of the conditions*

(i) $2^\kappa=2^\rho$ *for some* $\rho<\kappa$,

(ii) $2^{\underset{\cdot}{\kappa}} = \kappa > \mathrm{cf}(\kappa)$,

and

(iii) $\mathrm{cf}(\kappa) \not\rightarrow (\mathrm{cf}(\kappa), \mathrm{cf}(\kappa))^2$

ensures that

$$2^{\underset{\cdot}{\kappa}} \not\rightarrow (\kappa, \kappa)^2 \tag{12}$$

holds.

Proof. The result follows from (19.14), Corollary 21.3, or (9) with $\gamma \doteq 0$ according as (i), (ii), or (iii) holds.

If $2^{\underset{\cdot}{\kappa}} = \kappa$ and κ is regular, then the problem whether (12) holds is a question concerning large cardinals and will be discussed later. If $2^{\underset{\cdot}{\kappa}}$ is singular, then the above corollary gives a necessary and sufficient condition for (12) to hold. Namely, a theorem of Shelah (see Corollary 27.4 below) says that if $2^{\underset{\cdot}{\kappa}} > \kappa$ and κ is singular, $2^\rho < 2^{\underset{\cdot}{\kappa}}$ for every $\rho < \kappa$, and $\mathrm{cf}(\kappa) \rightarrow (\mathrm{cf}(\kappa))^2_2$, then $2^{\underset{\cdot}{\kappa}} \rightarrow (\kappa)^2_2$.

Next we turn to the same type of problem involving partitions of length three.

Corollary 21.6. *Assume* $\kappa \geq \omega$. *Then*

$$2^{\underset{\cdot}{\kappa}} \not\rightarrow (\kappa, \kappa, \mathrm{cf}(\kappa)^+)^2 , \tag{13}$$

and if $\kappa \not\rightarrow (\kappa, \kappa)^2$ *then*

$$2^{\underset{\cdot}{\kappa}} \not\rightarrow (\kappa, \kappa, \kappa)^2 . \tag{14}$$

Proof. (13) follows from (9) and the trivial relation $\mathrm{cf}(\kappa) \not\rightarrow (\mathrm{cf}(\kappa), \mathrm{cf}(\kappa), \mathrm{cf}(\kappa)^+)^2$. If $\mathrm{cf}(\kappa) < \kappa$ then (14) follows from (13), and if $\mathrm{cf}(\kappa) = \kappa$, then (14) follows from case (iii) of the preceding corollary.

22. ADDITION OF NEGATIVE PARTITION RELATIONS FOR $r \geq 3$

If we are given pairwise disjoint sets A_ξ, $\xi < \tau$, and colorings $f: [\tau]^3 \rightarrow \gamma$ and $f_\xi: [A_\xi]^3 \rightarrow \gamma$, then we can again define a coloring f' on the triples of the set $A = \bigcup_{\xi < \tau} A_\xi$ in a way analogous to what we did in the preceding section. An additional problem is what color to give to triples $\{xyz\} \in [A]^3$ such that $x, y \in A_\xi$ and $z \in A_\eta$ for some $\xi, \eta < \tau$ with $\xi \neq \eta$. If we decide to give them a new color, say γ, then the advantage of this is that a homogeneous set of color ν with $\nu < \tau$ is either included in a single A_ξ or is a transversal of the sequence $\langle A_\xi : \xi < \tau \rangle$. This means that we now have much better estimates for the sizes of homogeneous sets of color $< \gamma$, while we have an additional worry to keep down the size of homogeneous sets of the new color. Usually, the new color will be lumped together with some of the old ones. Our main result based on these considerations is

THEOREM 22.1. *Let $r \geq 3$ be an integer. Assume that*

$$\kappa_\xi \nrightarrow (\lambda_\nu)^r_{\nu<\gamma} \tag{1}$$

for any $\xi < \tau$ and

$$\tau \nrightarrow (\lambda'_0, (\lambda_\nu)_{1 \leq \nu < \gamma})^r, \tag{2}$$

where $\gamma \geq 2$, $\lambda'_0 \geq r+1$, and $\lambda_\nu \geq r-1$ for any $\nu < \gamma$. Put $\kappa = \sum_{\xi<\tau} \kappa_\xi$. Then

$$\kappa \nrightarrow (\lambda_0 + \lambda'_0 - 2, (\lambda_\nu)_{1 \leq \nu < \gamma})^r \tag{3}$$

provided $\lambda_1 \geq \omega$.

Assume $\lambda'_0 = \lambda_0$ from now on, i.e., assume

$$\tau \nrightarrow (\lambda_\nu)^r_{\nu<\gamma} \tag{4}$$

instead of (2). *Then*

$$\kappa \nrightarrow (\lambda_\nu)^r_{\nu<\gamma} \tag{5}$$

holds if either $\lambda_0, \lambda_1 \geq \omega$ or $r \geq 4$ and $\lambda_0 \geq \omega$, and we have

$$\kappa \nrightarrow ((\lambda_\nu)_{\nu<\gamma}, r+1)^r \tag{6}$$

provided $\lambda_0 \geq \omega$.

PROOF. Let A_ξ, $\xi < \tau$, be pairwise disjoint sets such that $|A_\xi| = \kappa_\xi$, and let $A = \bigcup_{\xi<\tau} A_\xi$. Consider the following subsets of $[A]^r$:

$$K_0 = \{u \in [A]^r : \forall \xi < \tau [|A_\xi \cap u| \leq 1]\},$$

$$K_1 = \{u \in [A]^r : \exists \xi < \tau [u \subseteq A_\xi]\},$$

$$K_2 = \{u \in [A]^r : \exists \xi, \eta < \tau [\xi < \eta \,\&\, |A_\xi \cap u| = 2 \,\&\, |A_\eta \cap u| = r-2]\},$$

and

$$K_3 = [A]^r \setminus (K_0 \cup K_1 \cup K_2).$$

We are going to establish the following simple

CLAIM. *Let $X \subseteq A$ and $|X| \geq r$.* a) *If $[X]^r \subseteq K_0 \cup K_1$, then either $[X]^r \subseteq K_0$ or $[X]^r \subseteq K_1$.* b) *Assume $r=3$ and $[X]^3 \subseteq K_0 \cup K_1 \cup K_2$. Then there is a $\xi < \tau$ such that $|X \cap A_\eta| \leq 1$ for all $\eta < \tau$ with $\eta \neq \xi$.* c) *Assume $r \geq 4$ and $[X]^r \subseteq K_0 \cup K_1 \cup K_2$. Then either $[X]^r \subseteq K_0$ or $[X]^r \subseteq K_1$ or $|X| = r$.* d) *If $[X]^r \subseteq K_2$, then $|X| = r$.* e) *Assume $[X]^r \subseteq K_0 \cup K_1 \cup K_3$. Then there is a $\xi < \tau$ such that $|X \cap A_\eta| \leq r-2$ for all $\eta < \tau$ with $\eta \neq \xi$.*

PROOF OF CLAIM. *Ad* a). Supposing $[X]^r \nsubseteq K_0$, there are a $\xi < \tau$ and distinct $x, y \in A_\xi \cap X$. Supposing further, that $[X]^r \nsubseteq K_1$, we have $X \nsubseteq A_\xi$; pick a

$z \in X \setminus A_\xi$. Then, noting that we assumed $r \geq 3$ in the theorem, an r-element subset $\{x, y, z, \ldots\}$ of X belongs neither to K_0 nor to K_1, which is a contradiction.

Ad b). Assume, on the contrary, that x, y, z, and t are distinct elements of X with $x, y \in A_\xi$ and $z, t \in A_\eta$, where $\xi < \eta$. Then $\{xzt\} \in K_3$, which is again a contradiction.

Ad c). Assume that $u \in [X]^r$, $u \in K_2$, and $x \in X \setminus u$. Let $\xi, \eta, \vartheta < \tau$ be such that $\xi < \eta$, $u \cap A_\xi \neq 0$, $u \cap A_\eta \neq 0$, and $x \in A_\vartheta$. Let s and t be such that $s \in u \cap A_\xi$ and $t \in u \cap A_\eta$. Then we have $(u \cup \{x\}) \setminus \{s\} \in [X]^r \cap K_3$ or $(u \cup \{x\}) \setminus \{t\} \in [X]^r \cap K_3$ according as $\vartheta \neq \xi$ or $\vartheta \neq \eta$ (at least one of these holds since $\xi < \eta$); the assumption $r \geq 4$ is used in the second case, since in case $r = 3$ and $\vartheta = \xi$ we would obviously have $(u \cup \{x\}) \setminus \{t\} \in K_1$. We again obtained a contradiction.

Ad d). The assertion follows from c) if $r \geq 4$, so assume $r = 3$. Let $x_i \in A_{\xi i}$, $i < 4$, be distinct elements of X, where $\xi_0 \leq \xi_1 \leq \xi_2 \leq \xi_3$. As $\{x_0 x_1 x_2\} \in K_2$ and $\{x_0 x_1 x_3\} \in K_2$, we have $\xi_0 = \xi_1 < \xi_2, \xi_3$. Hence $\{x_0 x_2 x_3\} \notin K_2$, a contradiction.

Ad e). Assume, on the contrary, that ξ and η are such that $\xi < \eta < \tau$ and $|X \cap A_\xi|$, $|X \cap A_\eta| \geq r - 1$. Noting that we assumed $r \geq 3$ in the theorem, let $u \subseteq X \cap A_\xi$ be a set containing two elements, and $v \subseteq X \cap A_\eta$, a set of $r - 2$ elements. Then $u \cup v \in [X]^r \cap K_2$, which is a contradiction. The proof of our Claim is complete.

We now return to the proof of the theorem. Let $f_\xi \colon [A_\xi]^r \to \gamma$ and $f \colon [\tau]^r \to \gamma$ be colorings that verify the relations in (1) and (2). Define the colorings $g \colon [A]^r \to \gamma$ and $g' \colon [A]^r \to \gamma \dot{+} 1$ as follows: given any $u \in [A]^r$, put

$$g(u) = g'(u) = f(\{\xi < \tau \colon A_\xi \cap u \neq 0\})$$

if $u \in K_0$,

$$g(u) = g'(u) = f_\xi(u)$$

if $u \in K_1$, where $\xi < \tau$ is the unique ordinal for which $u \subseteq A_\xi$,

$$g(u) = 0 \qquad \text{and} \qquad g'(u) = \gamma$$

if $u \in K_2$ and, finally,

$$g(u) = g'(u) = 1$$

if $u \in K_3$.

We claim that g verifies (3); and, moreover, g verifies (5) and g' verifies (6), if we exchange the roles of λ_0 and λ_1 in the assumptions (i.e., g verifies (5) if either $\lambda_0, \lambda_1 \geq \omega$ or $r \geq 4$ and $\lambda_1 \geq \omega$, and g' verifies (6) if $\lambda_1 \geq \omega$. These exchanges of roles are necessary only in order to enable us to treat these relations in a uniform way).

To show this, let X first be a homogeneous set of color 0 with respect to g. We claim that, assuming (1) and (2) (and not (4)), $|X| < \lambda_0 + \lambda_0' - 2$ if $r = 3$ and $|X| < \max\{\lambda_0, \lambda_0'\}$ if $r > 3$. To see this, observe that we have

$$[X]^r \subseteq K_0 \cup K_1 \cup K_2,$$

and we may of course assume that $|X|>r$. Consider first the case $r=3$. Then Claim b) implies that there is a $\xi<\tau$ such that

$$|X \cap A_\eta| \leq 1 \tag{7}$$

for every $\eta<\tau$ with $\eta \neq \xi$, and here we may obviously assume that

$$|X \cap A_\xi| \neq 0 \,. \tag{8}$$

$X \cap A_\xi$ is homogeneous of color 0 with respect to f_ξ, and so we have

$$|X \cap A_\xi| \leq \lambda_0 - 1 \,. \tag{9}$$

The set $\{\eta<\tau\colon X \cap A_\eta \neq 0\}$ is homogeneous of color 0 with respect to f: hence its cardinality is $<\lambda_0'$. Thus, noting that ξ also belongs to this set by (8), we obtain by (7) that

$$|X \setminus A_\xi| = |\{\eta<\tau\colon X \cap A_\eta \neq 0 \,\&\, \eta \neq \xi\}| < \lambda_0' - 1 \,.$$

Combining this with (9), we can conclude that $|X|<\lambda_0+\lambda_0'-2$, which we wanted to show.

Consider now the case $r \geq 4$. Then either $[X]^r \subseteq K_0$ or $[X]^r \subseteq K_1$ holds according to Claim c). In the first case $|X|<\lambda_0'$, as the set $\{\xi<\tau\colon |X \cap A_\xi| \neq 0\}$ is a homogeneous set of color 0 with respect to f, and in the second case $|X|<\lambda_0$, as X is homogeneous with respect to f_ξ, where $\xi<\tau$ is the unique ordinal for which $X \subseteq A_\xi$. Hence $|X|<\max\{\lambda_0, \lambda_0'\}$. So far we have proved that g verifies the partition relation

$$\kappa \nrightarrow (\lambda_0+\lambda_0'-2, \ldots)^r \tag{10}$$

if $r=3$ (and also if $r>3$, see below), and

$$\kappa \nrightarrow (\max\{\lambda_0, \lambda_0'\}, \ldots)^r \tag{11}$$

if $r \geq 4$ (note that (10) is valid also in case $r>3$ by (11), since $\lambda_0+\lambda_0'-2 \geq \max\{\lambda_0, \lambda_0\}$ as we assumed $\lambda_0, \lambda_0' \geq r+1>2$).

Let now X be a homogeneous set of color 1 with respect to g. Then $[X]^r \subseteq K_0 \cup K_1 \cup K_3$. Claim e) implies that there is a $\xi<\tau$ such that

$$|X \cap A_\eta| \leq r-2 \tag{12}$$

for every $\eta<\tau$ with $\eta \neq \xi$. $X \cap A_\xi$ is a homogeneous set of color 1 with respect to f_ξ, and so

$$|X \cap A_\xi| < \lambda_1 \,, \tag{13}$$

since f_ξ verifies (1). Write

$$H = \{\eta<\tau\colon \eta \neq \xi \,\&\, A_\eta \cap X \neq 0\} \,;$$

then H is a homogeneous set of color 1 with respect to f. Therefore (2) (or (4)) implies that $|H| < \lambda_1$, and so we obtain by using (12) that $|X \setminus A_\xi| \leq (r-2) \cdot |H| < \lambda_1$, as λ_1 was assumed to be infinite in all cases (note that we changed the original assumptions from $\lambda_0 \geq \omega$ to $\lambda_1 \geq \omega$). This together with (13) implies that $|X| < \lambda_1$. So we have shown that g verifies the relation

$$\kappa \nrightarrow (\cdot, \lambda_1, \ldots)^r; \tag{14}$$

note that this relation depends on the assumption $\lambda_1 \geq \omega$.

Let now $v \geq 2$ and assume that X is a homogeneous set of color v with respect to g. Then $[X]^r \subseteq K_0 \cup K_1$, and so Claim a) implies that either $[X]^r \subseteq K_0$ or $[X]^r \subseteq K_1$. We have $|X| < \lambda_v$ in both cases. In fact, in the first case $\{\xi < \tau: A_\xi \cap X\} \neq 0$ is a homogeneous set of color v with respect to f, and in the second case $X \subseteq A_\xi$ for some $\xi < \tau$, and then X is a homogeneous set of color v with respect to f_v. So we have just shown that g verifies the relation

$$\kappa \nrightarrow (\cdot, \cdot, (\lambda_v)_{2 \leq v < \gamma})^r. \tag{15}$$

Putting (10), (14), and (15) together, we obtain (3). If $\lambda_0' = \lambda_0 > \omega$ and $\lambda_1 > \omega$, then (5) follows from (3) since $\lambda_0 + \lambda_0' - 2 = \lambda_0$ holds in that case. If $r \geq 4$ and $\lambda_1 \geq \omega$ (note the exchange of roles of λ_0 and λ_1 in the assumption), then (11), (14), and (15) give (5).

We are now going to show that g' verifies (6) if we assume $\lambda_1 \geq \omega$ instead of $\lambda_0 \geq \omega$. Suppose $\lambda_0 = \lambda_0'$. It follows from our considerations above that g' verifies the relation

$$\kappa \nrightarrow ((\lambda_v)_{v < \gamma}, \cdot)^r. \tag{16}$$

In fact, since for any $u \in [A]^r$ and for any v with $1 \leq v < \gamma$ we have $g'(u) = v$ exactly if $g(u) = v$, we have the same bounds for the sizes of homogeneous sets of color v with respect to g' and g provided $1 \leq v < \gamma$. As for the case $v = 0$, this is now analogous to the case $2 \leq v < \gamma$, and the same argument gives that a set homogeneous of color 0 with respect to g' has cardinality $< \lambda_0 = \lambda_0'$.

Let X be a homogeneous set of color γ with respect to g'. Then $[X]^3 \subseteq K_2$, and so Claim d) implies $|X| \leq r$. This together with (16) means that g' verifies (6). The proof is complete.

We do not list corollaries here, just mention one application and a related problem. We shall later show that

$$2^{\aleph_n} \nrightarrow (4, \aleph_n)^3$$

holds for all $n < \omega$ (cf. (24.6)); hence we have

$$2^{\aleph_n} \nrightarrow (4, \aleph_\omega)^3 \tag{17}$$

a fortiori. Note that

$$\aleph_0 \nrightarrow (4, \aleph_\omega)^3 \tag{18}$$

holds trivially. Using the above theorem with (17) and (18) replacing (1) and (2), we obtain that

$$2^{\aleph_\omega} \nrightarrow (5, \aleph_\omega)^3 \tag{19}$$

according to (3).

Problem 22.2. Does the relation

$$2^{\aleph_\omega} \nrightarrow (4, \aleph_\omega)^3 \tag{20}$$

hold?

This is certainly true if $2^{\aleph_\omega} = 2^{\aleph_n}$ for some $n < \omega$ or $2^{\aleph_\omega} = \aleph_\omega$ (see Corollaries 25.2 and 25.3 below).

23. MULTIPLICATION OF NEGATIVE PARTITION RELATIONS IN CASE $r \geq 3$

Lemma 16.1, the Stepping-up Lemma, gives us a method to obtain positive relations for $r \geq 3$. This is the only method that we know for proving positive relations in the case $r \geq 3$, except if the cardinality of the set to be partitioned is countable or a large inaccessible cardinal (partition relations for large cardinals will be discussed in Chapter VII, especially in Sections 29 and 34). One would like therefore to show that the converse of Lemma 16.1 is true, i.e., that (16.2) there implies (16.1), possibly under some reasonable restrictions. This would mean, say, that the relation $\kappa \nrightarrow (\lambda_\xi)_{\xi<\tau}^{r-1}$ implies the relation $(2^\kappa)^+ \nrightarrow (\lambda_\xi + 1)_{\xi<\tau}^r$ if $\kappa \geq \omega$ (note that $+$ means cardinal addition; here we are concerned only with relations involving cardinals). This implication is, however, not provable, as is shown by the following example: Assume that $2^{\aleph_0} = \aleph_2$ and $2^{\aleph_1} = 2^{\aleph_2} = \aleph_3$. Then we have $(2^{2^{\aleph_0}})^+ = \aleph_4 \to (\aleph_1, \aleph_1)^3$ according to (17.32), i.e., $(2^{\aleph_3})^+ = \aleph_4 \to (\aleph_1, \aleph_1)^3$, while $\aleph_2 = 2^{\aleph_0} \nrightarrow (\aleph_1, \aleph_1)^2$ according to (19.14).

The correct formulation is suggested by (19.14). This says that $2^\lambda \nrightarrow (\lambda^+, \lambda^+)^2$ for $\lambda \geq \omega$, and this is derived from the trivial relation $\lambda \nrightarrow (\lambda^+, \lambda^+)^1$. Analogously, we can expect that if $\kappa \geq \omega$ and $\kappa \nrightarrow (\lambda_\xi)_{\xi<\tau}^{r-1}$, then $2^\kappa \nrightarrow (\lambda_\xi + 1)_{\xi<\tau}^r$. We shall indeed be able to show this under fairly general circumstances, though we shall not be able to give a completely unified treatment; we shall, however, have no difficulty if, say, at least two of the λ_ξ's are infinite and one of these is regular. In the present section we prove some preliminary technical lemmas, and the main results will come in the next section. Our preparation for these results here will consist in defining a partition of $[{}^\kappa 2]^r$ in a canonical way if we are given a partition of $[\kappa]^{r-1}$, and then investigating the properties of this new partition. To this end, we shall first consider the set ${}^\kappa 2 = \{f\colon f \text{ is a function } \& \operatorname{dom}(f) = \kappa \;\&\; \operatorname{ra}(f) \subseteq 2\}$.

Let $\kappa > 0$ be a cardinal and $r \geq 3$ an integer fixed throughout the discussion below.

DEFINITION 23.1. (i) Let $\prec$ be the lexicographic ordering of ${}^{\kappa}2$, i.e., if $<_2$ denotes the usual ordering of the set $2 = \{0, 1\}$ then put

$$\prec = \text{lex} \langle <_2 : \xi < \kappa \rangle.$$

Let, further, $<^*$ be a fixed wellordering of ${}^{\kappa}2$.

(ii) For any $x, y \in {}^{\kappa}2$ with $x <^* y$ write $\eta(x, y) = 0$ if $x \prec y$ and $\eta(x, y) = 1$ if $y \prec x$. Put $\{xy\} \in T_i$ if $\eta(x, y) = i$ $(i = 0, 1)$. $\langle T_0, T_1 \rangle$ is a **Sierpiński** partition of $\langle {}^{\kappa}2, \prec \rangle$ which establishes the relation $2^{\kappa} \nrightarrow (\kappa^+, \kappa^+)^2$ provided $\kappa \geq \omega$, according to (19.14).

(iii) Given a sequence of elements $x_0 <^* x_1 <^* \ldots <^* x_{r-1}$ of ${}^{\kappa}2$, put

$$\eta(\{x_0, \ldots, x_{r-1}\}) = \langle \eta(x_0, x_1), \ldots, \eta(x_{r-2}, x_{r-1}) \rangle. \tag{1}$$

If $1 \leq s \leq r-1$ and $k_0, \ldots, k_{s-1}$ are integers < 2, then write

$$K(k_0, \ldots, k_{s-1}) = \{u \in [{}^{\kappa}2]^r : \eta(u) \restriction s = \langle k_0, \ldots, k_{s-1} \rangle\}. \tag{2}$$

E.g. $K(0, 1, 0)$ is defined in case $r \geq 4$ and is the set of all $\{xyvz \ldots\} \in [{}^{\kappa}2]^r$ such that $x <^* y <^* v <^* z <^* \ldots$ and $x \prec y \succ v \prec z$. We put

$$K_i = K(i, \ldots, i) \tag{3}$$

with $i = 0$ or 1, where the number of arguments is $r-1$ (i.e., the largest possible), and

$$K = K_0 \cup K_1. \tag{4}$$

(iv) For any sequence of elements $x_0 <^* x_1 <^* \ldots <^* x_{r-1}$ of ${}^{\kappa}2$ we write

$$\delta(\{x_0, \ldots, x_{r-1}\}) = \langle x_0 \cap x_1, \ldots, x_{r-2} \cap x_{r-1} \rangle; \tag{5}$$

the operation $x \cap y$, the first discrepancy of x and y, was described in Definition 19.1, and it assigns to the functions x and y the least ordinal at which their values differ. Clearly, $x \cap y < \kappa$ for $x \neq y$.

(v) For any two distinct ordinals δ_1 and δ_2, write $\zeta(\delta_0, \delta_1) = 0$ if $\delta_0 < \delta_1$ and $\zeta(\delta_0, \delta_1) = 1$ if $\delta_1 < \delta_0$. If $\langle \delta_0, \ldots, \delta_{r-2} \rangle$ is a sequence of ordinals such that

$$\delta_i \neq \delta_{i+1} \tag{6}$$

for any $i < r-2$, then put

$$\zeta(\langle \delta_0, \ldots, \delta_{r-2} \rangle) = \langle \zeta(\delta_0, \delta_1), \ldots, \zeta(\delta_{r-3}, \delta_{r-2}) \rangle. \tag{7}$$

(vi) For any integer s with $1 \leq s \leq r-2$ and any sequence $\langle k_0, \ldots, k_{s-1} \rangle$ of zeros and ones, write

$$P(k_0, \ldots, k_{s-1}) = \{u \in K : \zeta(\delta(u)) \restriction s = \langle k_0, \ldots, k_{s-1} \rangle\}. \tag{8}$$

Here K was defined in (4). Note that $\zeta(\delta(u))$ is always defined, i.e., the inequalities analogous to those in (6) are satisfied. For, given any $x_0, x_1, x_2 \in {}^\kappa 2$ with either $x_0 \prec x_1 \prec x_2$ or $x_0 \succ x_1 \succ x_2$, we have

$$x_0 \cap x_1 \neq x_1 \cap x_2 . \tag{9}$$

In fact, assuming $\xi = x_0 \cap x_1 = x_1 \cap x_2$, the inequalities $x_0 \prec x_1 \prec x_2$ would mean that $x_0(\xi) < x_1(\xi) < x_2(\xi)$, which is impossible since $x_i(\xi) = 0$ or 1 ($i = 0, 1, 2$); $x_0 \succ x_1 \succ x_2$ is equally impossible.

To illustrate the symbol just introduced, $P(0, 1, 0)$ is defined for $r \geq 5$, and it denotes the set of all $\{x_0, \ldots, x_{r-1}\} \subseteq {}^\kappa 2$ with $x_0 <^* x_1 <^* \ldots <^* x_{r-1}$ such that (a) either $x_0 \prec \ldots \prec x_{r-1}$ or $x_0 \succ \ldots \succ x_{r-1}$ and (b) $x_0 \cap x_1 < x_1 \cap x_2 > x_2 \cap x_3 < < x_3 \cap x_4$.

Given $i = 0$ or 1, write

$$P_i = P(i, \ldots, i), \tag{10}$$

where the number or arguments of P is $r - 2$, i.e., the largest possible. Put $P = = P_0 \cup P_1$.

(vii) Given any partition $I = \langle I_\xi : \xi < \tau \rangle$ of $[\kappa]^{r-1}$, we define a partition $I^* = \langle I_\xi^* : \xi < \tau \rangle$ of P_0 by putting

$$I_\xi^* = P_0 \cap \{u \in [{}^\kappa 2]^r : \delta(u) \in I_\xi\} . \tag{11}$$

The idea is that the set $[{}^\kappa 2]^r \setminus P_0$ is small under reasonable assumptions; hence I^* gives a partition of almost the whole of $[{}^\kappa 2]^r$. So, if we are given a partition I verifying the relation $\kappa \nrightarrow (\lambda_\xi)_{\xi<\tau}^{r-1}$, then I^* can be slightly changed so that it verify the relation $2^\kappa \nrightarrow (\lambda_\xi + 1)_{\xi<\tau}^r$, provided the λ_ξ's satisfy certain conditions (e.g. if $\lambda_0, \lambda_1 \geq \omega$ and λ_0 is regular). The rest of this section is devoted to proving simple technical lemmas about the concepts introduced above. The main results will be given in the next section. As the lemmas coming now will inevitably be unexciting, the reader may choose to skip them and check their proofs only when they are referred to.

$r \geq 3$ will be assumed throughout the lemmas. We start with the following simple result:

LEMMA 23.2. *Assume* $2 \leq s < r$, *and* $[X]^r \subseteq K(k_0, \ldots, k_{s-1})$ *for some* $k_0, \ldots, k_{s-1} < 2$. *If* $|X| \geq r + 1$ *then* $k_0 = \ldots = k_{s-1}$.

PROOF. We prove the assertion by showing that, for any $i < s - 1$, $k_i = 0$ holds if and only if $k_{i+1} = 0$ holds. To this end, let $x_0 <^* \ldots <^* x_r$ be elements of X; then $\{x_0, \ldots, x_{r-1}\} \in K(k_0, \ldots, k_{s-1})$ and $\{x_1, \ldots, x_r\} \in K(k_0, \ldots, k_{s-1})$. According to the former relation, $k_{i+1} = 0$ is equivalent to $x_{i+1} \prec x_{i+2}$, while this is equivalent to $k_i = 0$ according to the latter one.

LEMMA 23.3. *Assume that either* a) $[X]^r \subseteq K$ *or* b) $r \geq 4$, $|X| \geq r+1$, *and* $[X]^r \subseteq K(0, 1, 0) \cup K$. *Then* $[X]^r \subseteq K_0$ *or* $[X]^r \subseteq K_1$.

PROOF. Assume, on the contrary, that there are $u_0, u_1 \in [X]^r$ such that $u_0 \notin K_0$, $u_1 \notin K_1$. We claim that $|X| \geq r+1$ holds. In fact, we assumed this in case b). In case a), we have $u_0 \in K_1$ and $u_1 \notin K_1$, and so $u_0 \neq u_1$; as $u_0, u_1 \subseteq X$ and $|u_0| = |u_1| = r$, we indeed must have $|X| \geq r+1$ in this case as well. Hence $X \setminus u_0$ is not empty; pick an element x from this set and write $u_0 \cup u_1 \cup \{x\} = \{x_0, \ldots, x_{n-1}\}$, where $x_0 <^* x_1 <^* \ldots <^* x_{n-1}$. Note that $n \geq r+1$. We distinguish two cases:

Case 1. $\{x_i, \ldots, x_{i+r-1}\} \in K$ for all $i \leq n-r$. Then, writing $\eta_j = \eta(x_j, x_{j+1})$ for $j < n-2$, we have $\eta_i = \ldots = \eta_{i+r-2}$ for all $i \leq n-r$, and so $\eta_0 = \eta_1 = \ldots = \eta_{n-2}$, as $r-2>0$. Hence $u_1, u_2 \in K_{\eta_0}$, which contradicts the choice of u_0 and u_1.

Case 2. $\{x_i, \ldots, x_{i+r-1}\} \in K(0, 1, 0\}$ holds for some $i \leq n-r$. If $i=0$ then $\{x_{i+1}, \ldots, x_{i+r}\} \notin K(0, 1, 0) \cup K$, and if $i \neq 0$ then $\{x_{i-1}, \ldots, x_{i+r-2}\} \notin K(0, 1, 0) \cup K$. This is a contradiction, completing the proof.

LEMMA 23.4. *Let* $r \geq 4$ *and assume* $[X]^r \subseteq P$. *Then either* $[X]^r \subseteq P_0$ *or* $[X]^r \subseteq P_1$.

PROOF. We have $[X]^r \subseteq P \subseteq K$; hence $[X]^r \subseteq K_0$ or $[X]^r \subseteq K_1$, by the preceding lemma. Assume that the assertion to be proved fails, and let $u_0, u_1 \in [X]^r$ such that $u_0 \notin P_0$ and $u_1 \notin P_1$. Write $u_0 \cup u_1 = \{x_0, \ldots, x_{n-1}\}$ where $x_0 <^* \ldots <^* x_{n-1}$. Note that

$$x_0 \prec \ldots \prec x_{n-1} \tag{12}$$

or

$$x_0 \succ \ldots \succ x_{n-1} \tag{13}$$

according as $[X]^r \in K_0$ or $[X]^r \in K_1$. Write $\zeta_j = \zeta(x_j \cap x_{j+1}, x_{j+1} \cap x_{j+2})$ for $j < n-2$ (ζ was defined in Definition 23.1(v) above; note that (9) guarantees that ζ_j is meaningful); the relation $\{x_i, \ldots, x_{i+r-1}\} \in P$, valid for any $i \leq n-r$, implies that $\zeta_i = \ldots = \zeta_{i+r-3}$. Hence $\zeta_0 = \ldots = \zeta_{n-3}$, as $r-3 \geq 1$. We claim that we have

$$\zeta(x_i \cap x_j, x_j \cap x_k) = \zeta_0 \tag{14}$$

for any i, j, k with $i<j<k<n$. In fact, assume e.g. that $\zeta_0 = 0$, i.e., that $x_0 \cap x_1 < x_1 \cap x_2 < \ldots < x_{n-2} \cap x_{n-1}$; then we have to show that

$$x_i \cap x_j < x_j \cap x_k. \tag{15}$$

Here $x_i \cap x_j \leq x_i \cap x_{i+1}$ in view of (12) or (13), and

$$x_j \cap x_k \geq \min \{x_l \cap x_{l+1} : j \leq l < k\}$$

by a repeated application of (19.2); hence (15), and so (14), follows in case $\zeta_0 = 0$.

The case $\zeta_0 = 1$ can be dealt with similarly. (14) implies that $u_0, u_1 \in P_{\zeta_0}$, which is a contradiction. The proof is complete.

LEMMA 23.5. *Let $X \subseteq {}^{\kappa}2$, and assume $|X| \geq \omega$. Then there is a set $X' \subseteq X$ with $|X'| = |X|$ such that $[X']^r \subseteq K_0$ or $[X']^r \subseteq K_1$ provided at least one of the following conditions holds:* (i) $[X]^r \cap K(0, 1) = 0$, (ii) $[X]^r \cap K(1, 0) = 0$, *or* (iii) $r \geq 4$ *and* $[X]^r \cap K(0, 1, 0) = 0$.

PROOF. Let $\lambda = |X|$. We may suppose that tp $\langle X, <^* \rangle = \lambda$. Assuming that the assertion of the lemma is false, we are first going to establish the following.

CLAIM. *There are elements $x_0 <^* x_1 <^* x_2 <^* x_3$ of X such that $x_0 \prec x_1 \succ x_2 \prec x_3$.*

PROOF OF CLAIM. For every $x \in X$, there are $y, z \in X$ and $y', z' \in X$ such that $x \leq^* y <^* z$, $x \leq^* y' <^* z'$, $y \prec z$ and $y' \succ z'$; in fact, if $x \in X$ did not satisfy these requirements, then we could take $X' = \{x' \in X : x \leq^* x'\}$ to verify the assertion of the lemma. Now let x_0 and z_1 be elements of X with $x_0 <^* z_1$, and $x_0 \prec z_1$. Choose $y_1, z_2 \in X$ with $z_1 \leq^* y_1 <^* z_2$ and $y_1 \succ z_2$. Denoting by $\min_\prec$ and $\max_\prec$ the minimum and maximum, respectively, of a set in the ordering $\prec$, put $x_1 = \max_\prec \{z_1, y_1\}$. Pick $y_2, x_3 \in X$ with $z_2 \leq^* y_2 <^* x_3$ and $y_2 \prec x_3$, and write $x_2 = \min_\prec \{z_2, y_2\}$. Clearly, we have $x_0 \prec x_1 \succ x_2 \prec x_3$. Our claim is established.

Let now $x_i \in X$, $i < 4$ be such as described in the above Claim, and choose pairwise distinct $x_i \in X$ for $4 \leq i \leq r$ with $x_3 <^* x_i$. Then $\{x_i : i \leq r-1\} \in K(0, 1)$, violating (i), $\{x_i : 1 \leq i \leq r\} \in K(1, 0)$, violating (ii), and, in case $r \geq 4$, $\{x_i : i \leq r-1\} \in K(0, 1, 0)$, violating (iii). These contradictions prove the lemma.

LEMMA 23.6. *Let $X \subseteq {}^{\kappa}2$, assume the cardinality of X is regular, and $[X]^r \subseteq K_0$ or $[X]^r \subseteq K_1$. Then there is an $X' \subseteq X$ with $|X'| = |X|$ and $[X']^r \subseteq P_0$.*

PROOF. Write $|X| = \lambda$. As $[X]^r \subseteq K_0$ or K_1, X is wellordered by $\prec$ or $\succ$; hence Lemma 19.3 (and the remark immediately afterwards) implies that there exists a strictly $<^*$-increasing sequence $\langle x_\alpha : \alpha < \lambda \rangle$ of elements of X and a nondecreasing sequence $\langle \xi_\alpha : \alpha < \lambda \rangle$ of ordinals such that $x_\alpha \cap x_\beta = \xi_\alpha$ for any α, β with $\alpha < \beta < \lambda$. Note that $\langle \xi_\alpha : \alpha < \lambda \rangle$ is actually strictly increasing in view of (9). For any $\alpha_0 < \alpha_1 < \ldots < \alpha_{r-1} < \lambda$, the sequence

$$\delta(\{x_{\alpha_0}, \ldots, x_{\alpha_{r-1}}\}) = \langle \xi_{\alpha_0}, \ldots, \xi_{\alpha_{r-1}} \rangle$$

is increasing, and so $[\{x_\alpha : \alpha < \lambda\}]^r \in P_0$, which means that the assertion of the lemma holds with $X' = \{x_\alpha : \alpha < \lambda\}$.

LEMMA 23.7. *Let $X \subseteq {}^{\kappa}2$, $2 \leq s \leq r-2$, and assume $|X| \geq r+1$ and $[X]^r \subseteq P(k_0, \ldots, k_{s-1})$ with some $k_0, \ldots, k_{s-1} < 2$. Then $k_0 = \ldots = k_{s-1}$.*

PROOF. Pick arbitrary elements $x_0<^*\ldots<^*x_r$ of X. Then the relations $\{x_0,\ldots,x_{r-1}\}\in P(k_0,\ldots,k_{s-1})$ and $\{x_1,\ldots,x_{r-1}\}\in P(k_0,\ldots,k_{s-1})$ show that $k_i=k_{i+1}$ for any i with $0\leq i<s-1$, which completes the proof. Note that the assertion is vacuous for $r=3$.

LEMMA 23.8. *If* $[X]^r\subseteq P_1$, *then* X *is finite.*

PROOF. Assuming the contrary, let $\langle x_n: n<\omega\rangle$ be a $<^*$-increasing sequence of elements of X. Then, by the definition of P_1, we have $x_n\cap x_{n+1}>x_{n+1}\cap x_{n+2}$ for all $n<\omega$, i.e., we have an infinite decreasing sequence of ordinals, which is a contradiction.

LEMMA 23.9. *Let* $r\geq 4$, *let* $X\subseteq{}^{\kappa}2$ *be infinite, and assume* $[X]^r\subseteq K_0$ *or* $[X]^r\subseteq K_1$. *Then there is an* $X'\subseteq X$ *with* $|X'|=|X|$ *and* $[X']^r\subseteq P_0$, *provided at least one of the following conditions holds*: (i) $[X]^r\cap P(0,1)=0$, (ii) $[X]^r\cap P(1,0)=0$, (iii) $r\geq 5$ *and* $[X]^r\cap P(0,1,0)=0$, *or* (iv) $r\geq 5$ *and* $[X]^r\cap P(1,0,1)=0$.

PROOF. We may assume that $\mathrm{tp}\langle X,<^*\rangle=|X|$. First we establish the following

CLAIM. *If* $x_0<^*x_1<^*\ldots<^*x_{s-1}$ *are elements of* X such that

$$\zeta(x_i\cap x_{i+1},x_{i+1}\cap x_{i+2})\neq\zeta(x_{i+1}\cap x_{i+2},x_{i+2}\cap x_{i+3})\tag{16}$$

for every $i\leq s-4$, *then* $s\leq 5$.

In fact, if $r\geq s$, then choose further elements $(x_{s-1}<^*)x_s<^*\ldots<^*x_r$ of X. Then (16) implies that there is a $j=0$ or 1 for which

$$\{x_i:\ i<r\}\in P(j,1-j)$$

and

$$\{x_i:\ 1\leq i\leq r\}\in P(1-j,j)$$

hold provided $r\geq 4$ and $s\geq 5$, and it even implies

$$\{x_i: i<r\}\in P(j,1-j,j)$$

and

$$\{x_i: 1\leq i\leq r\}\in P(1-j,j,1-j)$$

with $j=0$ or 1, provided $r\geq 5$ and $s\geq 6$. These relation violate (i), (ii) and (iii), (iv) in turn, and so $s\geq 6$ is certainly impossible. Our Claim is established.

Let now $s\leq 5$ be the largest possible value in the Claim above (note that $s\geq 3$, since (16) vacuously holds if $s=3$), and choose $x_0<^*\ldots<^*x_{s-1}$ satisfying (16).

Write $x = x_{s-3}$, $y = x_{s-2}$, and $z = x_{s-1}$. According as a) $x \cap y > y \cap z$ or b) $x \cap y < y \cap z$ holds, the maximality of s implies that we have

$$\text{not} \quad x \cap y > y \cap z_0 < z_0 \cap z_1 \tag{17}$$

or

$$\text{not} \quad x \cap y < y \cap z_0 > z_0 \cap z_1 \tag{18}$$

for any $z_0, z_1, z_2 \in X$ with $z \leq^* z_0 <^* z_1 <^* z_2$. Observe that we always have

$$y \cap z_0 \neq z_0 \cap z_1$$

here in view of (9), since either $y \prec z_0 \prec z_1$ or $y \succ z_0 \succ z_1$ holds by our assumption that either $[X]^r \in K_0$ or $[X]^r \in K_1$. When using (17) or (18), we shall always take this inequality into account, without explicitly saying so. We are also going to use repeatedly (19.2) and the remark on equality there.

We show that case a) is impossible. In fact, assuming that a) holds, for any $z_0 \in X$ with $z <^* z_0$ (17) implies that

$$x \cap y > y \cap z > z \cap z_0 = y \cap z_0 \ ;$$

the last equality here follows from (19.2). For any $z_1 \in X$ with $z_0 <^* z_1$, we obtain similarly that

$$x \cap y > y \cap z_0 > z_0 \cap z_1 = y \cap z_1 \ .$$

Hence $y \cap z'$ strictly decreases as $z' \in X$ increases in the ordering $<^*$ provided $z <^* z'$; this would give us an infinite decreasing sequence of ordinals, which is absurd.

Assume therefore b). Then, for any $z_0 \in X$ with $z <^* z_0$, (18) implies that

$$x \cap y < y \cap z < z \cap z_0 \ ;$$

it follows from here by (19.2) that $y \cap z_0 = y \cap z$. Thus, for any $z_1 \in X$ with $z_0 <^* z_1$ we obtain, similarly,

$$x \cap y < y \cap z_0 < z_0 \cap z_1 \ . \tag{19}$$

Again, we have $x \cap y = x \cap z_0$ by (19.2). Hence we cannot have

$$x \cap z_0 < z_0 \cap z_1 > z_1 \cap z_2$$

for any $z_2 \in X$ with $z_1 <^* z_2$, in view of the maximality of s. The first inequality cannot fail here by virtue of (19) and the equality $x \cap y = x \cap z_0$ mentioned afterwards. Thus

$$z_0 \cap z_1 < z_1 \cap z_2 \ ;$$

here $z_0, z_1, z_2 \in X$ were arbitrarily chosen such that $z <^* z_0 <^* z_1 <^* z_2$; therefore $[\{z' \in X : z <^* z'\}]^r \in P_0$, i.e., the assertion holds with $X' = \{z' \in X : z <^* z'\}$. The proof is complete.

LEMMA 23.10. *Let* $r \geq 5$, $X \subseteq {}^{\kappa}2$, $|X| \geq r+1$, *and* $i=0$ *or* 1, *and assume that*

$$[X]^r \subseteq K_i \cap (P \cup P(0, 1, 0)). \tag{20}$$

Then $[X]^r \subseteq P_0$ *or* $[X]^r \subseteq P_1$.

PROOF. Assume $u, v \in [X]^r$ are such that $u \notin P_0$ and $v \notin P_1$, and let $x \in X \setminus u$; write

$$u \cup v \cup \{x\} = \{x_l : l < n\},$$

where $x_0 <^* x_1 <^* \ldots <^* x_{n-1}$. Clearly, $n \geq r+1$. We distinguish three cases according as $\{x_l : l < r\}$ belongs to P_0, P_1, or $P(0, 1, 0)$.

a) Assume $\{x_l : l < r\} \in P_0$. Then

$$x_j \cap x_{j+1} < x_{j+1} \cap x_{j+2} \tag{21}$$

for any $j < r-2$; we claim that this holds for any $j < n-2$. Assume, on the contrary, that $j < n-2$ is the least integer for which (21) fails; as pointed out just before, $j \geq r-2$ holds then. We have

$$x_{j-r+3} \cap x_{j-r+4} < x_{j-r+4} \cap x_{j-r+5} < x_{j-r+5} \cap x_{j-r+6}$$

in view of $r \geq 5$, and so $\{x_l : j-r+3 \leq l < j+3\} \in P(0, 0)$; hence (20) implies that we actually have

$$\{x_i : j-r+3 \leq i < j+3\} \in P_0.$$

So $x_j \cap x_{j+1} < x_{j+1} \cap x_{j+2}$, which establishes (21) for any $j < n-2$. Using (19.2), for any $l < j < k < n$ we obtain that

$$x_l \cap x_j = x_l \cap x_{l+1} < x_j \cap x_{j+1} = x_j \cap x_k,$$

and so $[\{x_l : l < n\}]^r \subseteq P_0$. This contradicts the choice of u.

b) Assume $\{x_l : l < r\} \in P_1$. Then we can conclude $[\{x_l : l < n\}]^r \subseteq P_1$ in a similar way, which contradicts the choice of v.

c) Assume $\{x_l : l < r\} \in P(0, 1, 0)$. Using $n \geq r+1$, we can see that $\{x_l : 1 \subseteq l \subseteq r\} \in P(1, 0)$, which contradicts (20). The proof is complete.

The next lemma sums up results that we have essentially verified above already. We state it only because it will be convenient to refer to it in the proof of the lemma afterwards.

LEMMA 23.11. *Let* $X \subseteq {}^{\kappa}2$, *and assume* $|X| \geq r$ *and* $[X]^r \subseteq P_s$, *where* $s=0$ *or* 1. *Then* $[X]^r \subseteq K_0$ *or* $[X]^r \subseteq K_1$. *Moreover, if* $x <^* y <^* z$ *are elements of* X, *then* $x \cap y = x \cap z < y \cap z$ *or* $x \cap y > x \cap z = y \cap z$ *according as* $s=0$ *or* 1.

PROOF. As $P_s \subseteq K$, the assertion that $[X]^r \subseteq K_0$ or K_1 is given in Lemma 23.3.a above. To show the second assertion, let $x_0 <^* \ldots <^* x_{r-1}$ be elements of X such that $x=x_i$, $y=x_j$, and $z=x_k$ for some $i<j<k$. Considering e.g. the case $s=1$, we have $\{x_0, \ldots, x_{r-1}\} \in P_1$ and so, using (19.2), we can see that

$$x_i \cap x_j = x_{j-1} \cap x_j > x_{k-1} \cap x_k = x_i \cap x_k = x_j \cap x_k ,$$

which completes the proof.

The following result establishes a relation between the sizes of homogeneous sets with respect to the partitions I and I^*, described in Definition 23.1 (vii). There is a slight difference however; here, in the definition of A^* we may also take P_1 instead of P_0:

LEMMA 23.12. *Let* $A \subseteq [{}^\kappa 2]^{r-1}$, $s=0$ *or* 1, *and put*

$$A^* = \{u \in P_s : \delta(u) \in A\} .$$

Assume $[X]^r \subseteq A^*$, $X \neq 0$. *Then there is a* $D \subseteq \kappa$ *with* $|D|+1=|X|$ *such that* $[D]^{r-1} \subseteq A$.

PROOF. We may assume that $|X| \geq r$. Write $X=\{x_\xi : \xi<\alpha\}$, where $x_\xi <^* x_\eta$ whenever $\xi<\eta<\alpha$, and, for any ξ with $\xi \dotplus 1<\alpha$ put

$$\delta_\xi = x_\xi \cap x_{\xi \dotplus 1} .$$

Set

$$D=\{\delta_\xi : \xi \dotplus 1<\alpha\} .$$

Given any $\xi_0<\ldots<\xi_{r-2}$ satisfying $\xi_{r-2} \dotplus 1<\alpha$, we have to prove that

$$\{\delta_{\xi_i} : i<r-1\} \in A .$$

Write $\xi_{r-1}=\xi_{r-2} \dotplus 1$. In case $s=0$, the preceding lemma implies that

$$\{\delta_{\xi_i} : i<r-1\} = \{x_{\xi_i} \cap x_{\xi_i \dotplus 1} : i<r-1\} =$$
$$= \{x_{\xi_i} \cap x_{\xi_{i+1}} : i<r-1\} = \delta(\{x_{\xi_i} : i<r\}) ,$$

and this set belongs to A, as $\{x_{\xi_i} : i<r\} \in A^*$. In case $s=1$, the same lemma implies that

$$\{\delta_{\xi_i} : i<r-1\} = \{x_{\xi_i} \cap x_{\xi_i \dotplus 1} : i<r-1\} =$$
$$= \{x_{\xi_0} \cap x_{\xi_0 \dotplus 1}\} \cup \{x_{\xi_{i-1} \dotplus 1} \cap x_{\xi_i \dotplus 1} : 1 \leq i<r-1\} =$$
$$= \delta(\{x_{\xi_0}\} \cup \{x_{\xi_i \dotplus 1} : i<r-1\}) \in A ,$$

as $\{x_{\xi_0}\} \cup \{x_{\xi_i \dotplus 1} : i<r-1\} \in A^*$. The proof is complete.

For the discussion of the finite case of the ordinary partition relation, we shall need a simple lemma, our last result in this section. In order to state this lemma, we introduce the following notation:

$$P_{01} = \bigcup_{\vec{k}} P(\vec{k}), \tag{22}$$

where $\vec{k}$ runs over all 0-1 sequences of length $r-2$ that begin with a 0 and contain at least one 1, and

$$P_{10} = \bigcup_{\vec{l}} P(\vec{l}) \tag{23}$$

where $\vec{l}$ runs over all 0-1 sequences of length $r-2$ that begin with a 1 and contain at least one 0. Note that we obviously have

$$K = P \cup P_{01} \cup P_{10}, \tag{24}$$

and the sets on the right-hand side are pairwise disjoint. Our last result here is

LEMMA 23.13. *Assume $r \geq 4$. Let m be an integer, and let $X \subseteq K$ be a set with $|X| = 2m$. Then there is a set $X' \subseteq X$ with $|X'| \geq m+1$ and $[X']^r \subseteq P_j$ for $j=0$ or 1 provided at least one of the following conditions holds:* (i) $[X]^r \cap P_{10} = 0$, *or* (ii) $[X]^r \cap P_{01} = 0$.

PROOF. Write $X = \{x_i : i < 2m\}$, where $x_0 <^* \ldots <^* x_{2m-1}$. Assume e.g. (i), and write $\delta_i = x_i \cap x_{i+1}$. Put

$$i_0 = \min\{i < 2m-1 : \delta_i > \delta_{i+1}\}$$

if the set on the right-hand side is not empty, and put $i_0 = 2m-2$ otherwise. Write $X_0 = \{x_i : i \leq i_0\}$ and $X_1 = \{x_i : i_0 \leq i < 2m\}$. We claim that we have

$$[X_0]^r \subseteq P_0 \qquad \text{and} \qquad [X_1]^r \subseteq P_1 . \tag{25}$$

In fact, we have $x_i \cap x_{i+1} = \delta_i < \delta_{i+1} = x_{i+1} \cap x_{i+2}$ for any $i < i_0 - 1$ by the definition of i_0 (note that $\delta_i = \delta_{i+1}$ cannot happen in view of (9)), hence $x_i \cap x_j < < x_j \cap x_k$ holds whenever $i < j < k < i_0$ by virtue of (19.2). This establishes the first relation in (25). To prove the second one, observe that $[X_1]^r$ is empty, unless $|X_1| \geq r$; hence we may assume that this latter inequality holds. Then we have

$$\delta_i > \delta_{i+1} \qquad \text{whenever} \qquad i_0 \leq i < 2m-1 . \tag{26}$$

Assume the contrary; then there is a least i with $i_0 \leq i \leq 2m-1$ such that $\delta_i < \delta_{i+1}$ (cf. (9) again), and $i > i_0$ here by the definition of i_0. Taking an $Y \in [X_1]^r$ with $x_{i_0}, x_i, x_{i+1}, x_{i+2} \in Y$, we can see by (19.2) that $Y \in P_{10}$. This contradicts (i). Hence (26) must indeed hold in case $|X_1| \geq r$. Thus the second relation in (25) also follows by (19.2). As $|X_0| + |X_1| = 2m+1$, we must have $|X_j| \geq m+1$ for $j=0$ or 1. Put $X' = X_j$ with this j. The proof of our lemma is complete in case assumption (i) holds. The handling of assumption (ii) is entirely analogous.

24. THE NEGATIVE STEPPING-UP LEMMA

Relying upon the results of the preceding section, we are now in a position to prove the following lemma, which is a kind of converse to the Stepping-up Lemma (Lemma 16.1). Note that + always means cardinal addition.

LEMMA 24.1. (Negative Stepping-up Lemma). *Let* $r \geq 3$ *be an integer, let* $\kappa, \tau > 0$ *and* $\lambda_\xi \geq r$ $(\xi < \tau)$ *be cardinals, and assume that*

$$\kappa \nrightarrow (\lambda_\xi)^{r-1}_{\xi<\tau}. \tag{1}$$

Then we have

$$2^\kappa \nrightarrow (\lambda_\xi + 1)^r_{\xi<\tau} \tag{2}$$

provided at least one of the conditions a), b), c), d) *or* e) *holds*:

a) $\tau \geq 2$, $\kappa, \lambda_0, \lambda_1 \geq \omega$ *and* λ_0 *is regular*;

b) $\tau \geq 2$, $\kappa, \lambda_0 \geq \omega$, λ_0 *is regular, and* $r \geq 4$;

c) $\tau \geq 2$, $\kappa, \lambda_0, \lambda_1 \geq \omega$ *and* $r \geq 4$;

d) $\tau \geq 2$, $\kappa, \lambda_0 \geq \omega$ *and* $r \geq 5$;

e) $\kappa \geq \omega$ *and* $\lambda_\xi < \omega$ *for all* $\xi < \tau$.

Moreover, without assuming any of a), b), c), d) *or* e), *we have*

$$2^\kappa \nrightarrow ((\lambda_\xi + 1)_{\xi<\tau}, (r+1)_m)^r \tag{3}$$

with $m = 2^{r-1} + 2^{r-2} - 4$ *provided* $r \geq 4$, *and*

$$2^\kappa \nrightarrow ((\lambda_\xi + 1)_{\xi<\tau}, r+1)^r \tag{4}$$

provided $\lambda_0 \geq \omega$ *is regular, and*

$$2^\kappa \nrightarrow ((\lambda_\xi + 1)_{\xi<\tau}, (\lambda_\xi + 1)_{\xi<\tau}, 4, 4)^3; \tag{5}$$

for this last relation, one has to assume (1) *with* $r = 3$.

The main interest of (4) lies in the case $r = 3$. In fact, (4) follows from (2) under condition b) if $r \geq 4$ and $\tau \geq 2$. If $\tau = 1$ and $\kappa \geq \omega$ then (4) becomes

$$2^\kappa \nrightarrow (\kappa^+, r+1)^r \qquad (r \geq 3). \tag{6}$$

PROOF. Assume that $I = \langle I_\xi \colon \xi < \tau \rangle$ is a partition of $[\kappa]^{r-1}$ verifying (1). Recall that we defined I^*_ξ as

$$I^*_\xi = \{u \in P_0 \colon \delta(u) \in I_\xi\} \tag{7}$$

in (23.11) for any $\xi < \tau$; here and below we shall freely use the concepts introduced in Definition 23.1.

PROOF OF (2). We shall define a partition $J = \langle J_\xi \colon \xi < \tau \rangle$ of $[{}^\kappa 2]^r$ verifying (2) and depending on whether a), b), c) or d) hold; e) requires a different approach.

We shall fix a $\xi<\tau$, and $X\subseteq{}^{\kappa}2$ will denote a set such that

$$[X]^r\subseteq J_\xi. \tag{8}$$

We shall have to prove that

$$|X|<\lambda_\xi+1. \tag{9}$$

We may assume to this end that $X\neq 0$. Observe that if

$$[X]^r\subseteq I^*_\xi \tag{10}$$

holds instead of (8), then (9) is satisfied. In fact, Lemma 23.12 implies that there is a $D\subseteq\kappa$ with $|D|+1=|X|$ and $[D]^{r-1}\subseteq I_\xi$; as the partition I verifies the relation in (1), we obtain that $|D|<\lambda_\xi$, and so (9) follows. As we shall invariably put

$$J_\xi=I^*_\xi \tag{11}$$

if $2\leq\xi<\tau$, we shall have to establish (9) only in case $\xi=0$ or 1. We shall have

$$J_\xi\cap P_0=I^*_\xi \tag{12}$$

also for $\xi=0, 1$. We shall assume that (9) fails in the remaining cases, i.e., that

$$|X|=\lambda_\xi+1\qquad(\xi=0\text{ or }1). \tag{13}$$

Ad a). Set

$$J_\xi=I^*_\xi\qquad\text{if}\qquad 2\leq\xi<\tau,$$

$$J_1=K(0,1)\cup I^*_1,$$

and

$$J_0=[{}^{\kappa}2]^r\setminus\bigcup_{1\leq\xi<\tau}J_\xi.$$

It is easy to verify that (12) holds.

To prove (9), suppose first that $\xi=0$, i.e., that $[X]^r\subseteq J_0$. In this case $[X]^r\cap K(0,1)=0$; noting that $\lambda_0\geq\omega$, and so X is infinite by (13), we can conclude from Lemma 23.5(i) that there exists an $X'\subseteq X$ with $|X'|=|X|$ and $[X']^r\subseteq K_0$ or $[X']^r\subseteq K_1$. As $|X'|=\lambda_0+1=\lambda_0$ is regular, Lemma 23.6 now implies that there is an $X''\subseteq X'$ with $|X''|=|X'|$ $(=|X|=\lambda_0+1)$ and $[X'']^r\subseteq P_0$; hence (12) implies that $[X'']^r\subseteq I^*_0$. So, with X'' replacing X we have (10), and so (9) as well; i.e., $\lambda_0+1=|X''|<\lambda_0+1$, which is a contradiction.

Suppose now $\xi=1$. We have $[X]^r\cap K(1,0)=0$, and so, noting that $|X|=$ $=\lambda_1+1\geq\omega$, Lemma 23.5(ii) implies that there is an $X'\subseteq X$ with $|X'|=|X|$ and $[X']^r\subseteq K_0$ or K_1. Then $[X']^r\cap K(0,1)=0$, and so $[X']^r\subseteq I^*_1$; therefore (10), and (9) hold with X' replacing X. But then $\lambda_1+1=|X'|<\lambda_1+1$, which is again a contradiction.

Ad b). Set

$$J_\xi = I_\xi^* \quad \text{if} \quad 2 \le \xi < \tau,$$

$$J_1 = K(0,1,0) \cup I_1^*,$$

and

$$J_0 = [{}^\kappa 2] \setminus \bigcup_{1 \le \xi < \tau} J_\xi .$$

It is easy to see that (12) is satisfied.

Suppose $\xi = 0$. We have $[X]^r \cap K(0,1,0) = 0$, so Lemma 23.5(iii) and Lemma 23.6 imply in view of the regularity of $|X| = \lambda_0 + 1 = \lambda_0$ that there exists an $X' \subseteq X$ with $|X'| = |X|$ and $[X']^r \subseteq P_0$. Then $[X']^r \subseteq I_0^*$, and so we have $\lambda_0 + 1 = |X'| < \lambda_0 + 1$ by the implication (10)⇒(9), which is a contradiction.

Suppose $\xi = 1$. Noting that $|X| = \lambda_1 + 1 \ge r + 1$, $r \ge 4$, and $[X]^r \subseteq J_1 \subseteq K(0,1,0) \cup I_1^* \subseteq K(0,1,0) \cup K$, we obtain by Lemma 23.3b that $[X]^r \subseteq K$; therefore $[X]^r \subseteq I_1^*$, as $K \cap K(0,1,0) = 0$. So (10) holds, and this implies (9), i.e., that $\lambda_1 + 1 = |X| < \lambda_1 + 1$, which is a contradiction.

Ad c). Set

$$J_\xi = I_\xi^* \quad \text{if} \quad 2 \le \xi < \tau,$$

$$J_1 = K(0,1) \cup P(0,1) \cup I_1^*,$$

and

$$J_0 = [{}^\kappa 2]^r \setminus \bigcup_{1 \le \xi < \tau} J_\xi .$$

It is easy to prove that (12) is satisfied.

Suppose first that $\xi = 0$. Noting that $|X| = \lambda_0 + 1 \ge \omega$ and $[X]^r \cap K(0,1) = 0$, Lemma 23.5(i) entails that there is an $X' \subseteq X$ with $|X'| = |X| = \lambda_0 + 1$ and $[X']^r \subseteq K_0$ or K_1. Observing that $[X]^r \cap P(0,1) = 0$, i.e., $[X']^r \cap P(0,1) = 0$, we obtain by Lemma 23.9(i) that there exists an $X'' \subseteq X'$ with $|X''| = |X'| = \lambda_0 + 1$ and $[X'']^r \subseteq P_0$. Then $[X'']^r \subseteq I_0^*$, and so we have $\lambda_0 + 1 = |X''| < \lambda_0 + 1$ in view of the implication (10)⇒(9), which is a contradiction.

Suppose now $\xi = 1$. Noting that $|X| = \lambda_1 + 1 \ge \omega$ and $[X]^r \subseteq K \cup K(0,1)$, i.e., that $[X]^r \cap K(1,0) = 0$, we can conclude by Lemma 23.5(ii) that there is a set $X' \subseteq X$ with $|X'| = |X| = \lambda_1 + 1$ and $[X']^r \subseteq K_0$ or K_1. Then it follows from the assumption $[X]^r \subseteq J_1$ that $[X']^r \subseteq P(0,1) \cup I_1^*$, i.e., that $[X']^r \cap P(1,0) = 0$; hence Lemma 23.9(ii) implies that there exists an $X'' \subseteq X$ with $|X''| = |X'| = \lambda_1 + 1$ and $[X'']^r \subseteq P_0$. Then $[X'']^r \subseteq I_1^*$, i.e., $\lambda_1 + 1 = |X''| < \lambda_1 + 1$ by the implication (10)⇒(9), which is a contradiction.

Ad d). Set

$$J_\xi = I_\xi^* \quad \text{if} \quad 2 \le \xi < \tau,$$

$$J_1 = K(0,1,0) \cup P(0,1,0) \cup I_1^*,$$

and

$$J_0=[{}^{\kappa}2]\setminus \bigcup_{1\le\xi<\tau} J_\xi .$$

It is easy to see that (12) is again satisfied.

Suppose first that $\xi=0$. Then $|X|=\lambda_0+1\ge\omega$ and $[X]^r\cap K(0,1,0)=0$; so, by Lemma 23.5(iii), there is an $X'\subseteq X$ with $|X'|=|X|=\lambda_0+1$ and $[X']^r\subseteq K_0$ or K_1. As $|X'|\cap P(0,1,0)=0$ by $[X']^r\subseteq[X]^r\subseteq J_0$, Lemma 23.9(iii) entails the existence of an $X''\subseteq X$ with $|X''|=|X'|=\lambda_0+1$ and $[X'']^r\subseteq P_0$. Hence $[X'']^r\subseteq I_0^*$ by (12), and so it follows from the implication (10)⇒(9) that $\lambda_0+1=|X''|<\lambda_0+1$, which is a contradiction.

Suppose $\xi=1$ now. Then $|X|=\lambda_1+1\ge r+1$ and $[X]^r\subseteq J_1\subseteq K(0,1,0)\cup K$, and so actually $[X]^r\subseteq K_0$ or K_1 by Lemma 23.3b. Hence $[X]^r\subseteq$ $\subseteq P(0,1,0)\cup I_1^*\subseteq P(0,1,0)\cup P_0$, and so $[X]^r\subseteq P_0$ by Lemma 23.10. This means that $[X]^r\subseteq I_1^*$, i.e., that (10) holds; hence (9) also holds, i.e., $\lambda_0+1=$ $=|X|<\lambda_0+1$, which is a contradiction.

Ad e). The proof of this case does not directly rely on the preparations made above; rather, it makes simple use of relations (3) and (5) to be proved below. We have $\tau\ge\omega$ by Ramsey's theorem in virtue of the assumptions made on κ and the λ_ξ's in this case. In case $r\ge4$ (3) says that $2^\kappa\nrightarrow((r+1)_m,(\lambda_\xi+1)_{\xi<\tau})^r$, where m is finite, but this is stronger than (2), as $\tau\ge\omega$ implies that there is a rearrangement $\langle\lambda'_\xi:\xi<\tau\rangle$ of the sequence $\langle\lambda_\xi:\xi<\tau\rangle$ such that $\lambda'_{m\dotplus\xi}\ge\lambda_\xi$ for all $\xi<\tau$ and, of course, $\lambda_\xi\ge r$ for $\xi<m$ (cf. Monotonicity Property b) in Subsection 9.6). In case $r=3$ (1) also implies $2^\tau\ge\kappa$ in view of Corollary 17.5 with $\lambda=2$ and $\rho=\tau$, and so we have $\kappa\nrightarrow(3)^2_\tau$ by (19.17). Replacing (1) with this relation, the relation corresponding to (5) becomes $2^\kappa\nrightarrow((4)_\tau,(4)_\tau,4,4)^3$, i.e., $2^\kappa\nrightarrow(4)^3_\tau$; this is again stronger than (2), as $r=3$ now. This completes the proof of (2).

Proof of (3). Let $\vec{k}_i$, $i<2^{r-1}-2$, and $\vec{l}_j$, $j<2^{r-2}-2$, be 1-1 enumerations of all nonconstant 0-1 sequences of length $r-1$ and $r-2$, respectively. Put

$$J_\xi=\{u\in P:\delta(u)\in I_\xi\}\quad\text{for}\quad \xi<\tau,$$

$$J_{\tau\dotplus i}=\{u\in[{}^{\kappa}2]^r:\eta(u)=\vec{k}_i\}\quad\text{for}\quad i<2^{r-1}-2,$$

and

$$J_{\tau\dotplus(2^{r-1}-2)\dotplus j}=\{u\in K:\zeta(\delta(u))=\vec{l}_j\}\quad\text{for}\quad j<2^{r-2}-2.$$

Clearly, $\langle J_\xi:\xi<\tau\dotplus(2^{r-1}+2^{r-2}-4)\rangle$ is a partition of $[{}^{\kappa}2]^r$. Let $X\subseteq{}^{\kappa}2$ be a nonempty set such that $[X]^r\subseteq J_\xi$ for some $\xi<\tau+(2^{r-1}+2^{r-2}-4)$.

Assume first that $\xi<\tau$. Noting that $r\ge4$ and $[X]^r\subseteq P$, we have $[X]^r\subseteq P_s$ for $s=0$ or 1 by Lemma 23.4. (This means that the definition of J_ξ here is analogous to that of I_ξ^* in (7), though here we may have $s=1$ as well; this will not make much difference.) Lemma 23.12 implies that there is a $D\subseteq\kappa$ with $|D|+1=|X|$ and $[D]^{r-1}\subseteq I_\xi$; as the partition I was assumed to verify (1), we have $|D|<\lambda_\xi$, i.e., $|X|<\lambda_\xi+1$, which we wanted to show.

If $\tau\leq\xi<\tau\dot{+}(2^{r-1}-2)$, then Lemma 23.2 implies $|X|<r+1$, and if $\tau\dot{+}(2^{r-1}-2)\leq\xi<\tau\dot{+}(2^{r-1}+2^{r-2}-4)$, then Lemma 23.7 implies $|X|<r+1$. This completes the proof of (3).

Proof of (4). Write

$$J_\tau=K(0,1),$$

$$J_\xi=I_\xi^* \qquad \text{for} \qquad 1\leq\xi<\tau$$

and

$$J_0=[{}^\kappa 2]^r\setminus\bigcup_{1\leq\xi<\tau} J_\xi.$$

Assume $[X]^r\subseteq J_\xi$ for some $\xi\leq\tau$. If $\xi=\tau$, then $|X|<r+1$ by Lemma 23.2. If $1\leq\xi<\tau$, then $|X|<\lambda_\xi+1$ in view of the implication (10)⇒(9). If, finally, $\xi=0$, then $[X]^r\cap K(0,1)=0$; as the regularity of λ_0 is assumed here, we can argue exactly as we did in the proof of (2) under the assumption of a) in case $\xi=0$; we obtain $|X|<\lambda_0+1$.

Proof of (5). Note that $r=3$ in the present case, and so we have $K=P_0\cup P_1$; hence

$$[{}^\kappa 2]^3=K(0,1)\cup K(1,0)\cup P_0\cup P_1. \tag{14}$$

Write

$$J_{\tau\dot{+}\tau\dot{+}1}=K(1,0),$$

$$J_{\tau\dot{+}\tau}=K(0,1),$$

$$J_{\tau\dot{+}\xi}=\{u\in P_1:\delta(u)\in I_\xi\},$$

and

$$J_\xi=\{u\in P_0:\delta(u)\in I_\xi\},$$

where $\xi<\tau$. It follows from (14) that $\langle J_\eta:\eta<\tau\dot{+}\tau\dot{+}2\rangle$ is a partition of $[{}^\kappa 2]^3$.

Assume $[X]^r\subseteq J_\eta$ for some $\eta<\tau\dot{+}\tau\dot{+}2$. If $\eta<\tau$, then $|X|<\lambda_\eta+1$ by the implication (10)⇒(9). If $\eta=\tau\dot{+}\xi$ for some $\xi<\tau$, then we have $|X|<\lambda_\xi+1$ by an analogous argument. In fact, assuming $|X|=\lambda_\xi+1$, there is a $D\subseteq\kappa$ with $[D]^{r-1}\subseteq I_\xi$ and $|D|+1=|X|$ by Lemma 23.12; as the partition I verifies the relation in (1), we have $|D|<\lambda_\xi$, i.e., $|X|<\lambda_\xi+1$. If $\eta=\tau\dot{+}\tau$ or $\eta=\tau\dot{+}\tau\dot{+}1$, then $|X|<r+1=4$ by Lemma 23.2. The proof of Lemma 24.1 is complete.

Our next, and last, result in this section is a finite negative stepping-up lemma which will be useful in the section after the next one.

LEMMA 24.2. *Let $r \geq 3$, $n, s > 0$, and $m_i \geq r$ $(i < s)$ be integers, and assume that*

$$n \nrightarrow (m_i)^{r-1}_{i<s} \tag{15}$$

holds. Then we have

$$2^n \nrightarrow ((m_i+1)_{i<s}, (m_i+1)_{i<s})^3 \tag{16}$$

in case $r = 3$, and

$$2^n \nrightarrow (2m_0, 2m_1, (m_i+1)_{2 \leq i < s})^r \tag{17}$$

in case $r \geq 4$ and $s \geq 2$.

PROOF. Write $\kappa = n$, and use the notations introduced in the preceding section. As κ is finite, the lexicographical ordering $\prec$ of ${}^\kappa 2$ is a wellordering now, and so we may assume that the wellordering $<^*$ of ${}^\kappa 2$ is the same as $\prec$ (in fact, it is unlikely that we could gain anything by choosing $<^*$ different from $\prec$); we have

$$[{}^\kappa 2] = K_0 \tag{18}$$

then. Assume we are given a partition $I = \langle I_i : i < s \rangle$ of $[\kappa]^{r-1} = [n]^{r-1}$ verifying (15). To prove (16), suppose $r = 3$ and choose J_i and J_{s+i} $(i < s)$ analogously as in the proof of (5), with i and s replacing ξ and τ, respectively; i.e., put

$$J_i = \{u \in P_0 : \delta(u) \in I_i\}$$

and

$$J_{s+i} = \{u \in P_1 : \delta(u) \in I_i\}$$

for each $i < s$. In contrast to the situation in the proof of (5), it follows now from (14) that $J = \langle J_i : i < 2s \rangle$ is a partition of $[{}^n 2] = [{}^\kappa 2]$, as the sets $K(0, 1)$ and $K(1, 0)$ are now empty in view of (18). The same argument as in the proof of (5) with m_i replacing λ_i gives that there is no homogeneous set of class i or $s+i$ that has cardinality $m_i + 1$ for any $i < s$. Thus (16) is established.

To prove (17), assume $r \geq 4$ and $s \geq 2$. Define I'_i, $i < s$, with a slight modification of (7) above as

$$I'_i = \{u \in P : \delta(u) \in I_i\}.$$

Using the notations introduced in (23.22) and (23.23), put

$$J_0 = P_{01} \cup I'_0,$$

$$J_1 = P_{10} \cup I'_1$$

and

$$J_i = I'_i \quad \text{if} \quad 2 \leq i < s$$

(note that we assumed $s \geq 2$). Then $J = \langle J_i : i < s \rangle$ is a partition of $[{}^n 2]^r = [{}^\kappa 2]^r$ in view of (18) and (23.24). Let now $X \subseteq {}^n 2$ be a set such that $[X]^r \subseteq J_i$ holds for some $i < s$. Assume $i \geq 2$ first. Then $X \subseteq P$, and so $X \subseteq P_j$ for $j = 0$ or 1 by Lemma

23.4, since we assumed $r \geq 4$. Lemma 23.12 now implies that there is a $D \subseteq n$ with $|D|+1=|X|$ and $[D]^{r-1} \subseteq I_i$. As the partition I verifies (15), we have $|D|<m_i$, and so $|X|<m_i+1$. Hence I verifies the relation

$$2^n \nrightarrow (\cdot, \cdot, (m_i+1)_{2 \leq i<s})^r_s.$$

Assume now e.g. $i=0$, and suppose that $|X| \geq 2m_0$. $[X]^r \subseteq J_0$ implies that $[X]^r \cap P_{10}=0$, and so, noting that we assumed $r \geq 4$, Lemma 23.13 with assumption (i) ensures the existence of a set $X' \subseteq X$ with $|X'| \geq m_0+1$ and $[X']^r \in P_j$ for $j=0$ or 1. Then $[X']^r \subseteq P_j \cap J_0$, and so there is a D with $[D]^{r-1} \subseteq I_0$ such that $|X'|=|D|+1$, i.e., such that $|D| \geq m_0$, according to Lemma 23.12. This contradicts our assumption that I verifies (15). Hence we must have $|X|<2m_0$. In case $i=1$ we obtain analogously that $|X|<2m_1$. Thus (17) is established; the proof is complete.

25. SOME SPECIAL NEGATIVE PARTITION RELATIONS FOR $r \geq 3$

We are going to prove two more theorems not covered by the results in the preceding section. The first result is a kind of negative stepping up:

THEOREM 25.1. *Let κ, $\lambda \geq \omega$ and $\tau \geq 2$ be cardinals, and let $r \geq 2$ be an integer. Assume that*

$$\kappa \nrightarrow (\lambda)^r_\tau. \tag{1}$$

Then

$$\kappa \nrightarrow (\lambda, (r+2)_{\tau-1})^{r+1} \tag{2}$$

and, moreover

$$\kappa \nrightarrow (\lambda, \tau+r)^{r+1} \tag{3}$$

provided $\tau<\omega$, and

$$\kappa \nrightarrow (\lambda, r+3)^{r+2} \tag{4}$$

provided $\tau<\lambda$.

PROOF. Let $f:[\kappa]^r \to \tau$ be a coloring verifying relation (1).

PROOF OF (2). We are going to define a partition $\langle I_\xi : \xi<\tau \rangle$ verifying (2) as follows: given ordinals $x_0<x_1<\ldots<x_r<\kappa$, we put

$$\{x_i : i \leq r\} \in I_0 \quad \text{iff} \quad f(\{x_i : i<r\}) \geq f(\{x_i : 1 \leq i \leq r\}),$$

and

$$\{x_i : i \leq r\} \in I_{1+\xi} \quad \text{iff} \quad \xi=f(\{x_i : i<r\})<f(\{x_i : 1 \leq i \leq r\}),$$

where $1\dot{+}\xi<\tau$ (i.e., $\xi<\tau-1$, as τ is a cardinal). Assume that this partition does not verify (2).

Suppose first that $[X]^{r+1}\subseteq I_0$ and $|X|=\lambda$; we may assume tp $\langle X, <\rangle=\lambda$ here. Choose elements $x_0<x_1<\ldots<x_{r-1}$ of X such that $f(\{x_i: i<r\})=\mu$ is least possible. We claim that

$$f(u)=\mu \tag{5}$$

holds for any $u\in[\{x\in X: x_{r-1}<x\}]^r$. In fact, assume, on the contrary, that we have $f(u)>\mu$ for some u in this set, and write $u=\{x_i: r\leq i<2r\}$, where $x_r<\ldots<x_{2r-1}$. Then $f(\{x_i: j\leq i<j+r\})<f(\{x_i: j<i\leq j+r\})$ holds with some $j<r-1$. This contradicts the relation $\{x_i: j\leq i\leq j+r\}\in([X]^{r+1}\subseteq)I_0$, establishing (5). So $\{x\in X: x_{r-1}<x\}$ is a homogeneous set of color μ with respect to the coloring f; hence, noting that f establishes relation (1), we can see that this set has cardinality $<\lambda$. This contradicts the assumption that tp $\langle X, <\rangle=\lambda$.

Suppose now that $|X|=r+2$ and $[X]^{r+1}\subseteq I_{1+\xi}$ for some $\xi<\tau-1$. Let x_i, $i<r+2$, be an enumeration of the elements of X in increasing order. Then $\{x_i: i<r+1\}\in I_{1\dot{+}\xi}$, and so

$$\xi=f(\{x_i: i<r\})<f(\{x_i: 1\leq i\leq r\})$$

and $\{x_i: 1\leq i<r+2\}\in I_{1\dot{+}\xi}$, and so

$$\xi=f(\{x_i: 1\leq i\leq r\})<\ldots,$$

which is a contradiction, proving (2).

Proof of (3). Set

$$I_1' = \bigcup_{\xi<\tau-1} I_{\xi\dot{+}1},$$

where $I=\langle I_\xi: \xi<\tau\rangle$ is the partition defined just before in the proof of (2), i.e., if $x_0<x_1<\ldots<x_r<\kappa$, then

$$\{x_i: i\leq r\}\in I_1' \quad \text{iff} \quad f(\{x_i: i<r\})<f(\{x_i: 1\leq i\leq r\}).$$

We claim that the partition $\langle I_0, I_1'\rangle$ verifies the relation in (3). We saw in the proof of (2) that this partition verifies the relation

$$\kappa\not\rightarrow(\lambda, \cdot)^{r+1}.$$

Assume, therefore, on the contrary, that $|X|=\tau+r$ and $[X]^{r+1}\subseteq I_1'$. Noting that τ is assumed to be finite here, let x_i, $i<\tau+r$, be an enumeration of the elements of X in increasing order. We have $\{x_i: j\leq i\leq j+r\}\in I_1'$, and so

$$f(\{x_i: j\leq i<j+r\})<f(\{x_i: j<i\leq j+r\})$$

for any $j<\tau$. Hence $f(\{x_i: \tau\leq i<\tau+r\})\geq\tau$, which is a contradiction, since ra $(f)\subseteq\tau$. (3) is established.

Proof of (4). We define a partition $J=\langle J_0, J_1\rangle$ verifying (4). For any ordinals $x_0<x_1<\ldots<x_{r+1}<\kappa$ put

$$\{x_i: i\leq r+1\}\in J_1 \quad \text{iff}$$

$$f(\{x_i: i<r\})<f(\{x_i: 1\leq i<r+1\})\geq f(\{x_i: 2\leq i<r+2\}),$$

and write $J_0=[\kappa]^{r+2}\setminus J_1$. Assume that J does not verify (4).

Suppose first that there is an $X\subseteq\kappa$ with $[X]^{r+2}\subseteq J_0$ and tp $\langle X, <\rangle=\lambda$. Let $\langle x_\alpha: \alpha<\lambda\rangle$ be an enumeration of the elements of X in increasing order, and write

$$g(\alpha)=f(\{x_{\alpha\dot{+}i}: i<r\}).$$

If $g(\alpha)\geq g(\beta)$ whenever $\alpha<\beta<\lambda$, then g is eventually constant, i.e., there is an $\alpha_0<\lambda$ and a $\mu<\tau$ such that $g(\alpha)=\mu$ whenever $\alpha_0\leq\alpha<\lambda$; then $\{x_\alpha: \alpha_0\dot{+}r\leq\alpha<\lambda\}$ is a homogeneous set of color μ with respect to f. In fact, assuming the contrary, let $\alpha_0\dot{+}r\leq\gamma_r<\ldots<\gamma_{2r-1}$ be ordinals such that $f(\{x_{\gamma_r}, \ldots, x_{\gamma_{2r-1}}\})=\mu'\neq\mu$, and write $\gamma_j=\alpha_0\dot{+}j$ and $\gamma_{2r+j'}=\gamma_{2r-1}\dot{+}j'\dot{+}1$ for $j<r$, $j'<r+1$. Writing $h(j)=f(\{x_{\gamma_j}, \ldots, x_{\gamma_{j+r-1}}\})$, we have $h(0)=h(2r)=h(2r+1)=\mu$ and $h(r)=\mu'\neq\mu$. So $h(j)<h(j+1)\geq h(j+2)$ for some $j\leq 2r-1$. Then $\{x_{\gamma_j}, \ldots, x_{\gamma_{j+r+1}}\}\in J_1$, contradicting the assumption that $[X]^{r+2}\subseteq J_0$. Hence $\{x_\alpha: \alpha_0\dot{+}r\leq\alpha<\lambda\}$ is indeed a homogeneous set of color μ with respect to f, as claimed, which contradicts our assumption that the coloring f verifies the relation in (1). So we must have $g(\alpha)<g(\beta)$ for some α and β with $\alpha<\beta<\lambda$. We cannot have $g(\gamma)<g(\delta)$ whenever $\beta\leq\gamma<\delta<\lambda$, since, noting that ra $(g)\subseteq$ ra $(f)\subseteq\tau$, this would contradict the assumption $\tau<\lambda$. Hence there must be ordinals γ, δ with $\beta\leq\gamma<\delta<\lambda$ such that $g(\alpha)<g(\gamma)\geq g(\delta)$. Let now $y_j, j<n$, be an enumeration of the elements of the set $\bigcup_{i<r}\{x_{\alpha\dot{+}i}, x_{\gamma\dot{+}i}, x_{\delta\dot{+}i}\}$ in increasing order (clearly, $n\leq 3r$). Then $y_0=x_\alpha$, and writing $y_k=x_\gamma$ and $y_l=x_\delta$, we have $0<k<l$ and

$$g(\alpha)=f(\{y_i: i<r\})<g(\gamma)=f(\{y_{k+i}: i<r\})\geq$$
$$\geq g(\delta)=f(\{y_{l+i}: i<r\});$$

so there must be a $j\leq l-2$ such that

$$f(\{y_i: j\leq i<j+r\})<f(\{y_i: j+1\leq i<j+r+1\})\geq$$
$$\geq f(\{y_i: j+2\leq i<j+r+2\}).$$

Then $\{y_i: j\leq i<j+r+2\}\in J_1$, which contradicts our assumption.

Suppose now there is an $X \subseteq \kappa$ with $[X]^{r+1} \subseteq J_1$ and $|X| = r+3$. Let $x_i, i < r+3$ be an enumeration of the elements of X in increasing order. Then $\{x_i : i < r+2\} \in J_1$, and so

$$f(\{x_i : 1 \leq i < r+1\}) \geq f(\{x_i : 2 \leq i < r+2\},$$

and $\{x_i : 1 \leq i < r+3\} \in J_1$, which implies

$$f(\{x_i : 1 \leq i < r+1\}) < f(\{x_i : 2 \leq i < r+2\}).$$

This is a contradiction, completing the proof of the theorem.

We now turn to a discussion of the corollaries of the above result.

Corollary 25.2. *Let* $\kappa \geq \omega$. *Then*

$$2^\kappa \nrightarrow (\kappa^+, 4)^3.$$

Proof. We have $2^\kappa \nrightarrow (\kappa^+)^2_2$ by (19.14), and so the result follows from (3).

Corollary 25.3. *Let* κ *be singular. Then*

$$\kappa \nrightarrow (\kappa, 4)^3.$$

Proof. We have $\kappa \nrightarrow (\kappa)^2_2$ for singular κ according to Corollary 21.3, and so the result follows from (3).

Corollary 25.4. *Assume* $\kappa \geq \omega$. *Then any of the conditions*

(i) $2^{\underline{\kappa}} = 2^\rho$ *for some* $\rho < \kappa$,

(ii) $2^{\underline{\kappa}} = \kappa > \mathrm{cf}(\kappa)$,

(iii) $\mathrm{cf}(\kappa) \nrightarrow (\mathrm{cf}(\kappa))^2_2$

ensures that

$$2^{\underline{\kappa}} \nrightarrow (\kappa, 4)^3$$

holds.

Proof. We have $2^{\underline{\kappa}} \nrightarrow (\kappa)^2_2$ under the above conditions by Corollary 21.5, and so the result follows from (3).

Corollary 25.5. *Let* $\kappa \geq \omega$ *be such that* $\kappa \nrightarrow (\kappa)^2_2$ *Then*

$$2^{\underline{\kappa}} \nrightarrow (\kappa, 4, 4)^3, \tag{6}$$

$$2^{\underline{\kappa}} \nrightarrow (\kappa, 5)^3, \tag{7}$$

and, furthermore,

$$2^{\underline{\kappa}} \nrightarrow (\kappa, r+1)^r \tag{8}$$

for any $r \geq 4$.

PROOF. We have $2^\kappa \nrightarrow (\kappa)_3^2$ for κ with $\kappa \nrightarrow (\kappa)_2^2$ according to (21.14), and so (2) implies (6), (3) implies (7), and (4) implies

$$2^\kappa \nrightarrow (\kappa, 5)^4 .$$

This confirms (8) in case $r=4$. In case $r>4$, (8) follows from here by a direct application of result a) in Subsection 9.7 (this result is a monotonicity property in the superscript of the ordinary partition relation).

We do not know whether (7) can be strengthened to $2^\kappa \nrightarrow (\kappa, 4)^3$. The simplest problem that remains open is whether

$$2^{\aleph_\omega} \rightarrow (\aleph_\omega, 4)^3$$

holds if $2^{\aleph_n} < 2^{\aleph_\omega} > \aleph_\omega$ for every $n < \omega$. We have already stated this problem as Problem 22.2 above.

The second theorem we prove is a kind of pseudo negative stepping-up, that is, it gives us a negative partition relation for $r=3$ that should be obtained from a relation for $r=2$, except that this latter fails to satisfy certain conditions, and so the lemma of the preceding section is not applicable.

THEOREM 25.6. *Assume* $\kappa \geq \omega$, $2^\rho \geq \mathrm{cf}(\kappa)$, $\kappa = \sum_{\nu < \mathrm{cf}(\kappa)} \kappa_\nu$, $\lambda_\xi \geq 4$ *for any* $\xi < \tau$, *and*

$$2^{\kappa_\nu} \nrightarrow (\lambda_\xi)_{\xi<\tau}^3 \tag{9}$$

for any $\nu < \mathrm{cf}(\kappa)$. *Then*

$$2^\kappa \nrightarrow ((\lambda_\xi)_{\xi<\tau}, (4)_\rho)^3 . \tag{10}$$

Though we did not assume here that κ is singular, the case when κ is regular contains nothing new. In fact, we have

$$2^\rho \nrightarrow (3)_\rho^2$$

according to (19.17), and so

$$2^{2^\rho} \nrightarrow (4)_\rho^3 \tag{11}$$

for any $\rho \geq \omega$ by (24.5). If κ is regular and the assumptions of the above theorem hold, then $2^{2^\rho} \geq 2^\kappa$, and so (10) is satisfied.

PROOF. Let

$$I^\nu = \langle I_\xi^\nu : \xi < \tau \rangle$$

be a partition of $[2^{\kappa_\nu}]^3$ verifying relation (9), and let

$$h: \mathrm{cf}(\kappa) \rightarrow {}^\rho 2$$

be a 1-1 mapping. Write

$$A = \times_{\nu < \mathrm{cf}(\kappa)} 2^{\kappa_\nu};$$

clearly, $|A| = 2^\kappa$. Let $\prec$ be the lexicographic ordering of A, i.e., let

$$\prec = \text{lex} \langle <_\nu : \nu < \text{cf}(\kappa) \rangle ,$$

where $<_\nu$ is the natural wellordering of the cardinal 2^{κ_ν}. We are going to define a partition

$$I = \langle I_\xi : \xi < \tau \dot{+} \rho \dot{+} \rho \rangle$$

of $[A]^3$ verifying (10), or, rather, the equivalent relation

$$2^\kappa \nrightarrow ((\lambda_\xi)_{\xi<\tau}, (4)_{\rho \dot{+} \rho})^3 . \tag{12}$$

To this end, consider an arbitrary element $\{f_0, f_1, f_2\}$ of $[A]^3$, where $f_0 \prec f_1 \prec f_2$. If $f_0 \cap f_1 = f_1 \cap f_2 = \nu (< \text{cf}(\kappa))$, then, observing that $f_0(\nu)$, $f_1(\nu)$, and $f_2(\nu)$ are pairwise distinct, put

$$\{f_0, f_1, f_2\} \in I_\xi ,$$

where ξ is such that

$$\{f_0(\nu), f_1(\nu), f_2(\nu)\} \in I_\xi^\nu .$$

If $\nu_0 = f_0 \cap f_1 \neq f_1 \cap f_2 = \nu_1$, then put

$$\{f_0, f_1, f_2\} \in I_{\alpha, i}$$

if $\alpha < \rho$ and $i < 2$ are such that

$$i = h(\nu_0)(\alpha) \neq h(\nu_1)(\alpha) \, (= 1 - i) .$$

For any $\alpha < \rho$ write

$$I_{\tau \dot{+} \alpha} = I_{\alpha,0} \qquad \text{and} \qquad I_{\tau \dot{+} \rho \dot{+} \alpha} = I_{\alpha,1} .$$

This completes the definition of the (not necessarily disjoint) partition I, since it is obvious that $\bigcup_{\xi < \tau \dot{+} \rho \dot{+} \rho} I_\xi = [A]^3$. We claim that I verifies (12).

Assume the contrary. Then there is a $\xi < \tau \dot{+} \rho \dot{+} \rho$ and an $X \subseteq A$ such that $[X]^3 \subseteq I_\xi$, and $|X| \geq \lambda_\xi (\geq 4)$ in case $\xi < \tau$, and $|X| \geq 4$ otherwise. Consider first the case $\xi < \tau$. We claim that in this case there is a $\nu < \text{cf}(\kappa)$ such that

$$f \cap g = \nu \tag{13}$$

for any two distinct $f, g \in X$. In fact, we have to show that $f \cap g = f' \cap g'$ whenever $f, g, f', g' \in X$ and $f \neq g$, $f' \neq g'$. Let $f_0 \prec f_1 \prec f_2 \prec f_3$ be elements of X such that $\{f, g, f', g'\} \subseteq \{f_0, f_1, f_2, f_3\}$. Then $\{f_0, f_1, f_2\}, \{f_1, f_2, f_3\} \in I_\xi$, and so $f_0 \cap f_1 = f_1 \cap f_2 = f_2 \cap f_3 = \nu$ for some ν. Then $f_0 \upharpoonright \nu = f_1 \upharpoonright \nu = f_2 \upharpoonright \nu = f_3 \upharpoonright \nu$ and $f_0(\nu) < f_1(\nu) < f_2(\nu) < f_3(\nu)$. Hence $f_i \cap f_j = \nu$ for any i, j with $i < j < 4$, and so $f \cap g = f' \cap g' = \nu$, verifying (13). $[X]^3 \subseteq I_\xi$ and (13) together mean that

$$[\{f(\nu) : f \in X\}]^3 \subseteq I_\xi^\nu .$$

As I^ν verifies relation (9), we have

$$|\{f(\nu): f\in X\}|<\lambda_\xi .$$

Noting that $f(\nu)\neq g(\nu)$ for any two distinct $f, g\in X$ by (13), this implies that $|X|<\lambda_\xi$, which contradicts the assumption $|X|\geq\lambda_\xi$.

Consider now the case $\xi\geq\tau$. Then $[X]^3\subseteq I_{\alpha,i}$ for some $\alpha<\tau$ and $i<2$. Let $f_0\prec f_1\prec f_2\prec f_3$ be elements of X, write $\nu_0=f_0\cap f_1$, $\nu_1=f_1\cap f_2$, and $\nu_2=f_2\cap f_3$. Then the relation $\{f_0, f_1, f_2\}\in I_{\alpha,i}$ means that

$$i=h(\nu_0)(\alpha)\neq h(\nu_1)(\alpha),$$

and $\{f_1, f_2, f_3\}\in I_{\alpha,i}$ means that

$$i=h(\nu_1)(\alpha)\neq \ldots,$$

which is a contradiction. The proof is complete.

Corollary 25.7. *Let* $\kappa\geq\omega$ *and* $2^\rho\geq\operatorname{cf}(\kappa)$. *Then*

$$2^\kappa\nrightarrow(\kappa, (4)_\rho)^3 .$$

Proof. If κ is regular, then $2^\rho\geq\kappa$, and so $2^\kappa\nrightarrow(4)^3_\rho$ according to (11). Hence we may assume that κ is singular; then $\kappa>\omega$ in particular. Write $\kappa=\sum_{\nu<\operatorname{cf}(\kappa)}\kappa_\nu$, where $\omega\leq\kappa_\nu<\kappa$. We have $2^{\kappa_\nu}\nrightarrow(\kappa^+, 4)^3$ by Corollary 25.2, and so we have $2^{\kappa_\nu}\nrightarrow(\kappa, 4)^3$ a fortiori. Hence $2^\kappa\nrightarrow(\kappa, 4, (4)_\rho)^3$, i.e., $2^\kappa\nrightarrow(\kappa, (4)_\rho)^3$, follows from the theorem just proved.

Remark. We do not know whether the condition $2^\rho\geq\operatorname{cf}(\kappa)$, which is equivalent to $\operatorname{cf}(\kappa)\nrightarrow(3)^2_\rho$ (cf. (17.32) and (19.17)), can be replaced by the weaker condition $\operatorname{cf}(\kappa)\nrightarrow(\operatorname{cf}(\kappa), (3)_\rho)^2$ if κ is singular. (This can certainly be done if κ is regular in view of (24.4)). This leads to the following unsolved problem:

Problem 25.8. Assume GCH. Does then

$$\aleph_{\omega_{\omega+1}+1}\nrightarrow(\aleph_{\omega_{\omega+1}}, (4)_{\aleph_0})^3 \tag{14}$$

hold?

Note that, assuming GCH, we have

$$\aleph_{\omega+1}\nrightarrow(\aleph_{\omega+1}, (3)_{\aleph_0})^2 \tag{15}$$

by Theorem 20.2, which implies

$$\aleph_{\omega_{\omega+1}}\nrightarrow(\aleph_{\omega_{\omega+1}}, (3)_{\aleph_0})^2 \tag{16}$$

according to Corollary 21.2, but Lemma 24.1 is not applicable here to deduce (14).

COROLLARY 25.9. *Let $\kappa \geq \omega$ and $2^\rho \geq \mathrm{cf}(\kappa)$, and assume that $2^\sigma < 2^{\underset{\cdot}{\kappa}}$ for any cardinal $\sigma < \kappa$. Then*

$$2^{2^{\underset{\cdot}{\kappa}}} \nrightarrow (\kappa, \kappa, (4)_\rho)^3. \tag{17}$$

PROOF. Our assumptions imply in view of Theorem 6.10.d) and c) that $\mathrm{cf}(2^{\underset{\cdot}{\kappa}}) = \mathrm{cf}(\kappa)$ and there are $\kappa_\nu < \kappa$ $(\nu < \mathrm{cf}(\kappa))$ such that $2^{\underset{\cdot}{\kappa}} = \sum_{\nu < \mathrm{cf}(\kappa)} 2^{\kappa_\nu}$. We have $2^{\kappa_\nu} \nrightarrow (\kappa_\nu^+, \kappa_\nu^+)^2$ according to (19.14), and so Lemma 24.1.a) implies $2^{2^{\kappa_\nu}} \nrightarrow (\kappa_\nu^+, \kappa_\nu^+)^3$; therefore we have $2^{2^{\kappa_\nu}} \nrightarrow (\kappa, \kappa)^3$ a fortiori. Using Theorem 25.6 now, we obtain the desired result.

To give an illustration for the role of this result, assume GCH now. Then

$$\aleph_1 \nrightarrow (\aleph_1, \aleph_1, (3)_{\aleph_0})^2$$

according to (19.14) and so

$$\aleph_{\omega_1} \nrightarrow (\aleph_{\omega_1}, \aleph_{\omega_1}, (3)_{\aleph_0})^2$$

by Corollary 21.2. The Negative Stepping-up Lemma (Lemma 24.1) is not applicable here, and fails to confirm the expected relation

$$2^{\aleph_{\omega_1}} = \aleph_{\omega_1+1} \nrightarrow (\aleph_{\omega_1}, \aleph_{\omega_1}, (4)_{\aleph_0})^3.$$

And yet, this relation follows from (17) with $\kappa = \aleph_{\omega_1}$. This is the state of affairs which we referred to as the pseudo stepping-up character of Theorem 25.6. There still remain gaps, however, which seem to be filled neither by stepping up nor by pseudo stepping up, as illustrated by Problem 25.8 above.

26. THE FINITE CASE OF THE ORDINARY PARTITION RELATION

The aim of this section is to prove the main known results concerning the partition relation $n \to (m)_s^r$, where m, n, r, and s are integers. The reason we do this at this place in the book is that the derivation of some negative results extensively relies on Lemma 24.2, a negative stepping-up lemma for the finite case whose proof heavily exploited the methods described in Sections 23 and 24, while many of the other proofs are elementary and so can be given at any place. For our discussion below, it will be convenient to extend the meaning of the partition relation

$$n \to (c_i)_{i<s}^r \tag{1}$$

for the case when the c_i's are allowed to denote arbitrary positive real numbers.

Define the meaning of this relation as the same as that of the relation

$$n \to (c'_i)^r_{i<s}$$

where, for each $i<s$, c'_i is the least integer $\geq c_i$. This is a natural way of defining the relation in (1). In fact, with this definition, (1) holds if and only if given any coloring $f: [n]^r \to s$, there is an $i<s$ such that there is a set of cardinality $\geq c_i$ that is homogeneous of color i with respect to f. The first result we prove is due to Erdős [1947] and uses a simple probabilistic argument.

THEOREM 26.1. *There is a positive quantity $c_{r,s}$ depending only on r and s such that*

$$n \not\to (c_{r,s} (\log n)^{\frac{1}{r-1}})^r_s \tag{2}$$

holds for any integers n, r, $s \geq 2$. In case $r=2$ and $s \geq 2$ we have

$$n \not\to (2 \log n / \log s)^2_s \tag{3}$$

for all integers $n \geq s^s$.

PROOF. Let k be an integer, and consider the chances for the existence of a k-element homogeneous set. Denoting by N the number of all colorings $f: [n]^r \to s$, we obviously have

$$N = s^{\binom{n}{r}}. \tag{4}$$

Given a fixed k-element set $X \subseteq n$ and a fixed color $i<s$, the number of those colorings $f: [n]^r \to s$ with respect to which X is a homogeneous set of color i is obviously $s^{\binom{n}{r}-\binom{k}{r}}$. Letting X and i run here, we obtain for the number N_k of those colorings $f: [n]^r \to s$ with respect to which there is a k-element homogeneous set that

$$N_k \leq \binom{n}{k} \cdot s \cdot s^{\binom{n}{r}-\binom{k}{r}}, \tag{5}$$

since $\binom{n}{k}$ and s are the number of possibilities for X and i, respectively. (We do not necessarily have equality here since a coloring may have several k-element homogeneous sets.) If $N_k < N$, then there is a coloring $f: [n]^r \to s$ with respect to which there is no k-element homogeneous set; i.e.,

$$n \not\to (k)^r_s \tag{6}$$

holds then. By (4) and (5), this is certainly the case if

$$\binom{n}{k} \cdot s \cdot s^{\binom{n}{r}-\binom{k}{r}} < s^{\binom{n}{r}}.$$

This is equivalent to

$$\binom{n}{k} < s^{\binom{k}{r}-1}; \tag{7}$$

using the estimate $n^k/k! > \binom{n}{k}$, we can see that this certainly holds if

$$n^k \leq k!\, s^{\binom{k}{r}-1}, \tag{8}$$

i.e., if

$$k \log n \leq \log k! + \left(\binom{k}{r} - 1\right) \log s. \tag{9}$$

What we have proved so far is that (8) or (9) implies (6). To show (2), observe that $\binom{k}{r} - 1 > c'_{r,s}\, k^r$ holds for all integers $k > r$ with some positive quantity $c'_{r,s}$ depending only on r and s. Assuming $k > r$, (9) therefore follows from the inequality

$$k \log n \leq \log k! + c'_{r,s}\, k^r \log s,$$

and so a fortiori from

$$k \log n \leq c'_{r,s}\, k^r \log s;$$

this latter is equivalent to

$$k \geq (c'_{r,s} \log s)^{\frac{1}{1-r}} (\log n)^{\frac{1}{r-1}}. \tag{10}$$

As we assumed $n \geq 2$, the inequality

$$k \geq c_{r,s} (\log n)^{\frac{1}{r-1}} \tag{11}$$

with a suitable positive $c_{r,s}$ ensures both that (10) holds and that $k > r$. Hence (9), and so (6), hold for any integer k satisfying (11). This establishes (2).

To prove (3), we go back to (9) again. For $r = 2$ this means that

$$k \log n \leq \log k! + (k(k-1)/2 - 1) \log s, \tag{12}$$

and we claim that this holds in case

$$s \log s \leq \log n \leq \frac{k}{2} \log s. \tag{13}$$

Using the second inequality here, (12) certainly holds if

$$\frac{k^2}{2} \log s \leq \log k! + (k(k-1)/2 - 1) \log s,$$

i.e., if

$$(k/2+1)\log s \leq \log k!.$$

As $s \leq k/2$ according to (13) and $s \geq 2$ by our assumptions, it is enough to show for this that

$$(k/2+1)\log \frac{k}{2} \leq \log k!$$

holds provided $k(\geq 2s) \geq 4$. We have

$$\log k! = \sum_{i=2}^{k} \log i > \int_1^k \log t\, dt = k \log k - k + 1,$$

and so it is sufficient to show that

$$(k/2+1)\log k/2 \leq k \log k - k + 1,$$

i.e., that

$$(k/2+1)(\log k - \log 2) \leq k \log k - k + 1$$

holds for $k \geq 4$. This is equivalent to

$$(k/2-1)(2-\log 2) \leq (k/2-1)\log k + 2\log 2 - 1,$$

and noting that $2\log 2 - 1 = \log \frac{4}{e} > 0$, it is enough to show for this that

$$(k/2-1)(2-\log 2) \leq (k/2-1)\log k,$$

i.e.,

$$2 - \log 2 \leq \log k$$

holds for $k \geq 4$. This last inequality can simply be checked by computation; hence (13) implies (12), i.e., (9) with $r=2$, which means that (6) holds with $r=2$ in case (13) is satisfied, i.e., in case $k \geq 2 \log n/\log s$ and $n \geq s^s$. This establishes (3). The proof of the theorem is complete.

Using a sieve method due to L. Lovász, J. Spencer [1975] was able to obtain a somewhat better estimate for N_k in case $r=2$ than given in (5), and thereby he proved that even the inequality

$$n^{k-2} \leq \frac{k!}{2} s^{(k+1)(k-2)/2} \tag{14}$$

entails the relation $n \not\to (k)_s^2$. This is a slight improvement upon the implication (8)⇒(6) in case $r=2$. In fact, noting that $\sqrt[k-2]{k!}$ and $\sqrt[k]{k!}$ asymptotically equal k/e, (14) roughly means that

$$n \leq \frac{k}{e} s^{\frac{k+1}{2}},$$

while (8) requires a little more, since in case $r=2$ it means about as much as

$$n \leq \frac{k}{e} s^{\frac{k-1}{2}}$$

for large k.

Relation (2) above is interesting only in case $s=2$ since, as is easily seen, the case $s>2$ easily follows from this by the Omission Rule (see Subsection 9.3) with $c_{r,s}=c_{r,2}$. As we shall immediately see, the case $r\geq 4$ is again of no interest, since we can derive stronger results then. Therefore we think that it is worth formulating the important particular cases of (2):

COROLLARY 26.2. *There are positive constants c_2 and c_3 such that the relations*

$$n \nrightarrow (c_2 \log n)_2^2 \tag{15}$$

and

$$n \nrightarrow (c_3 \sqrt{\log n})_2^3 \tag{16}$$

hold for every integer $n \geq 2$.

For an integer k let $\log_k$ denote the k times iterated logarithm, i.e., put

$$\log_0 x = x \qquad \text{and} \qquad \log_{i+1} x = \log \log_i x \tag{17}$$

for each integer i and every large enough real number x. We have the following result, which throws some light upon the situation also in case $r\geq 4$.

THEOREM 26.3 (Erdős–Hajnal–Rado [1965]). *Let $r\geq 3$. There are positive real numbers c_r, d_r, and n_r depending only on r such that*

$$n \nrightarrow (c_r \sqrt{\log_{r-2} n})_2^r \tag{18}$$

and

$$n \nrightarrow (d_r \log_{r-1} n)_4^r \tag{19}$$

hold for every integer $n \geq n_r$.

PROOF. The result easily follows from the above corollary with the aid of Lemma 24.2. As for (18), this is valid for $r=3$ according to (15). Using induction on r, assume that we have

$$n' \nrightarrow (c_{r-1} \sqrt{\log_{r-3} n'})_2^{r-1},$$

where $r>3$ and n' is such that $2^{n'-1} < n \leq 2^{n'}$. Then

$$2^{n'} \nrightarrow (2c_{r-1} \sqrt{\log_{r-3} n'})_2^r$$

holds according to (24.17). As $n>2^{n'-1}$, we have, say,

$$\log_{r-3} n' \leq 2 \log_{r-2} n$$

if n is large enough. Since we also have $n \leq 2^{n'}$, we obtain (18) from here with $c_r = 2\sqrt{2}\, c_{r-1}$.

As for (19), we first prove this in case $r=3$. Assume that n is large enough, e.g. $n > 20$ ($> e^e$), and choose n' such that $2^{n'-1} < n \leq 2^{n'}$. We have

$$n' \nrightarrow (c_2 \log n')^2_2$$

according to (15) above. (24.16) now implies that

$$2^{n'} \nrightarrow (c_2 \log n' + 1)^3_4.$$

As $2^{n'-1} < n$, we have $\log n' < 2 \log\log n$ for, say, $n > 20$. Thus (19) follows from here for $r=3$ with an appropriate $d_3 > 0$. Using induction on r as in the proof of (18), it is now easy to establish (19) for an arbitrary $r \geq 3$ with the aid of (24.17). The proof of the theorem is complete.

Let $\varepsilon > 0$ be an arbitrary real number. One can easily see from the above proof that there are integers $n_{r,\varepsilon}$ depending only on r and ε such that (18) and (19) hold with

$$c_r = (1+\varepsilon)\, 2^{r-3} c_3 \qquad \text{and} \qquad d_r = (1+\varepsilon)\, 2^{r-3} c_2 \tag{20}$$

provided $n \geq n_{r,\varepsilon}$. (For c_2 and c_3 see (15) and (16). $c_2 \leq 2/\log 2$ holds according to (3).)

We now turn to a discussion of positive relations. To this end we shall need the concept of end-homogeneous sets, which was introduced in Section 15. We first prove the following

LEMMA 26.4. *Let n, k, r, and s be integers, and assume that $k > r \geq 2$ and $s \geq 2$. Given a coloring $f\colon [n]^r \to s$, the existence of a k-element end-homogeneous set is guaranteed by the condition*

$$n \geq r + \sum_{i=r-1}^{k-2} s^{\binom{i}{r-1}}. \tag{21}$$

Hence, a fortiori, it is guaranteed by the condition

$$n \geq r + s \tag{22}$$

in case $k = r+1$, and by the condition

$$n \geq 2s^{\binom{k-2}{r-1}} \tag{23}$$

in case $k > r+1$.

PROOF. We are first going to deal with conditions (22) and (23). It is obvious that in case $k = r+1$ (21) and (22) are equivalent. We are going to show that in case $k > r+1$ (23) implies (21); in fact, writing $l = k-2$ and $m = r-1$, we are going to show that the inequality

$$m+\sum_{i=m}^{l} s^{\binom{i}{m}} < 2s^{\binom{l}{m}} \tag{24}$$

holds for every $l>m$ provided $m\geq 1$ and $s\geq 2$. We use induction on l. As for the case $l=m+1$, the above inequality simply says that

$$m+s+s^{m+1}<2s^{m+1},$$

i.e., that

$$m\leq s^{m+1}-s,$$

and this does in fact hold, as

$$s^{m+1}-s=\int_1^{m+1} s^t \log s\, dt \geq ms \log s \geq m\, 2 \log 2 > m,$$

because we assumed $s\geq 2$. Assume now that $l>m+1$ and that (24) holds with $l-1$ replacing l. Then we have

$$m+\sum_{i=m}^{l} s^{\binom{i}{m}} < 2s^{\binom{l-1}{m}} + s^{\binom{l}{m}}.$$

Hence, to show (24) we only have to verify that

$$2s^{\binom{l-1}{m}} \leq s^{\binom{l}{m}}.$$

This, however, obviously holds, since in case $l-1>m\geq 1$ and $s\geq 2$ we have

$$2s^{\binom{l-1}{m}} \leq s\cdot s^{\binom{l-1}{m}} = s^{\binom{l-1}{m}+1} \leq s^{\binom{l}{m}}.$$

Thus (24) is established, and so we have also proved the implication (23)⇒(21).

We now turn to the proof of the main part of the lemma. For this purpose assume that (21) holds, and let $f\colon [n]^r \to s$ be an arbitrary coloring; we shall then have to find an end-homogeneous set of k elements. In order to do this, we shall in effect perform a tree argument using the canonical partition tree associated with the coloring f (see Sections 15 and 18). In practice, however, we shall not construct the whole tree here, as we shall be able to find a long enough branch without bothering about the whole tree.

We are going to construct an end-homogeneous set

$$X=\{x_j\colon j<k\}\subseteq n,$$

where $x_0<\ldots<x_{k-1}<n$. To this end, we shall also have to define the sets S_j for $j\leq k-2$; intuitively, S_j will be the set of the potential successors of the sequence $\langle x_i\colon i<j\rangle$. Put $x_j=j$ and $S_j=\{i\colon j\leq i<n\}$ for $j\leq r-2$. Let $r-1\leq j\leq k-2$ and

assume that S_i and x_i have already been constructed for all $i<j$ such that

$$x_i = \min S_i \tag{25}$$

holds, and we have

$$f(Z \cup \{x_i\}) = f(Z \cup \{y\}) \tag{26}$$

whenever $y \in S_i$, $Z \subseteq \{x_l : l<i\}$, and $|Z| = r-1$. Assume, further, that

$$|S_{j-1}| > \sum_{i=j-1}^{k-2} s^{\binom{i}{r-1} - \binom{j-1}{r-1}}. \tag{27}$$

Note that (25) obviously holds in case $i<r-1$, and (26) holds vacuously in this case. Finally, (27) also holds with $j=r-1$. In fact, we have $|S_{r-2}| = n-r+2$ and $\binom{r-2}{r-1} = 0$, and so (27) says that

$$n-r+2 > \sum_{i=r-2}^{k-2} s^{\binom{i}{r-1}} = 1 + \sum_{i=r-1}^{k-1} s^{\binom{i}{r-1}},$$

which is true in view of (21).

Define the equivalence relation $\equiv_j$ on $S_{j-1} \setminus \{x_{j-1}\}$ by putting $y \equiv_j z$ for any $y, z \in S_{j-1} \setminus \{x_{j-1}\}$ satisfying

$$f(Z \cup \{x\}) = f(Z \cup \{y\}) \tag{28}$$

whenever $Z \subseteq \{x_i : i<j\}$ and $|Z| = r-1$. Observe that this equality always holds if $x_{j-1} \notin Z$ in view of (26) with $i=j-1$. Hence, noting that there are at most $\binom{j-1}{r-2}$ possibilities for Z satisfying $x_{j-1} \in Z$, we can see that there are at most $s^{\binom{j-1}{r-2}}$ equivalence classes under the relation $\equiv_j$. Thus, using (27), we can see that there is at least one equivalence class of cardinality $\geq$

$$(|S_{j-1}|-1)\, s^{-\binom{j-1}{r-2}} > \sum_{i=j}^{k-2} s^{\binom{i}{r-1} - \binom{j-1}{r-1} - \binom{j-1}{r-2}} =$$

$$= \sum_{i=j}^{k-2} s^{\binom{i}{r-1} - \binom{j}{r-1}}.$$

Choose this equivalence class as S_j. Then (27) holds with j replacing $j-1$. Putting $x_j = \min S_j$, (25) and (26) will also hold with $i=j$. In fact, (25) needs no comment, and (26) follows from the fact that S_j is an equivalence class under the relation $\equiv_j$.

So far we have constructed x_i and S_i for $i \leq k-2$. Noting that we have $|S_{k-2}| \geq 2$ in view of (27), let x_{k-1} be arbitrary such that

$$x_{k-1} \in S_{k-2} \setminus \{x_{k-2}\}. \tag{29}$$

This finishes the construction of the set $X = \{x_i: i<k\}$. It is easy to see from (25), (26) (both valid for $i \leq k-2$), and (29) that X is an end-homogeneous set with respect to f. The proof is complete.

To prove now the existence of large homogeneous sets, we need an analogue of Lemma 15.1. That lemma is not directly applicable here since it concerns colorings $f: [\alpha]^{<\omega} \to \tau$, which are of no use in the finite case. We have, however, the following variant of that lemma:

Lemma 26.5. *Let n, r, s, k and l be positive integers, and let $f: [n]^r \to s$ be a coloring. Assume that there is a k-element end-homogeneous set with respect to f and that we have*

$$k-1 \;\to\; (l-1)_s^{r-1}. \tag{30}$$

Then there is an l-element homogeneous set for f. Hence if there is a k-element end-homogeneous set for any coloring $f: [n]^r \to s$, then (30) implies

$$n \to (l)_s^r. \tag{31}$$

The proof is omitted, since it is an exact copy of the proof of (15.5) in Lemma 15.1 with n, s, $k-1$, and $l-1$ replacing α, τ, β, and v_ξ ($\xi < \tau$), respectively.

We have now made all the necessary preparations for proving positive ordinary partition relations. Noting that, as above, $\log_k$ denotes the k-times iterated logarithm (cf. (17) above), we have the following:

Theorem 26.6. *Let r, $s \geq 2$ be integers. Then there are positive quantities $c^*_{r,s}$ and $n_{r,s}$ depending only on r and s such that*

$$n \to (c^*_{r,s} \log_{r-1} n)_s^r \tag{32}$$

holds for all integers $n \geq n_{r,s}$. In particular, we have

$$n \to \left(\frac{\log n}{s \log s}\right)_s^2 \tag{33}$$

for all integers $n \geq 3$.

Proof. We shall first establish (33). Let $f: [n]^2 \to s$ be an arbitrary coloring. Taking the trivial relation

$$(l-2)\, s+1 \to (l-1)_s^1$$

into account, the preceding lemma implies that an end-homogeneous set of

$k=(l-2)s+2$ elements ensures the existence of an l-element homogeneous set. Noting that $r=2$ in the present case, (23) above becomes

$$n \geq 2s^{(l-2)s} \tag{34}$$

with this value for k; according to Lemma 26.4, this inequality guarantees the existence of a k-element end-homogeneous set, and so that of an l-element homogeneous set, provided $k \geq 3$; this latter holds in case $l \geq 3$. Since a two-element set is always homogeneous, (34) guarantees the existence of an l-element homogeneous set even without the assumption $l \geq 3$. (34) can be written as

$$\log n \geq \log 2 - s \log s + (l-1) s \log s.$$

As $s \geq 2$, we have $\log 2 < 2 \log 2 \leq s \log s$, and so this inequality certainly holds if

$$\log n \geq (l-1) s \log s,$$

i.e., if

$$l \leq \frac{\log n}{s \log s} + 1.$$

This inequality holds with the least integer l not smaller than $\dfrac{\log n}{s \log s}$, and so we have $n \to (l)_s^2$ with this l. Thus we have established (33).

We now turn to the proof of (32). We use induction on r. In case $r=2$, (32) is confirmed by (33) with $c_{r,s}^* = \dfrac{1}{s \log s}$. Assume $r>2$ now, and let $f\colon [n]^r \to s$ be a coloring. In order to have a k-element end-homogeneous set with $k>r+1$, it is enough according to (23) in Lemma 26.4 that we have

$$n \geq 2s^{\binom{k-2}{r-1}}.$$

This certainly holds if $n \geq 2s^{(k-2)^{r-1}}$, and so, a fortiori, if $n \geq s^{k^r}$, as $s \geq 2$. In other words, this latter inequality, that is, the inequality

$$k \leq (\log n / \log s)^{1/r} \tag{35}$$

ensures the existence of a k-element end-homogeneous set, provided $k>r+1$. As we have

$$k-1 \to (c_{r-1,s}^* \log_{r-2}(k-1))_s^{r-1}$$

for $k-1 \geq n_{r-1,s}$ by the induction hypothesis, the preceding lemma implies that (35) ensures the existence of a set of cardinality $\geq c_{r-1,s}^* \log_{r-2}(k-1)+1$ that is homogeneous with respect to f provided $k \geq r+1$ and $k > n_{r-1,s}$. Taking here the largest integer k satisfying (35), we can see that this number is greater than

$$\frac{2}{3r}c^*_{r-1,s}(\log_{r-1} n-\log_2 s)\geq\frac{1}{2r}c^*_{r-1,s}\log_{r-1} n$$

provided n is greater than an integer $n_{r,s}$ depending on r and s. This proves (32) with $c^*_{r,s}=\frac{1}{2r}c^*_{r-1,s}$. It is easy to see that if $\varepsilon>0$ then we can have $c^*_{r,s}=$ $=(2-\varepsilon)c^*_{2,s}/r!=\frac{2-\varepsilon}{r!\,s\log s}$ in (32), provided n is large enough depending on r, s, and ε. The proof is complete.

There is a different approach to the case $r=2$ of the above theorem. In fact, we have the following simple result due to P. Erdős and G. Szekeres [1935], which was rediscovered in Greenwood–Gleason [1955]:

THEOREM 26.7. *Let k and n be integers ≥ 2. Then we have*

$$\binom{n+k-2}{k-1}\rightarrow(k,n)^2. \tag{36}$$

PROOF. In case k or n equals 2 this simply says that $n\rightarrow(2,n)^2$ or $k\rightarrow(k,2)^2$, both of which obviously hold. Using induction, assume therefore that $k,n>2$, and that (36) holds if k or n is replaced with any $k'<k$ or $n'<n$, respectively. Let X be a set of cardinality $\binom{n+k-2}{k-1}$ and let $f\colon[X]^2\rightarrow 2$ be a coloring. Fix $z\in X$ arbitrarily, and write

$$X_0=\{x\in X\setminus\{z\}:f(\{xz\})=0\}$$

and

$$X_1=\{x\in X\setminus\{z\}:f(\{xz\})=1\}.$$

In view of the equality

$$\binom{n+k-2}{k-1}=\binom{n+k-3}{k-2}+\binom{n+k-3}{k-1}$$

we must have either

$$|X_0|\geq\binom{n+k-3}{k-2} \tag{37}$$

or

$$|X_1|\geq\binom{n+k-3}{k-1}. \tag{38}$$

If (37) holds, then the induction hypothesis implies that there is a subset Y_0 or Y_1 of X_0 of cardinality $k-1$ or n that is homogeneous of color 0 or 1, respectively, with respect to f. In the second case we are ready, and so are we also in the first case, since then $Y_0 \cup \{z\}$ is a homogeneous set of cardinality k and of color 0. Similarly, if (38) holds, then there is a subset Y_0 or Y_1 of X_1 of cardinality k or $n-1$ of X_1 that is homogeneous of color 0 or 1, respectively. In the first case we are ready, and so are we in the second case, since then $Y_1 \cup \{z\}$ is a homogeneous set of cardinality n and color 1. The proof is complete.

In case $n=k=l+1$ (36) gives that

$$\binom{2l}{l} \to (l+1)^2_2 . \tag{39}$$

holds for any integer $l \geq 1$. As we have

$$\binom{2l}{l} = 2^l \cdot \frac{1 \cdot 3 \cdot \ldots \cdot (2l-1)}{l!} \leq 2^l \cdot \frac{4 \cdot 6 \cdot \ldots \cdot 2l}{l!} = 2^{2l-1} \tag{40}$$

(equality above holds only in case $l=1$), this means that

$$2^{2l-1} \to (l+1)^2_2 \tag{41}$$

holds for any integer $l \geq 1$. One can easily derive (33) with $s=2$ from here. Note that (40) is not the best estimate for $\binom{2l}{l}$. It is well known that

$$\frac{4^l}{2\sqrt{l}} \leq \binom{2l}{l} < \frac{4^l}{\sqrt{2l+1}}$$

holds for all integers $l \geq 1$. Hence (41) can be improved to

$$4^l/\sqrt{2l+1} \to (l+1)^2_2 . \tag{42}$$

This is, however, insufficient to make any improvement in the constant factor $\frac{1}{2 \log 2}$ in (33) with $s=2$. Note the discrepancy in the constant factors in (3) and (33). Putting

$$r^2_2(n) = \sup \{x : x > 0 \text{ real } \& \; n \to (x \log n)^2_2\} ,$$

(3) and (33) with $s=2$ gives that

$$1/\log 4 \leq \liminf_{n \to \infty} r^2_2(n) \leq \limsup_{n \to \infty} r^2_2(n) \leq 2/\log 2 . \tag{43}$$

It would be interesting to know how much, if at all, this inequality can be improved. A more intriguing problem concerns the case $r=3$:

PROBLEM 26.8. Does there exist a positive quantity c such that

$$n \nrightarrow (c \log \log n)_2^3 \tag{44}$$

holds for every large enough integer n?

Compare this problem with (16), and with the case $r=3$ of (19) and (32).

CHAPTER VI

THE CANONIZATION LEMMAS

Given a coloring $f:[A]^r \to \tau$, the canonization lemmas ensure under certain assumptions that a 'large' subset of A can be represented as a disjoint union in such a way that the color of an r-tuple depends only on the summands to which the elements of this r-tuple belong, and not on the elements themselves. These lemmas will be very useful in establishing positive partition relations for singular cardinals, especially in the case $r=2$ of the ordinary partition relation, but they also have important applications in set theoretical topology (see Section 43) and in the theory of set mappings of type >1 (see Section 45).

27. SHELAH'S CANONIZATION

The main purpose of this section is to show that, as mentioned earlier, Corollary 21.5 indeed gives a necessary and sufficient condition for $2^{\kappa} \to (\kappa, \kappa)^2$ to hold in case 2^{κ} is singular (see Corollary 27.4 below). We shall actually do something more general here, as will be seen from the explanation given after the following

DEFINITION 27.1. Let $\langle A_\xi : \xi < \vartheta \rangle$ be a sequence of pairwise disjoint sets, write $A = \bigcup_{\xi < \vartheta} A_\xi$, and let $f:[A']^r \to \tau$ be a coloring, where $A' \supseteq A$. We say that $\langle A_\xi : \xi < \tau \rangle$ is *canonical* with respect to f if $f(u) = f(v)$ whenever $u, v \in [A]^r$ and $|u \cap A_\xi| = |v \cap A_\xi|$ for every $\xi < \vartheta$. We say that $\langle A_\xi : \xi < \vartheta \rangle$ is *weakly canonical* with respect to f if $f(u) = f(v)$ whenever $u, v \in [A]^r$ and $|u \cap A_\xi| = |v \cap A_\xi| \leq 1$ for every $\xi < \vartheta$.

Given a sequence $\langle A_\xi : \xi < \vartheta \rangle$ of pairwise disjoint sets and a coloring $f:[\bigcup_{\xi < \vartheta} A_\xi]^r \to \tau$, we shall be able to construct, under certain general conditions, another sequence $\langle B_\xi : \xi < \vartheta \rangle$ such that $B_\xi \subseteq A_\xi$, B_ξ is 'fairly large', and $\langle B_\xi : \xi < \vartheta \rangle$ is canonical or weakly canonical with respect to f. The construction of such a sequence $\langle B_\xi : \xi < \vartheta \rangle$ will be called canonization. It seems plausible that the canonicity of $\langle B_\xi : \xi < \vartheta \rangle$ with respect to f will facilitate the construction of a large homogeneous subset of $\bigcup_{\xi < \vartheta} B_\xi$. We shall soon be able to give an illustration confirming this. The main result of this section is Shelah's Canonization Lemma (cf. Shelah [1975]), which is an improvement for $r=2$ of the original canonization lemma of Erdős, Hajnal, and Rado (Lemma 28.1):

LEMMA 27.2 (Shelah's Canonization Lemma). *Let* $\langle \kappa_\xi : \xi < \vartheta \rangle$ *be a strictly increasing sequence of cardinals,* $\langle A_\xi : \xi < \vartheta \rangle$ *a sequence of pairwise disjoint sets,* $F_\xi \subseteq \mathscr{P}(A_\xi)$ *and, writing* $A = \bigcup_{\xi < \vartheta} A_\xi$, *let* $f : [A]^2 \to \tau$ *be a coloring. Assume that we have*

$$2^{\kappa_\xi} < 2^{\kappa_\eta} \tag{1}$$

for any $\xi < \eta < \vartheta$,

$$\kappa_0 \geq \vartheta,\ \tau,\ \omega, \tag{2}$$

$$|A_\xi| > 2^{\kappa_\xi} \tag{3}$$

for any $\xi < \vartheta$ *and, moreover,*

$$\text{for any } \xi < \vartheta \text{ and for any } X \subseteq A_\xi \text{ with } |X| = (2^{\kappa_\xi})^+ \\ \text{there is an } Y \in F_\xi \text{ such that } Y \subseteq X \text{ and } |Y| = \kappa_\xi^+ . \tag{4}$$

Then there are $B_\xi \subseteq A_\xi$, $\xi < \vartheta$, *such that*

$$B_\xi \in F_\xi, \quad |B_\xi| = \kappa_\xi^+ , \tag{5}$$

and $\langle B_\xi : \xi < \vartheta \rangle$ *is weakly canonical with respect to* f.

To illustrate the role of F_ξ in the above lemma, we mention a corollary; we shall, however, not need this corollary for future references. The real illustration of the role of the above lemma will be Corollary 27.4 below.

COROLLARY 27.3. *Let* κ_ξ, A_ξ, *and* f *be as in the above lemma, and assume* (1), (2), *and* (3). *Then there are* $B_\xi \subseteq A_\xi$ *with* $|B_\xi| = \kappa_\xi^+$ *such that the sequence* $\langle B_\xi : \xi < \vartheta \rangle$ *is canonical with respect to* f.

PROOF. Use the above lemma with

$$F_\xi = \{X \subseteq A_\xi : X \text{ is homogeneous with respect to } f\};$$

then (4) holds in view of the relation $(2^{\kappa_\xi})^+ \to (\kappa_\xi^+)^2_\tau$ (note that $\tau \leq \kappa_0 \leq \kappa_\xi$; cf. (17.32)). We obtain $B_\xi \subseteq A_\xi$ with $|B_\xi| = \kappa_\xi^+$ and $B_\xi \in F_\xi$ such that $\langle B_\xi : \xi < \vartheta \rangle$ is weakly canonical with respect to f. But then the relation $B_\xi \in F_\xi$ implies that this sequence is canonical with respect to f.

PROOF OF LEMMA 27.2. We may assume that

$$|A_\xi| = (2^{\kappa_\xi})^+ . \tag{6}$$

Let $\xi<\vartheta$ and $Z\subseteq\bigcup_{\alpha<\xi} A_\alpha$. Define the equivalence relation $\equiv_{\xi,Z}$ on A_ξ as follows: for any $x, y\in A_\xi$,

$$x\equiv_{\xi,Z} y \qquad \text{iff} \qquad \forall z\in Z[f(\{zx\})=f(\{zy\})]\,.$$

Call the element x of A_ξ good with respect to Z if there are at least $(2^{\kappa_\xi})^+$ elements y of A_ξ such that $x\equiv_{\xi,Z}y$. Call x good if it is good with respect to any $Z\subseteq\bigcup_{\alpha<\xi} A_\alpha$ with $|Z|\le\kappa_\xi$. We claim that there is a good $x\in A_\xi$ for any $\xi<\vartheta$. In fact, for a fixed $Z\subseteq\bigcup_{\alpha<\xi} A_\xi$ with $|Z|\le\kappa_\xi$, the number of equivalence classes under the equivalence relation $\equiv_{\xi,Z}$ is $\le 2^{\kappa_\xi}$. An element of A_ξ which is bad (i.e., not good) with respect to Z is contained in an equivalence class of cardinality $\le 2^{\kappa_\xi}$; hence, the number of bad elements with respect to Z is $\le 2^{\kappa_\xi}$. As $|\bigcup_{\alpha<\xi} A_\alpha|\le 2^{\kappa_\xi}$ in view of (1) and (6), there are at most $(2^{\kappa_\xi})^{\kappa_\xi}=2^{\kappa_\xi}$ possibilities for Z; therefore, all but 2^{κ_ξ} elements of A_ξ are good.

For any $\xi<\vartheta$, let $a_\xi\in A_\xi$ be good. We are going to define the sets $B_\xi\subseteq A_\xi$, $\xi<\vartheta$, by transfinite recursion such that (5) holds,

$$f(\{x_\xi a_\eta\})=f(\{y_\xi a_\eta\}) \tag{7}$$

for any $\xi<\eta<\vartheta$ and $x_\xi, y_\xi\in B_\xi$, and

$$f(\{x_\xi x_\zeta\})=f(\{x_\xi a_\zeta\}) \tag{8}$$

for any $\xi<\zeta<\vartheta$, $x_\xi\in B_\xi$, and $x_\zeta\in B_\zeta$. Fix $\alpha<\vartheta$, and suppose that the sets $B_\xi\subseteq A_\xi$ have already been constructed for all $\xi<\alpha$ in such a way that (5), (7), and (8) hold for all $\xi,\zeta<\alpha$ and $\eta<\vartheta$ (i.e., $\eta\ge\alpha$ is allowed here). We are about to define B_α. To this end, put

$$C_\alpha=\{x\in A_\alpha : x\equiv_{\alpha,\bigcup_{\xi<\alpha} B_\xi} a_\alpha\}\,;$$

as a_α is good and $|\bigcup_{\xi<\alpha} B_\xi|\le 2^{\kappa_\alpha}$ by (1), we have

$$|C_\alpha|>2^{\kappa_\alpha}\,. \tag{9}$$

Define the equivalence relation $\sim_\alpha$ on C_α as follows: for any $x, y\in C_\alpha$ put

$$x\sim_\alpha y \qquad \text{iff} \qquad \forall\beta[\alpha<\beta<\vartheta\Rightarrow f(\{xa_\beta\})=f(\{ya_\beta\})]\,.$$

There are at most $\tau^{|\vartheta|}$ equivalence classes under the relation $\sim_\alpha$; note that we have $\tau^{|\vartheta|}<2^{\kappa_0}\le 2^{\kappa_\alpha}$ according to (2). So, by (9), there is an equivalence class C'_α under the relation $\sim_\alpha$ that has cardinality $(2^{\kappa_\alpha})^+$. Let $B_\alpha\subseteq C'_\alpha$ be a set of cardinality κ_α^+ such that $B_\alpha\in F_\alpha$; there is such a B_α by (4). (5) clearly holds with $\xi=\alpha$. (7) also holds with $\xi=\alpha$, since $x_\alpha, y_\alpha\in B_\alpha\subseteq C'_\alpha$ implies that $x_\alpha\sim_\alpha y_\alpha$, and so indeed $f(\{x_\alpha a_\eta\})=f(\{y_\alpha a_\eta\})$ whenever $\alpha<\eta<\vartheta$. To see that (8) is valid with $\zeta=\alpha$, one has only to observe that $x_\alpha\in B_\alpha\subseteq C_\alpha$ implies that $x_\alpha\equiv_{\alpha,\bigcup_{\xi<\alpha}B_\xi} a_\alpha$, and so $f(\{x_\xi a_\alpha\})=f(\{x_\xi x_\alpha\})$ for any $\xi<\alpha$ and $x_\xi\in B_\xi$. The definition of the sequence $\langle B_\xi:\xi<\vartheta\rangle$ is complete. (7) and (8) imply that this sequence is weakly canonical with respect to f, which we wanted to prove.

COROLLARY 27.4 (Shelah [1975]). *Let κ be a singular cardinal such that $\kappa<2^\kappa$ and $2^\rho<2^\kappa$ for any $\rho<\kappa$, and assume that*

$$cf(\kappa)\to(\mathrm{cf}(\kappa))^2_2 . \tag{10}$$

Then we have

$$2^\kappa\to((\kappa)_2, (\mathrm{cf}(\kappa))_\tau)^2 \tag{11}$$

for any $\tau<\mathrm{cf}(\kappa)$.

This result confirms that Corollary 21.5 gives a necessary and sufficient condition for $2^\kappa\to(\kappa,\kappa)^2$ to hold in case κ is singular, namely, that $2^\kappa>\kappa$ and $2^\rho<2^\kappa$ for any $\rho<\kappa$.

PROOF. We shall establish (11) under the assumption

$$\mathrm{cf}(\kappa)\to(\mathrm{cf}(\kappa))^2_{2\dot{+}\tau} ; \tag{12}$$

it will follow from Theorems 29.1 and 29.5 below that (12) is equivalent to (10). Assume there is a coloring $f:[2^\kappa]^2\to 2\dot{+}\tau$ that refutes the relation in (11). By Theorem 6.10.c(ii), there are cardinals κ_ξ, $\xi<\mathrm{cf}(\kappa)$, such that $\kappa_\xi\geq\tau, \mathrm{cf}(\kappa)$,

$$2^\kappa=\sum_{\xi<\mathrm{cf}(\kappa)} 2^{\kappa_\xi} , \tag{13}$$

and

$$\kappa<2^{\kappa_\xi}<2^{\kappa_\eta}<2^\kappa \tag{14}$$

whenever $\xi<\eta<\mathrm{cf}(\kappa)$. Let $\langle A_\xi:\xi<\mathrm{cf}(\kappa)\rangle$ be a disjoint partition of 2^κ such that $|A_\xi|=(2^{\kappa_\xi})^+$, and, for any $\xi<\mathrm{cf}(\kappa)$, let

$$F_\xi=\{X_0\cup X_1: X_0, X_1\subseteq A_\xi \,\&\, |X_0|=|X_1|=\kappa_\xi^+ \,\&$$
$$\& \, f``[X_0]^2=\{0\} \,\&\, f``[X_1]^2=\{1\}\} .$$

Observe that (4) holds with this F_ξ. In fact, fixing $\xi<\mathrm{cf}(\kappa)$, we have to show that, for $i=0,1$ and for any set $Z\subseteq A_\xi$ of cardinality $(2^{\kappa_\xi})^+$, there is a subset of cardinality κ_ξ^+ of Z that is homogeneous of color i with respect to the coloring f. Assume that this is not the case e.g. for $i=1$. Then considering the relation $(2^{\kappa_\xi})^+\to((2^{\kappa_\xi})^+, (\kappa_\xi^+)_{1\dot{+}\tau})^2$, which follows from Corollary 17.5 since $\tau\leq\kappa_\xi$, we obtain that Z either has a homogeneous subset of color 0 that has cardinality

$(2^{\kappa_\xi})^+ > \kappa$, or it has a homogeneous subset of color ζ for some ζ with $2 \le \zeta < 2 \dot{+} \tau$ that has cardinality $\kappa_\xi^+ > \mathrm{cf}(\kappa)$. This, however, contradicts our assumption that f refutes the relation in (11).

Thus (4) holds, and so we can use Lemma 27.2. We obtain B_ξ, $\xi < \mathrm{cf}(\kappa)$. As $B_\xi \in F_\xi$ by (5), B_ξ has the form $X_0^\xi \cup X_1^\xi$, where

$$|X_i^\xi| = \kappa_\xi^+ \qquad \text{and} \qquad f``[X_i^\xi]^2 = \{i\} \qquad (i = 0, 1). \tag{15}$$

Define $\bar{f}: [\mathrm{cf}(\kappa)]^2 \to 2 \dot{+} \tau$ by putting

$$\bar{f}(\{\xi\eta\}) = f(\{x_\xi x_\eta\}) \tag{16}$$

for any ξ, η with $\xi < \eta < \mathrm{cf}(\kappa)$, where $x_\xi \in B_\xi$ and $x_\eta \in B_\eta$; as $\langle B_\xi : \xi < \mathrm{cf}(\kappa) \rangle$ is weakly canonical with respect to f, the value of $\bar{f}(\{\xi\eta\})$ does not depend on the particular choice of x_ξ and x_η. (12) implies that there is a set $H \subseteq \mathrm{cf}(\kappa)$ of cardinality $\mathrm{cf}(\kappa)$ that is homogeneous with respect to $\bar{f}$; let ζ be the color of this homogeneous set, i.e., let $\{\zeta\} = \bar{f}``[H]^2$ $(\zeta < 2 \dot{+} \tau)$. If $\zeta = 0$ or 1, then putting

$$X = \bigcup_{\xi \in H} X_\zeta^\xi ,$$

it is easy to see by (15) and (16) that X is a homogeneous set of color ζ and $|X| = \sum_{\xi \in H} \kappa_\xi^+$; as $|H| = \mathrm{cf}(\kappa)$, this is equal to κ in view of (13) and (14). So $|X| = \kappa$, which contradicts our assumption that f refutes (11). If $2 \le \zeta < 2 \dot{+} \tau$ then, pick arbitrary elements x_ξ from each B_ξ, $\xi < \mathrm{cf}(\kappa)$, and put $X = \{x_\xi : \xi \in H\}$. It is easy to see by (16) that X is a homogeneous set of cardinality $\mathrm{cf}(\kappa)$ and of color ζ, where $2 \le \zeta < 2 \dot{+} \tau$ now. This again contradicts our assumption that f refutes (11). The proof is complete.

The next corollary can be derived from Corollary 27.3 above, but Lemma 28.1 with $r = 2$, which is a sligthly weaker assertion given in the next section, would also suffice.

Corollary 27.5. *Let κ be a singular cardinal such that $2^\rho < 2^\kappa$ for all $\rho < \kappa$, let τ be an arbitrary cardinal, let ν_α, $\alpha < \tau$, be ordinals, and assume that*

$$cf(\kappa) \to (\mathrm{cf}(\kappa), (\nu_\xi)_{\xi < \tau})^2 . \tag{17}$$

Then

$$2^\kappa \to (\kappa, (\nu_\xi)_{\xi < \tau})^2 . \tag{18}$$

Proof. Let $f: [2^\kappa]^2 \to 1 \dot{+} \tau$ be an arbitrary coloring, and assume that there is no homogeneous set of order type ν_ξ and of color $1 \dot{+} \xi$ for any $\xi < \tau$. We have to prove that then there is a homogeneous set of order type (or cardinality) κ and of color 0. To this end, observe that by Theorem 6.10.c(ii) there are cardinals κ_α,

$\alpha < \mathrm{cf}(\kappa)$, such that

$$\mathrm{cf}(\kappa) < \kappa_\alpha < \kappa,$$

$$2^\kappa = \sum_{\alpha < \kappa} 2^{\kappa_\alpha},$$

and, for any α, β with $\alpha < \beta < \mathrm{cf}(\kappa)$,

$$2^{\kappa_\alpha} < 2^{\kappa_\beta}.$$

Let $\langle A_\alpha : \alpha < \mathrm{cf}(\kappa)\rangle$ be a disjoint partition of 2^κ such that $|A_\alpha| = (2^{\kappa_\alpha})^+$. Observing that we may assume $\nu_\xi \geq 3$ for any $\xi < \tau$, and that (17) with this assumption implies $\tau < \mathrm{cf}(\kappa)$, we can use Corollary 27.3. We obtain that there are sets $B_\alpha \subseteq A_\alpha$, $\alpha < \mathrm{cf}(\kappa)$, with $|B_\alpha| = \kappa_\alpha^+$ such that $\langle B_\alpha : \alpha < \mathrm{cf}(\kappa)\rangle$ is canonical with respect to the coloring f. This means, in particular, that each B_α is a homogeneous set with respect to f. Noting that $|B_\alpha| = \kappa_\alpha^+ > \mathrm{cf}(\kappa)$ and $\nu_\xi \leq \mathrm{cf}(\kappa)$ by (17), we can conclude by the assumption on homogeneous sets at the beginning of the proof that each B_α has color 0, i.e., that

$$f``[B_\alpha]^2 = \{0\}. \tag{19}$$

Consider the coloring $\bar{f}: [\mathrm{cf}(\kappa)]^2 \to 1 \dot{+} \tau$ defined as follows: for any two distinct $\alpha, \beta < \mathrm{cf}(\kappa)$ we put $\bar{f}(\{\alpha\beta\}) = \gamma$ if and only if $f(\{xy\}) = \gamma$ for every $x \in B_\alpha$ and $y \in B_\beta$; note that this definition is sound in view of the canonicity of $\langle B_\alpha : \alpha < \mathrm{cf}(\kappa)\rangle$. There is no homogeneous set of order type ν_γ and of color $1 \dot{+} \gamma$ with respect to $\bar{f}$ for any $\gamma < \tau$, since this would imply the same for f, and there is no such homogeneous set with respect to f in view of the assumption made at the beginning of the proof. Hence (17) implies that there is an $X \subseteq \mathrm{cf}(\kappa)$ of cardinality $\mathrm{cf}(\kappa)$ that is homogeneous of color 0 with respect to $\bar{f}$. Then (19) implies that the set

$$Y = \bigcup_{\alpha \in X} B_\alpha$$

is a homogeneous set of color 0 with respect to f. Y has cardinality $\sum_{\alpha \in X} \kappa_\alpha = \kappa$, which completes the proof.

28. THE GENERAL CANONIZATION LEMMA

In the preceding section we introduced the concept of canonical sequence (see Definition 27.1) and proved a canonization lemma for $r = 2$ (Lemma 27.2). Here we are going to prove a canonization lemma for an arbitrary integer r, though in case $r = 2$ this will say somewhat less than Lemma 27.2. The General

Canonization Lemma will be used later in order to obtain some results in set-theoretical topology and about set mappings and the square bracket symbol (see Sections 43, 45, and 54, respectively), and not in the context of the ordinary partition relation. The reason for including this lemma here rather than at some later place lies more in its proof than in its applications, and also in the fact that in the preceding section we discussed a canonization lemma that had important applications for the discussion of the ordinary partition relation. In order to state the lemma in question, we recall that $\exp_n(\kappa)$, where n is an integer and κ a cardinal, is defined by the recursive equations

$$\exp_0(\kappa)=\kappa\,,\quad \exp_{n+1}(\kappa)=\exp_n(2^\kappa)\,. \tag{1}$$

LEMMA 28.1 (General Canonization Lemma; cf. Erdős–Hajnal–Rado [1965]). *Let $\langle\kappa_\xi:\xi<\vartheta\rangle$ be a strictly increasing sequence of cardinals, $\langle A_\xi:\xi<\vartheta\rangle$ a sequence of pairwise disjoint sets, write $A=\bigcup_{\xi<\vartheta}A_\xi$, and let $\tau\geq 2$ be a cardinal and $r\geq 1$ an integer. For any i with $1\leq i\leq r$ let*

$$f_i:[A]^i\to\exp_{r-i}(\tau)$$

be a coloring. Assume that

$$\exp_{\binom{r}{2}}(\kappa_\xi)<\exp_{\binom{r}{2}}(\kappa_\eta) \tag{2}$$

for any $\xi<\eta<\vartheta$,

$$\kappa_0\geq\tau^{|\vartheta|},\,\omega \tag{3}$$

and, for any $\xi<\vartheta$,

$$|A_\xi|\geq(\exp_{\binom{r}{2}}(\kappa_\xi))^+\,. \tag{4}$$

Then there are sets $B_\xi\subseteq A_\xi$, $\xi<\vartheta$, such that

$$|B_\xi|\geq\kappa_\xi^+ \tag{5}$$

and the sequence $\langle B_\xi:\xi<\vartheta\rangle$ is canonical with respect to the coloring f_i for each i with $1\leq i\leq r$.

The purpose of having several colorings in the above lemma instead of just one is technical; namely, in this way it will be possible to prove the result by a simple induction. Our main concern in applications will be the coloring f_r. In case $r=2$ we obtain a weaker version of Corollary 27.3; indeed, the assumption in (3) here requires more than that in (27.2).

PROOF. We may assume that

$$\tau=\tau^{|\vartheta|}\geq\omega\,; \tag{6}$$

indeed, the only assumption on τ is that in (3), and this remains valid if we replace τ by $\tau'=\max\{\tau^{|\vartheta|},\aleph_0^{|\vartheta|}\}$. Note also that the lemma is trivially true for

$r=1$; in this case $\binom{r}{2}=r(r-1)/2=0$. We use induction on r; let $r\geq 2$ and suppose that the lemma is valid with $r-1$ replacing r. First we construct a sequence $\langle \bar{A}_\xi : \xi<\vartheta\rangle$ such that $\bar{A}_\xi \subseteq A_\xi$,

$$|\bar{A}_\xi| = (\exp_{\binom{r-1}{2}}(\kappa_\xi))^+ , \tag{7}$$

and, writing

$$\bar{A} = \bigcup_{\xi<\vartheta} \bar{A}_\xi ,$$

the following holds:

$$\begin{gathered} \text{if } u, v \subseteq \bar{A} , \quad |u|=|v|=i\leq r , \\ \xi<\vartheta \text{ is the largest ordinal with } u\cap \bar{A}_\xi \neq 0 , \\ \text{and } u\setminus \bar{A}_\xi = v\setminus \bar{A}_\xi , \text{ then } f_i(u)=f_i(v) . \end{gathered} \tag{8}$$

Fix $\xi<\vartheta$, and suppose that $\bar{A}_\eta$ has already been constructed for any $\eta<\xi$ such that (7) and (8) are satisfied. To define $\bar{A}_\xi$, define colorings g_i on $[A_\xi]^i$ for any i with $1\leq i\leq r$ by putting:

$$g_i(u) = \{\langle v, f_{i+j}(u\cup v)\rangle : j=|v|\leq r-i \;\&\; v\subseteq \bigcup_{\eta<\xi} \bar{A}_\eta\} ;$$

note that, writing

$$\bar{A}^\xi = \bigcup_{\eta<\xi} \bar{A}_\eta , \tag{9}$$

we have

$$\operatorname{ra}(g_i) \subseteq \bigcup_{j\leq r-i} {}^{[\bar{A}^\xi]^j}(\operatorname{ra}(f_{i+j})) .$$

Note also that we have $|\bar{A}^\xi|\leq \exp_{\binom{r-1}{2}}(\kappa_\xi)$ by (7) and (9), since (2) implies that

$$\exp_l(\kappa_\xi) < \exp_l(\kappa_\eta) \tag{10}$$

whenever $\xi<\eta<\vartheta$ and $l<\binom{r}{2}$. Hence

$$|\operatorname{ra}(g_i)| \leq \sum_{j\leq r-i} (\exp_{r-i-j}(\tau))^{(\exp_{\binom{r-1}{2}}(\kappa_\xi))^j} =$$

$$= \exp_{r-i}(\tau) + \sum_{1\leq j\leq r-i} (\exp_{r-i-j}(\tau))^{\exp_{\binom{r-1}{2}}(\kappa_\xi)} .$$

In case $i=r$ this implies

$$|\operatorname{ra}(g_r)| \leq \tau \leq \kappa_\xi , \tag{11}$$

and in case $i<r$ we may disregard the first term if we write $\exp_{r-i}$ instead of $\exp_{r-i-j}$. We obtain

$$\begin{aligned}|\mathrm{ra}\,(g_i)| &\le \exp_{r-i}(\tau)^{\exp_{\binom{r-1}{2}}(\kappa_\xi)} = 2^{\exp_{r-i-1}(\tau)\cdot\exp_{\binom{r-1}{2}}\kappa_\xi} \le \\ &\le 2^{\exp_{\binom{r-1}{2}}(\kappa_\xi)} = \exp_{\binom{r-1}{2}+1}(\kappa_\xi),\end{aligned} \tag{12}$$

since we have $\tau \le \kappa_0 \le \kappa_\xi$ by (3) and (2), and $r-i-1 \le r \dot{-} 2 \le \binom{r-1}{2}$, as $r \ge 2$. By (11), (12), and Corollary 16.5, there is a subset $\bar{A}_\xi$ of A_ξ of cardinality

$$(\exp_{\binom{r}{2}-(r-1)}(\kappa_\xi))^+ = (\exp_{\binom{r-1}{2}}(\kappa_\xi))^+$$

that is simultaneously homogeneous with respect to the colorings g_i, $1 \le i \le r$. In view of the definition of the colorings g_i, this simply means that (8) holds. This concludes the construction of the sequence $\langle \bar{A}_\xi : \xi < \vartheta \rangle$.

We are about to construct the sequence $\langle B_\xi : \xi < \vartheta \rangle$ by using the induction hypothesis. To this end, define the colorings h_i on $[\bar{A}]^i$ for $1 \le i \le r-1$ as follows: for any $u \in [\bar{A}]^i$ put

$$h_i(u) = \{\langle j, \xi, f_{i+j}(u \cup v)\rangle : u \subseteq \bar{A}^\xi \;\&\; v \in [\bar{A}_\xi]^j \;\&\; 1 \le j < r-i\}. \tag{13}$$

($\bar{A}^\xi$ was defined in (9).) Observe that, by virtue of (8), $f_{i+j}(u \cup v)$ here depends only on u, j and ξ, and not on the particular choice of v (note that u determines i). The number of possibilities for $f_{i+j}(u \cup v)$ on the right-hand side is

$$\le |\mathrm{ra}\,(f_{i+j})| \le \exp_{r-i-j}(\tau) \le \exp_{r-i-1}(\tau).$$

Hence

$$|\mathrm{ra}\,(h_i)| \le (\exp_{r-i-1}(\tau))^{|\vartheta|\cdot r}.$$

In case $i = r-1$ this implies

$$|\mathrm{ra}\,(h_{r-1})| \le \tau^{|\vartheta|\cdot r} = \tau = \exp_{(r-1)-(r-1)}(\tau) \tag{14}$$

in virtue of (6). If $1 \le i < r-1$, then we get

$$\mathrm{ra}\,(h_i) \le 2^{|\vartheta|\cdot r\cdot\exp_{r-i-2}(\tau)} = 2^{\exp_{r-i-2}(\tau)} = \exp_{(r-1)-i}(\tau), \tag{15}$$

since $|\vartheta| \cdot r \le \tau \le \exp_{r-i-2}(\tau)$ by (6). In view of (7), (10), (14), (15), and the induction hypothesis, we can use the lemma to be proved with $r-1$, $\langle \bar{A}_\xi : \xi < \vartheta \rangle$, and h_i $(1 \le i \le r-1)$ replacing r, $\langle A_\xi : \xi < \vartheta \rangle$, and f_i $(1 \le i \le r)$, respectively. We obtain the sets $B_\xi \subseteq \bar{A}_\xi$, $\xi < \vartheta$, such that $|B_\xi| \ge \kappa_\xi^+$ and the sequence $\langle B_\xi : \xi < \vartheta \rangle$ is canonical with respect to each coloring h_i, $1 \le i \le r-1$. Then $\langle B_\xi : \xi < \vartheta \rangle$ is also canonical with respect to each f_k, $1 \le k \le r$. Indeed, in case $k = 1$ this simply means that each B_ξ is homogeneous with respect to f_1, and this is indeed so, as already $\bar{A}_\xi$ was homogeneous with respect to each f_k, $1 \le k \le r$, by (8); in case $1 < k \le r$, the canonicity of $\langle B_\xi : \xi < \vartheta \rangle$ with respect to f_k follows from the homogeneity of B_ξ, $\xi < \vartheta$, with respect to f_k, the canonicity of $\langle B_\xi : \xi < \vartheta \rangle$ with respect to the h_i's, and the remark made after (13), namely that, in (13), $f_{i+j}(u \cup v)$ depends only on u, j, and ξ, and not on v. The proof of the lemma is complete.

COROLLARY 28.2. *Let κ be a strong limit cardinal, i.e., such that $2^\rho < \kappa$ for any $\rho < \kappa$, and assume κ is singular. Let $r \ge 1$ be an integer, $\tau < \kappa$ a cardinal, and $f : [\kappa]^r \to \tau$ a coloring. Then there are pairwise disjoint sets $B_\xi \subseteq \kappa_\xi$, $\xi < \mathrm{cf}(\kappa)$, of cardinality $< \kappa$ such that $\bigcup_{\xi < \mathrm{cf}(\kappa)} B_\xi$ has cardinality κ and the sequence $\langle B_\xi : \xi < \mathrm{cf}(\kappa) \rangle$ is canonical with respect to f.*

PROOF. Writing $\vartheta = \mathrm{cf}(\kappa)$, there is a sequence $\langle \kappa_\xi : \xi < \vartheta \rangle$ of cardinals $< \kappa$ satisfying the requirements of Lemma 28.1 such that

$$\kappa = \sum_{\xi < \vartheta} \kappa_\xi = \sum_{\xi < \vartheta} (\exp_{\binom{r}{2}} \kappa_\xi)^+ .$$

Let $\langle A_\xi : \xi < \vartheta \rangle$ be a disjoint partition of κ such that $|A_\xi| = (\exp_{\binom{r}{2}}(\kappa_\xi))^+$. Applying Lemma 28.1, we obtain the sets B_ξ with the required properties. The proof is complete.

CHAPTER VII

LARGE CARDINALS

The ordinary partition relation shows remarkably different features for inaccessible and for accessible cardinals. A crucial role is played here by weakly compact inaccessible cardinals, which are the main topic of the next few sections. The chapter is concluded by a section concerning saturated ideals and partition properties of measurable cardinals.

29. THE ORDINARY PARTITION RELATION FOR INACCESSIBLE CARDINALS

The simplest instance of Ramsey's theorem says that $\omega \to (\omega)_2^2$ holds. The question arises whether an analogous relation holds for any other infinite cardinal replacing ω. A simple result in this direction is

THEOREM 29.1 ([Erdős–Hajnal–Rado [1965]). *If* $\kappa > \omega$ and $\kappa \to (\kappa)_2^2$ *holds, then* κ *is inaccessible.*

PROOF. Observe that κ must be strong limit, i.e., $2^\rho < \kappa$ must hold for any $\rho < \kappa$. In fact, if $2^\rho \geq \kappa$ for some $\rho < \kappa$, then $\kappa \not\geq (\kappa)_2^2$ holds in view of the relation $2^\rho \not\to (\rho^+)_2^2$ (see (19.14)). We have to show that κ must be regular as well. This too is confirmed by our earlier results (namely, Corollary 21.3), but it is worth presenting a direct proof. Assume κ is singular and let $\langle \alpha_\xi : \xi < \mathrm{cf}(\kappa) \rangle$ be an increasing continuous sequence with $\alpha_0 = 0$ (i.e., such that $\alpha_\xi = \bigcup_{\eta < \xi} \alpha_{\eta+1}$ for all $\xi < \mathrm{cf}(\kappa)$) tending to κ. We define a coloring $f: [\kappa]^2 \to 2$ verifying $\kappa \not\to (\mathrm{cf}(\kappa)^+, \kappa)^2$ as follows. For any $\mu, \nu < \kappa$ with $\mu < \nu$ put $f(\{\mu\nu\}) = 0$ if there is an α_ξ with $\mu < \alpha_\xi \leq \nu$, and put $f(\{\mu\nu\}) = 1$ otherwise. Now, if X is a homogeneous set of color 0, then $|X| \leq \mathrm{cf}(\kappa)$, as X has at most one element in any one of the intervals $[\alpha_\xi, \alpha_{\xi+1})$, and if X is a homogeneous set of color 1, then $|X| < \kappa$, as $X \subseteq [\alpha_\xi, \alpha_{\xi+1})$ for some $\xi < \mathrm{cf}(\kappa)$. The proof is complete.

The method used in obtaining the above theorem does not give any indication as to whether or not $\kappa \to (\kappa)_2^2$ holds for certain inaccessible cardinals. In their paper [1943], Erdős and Tarski even comtemplated the possibility that this might be independent of the usual axioms of set theory. A major breakthrough was achieved in 1961, when W. P. Hanf and A. Tarski proved that the relation

$\kappa \to (\kappa)^2_2$ fails for a large class of inaccessible cardinals, including the first one. Their result shows that if $\kappa \to (\kappa)^2_2$ holds, then κ must be a very large inaccessible cardinal that cannot be characterized in a 'constructive' way. This means, in particular, that it is consistent with the usual axioms of set theory (plus an axiom saying that there are 'many' inaccessible cardinals) that the relation $\kappa \to (\kappa)^2_2$ fails for any inaccessible cardinal κ, since we can cut down the universe of sets anywhere below the first cardinal κ (if there is any) for which $\kappa \to (\kappa)^2_2$ holds. This means that the assumption that there are inaccessible cardinals satisfying $\kappa \to (\kappa)^2_2$ can be considered a new axiom of set theory, in effect saying that the universe of sets cannot be cut down below a certain 'large' cardinal. Hence the name, introduced by K. Gödel [1947], of this type of axiom being a large cardinal axiom. There are many more, and much stronger, large cardinal axioms than the one we mentioned here. Their primary importance lies partly in the fact they have interesting consequences for 'small' sets, e.g. subsets of ω. The reader interested in large cardinals should turn to F. Drake's book [1974], which also has a good bibliography on the subject.

Many properties equivalent to $\kappa \nrightarrow (\kappa)^2_2$ of an inaccessible cardinal κ are given in the paper of Keisler and Tarski [1964]; these properties concern a large number of branches of mathematics such as infinitary logic, model theory, topology, the theory of Boolean algebras, representation theorems, separation principles, and so on. They all show the importance of the ordinary partition relation. It is beyond the scope of this book to reproduce all these properties; we shall nonetheless need a few of them. In the rest of this section we shall prove a number of implications between some of these properties, which will show that all problems concerning the relation $\kappa \to (\lambda_\xi)^r_{\xi<\tau}$ for inaccessible κ that cannot be decided with the aid of our earlier results can be reduced to the problem whether $\kappa \to (\kappa)^2_2$ holds. All these implications except (iv)$\Rightarrow$(v) below are stated in Silver [1971], where complete references are given. Thus when proving these implications, we shall only mention the authors' name, while omitting exact references. In order to formulate and prove these implications, we shall need some definitions; the first one concerns trees (for the definition of tree and related concepts see Section 13).

DEFINITION 29.2. Let κ be an infinite cardinal. An *Aronszajn κ-tree* is a tree of length κ that has no branch of length κ and is such that each of its levels has cardinality $<\kappa$. The cardinal κ is said to have the *tree property* if there is no Aronszajn κ-tree.

König's lemma (Lemma 10.3) says that ω has the tree property, and it is a well-known theorem of Aronszajn and Specker that there is an Aronszajn $\aleph_1$-tree (Theorem 52.1 below).

DEFINITION 29.3. (i) A nonempty set (of sets) F is called a *field of sets* if for every $x, y \in F$ we have $x \cup y \in F$ and $\bigcup F \setminus x \in f$ (so, in particular, 0 and $\bigcup F$ belong to F). For a cardinal κ, F is called ***κ-complete*** if in addition we have $\bigcup X \in F$ for any set $X \subseteq F$ of cardinality $<\kappa$.

(ii) Given an arbitrary set (of sets) G and a cardinal κ, a field of sets *κ-generated* by G is a κ-complete field of sets $F \subseteq \mathscr{P}(\bigcup G)$ such that it includes G and is minimal with respect to inclusion. Instead of '3-generated' one simply says '***generated***', and this is the same as 'ω-generated'.

Observe that in order to see that F is a κ-complete field of sets it is enough to show that $\bigcup F \setminus \bigcup X \in F$ whenever $X \subseteq F$ and $|X| < \kappa$.

LEMMA 29.4. *For any set G and for any regular cardinal κ, there is exactly one field F κ-generated by κ, and F has cardinality $\leq |G|^{\kappa} + 1$.*

PROOF. It is obvious that there is a unique field F κ-generated by G. In fact, we have

$$F = \bigcap \{F' \subseteq \mathscr{P}(\bigcup G) \colon G \subseteq F' \text{ and } F' \text{ is a } \kappa\text{-complete field of sets}\}.$$

(note that the set on the right-hand side here is not empty, since $\mathscr{P}(\bigcup G)$ belongs to it). In order to see that the assertion about the cardinality of F is valid we have to give another definition of F. To this end put $F_0 = G$, and if we have already defined F_β for any $\beta < \alpha$, where α is an arbitrary ordinal, then put

$$F_\alpha = \{\bigcup G \setminus \bigcup X \colon X \subseteq \bigcup_{\beta < \alpha} F_\beta \;\&\; |X| < \kappa\}. \tag{1}$$

We claim that F_κ is a κ-complete field of sets. By our remark after Definition 29.2, it is enough to show to this end that

$$\bigcup G \setminus \bigcup X \in F_\kappa \tag{2}$$

whenever $X \subseteq F_\kappa$ and $|X| < \kappa$. First we show that if $y \in F_\kappa$ then $y \in F_\xi$ for some $\xi < \kappa$. In fact, y has the form $\bigcup G \setminus \bigcup Y$ for some $Y \subseteq \bigcup_{\alpha < \kappa} F_\alpha$ with $|Y| < \kappa$. By the regularity of κ, we have $Y \subseteq \bigcup_{\alpha < \xi} F_\alpha$ for some $\xi < \kappa$; but then $y \in F_\xi$ by (1). To prove (2), note that $X \subseteq F_\alpha$ implies $X \subseteq \bigcup_{\xi < \kappa} F_\xi$ by the assertion just proved. Hence $\bigcup G \setminus \bigcup X \in F_\kappa$ by (1), and this is what we asserted in (2). Hence our claim that F_κ is a κ-complete field of sets is established. Now, on the one hand, F_κ obviously includes $G = F_0$ and, on the other, any field of sets including G must include all F_α, and so F_κ in particular. Hence F_κ equals the field F κ-generated by G (and $F_\alpha = F_\kappa$ for all $\alpha \geq \kappa$, but we shall not need this).

So $F = F_\kappa$, and all we have to do is to estimate the cardinality of F_κ. We consider the case $|G| \geq 2$ first. Writing $\lambda = |G|$, we then have to prove that $|F_\kappa| \leq \lambda^{\underline{\kappa}} + 1 = \lambda^{\underline{\kappa}}$. We have $|F_0| = |G| = \lambda$, and, obviously,

$$|F_\alpha| \leq \Big(\sum_{\beta<\alpha} |F_\beta| \Big)^{\underline{\kappa}}$$

holds for all $\alpha > 0$. Fix α with $0 < \alpha \leq \kappa$, and assume that $|F_\beta| \leq \lambda^{\underline{\kappa}}$ for all $\beta < \alpha$. Then, noting that κ was assumed to be regular, and using Theorems 6.10e and 6.3, we obtain that

$$|F_\alpha| \leq (\kappa \cdot \lambda^{\underline{\kappa}})^{\underline{\kappa}} = \sum_{\rho<\kappa} \kappa^{\rho} \cdot (\lambda^{\underline{\kappa}})^{\rho} = \sum_{\rho<\kappa} \kappa \cdot \kappa^{\rho} \cdot \lambda^{\underline{\kappa}} =$$

$$= \kappa \cdot \lambda^{\underline{\kappa}} \cdot \sum_{\rho<\kappa} \sum_{\sigma<\kappa} \sigma^{\rho} \leq \kappa \cdot \lambda^{\underline{\kappa}} \cdot \sum_{\rho<\kappa} \sum_{\sigma<\kappa} 2^{\sigma \cdot \rho} =$$

$$= \kappa \cdot \lambda^{\underline{\kappa}} \cdot 2^{\underline{\kappa}} = \lambda^{\underline{\kappa}},$$

where the last equality holds by our assumption $\lambda = |G| \geq 2$. Hence $|F_\alpha| \leq |G|^{\underline{\kappa}}$ for all α in case $|G| \geq 2$, and so $|F| \leq |G|^{\underline{\kappa}}$, proving the lemma in this case. If $|G| \leq 1$, then the field F κ-generated by G consists of the empty set and the only element (if any) of G, and so $|F| \leq |G| + 1 \leq |G|^{\underline{\kappa}} + 1$. This completes the proof.

Definition 29.5. (i) Given a field F of sets, an *ideal in* F is a set $I \subseteq F$ such that $X \in I$, $Y \subseteq X$, and $Y \in F$ imply $Y \in I$, and whenever $X, Y \in I$ we have $X \cup Y \in I$.

(ii) I is an ideal *over* F if I is an ideal in F and $\bigcup I = \bigcup F$. The ideal I in F is *proper* if $\bigcup F \notin I$.

(iii) For a cardinal κ, the ideal I in F is κ-*complete* if we have $\bigcup X \in I$ for any $X \subseteq I$ with $|X| < \kappa$ and $\bigcup X \in F$.

(iv) The ideal I in F is *prime* if it is proper and either $X \in I$ or $\bigcup F \setminus X \in I$ holds for any $X \in F$.

(v) The cardinal κ is said to have the *prime ideal property* if every proper κ-complete ideal I over a field F of sets with $\bigcup F = \kappa$, κ-generated by a set of cardinality $\geq \kappa$, is included in a κ-complete prime ideal over F (it is customary to say in this case that I can be *extended* to a κ-complete prime ideal over F).

Note that every ideal is ω-complete. Similarly as in the proof of Theorem 10.1, one can show that a proper ideal in a field F of sets is prime if and only if it is maximal with respect to inclusion among the proper ideals in F. As Zorn's lemma implies that every proper ideal can be extended to a maximal proper ideal (and this latter is ω-complete by our first remark), it follows that ω has the prime ideal property. We are now in a position to turn to the main result of this section.

This gives several properties of an inaccessible cardinal κ equivalent to $\kappa \not\to (\kappa)^2_2$. The proof that this property holds for many inaccessible cardinals (the Hanf–Tarski theorem mentioned above), as well as other equivalent properties will be given in Sections 30–33.

THEOREM 29.6 (Erdős–Tarski [1961], Hanf [1964a], Hajnal [1964]). *Let κ be an inaccessible cardinal or $\kappa = \omega$. Then the following properties are equivalent:*

(i) *κ does not have the prime ideal property.*

(ii) *κ does not have the tree property.*

(iii) *There is an ordered set $\langle A, \prec \rangle$ such that $|A| = \kappa$ and $\kappa, \kappa^* \not\leq \operatorname{tp} \langle A, \prec \rangle$.*

(vi) $\kappa \not\to (\kappa)^2_2$.

(v) $\kappa \not\to (\kappa, 4)^3$.

(vi) *There are an integer $r \geq 2$ and a cardinal $\tau < \kappa$ such that $\kappa \not\to (\kappa)^r_\tau$.*

(vii) *$\kappa \not\to \langle \kappa \rangle_\tau$ holds with some cardinal $\tau < \kappa$ (see* (15.2) *for the definition of the partition symbol used).*

Note that all these properties fail for $\kappa = \omega$. (i) by our remark preceding this theorem, (ii), as we remarked above, by Lemma 10.3, (iii) by obvious considerations, and (iv)–(vi) by Ramsey's theorem (Theorem 10.2), and as for (vii), the second proof of that theorem shows that $\omega \to \langle \omega \rangle_n$ holds for all integers n.

Hence we could disregard the case $\kappa = \omega$ in the proof below. Seemingly, this is indeed what we shall do; in actual fact, all our arguments below apply to the case $\kappa = \omega$ as well, and even if we may use a phrase like "since κ is an inaccessible cardinal", this could always be replaced by the longer phrase "since either $\kappa = \omega$ or κ is an inaccessible cardinal". So our proof below gives a "new" proof of Ramsey's theorem relying on the fact that κ does not have the tree property, essentially identical to the second proof of this theorem given in Section 10, with the difference that the proof here is more economical, since we now have the concept of partition tree at our disposal.

PROOF. We shall establish the following implications: (i)$\Rightarrow$(ii), (ii)$\Rightarrow$(i), and (ii)$\Rightarrow$(iii), (iii)$\Rightarrow$(iv), ..., (vii)$\Rightarrow$(ii). As will be clear in the proofs of these implications below, only the regularity, and not the inaccessibility of κ is needed in the proofs of (ii)$\Rightarrow$(i) and (ii)$\Rightarrow$(iii), and not even the regularity of κ is used in the proofs of (iii)$\Rightarrow$(iv), (iv)$\Rightarrow$(v), (v)$\Rightarrow$(vi), and (vi)$\Rightarrow$(vii). $\kappa \geq \omega$ is of course always assumed.

(i)$\Rightarrow$(ii). Assuming that κ has the tree property, we have to show that it also has the prime ideal property. To this end, let F be a field of sets with $\bigcup F = \kappa$ κ-generated by a set of cardinality $\leq \kappa$, and let I be a proper κ-complete ideal over F; we have to show that I can be extended to a κ-complete prime ideal over F. Observe that F has cardinality $\leq \kappa$ in view of the inaccessibility of κ by Lemma

29.4; in case $|F|<\kappa$ we would have $\bigcup F=\bigcup I\in I$ by the *κ-completeness* of I; this is, however, not the case, since I is proper. So $|F|=\kappa$; let $\langle A_\xi\colon \xi<\kappa\rangle$ be an enumeration of the elements of F. For every $\xi<\kappa$ let $x_\xi\in\bigcup F$ be such that

$$x_\xi\notin\bigcup\{A_\eta\colon \eta<\xi \,\&\, A_\eta\in I\}\,. \tag{3}$$

Note that the set on the right-hand side here belongs to I, and I is proper; so there is such an x_ξ. We define a function $f_\xi\colon \xi\to 2$ by putting

$$f_\xi(\eta)=0 \qquad \text{iff} \qquad x_\xi\notin A_\eta \tag{4}$$

for every $\eta<\xi$. The set of functions

$$T=\{f_\xi\restriction\alpha\colon \alpha\leq\xi<\kappa\}$$

ordered by inclusion is a tree of length κ, and it is obvious by the inaccessibility of κ that, for any $\alpha<\kappa$, the αth level of T, i.e., the set of functions in T whose domains are α, has cardinality $<\kappa$. Since we assumed that κ has the tree property, this implies that T has a branch of length κ. Let $b\subseteq T$ be such a branch, and let $f=\bigcup b$; clearly, f is a function from κ into 2, and $b=\{f\restriction\alpha\colon \alpha<\kappa\}\subseteq T$, i.e.,

$$\forall\alpha<\kappa\exists\xi[\alpha\leq\xi<\kappa \,\&\, f\restriction\alpha=f_\xi\restriction\alpha]\,. \tag{5}$$

We claim that the set

$$J=\{A_\eta\colon \eta<\kappa \,\&\, f(\eta)=0\}$$

is a κ-complete prime ideal extending I. In fact, J includes I, since if $A_\eta\in I$ then $f_\xi(\eta)=0$ for any ξ with $\eta<\xi<\kappa$ by (3) and (4), and so $f(\eta)=0$ by (5), i.e., $A_\eta\in J$. If $A_\zeta\in J$ and $A_\eta\subseteq A_\zeta$, then $A_\eta\in J$. To see this, choose an $\alpha<\kappa$ with $\zeta,\eta<\alpha$; by (5), there is a ξ with $\alpha\leq\xi<\kappa$ such that $f\restriction\alpha=f_\xi\restriction\alpha$. We have $f_\xi(\zeta)=f(\zeta)=0$ in view of $A_\zeta\in J$; hence (4) implies that $x_\xi\notin A_\zeta$. But then $x_\xi\notin A_\eta(\subseteq A_\zeta)$, and so $f(\eta)=f_\xi(\eta)=0$ again by (4), i.e., $A_\eta\in J$, as we wanted to show. J is a κ-complete ideal. Indeed, if $X\subseteq\kappa$, $|X|<\kappa$, and $A_\eta\in J$ for every $\eta\in X$, then we have to show that $\bigcup[A_\eta\colon \eta\in X\}\in J$. Note here that the field F is κ-complete (since it is κ-generated by a set), and so this set belongs to F; i.e., $\bigcup[A_\eta\colon \eta\in X\}=A_\gamma$ for some $\gamma<\kappa$. To see that $A_\gamma\in J$, let $\alpha<\kappa$ be an ordinal exceeding γ and all elements of X; in view of (5), there is a ξ with $\alpha\leq\xi<\kappa$ such that $f\restriction\alpha=f_\xi\restriction\alpha$. Our assumptions imply $f_\xi(\eta)=f(\eta)=0$ for all $\eta\in X$; hence $x_\xi\notin A_\eta$ by (4) if $\eta\in X$. So $x_\xi\notin A_\gamma= =\bigcup\{A_\eta\colon \eta\in X\}$; i.e., $f(\gamma)=f_\xi(\gamma)=0$ again by (4), and so $A_\gamma\in J$, and this is what we wanted to show. J is proper, i.e., $\bigcup F\notin J$. To verify this, let $\eta<\kappa$ be such that $A_\eta=\bigcup F(\in F)$, choose an α with $\eta<\alpha<\kappa$, and, in accordance with (5), let ξ be such that $\alpha\leq\xi<\kappa$ and $f\restriction\alpha=f_\xi\restriction\alpha$. Noting that $x_\xi\in\bigcup F=A_\eta$, (4) implies that $f(\eta)=f_\xi(\eta)=1$, i.e., $\bigcup F=A_\eta\notin J$ which we wanted to establish. J is a prime ideal, i.e., for any $\zeta,\eta<\kappa$ such that $A_\eta=\bigcup F\setminus A_\zeta$ we have either $A_\zeta\in J$ or $A_\eta\in J$. To see this, choose an α with $\zeta, \eta<\alpha<\kappa$ and, by (5), a ξ with $\alpha\leq\xi<\kappa$ and $f\upharpoonright\alpha=f_\xi\upharpoonright\alpha$. Then either $x_\xi\notin A_\zeta$ or $x_\xi\notin A_\eta$, i.e., either $f(\zeta)=f_\xi(\zeta)=0$ or $f(\eta)=f_\xi(\eta)=0$; in other words; we indeed have either $A_\zeta\in J$ or $A_\eta\in J$. Thus we have shown that J is a κ-complete prime ideal extending I, that is, κ has the prime ideal property.

(ii)⇒(i). (We have to assume only the regularity of κ here.) Assume κ is regular and has the prime ideal property. We have to show that κ has the tree property. To this end, let $\langle T, \prec\rangle$ be a tree of length κ such that each of its levels has cardinality $<\kappa$. Let F be the field of sets κ-generated by the set

$$G=\{\{x\}:x\in T\}\cup\{\{t\in T:x\prec t\}:x\in T\},$$

and put

$$I=\{X\in F:|X|<\kappa\}\ (=\{X\subseteq T:|X|<\kappa\}).$$

I is a κ-complete ideal over F by the regularity of κ, and I is proper, since the set $\bigcup F=T$ obviously has cardinality κ. As κ has the prime ideal property, there is a prime ideal J over F extending I.

We claim that the set

$$b=\{x\in T:\{t\in T:x\prec t\}\notin J\}$$

is a branch of length κ in T. As T was an arbitrary tree, this means that κ has the tree property, and this is what we want to establish. First we show that b is a chain, i.e., that $x\preceq y$ or $y\preceq x$ holds for any $x, y\in b$. In fact, if x any y are incomparable in the ordering $\prec$, then $\{t\in T:x\prec t\}$ and $\{t\in T:y\prec t\}$ are disjoint, and so one of these sets belongs to J, since J is a prime ideal; i.e., either $x\notin b$ or $y\notin b$. To see that b is a branch of length κ it is now sufficient to show that b intersects each level of T. Denoting, as usual, by $T[\xi]$ the ξth level of T, we have

$$T=\bigcup_{\beta\leq\alpha}T[\beta]\cup\bigcup_{x\in T[\alpha]}\{t\in T:x\prec t\}$$

for every $\alpha<\kappa$. There is a union of less than κ sets belonging to F on the right hand side here; so at least one of these sets does not belong to J, because otherwise we would have $\bigcup F=T\in J$. Now each $T[\beta]$ has cardinality $<\kappa$, and so $T[\beta]\in I\subseteq J$; hence $\{t\in T: x\prec t\}\notin J$, i.e., $x\in b$, holds for some $x\in T[\alpha]$, as we wanted to show.

(ii)⇒(iii). (Hanf [1964a]. We have to assume only the regularity of κ here.) Assume κ is regular and does not have the tree property, i.e., there is an Aronszajn κ-tree $\langle A, <_A\rangle$. Clearly, A has cardinality κ. We are going to define an ordering $\prec$ of A for which $\kappa, \kappa^*\not\leq\operatorname{tp}\langle A, \prec\rangle$. To this end, for any $x\in A$ and any $\xi\leq o(x)\ (=\operatorname{tp}\langle\{y\in A: y<_Ax\}, <_A\rangle$ by definition), let $x|\xi$ be the projection of x onto the ξth level $A[\xi]$ of A; i.e., let $x|\xi$ be the unique element of A such that $x|\xi\leq_Ax$ and $x|\xi\in A[\xi]$. For any $\xi<\kappa$, let $\prec_\xi$ be an arbitrary ordering of

$A[\xi]$. Define now the ordering $\prec$ as follows: for any two $x, y \in A$ that are incomparable in $<_A$, i.e., which are such that neither $x \leq_A y$ nor $y <_A x$ holds, put $x \prec y$ iff $x|\xi \prec_\xi y|\xi$ holds with the least ordinal ξ for which $x|\xi \neq y|\xi$. If x and y are comparable (but distinct), then we may put $x \prec y$ or $y \prec x$ arbitrarily as long as this will make $\prec$ into an ordering; this is obviously the case if we put a) always $x \prec y$ or b) always $y \prec x$ whenever $x <_A y$.

We have to show that $\kappa, \kappa^* \not\leq \mathrm{tp}\langle A, \prec\rangle$. Assume, on the contrary, that $X \subseteq A$ is such that $\langle X, \prec\rangle$ has order type κ (the case when $\langle X, \prec\rangle$ has order type κ^* is entirely analogous because we did not decide which of the alternatives a) and b) at the end of the preceding paragraph we chose). Let $\xi < \kappa$ be arbitrary. The set $T|\xi = \bigcup_{\alpha<\xi} T[\alpha]$ has cardinality $<\kappa$ and κ is regular, so the set $X \setminus T|\xi$ includes a final segment X^ξ of X. For any $y \in T[\xi]$ put

$$X^\xi_y = \{x \in X^\xi : x|\xi = y\}.$$

We have

$$X^\xi = \bigcup_{y \in T[\xi]} X^\xi_y,$$

and, moreover, the sets X^ξ_y do not interlace in the ordering $\prec$; that is, if $y, y' \in T[\xi]$ and $y \prec y'$, then, clearly, all elements of X^ξ_y precede all those of $X^\xi_{y'}$ in the ordering $\prec$. As $|T[\xi]| < \kappa$ and $\langle X^\xi, \prec\rangle$ has order type κ, this implies by the regularity of κ that there is a (unique) $y_\xi \in T[\xi]$ such that $X^\xi_{y_\xi}$ is a final segment of $\langle X^\xi, \prec\rangle$, and therefore of $\langle X, \prec\rangle$. Then $\{y_\xi : \xi < \kappa\}$ is a branch of length κ in $\langle A, <_A\rangle$; in fact, we have $y_\xi <_A y_\eta$ whenever $\xi < \eta < \kappa$, since $X^\xi_{y_\xi}$ and $X^\eta_{y_\eta}$ would be disjoint otherwise, which is impossible because they are final segments of the same ordered set. We have obtained a contradiction, since A is an Aronszajn κ-tree, and so it cannot have a branch of length κ.

(iii)$\Rightarrow$(iv). (No assumption on κ is needed here.) The relation $\kappa \not\to (\kappa)^2_2$ is established by the ordered set in (iii) (cf. Lemma 19.4, and also Definition 19.5).

(iv)$\Rightarrow$(v). (Hajnal [1964]. No assumption on κ is needed.) This implication follows directly from (25.2) in Theorem 25.1 with $r = \tau = 2$ and $\lambda = \kappa$.

(v)$\Rightarrow$(vi). (No assumption on κ is needed.) By one of the monotonicity properties (cf. Subsection 9.6b), $\kappa \not\to (\kappa, 4)^3$ implies $\kappa \not\to (\kappa)^3_2$.

(vi)$\Rightarrow$(vii). (No assumption on κ is needed.) Assuming that (vii) fails, we have $\kappa \to \langle\kappa\rangle_\lambda$ for all cardinals $\lambda < \kappa$. Fixing λ and observing that the relation $\kappa \to (\kappa)^0_\lambda$ holds vacuously, we can easily prove with the aid of (15.4) in Lemma 15.1 by induction on r that $\kappa \to (\kappa)^r_\lambda$ holds for all integers r, i.e., that (vi) also fails.

(vii)$\Rightarrow$(ii). (Erdős–Tarski [1961].) We shall use the inaccessibility of κ here. Assume that (ii) fails, i.e., that κ has the tree property. We are going to prove that (vii) also fails then, i.e., that $\kappa \to \langle\kappa\rangle_\lambda$ holds for all $\lambda < \kappa$. Fix $\lambda < \kappa$, and let $f: [\kappa]^{<\omega} \to \lambda$ be a coloring. Writing $E = \kappa$, let $\langle E, S, R, T\rangle$ be the canonical partition tree associated with f, as defined in the beginning of the proof of Lemma 15.2. (Alternatively, one could use the tree $\langle \kappa, \prec_f \rangle$ defined in Section 18). Observe that the relations in (15.8)–(15.16) remain valid in the present case, since we only used there that E is an ordinal (i.e., wellordered), and the assumption $E = (\lambda^\kappa)^+$ of that proof was used only later. We claim that for each $\alpha < \kappa$ the αth level $T[\alpha]$ of the tree $T = \langle T, \subset \rangle$ has cardinality $< \kappa$. Assume, on the contrary, that this is not the case and let $\xi < \kappa$ be the least ordinal for which $|T[\xi]| \geq \kappa$. Note that $\xi \neq 0$, as $T[0]$ has only one element (the empty set); $\xi = \alpha \dot{+} 1$ for some α is not possible, since then (15.16) implies that

$$|T[\xi]| = \sum_{h \in T[\alpha]} |\mathrm{ims}\,(h)| \leq \sum_{h \in T[\alpha]} \lambda^{|[|\alpha|+1]^{<\omega}|} < \kappa$$

by the inaccessibility of κ. Assume therefore that ξ is a limit ordinal. Then, noting that each element of $T[\xi]$ is the least successor of a path of length ξ in T, and each such path p has at most one least successor (namely, $\bigcup p$ or none), we can see that $|T[\xi]|$ is $\leq$ the number of paths of length ξ in T. As each such path intersects each of the levels $T[\alpha]$, $\alpha < \xi$, in exactly one element, their number is $\leq \prod_{\alpha < \xi} |T[\alpha]| < \kappa$ by the inaccessibility of κ. This contradicts our assumption that $|T[\xi]| \geq \kappa$; thus our claim that $|T[\alpha]| < \kappa$ holds for every $\alpha < \kappa$ is established. Note that $|R(h)| = 1$ for every $h \in T$ by (15.15), i.e., $|T| = \kappa$ by Lemma 14.2. Hence T must have length $\geq \kappa$. As κ was assumed to have the tree property, $T|\kappa$ cannot be an Aronszajn tree, i.e., it must have a branch $b = \{g_\alpha : \alpha < \kappa\}$ of length κ, where g_α denotes the αth element of b, i.e., $g_\alpha \in T[\alpha]$. Putting $\xi_\alpha = s(g_\alpha)$ (cf. (15.8)), it follows that $X = \{\xi_\alpha : 1 \leq \alpha < \kappa\}$ is an end-homogeneous set by arguments similar to the ones used at the end of Lemma 15.5. In fact, $\xi_\alpha < \xi_\beta$ whenever $\alpha < \beta < \kappa$, as $\xi_\alpha = \min S(g_\alpha)$ by (15.8), and $\xi_\beta \subseteq S(g_\beta) \subseteq S(g_\alpha) \setminus \{\xi_\alpha\}$. Hence, if $u \subseteq X$ is finite and $\max u = \xi_\alpha < \xi_\beta, \xi_\gamma$ (or $\alpha = 0$ in case u is empty), then we have $u \subseteq \{\xi_\nu : 1 \leq \nu \leq \alpha\}$ and $\xi_\beta, \xi_\gamma \in S(g_{\alpha+1})$, and so (15.14) implies that

$$f(u \cup \{\xi_\beta\}) = f_{g_{\alpha+1}}(u) = f(u \cup (\xi_\gamma\}),$$

which we wanted to show. Thus the relation $\kappa \to \langle \kappa \rangle_\lambda$ holds for any $\lambda < \kappa$, i.e., (vii) fails. This completes the proof of the implication (vii)$\Rightarrow$(ii), and so that of the theorem.

Remarks. The investigation of properties (i)–(vii) is also of interest in case κ is not inaccessible. As for (i), one can show that it holds for all uncountable accessible cardinals, i.e., that κ can have the prime ideal property only if it is inaccessible. As for (ii), it trivially holds if κ is singular. Specker proved under GCH that if λ is regular, then (ii) fails for $\kappa = \lambda^+$ (cf. Theorem 52.1 below). It is not known whether the regularity of λ is necessary here, e.g. if the same conclusion holds with $\lambda = \aleph_\omega$ (see Erdős–Tarski [1961]). J. Silver [1971] proved that if $2^{\aleph_0}$ is real-valued measurable (which is consistent relatively to ZFC+ 'there is a measurable cardinal' by a theorem of Solovay [1971]; see Section 34 for the definitions of measurability and real-valued measurability), then (ii) fails for $\kappa = 2^{\aleph_0}$. Noting that κ is regular in this case, this shows that (ii) and (iv) are not provably equivalent if we assume only that κ is regular, since (iv) holds for every accessible cardinal by Theorem 29.1.

30. WEAK COMPACTNESS AND A METAMATHEMATICAL APPROACH TO THE HANF–TARSKI RESULT

The present section describes a metamathematical method showing that the relation $\kappa \to (\kappa)^2_2$ fails for many inaccessible cardinals. As the tools used here differ radically from the combinatorial tools more common in this book, *the combinatorially oriented reader might skip this whole section:* he will find an alternative, purely combinatorial proof of the main result (Theorem 30.5) concerning the relation $\kappa \not\to (\kappa)^2_2$ in Sections 31–32. It would, however, not be entirely fair to suppress that the ideas in the next section have a metamathematical background. The idea that the question whether $\kappa \to (\kappa)^2_2$ holds for certain inaccessible cardinals might be approached by metamathematical methods occurred to Tarski; the actual proof, by these methods, that $\kappa \to (\kappa)^2_2$ fails for many inaccessible cardinals was obtained by Hanf. Afterwards, Keisler and Tarski, and later Erdős and Hajnal, found direct combinatorial proofs of this result.

As this section is only for readers having some interest in model theory, we omit the definitions of basic model-theoretical concepts. We start with the following

DEFINITION 30.1. Let $\mathscr{C}$ be a set of constants and $\mathscr{R}$ a set of (finitary) relation symbols. For a cardinal κ, the *language* $L_{\kappa\kappa}(\mathscr{C}, \mathscr{R})$ is defined as follows: the *atomic formulas* of $L_{\kappa\kappa}(\mathscr{C}, \mathscr{R})$ are strings of symbols of the form $R(\cdot, \cdot, \ldots, \cdot)$, where $R \in \mathscr{R}$, and the dots (whose number is n if R is an n-ary relation symbol) are to be replaced by constants (i.e., elements of $\mathscr{C}$) or variables (we suppose that there are κ variables, e.g. v_α for each $\alpha < \kappa$). The set of (well-formed) formulas of $L_{\kappa\kappa}(\mathscr{C}, \mathscr{R})$ are defined as members of the smallest set W such that

(i) every atomic formula belongs to W;

(ii) if φ belongs to W then so does its negation $\neg\varphi$;

(iii) if $\alpha < \kappa$ and each φ_ξ, $\xi < \alpha$, is an element of W then so is the disjunction

(iii) if $\alpha < \kappa$ and each φ_ξ, $\xi < \alpha$, is an element of W then so is the disjunction $\bigvee_{\xi<\alpha} \varphi_\xi$;

(iv) if $\alpha<\kappa$ and φ belongs to W then so does the quantification $(\exists x_\xi)_{\xi<\alpha}\varphi$, where x_ξ, $\xi<\alpha$, are variables.

It is clear that each formula φ of $L_{\kappa\kappa}(\mathscr{C}, \mathscr{R})$ is obtained via a transfinite repetition of steps (i)–(iv); the formulas obtained in the course of a nonredundant construction of φ are called *subformulas* of φ. A formula of the form given in (iv) is called an *existential formula*.

The language $L_{\kappa\kappa}(\mathscr{C}, \mathscr{R})$ is identified with the set of its formulas, i.e., we put $L_{\kappa\kappa}(\mathscr{C}, \mathscr{R})=W$ for the smallest set W described above. The language $L_{\kappa\kappa}$ will consist of all formulas belonging to $L_{\kappa\kappa}(\mathscr{C}, \mathscr{R})$ for some $\mathscr{C}$ and $\mathscr{R}$. (To explain subscripts in the notation $L_{\kappa\kappa}$, the first κ means that the length of a disjunction is $<\kappa$, and the second, that the length of a block of quantifiers is $<\kappa$; cf. (iii) and (iv) above, respectively. One often uses the language $L_{\kappa\lambda}$ for $\kappa\neq\lambda$.) Formulas without free variables will be called *sentences.* In writing formulas below, we shall also use symbols not occurring under (i)–(iv) above; thus we shall use the customary binary logical connectives (e.g. $\vee$, &, $\Rightarrow$, ...), infinitary conjuction $\bigwedge$, and universal quantification. The meanings of these signs will be obvious: e.g. $\bigwedge_{\xi<\alpha}$ abbreviates $\neg\bigvee_{\xi<\alpha}\neg$, and $(\forall x_\xi)_{\xi<\alpha}$ abbreviates $\neg(\exists x_\xi)_{\xi<\alpha}\neg$.

The formulas of $L_{\kappa\kappa}(\mathscr{C}, \mathscr{R})$ will be interpreted semantically in a fairly straightforward manner: Let $\mathfrak{A}=\langle A, c^{\mathfrak{A}}, R^{\mathfrak{A}}\rangle_{c\in\mathscr{C}, R\in\mathscr{R}}$ be a structure, and let φ be a formula of $L_{\kappa\kappa}(\mathscr{C}, \mathscr{R})$. For given elements a_ν, $\nu<\gamma$, of A as values of the free variables y_ν, $\nu<\gamma$, in φ, the truth value $[\![\varphi[a_\nu/y_\nu\colon \nu<\gamma]]\!]$ (which is always 0 or 1) of φ is defined as follows:

(i′) if φ is an atomic formula, then $[\![\varphi[a_\nu/y_\nu\colon \nu<\alpha]]\!]=1$ iff the relation $R^{\mathfrak{A}}$ corresponding to φ holds with the appropriate arguments, with a_ν substituted for y_ν;

(ii′) $[\![\neg\varphi[a_\nu/y_\nu\colon \nu<\gamma]]\!]=1-[\![\varphi[a_\nu/y_\nu\colon \nu<\gamma]]\!]$;

(iii′) $[\![\bigvee_{\xi<\alpha}\varphi_\xi[a_\nu/y_\nu\colon \nu<\gamma]]\!]=1$ iff $[\![\varphi_\xi[a_\nu/y_\nu\colon \nu<\gamma]]\!]=1$ for some $\xi<\alpha$;

(iv′) $[\![(\exists x_\xi)_{\xi<\alpha}\varphi[a_\nu/y_\nu\colon \nu<\gamma]]\!]=1$ iff $[\![(\varphi[b_\xi/x_\xi\colon \xi<\alpha])\,[a_\nu/y_\nu\colon \nu<\gamma]]\!]=1$

for some elements b_ξ, $\xi<\alpha$, of A.

A structure that satisfies a set S of sentences will be called a *model* of S. A set of sentences of $L_{\kappa\kappa}$ is said to be *consistent* if it has a model, and it is said to be *semantically consistent* if each of its subsets of cardinality $<\kappa$ is consistent. The following definition describes the key concept of this section:

DEFINITION 30.2. A cardinal κ is called *weakly compact* if every semantically consistent set of sentences of $L_{\kappa\kappa}$ is consistent.

ω is a weakly compact cardinal by the compactness theorem of finitary first order logic. According to results of Hanf [1964], an uncountable weakly compact cardinal must be a 'large' regular limit cardinal but it is not known whether it must be inaccessible. The importance of weakly compact cardinals for partition relations is shown by the following theorem:

THEOREM 30.3. *An inaccessible cardinal κ is weakly compact if and only if $\kappa \to (\kappa)^2_2$.*

PROOF. The result is clearly valid also for the case $\kappa = \omega$. Our proof does work for this case, and phrases like 'since κ is inaccessible' can always be replaced by longer phrases like 'since $\kappa = \omega$ or κ is inaccessible' in the arguments below.

'If' part. Assume $\kappa \to (\kappa)^2_2$. Let $S = \{\varphi_\xi : \xi < \kappa\}$ be a semantically consistent set of sentences of $L_{\kappa\kappa}(\mathscr{C}, \mathscr{R})$. As we have to consider only those constants and relations that occur in some φ_ξ, we may obviously assume that

$$|\mathscr{C}|, |\mathscr{R}| \leq \kappa. \tag{1}$$

Let

$$\mathfrak{A}_\xi = \langle A, c^{\mathfrak{A}_\xi}, R^{\mathfrak{A}_\xi} \rangle_{c \in \mathscr{C}, R \in \mathscr{R}}$$

be a model of the sentences φ_γ, $\gamma < \xi$. For each $\alpha < \kappa$, and each existential subformula $\chi = (\exists x_\nu)_{\nu < \eta} \psi$ of φ_α, add *Skolem functions* $f^{\chi,\xi}_\nu$, $\nu < \eta$; these are functions with the following property: Let $\{y_\mu : \mu < \vartheta_\chi\}$ be an enumeration of the set of free variables in χ; then each $f^{\chi,\xi}_\nu$ maps ${}^{\vartheta_\chi}A_\xi$ into A_ξ, where A_ξ is the underlying set of the structure $\mathfrak{A}_\xi$, and $f^{\chi,\xi}_\nu$ is such that for every $g \in {}^{\vartheta_\chi}A_\xi$ the equivalence

$$\begin{aligned} &\mathfrak{A}_\xi \models (\exists x_\nu)_{\nu < \eta}\, \psi[y_\mu / g(\mu) : \mu < \vartheta_\chi] \Leftrightarrow \\ &\Leftrightarrow \mathfrak{A}_\xi \models (\psi[x_\nu / f^{\chi,\xi}_\nu(g) : \nu < \eta])\, [y/g(\mu) : \mu < \vartheta_\chi] \end{aligned} \tag{2}$$

holds; in other words $f^{\chi,\xi}_\nu$ just picks a value whose existence is asserted by the existential quantifier.

Take the direct product

$$\mathfrak{A} = \langle A, c^{\mathfrak{A}}, f^\chi_\nu \rangle_{c \in \mathscr{C}, \chi, \nu}$$

of the structures $\langle A_\xi, c^{\mathfrak{A}_\xi}, f^{\chi,\xi}_\nu \rangle_{c \in \mathscr{C}, \chi, \nu}$ where χ and ν run over the domains described above. That is, $A = \mathsf{X}_{\xi < \kappa} A_\xi$, $c^{\mathfrak{A}}$ is the function $g \in A$ such that $g(\xi) = c^{\mathfrak{A}_\xi}$ for every $\xi < \kappa$, and the function $f^\chi_\nu : {}^{\vartheta_\chi}A \to A$ is such that for any function $h \in {}^{\vartheta_\chi}A$ and any ordinal $\xi < \vartheta_\chi$ we have

$$f^\chi_\nu(h)\,(\xi) = f^{\chi,\xi}_\nu(\{\langle \mu, h(\mu)\,(\xi) \rangle : \mu < \vartheta_\chi\}). \tag{3}$$

Let $\mathfrak{L} = \langle B, c^{\mathfrak{A}}, f^\chi_\nu \upharpoonright {}^{\vartheta_\chi}B \rangle_{c \in \mathscr{C}, \chi, \nu}$ be a substructure of $\mathfrak{A}$ that includes the set $\{c^{\mathfrak{A}} : c \in \mathscr{C}\}$ and is closed with respect to all functions f^χ_ν. Note that we may assume

$$|B| \le \kappa \tag{4}$$

in view of (1).

We shall take a suitable reduction by an ultrafilter of the model $\mathfrak{L}$ that satisfies all the sentences φ_ξ. To this end we shall need an ultrafilter or, alternatively, a prime ideal. We cannot define the prime ideal on the whole of $\mathscr{P}(\kappa)$, but, as we shall see, it is enough to define it on an appropriate κ-complete field of sets F. F will be defined as the field of sets κ-generated by the singletons $\{\xi\}$, $\xi < \kappa$, and the sets

$$T_{\psi[\vec{b}]} = \{\xi : \mathfrak{A}_\xi \models \psi[\vec{b}(\xi)/\vec{x}]\}, \tag{5}$$

where ψ runs over all subformulas of the formulas φ_α, $\alpha < \kappa$, and $\vec{b}$ runs over all possible sequences of elements of B whose length is equal to the length of the sequence $\vec{x}$ of free variables in ψ. Note that there are at most κ possibilities for ψ, as a formula of $L_{\kappa\kappa}$ has $\le \kappa$ subformulas (equality is possible only for singular κ; this is not the case at present), and by the inaccessibility of κ there are less than κ possibilities for $\vec{b}$, as $|B| \le \kappa$ by (4), and the sequence $\vec{b}$ consists of less than κ elements of B. Hence the field F is κ-generated by κ sets. Put

$$I = \{X \subseteq \kappa : |X| < \kappa\}.$$

Then I is a κ-complete ideal over F. According to the assumption $\kappa \to (\kappa)^2_2$, Theorem 29.6 implies that κ satisfies the prime ideal property. Hence there is a proper κ-complete prime ideal J over F that includes I (cf. Definition 29.5(iv)). We shall define a relation $R^{\mathfrak{L}}$ for each $R \in \mathscr{R}$ on the structure $\mathfrak{L}$; for each R and $g_0, g_1, \ldots \in B$ we stipulate that

$$\begin{gathered} R^{\mathfrak{L}}(g_0, g_1, \ldots) \quad \text{holds iff} \\ \{\xi < \kappa : R^{\mathfrak{A}_\xi}(g_0(\xi), g_1(\xi), \ldots) \quad \text{holds}\} \notin J. \end{gathered} \tag{6}$$

($R^{\mathfrak{L}}$ is a kind of subdirect ultraproduct of the relations $R^{\mathfrak{A}_\xi}$, $\xi < \kappa$.) Note that, by (5), the set on the right-hand side belongs to the field F provided $R(x_0, x_1, \ldots)$ with distinct variables $x_0, x_1, \ldots$ is a subformula of some φ_ξ, and this is an assumption we can easily afford to make (e.g. by supposing that the formulas $\exists x_0 x_1 \ldots [R(x_0, x_1, \ldots) \Leftrightarrow R(x_0, x_1, \ldots)]$ occur among the φ_ξ's).

We claim that for any subformula ψ of each sentence φ_ξ, $\xi < \kappa$, and for any substitution sequence $\vec{b}$ of elements of B for the free variable sequence $\vec{x}$ of ψ we have

$$\begin{gathered} \langle B, c^{\mathfrak{L}}, R^{\mathfrak{L}} \rangle_{c \in \mathscr{C}, R \in \mathscr{R}} \models \psi[\vec{b}/\vec{x}] \\ \text{iff} \qquad \{\xi : \mathfrak{A}_\xi \models \psi[\vec{b}(\xi)/\vec{x}]\} \notin J. \end{gathered} \tag{7}$$

Note that the set on the right-hand side here belongs to the field F in view of (5). The proof of this claim is like the standard proof of Łos's theorem; it proceeds by simple induction on the complexity of the formula ψ; the argument used in the induction steps depends on which of the forms given in (i)–(iv) of Definition 30.1 the formula ψ has. In case (i), i.e., when ψ is an atomic formula, the claim holds in view of the definition of the relations $R^{\mathfrak{L}}$ in (6); in case (ii) it follows from the induction hypothesis and the fact that J is a prime ideal (i.e., that for any $X \in F$ exactly one of the relations $X \in J$ and $\kappa \setminus X \in J$ hold); in case (iii) one has to exploit the κ-completeness of J; finally in case (iv), i.e., if ψ is an existential formula, then the Skolem functions f^{ψ}_{v} pick the elements of B whose existence is claimed by the existential quantifiers (it is here that the fact is used that B is closed with respect to these Skolem functions). This proves the claim in (7).

This claim is valid in particular for the sentences φ_{ξ}. But we have

$$\{\xi : \mathfrak{A}_{\xi} \models \varphi_{\xi}\} \supseteq \kappa \setminus (\xi \dotplus 1)$$

by the choice of the structure $\mathfrak{A}_{\xi}$. The set on the right-hand side does certainly not belong to J, as its complement belongs to $I \subseteq J$; hence, in view of (7), the structure $\langle B, c^{\mathfrak{L}}, R^{\mathfrak{L}} \rangle_{c \in \mathscr{C},\, R \in \mathscr{R}}$ is a model for all the sentences φ_{ξ}, $\xi < \kappa$. This settles the 'if' part.

REMARK. One should note that if equality occurs among the relation symbols in $\mathscr{R}$, then it is not necessarily true that equal elements in the obtained model are identical. This is a perfectly sound situation, but it may be inconvenient, since if one defines an isomorphism as a 1-1 mapping satisfying certain requirements, then two models, though essentially the same, may not be isomorphic just because in one of the models there are equal but nonidentical elements. To avoid this inconvenience, one may pick an element from each of the equivalence classes of B under the relation $=^{\mathfrak{L}}$ and throw away all the others.

'Only if' part. Assume κ is weakly compact and inaccessible. We shall then prove that κ satisfies the prime ideal property. The model that will be constructed for this purpose will, with minor modifications, be useful in other proofs as well in the remaining part of this section.

Let F be a κ-complete field of sets with $\bigcup F = \kappa$ that has cardinality κ (as κ is inaccessible, this is the same by virtue of Lemma 29.3 and Theorem 6.3 as saying that F is κ-generated by κ sets), and let I be a proper κ-complete ideal over F. We have to show that I can be extended to a κ-complete prime ideal J over F. To this end, consider the language

$$\mathscr{L} = L_{\kappa\kappa}(\{\underline{\alpha} : \alpha \leq \kappa\} \cup \{\underline{x} : x \in F\} \cup \{\pi\}, \{=, \in\}),$$

where the first set is the set of constants (intuitively, the constants $\underline{\alpha}$ and $\underline{x}$ denote the sets α and x, and π is a new constant symbol), and the second set is the set of relations. Let ZFC_n denote the conjunction of the first n axioms of Zermelo–Fraenkel set theory, and, for an integer n to be chosen later, let κ' be a cardinal $\geq \kappa$ such that $\langle V(\kappa'), =, \in \rangle \models ZFC_n$, where $V(\kappa')$ denotes the set of all sets of rank $<\kappa'$. The existence of such a κ' is well known. (In fact, we could choose $\kappa' = \kappa^+$ here and in the next two proofs, since no axiom of set theory will be needed in the model to be constructed that does not hold in $\langle V(\kappa^+), =, \in \rangle$, but it would be quite tedious to check this fact.)

We construct a semantically consistent set of sentences $S = S_0 \cup S_1 \cup S_2$ of $\mathscr{L}$ as follows:

a) Let S_0 be the set of sentences of $\mathscr{L}$ using only the constant symbols $\underline{\alpha}$, $\underline{x}$ $(\alpha \leq \kappa, x \in F)$ that are satisfied by $\langle V(\kappa'), =, \in \rangle$ if we interpret $\underline{\alpha}$ as α and $\underline{x}$ as x.

b) Let

$$S_1 = \{\pi \in \underline{\kappa}\}.$$

c) Let

$$S_2 = \{\pi \notin \underline{x} : x \in I\},$$

where I is the ideal to be extended.

It follows from the inaccessibility of κ that $|S| = \kappa$. It is easy to see that $S = S_0 \cup S_1 \cup S_2$ is semantically consistent provided n is large enough (so that the set-theoretical results used to establish this fact are valid in $\langle V(\kappa'), =, \in \rangle$, i.e., they are provable from ZFC_n). In fact, if we take a set $S' \subseteq S$ of cardinality $<\kappa$, then they are satisfied by $\langle V(\kappa'), =, \in \rangle$, provided we interpret $\underline{\alpha}$ as α, $\underline{x}$ as x, and the constant π as an ordinal $\pi' < \kappa$ such that

$$\pi' \notin x$$

holds for any $x \in I$ for which $\underline{x}$ occurs in $S' \cap S_2$. There is such a π' since there are less than κ of these x's, and so $\bigcup \{x : x \in I$ and $\underline{x}$ occurs in $S' \cap S_2\} \in I$, which implies that the set on the left-hand side does not equal κ.

Hence, by the weak compactness of κ, the set S of sentences has a model; denote it by $\mathfrak{A}$. Put

$$J = \{x \in F : \mathfrak{A} \models \pi \notin \underline{x}\}.$$

Then $I \subseteq J$ by the axioms in S_2. J is a κ-complete ideal over F. In fact, if $x, y \in F$ and $x \subseteq y$, then the sentence $\underline{x} \subseteq \underline{y}$ (which is the abbreviation of $\forall z[z \in \underline{x} \Rightarrow z \in \underline{y}]$; we shall freely use the customary set-theoretical abbreviations in formulas of $\mathscr{L}$ below) belongs to S_0, and so it holds in $\mathfrak{A}$. If now $y \in J$ then $\mathfrak{A} \models \pi \notin \underline{y}$, and so a fortiori $\mathfrak{A} \models \pi \notin \underline{x}$, i.e., $x \in J$. If x_α, $\alpha < \xi$, are elements of F for some $\xi < \kappa$, then $x = \bigcup_{\alpha < \xi} x_\alpha$ is also an element of F, and the sentence

$$\forall z[z \in \underline{x} \Leftrightarrow \bigvee_{\alpha < \xi} z \in \underline{x}_\alpha]$$

belongs to S_0, and so it is valid in $\mathfrak{A}$. If now $x_\alpha \in J$, i.e., $\mathfrak{A} \models \pi \notin \underline{x}_\alpha$, holds for all $\alpha < \xi$, then this sentence implies that $\mathfrak{A} \models \pi \notin \underline{x}$ also holds, i.e., we have $x \in J$. This shows that J is a κ-complete ideal over F. J is proper, since $\mathfrak{A} \models \pi \in \underline{\kappa}$ holds by the single sentence in S_1, and so $\bigcup F = \kappa \notin J$. Finally, J is a prime ideal. In fact, if $x, y \in F$ and $x \cap y = 0$, then $\underline{x} \cap \underline{y} = \underline{0}$ holds in $\mathfrak{A}$ by a sentence of S_0. Hence either $\pi \notin \underline{x}$ or $\pi \notin \underline{y}$ holds in $\mathfrak{A}$; i.e., either $x \in J$ or $y \in J$. The proof of the theorem is complete.

There are several properties of inaccessible cardinals that are equivalent to weak compactness; the reader interested in them should consult e.g. Silver's dissertation [1971]. Our main purpose here is to prove the Hanf–Tarski result saying that most of the 'small' inaccessible cardinals are not weakly compact. To this end we shall need the Mahlo operation M. (See Mahlo [1911]; the operation defined here differs slightly from what Keisler and Tarski [1964] call Mahlo operation.)

Definition 30.4. Let X be a class of ordinals. We define $M(X)$ by putting $\alpha \in M(X)$ iff either $\operatorname{cf}(\alpha) \leq \omega$ or $\alpha \setminus X$ is not stationary in α.

In other words, $M(X)$ contains the same limit ordinals as $\operatorname{nst}(On \setminus X)$, where On is the class of all ordinals, and the operation nst, the nonstationary points of a class of ordinals, was defined right after the announcement of Theorem 5.7.

The following theorem is essentially the Hanf–Tarski result (cf. Hanf [1964]), saying that many inaccessible cardinals (including the 'small' ones, in particular the first) are not weakly compact. We shall prove this theorem again in the Section 32 as Theorem 32.4, and we shall consider its corollaries there. Let AC denote the class of all accessible ordinals (i.e., of all ordinals which are not inaccessible cardinals). We have

Theorem 30.5. *Let κ be an inaccessible cardinal, and assume that there are sets $Z_\alpha \subseteq \kappa$, $\alpha < \kappa$, such that, writing*

$$X = AC \cup \bigcup_{\alpha < \kappa} [M(Z_\alpha) \setminus (\alpha \dot{+} 1)], \tag{8}$$

we have $\kappa \notin X$ and $\kappa \in M(X)$. Then κ is not weakly compact.

If each Z_α is taken to be the empty set, then this result says that the first inaccessible cardinal is not weakly compact; or, indeed, the class $M(AC)$ does not contain any weakly compact inaccessible cardinals.

Proof. ($\kappa > \omega$ is definitely needed now.) Assume, on the contrary, that κ is a weakly compact inaccessible cardinal. Let f be a regressive divergent function on the set $\kappa \setminus X$; such a function exists in view of Neumer's theorem (Theorem 5.3), as $\kappa \setminus X$ is nonstationary by our assumption $\kappa \in M(X)$. Let $Z\colon \kappa \to \mathscr{P}(\kappa)$ be the function such that $Z(\alpha)=Z_\alpha$ for every $\alpha<\kappa$. Consider the language

$$\mathscr{L}=L_{\kappa\kappa}(\{\underset{\sim}{\alpha}: \alpha\leq\kappa\}\cup\{\underset{\sim}{X}, \underset{\sim}{Z}, \underset{\sim}{f}, \pi\}, \{=, \in\}),$$

where π is a new constant symbol. Let $\kappa'>\kappa$ be a cardinal as in the second part of the proof of the preceding theorem, i.e., such that $\langle V(\kappa'), =, \in\rangle\models \mathrm{ZFC}_n$ for a large enough integer n. Consider the following set $S=S_0\cup S_1\cup S_2$ of sentences of $\mathscr{L}$:

a) S_0 is the set of all sentences of $\mathscr{L}$ not containing π that are satisfied by $\langle V(\kappa'), =, \in\rangle$ if we interpret $\underset{\sim}{\alpha}$, $\underset{\sim}{Z}$, and $\underset{\sim}{f}$, as α, Z, and f, respectively.

b) S_1 contains the single sentence $\pi\in\underset{\sim}{\kappa}$;

c) put

$$S_2=\{\underset{\sim}{\alpha}\in\pi: \alpha<\kappa\}.$$

As before, it is easy to see from the inaccessibility of κ that $|S|=\kappa$. (We remark that the theorem being proved has an analogue for the case when κ is not inaccessible but only a regular limit cardinal. In that case the set S_0 might have cardinality $>\kappa$; it will however turn out that only 'very few' of the sentences in S_0 are actually needed.) It is also easy to see that S is semantically consistent; in fact, if $S'\subseteq S$ has cardinality $<\kappa$, then all the sentences of S' are satisfied by $\langle V_{\kappa'}, =, \in\rangle$ if we interpret $\underset{\sim}{\alpha}$, $\underset{\sim}{Z}$, and $\underset{\sim}{f}$ as α, Z, and f, respectively, and π as an ordinal $<\kappa$ exceeding all ordinals α for which $\underset{\sim}{\alpha}$ occurs in the sentences of S'.

Hence, by the assumption that κ is weakly compact, S has a model, say $\mathfrak{A}=\langle A, =^{\mathfrak{A}}, \in^{\mathfrak{A}}\rangle$. $\mathfrak{A}$ satisfies the following sentence:

$$\forall xy[x=y\Leftrightarrow\forall z[z\in x\Leftrightarrow z\in y]\tag{9}$$

and

$$\neg(\exists x_k)_{k<\omega}\bigwedge_{k<\omega} x_{k+1}\in x_k,\tag{10}$$

as these sentences, being valid in $\langle V(\kappa'), \in, =\rangle$, belong to $S_0\subseteq S$. In fact, the first sentence is the Axiom of Extensionality, and the second one says that the membership relation is well-founded, which is an easy consequence of the Axiom of Regularity. Hence, by Mostowski's Collapsing Lemma (see Subsection 4.3), $\mathfrak{A}$ is isomorphic to a transitive model

$$\mathfrak{L}=\langle B, =, \in\rangle;$$

that is, B is a transitive set (i.e., $\forall x\in B[x\subseteq B]$), and, furthermore, $=$ and $\in$ are the true equality and membership relations, respectively. Naturally, $\mathfrak{L}$ is also a model of S.

We claim that we have

$$\underset{\sim}{\alpha}^{\mathfrak{L}}=\alpha\tag{11}$$

for every ordinal $\alpha<\kappa$. In fact, the sentence

$$\forall x[x\in \underline{\alpha}\Leftrightarrow \bigvee_{\beta<\alpha} x=\underline{\beta}]$$

belongs to S_0, and so it is valid in $\mathfrak{L}$. Hence if we assume that (11) is valid for any ordinal $\beta<\alpha$, then it is also valid for α by this formula. On the other hand

$$\kappa\in\underline{\kappa}^{\mathfrak{L}} \tag{12}$$

holds. Indeed, $\pi^{\mathfrak{L}}\in\underline{\kappa}^{\mathfrak{L}}$ by the sentence in S_1, and $\alpha=\underline{\alpha}^{\mathfrak{L}}\in\pi^{\mathfrak{L}}$ for every $\alpha<\kappa$ by the sentences in S_2 (and by (11)). Observe now that $\underline{\kappa}^{\mathfrak{L}}$ is an ordinal in $\mathfrak{L}$ (a sentence in S_0 says so), and so, by the transitivity of B, it is an ordinal in the real world. Hence (12) must hold. Note that, by (12) and the transitivity of B, we of course have $\kappa\in B$.

It is easy to see that

$$\kappa\notin AC^{\mathfrak{L}}, \tag{13}$$

i.e., κ is inaccessible in the sense of $\mathfrak{L}$. In fact, the only quantifiers in the sentence "κ is accessible" that are not bounded (i.e., not of form $\forall x\in y$ or $\exists x\in y$) are existential ones, and if the sets that could confirm the accessibility of κ do not exist in the real world, then they cannot exist in the model $\mathfrak{L}$ either.

Note that

$$\kappa\cap\underline{Z}^{\mathfrak{L}}(\underline{\alpha}^{\mathfrak{L}})=Z(\alpha)\,(=Z_\alpha) \tag{14}$$

holds for any $\alpha<\kappa$. In fact, $Z(\alpha)=Z_\alpha\subseteq\kappa$ by our assumptions, and one of the sentences

$$\underline{\beta}\in\underline{Z}(\underline{\alpha})\qquad\text{and}\qquad\underline{\beta}\notin\underline{Z}(\underline{\alpha})$$

(the one which is true in reality) belongs to $S_0(\subseteq S)$ for each $\beta<\kappa$. Hence (14) follows from (11).

It is also easy to see that

$$\kappa\notin M^{\mathfrak{L}}(\underline{Z}^{\mathfrak{L}}(\underline{\alpha}^{\mathfrak{L}}))\cap(\kappa\dot{+}1)=M^{\mathfrak{L}}(Z_\alpha)\cap(\kappa\dot{+}1) \tag{15}$$

holds for any $\alpha<\kappa$, where $M^{\mathfrak{L}}$ denotes the Mahlo operation in the sense of $\mathfrak{L}$. In fact, the equality here holds by (14), and $\kappa\in M^{\mathfrak{L}}(Z_\alpha)$ would mean that $\kappa\setminus Z_\alpha$ was not stationary in the sense of $\mathfrak{L}$. This would in turn mean in view of Neumer's theorem (Theorem 5.3) that, in $\mathfrak{L}$, there was a regressive divergent function on $\kappa\setminus Z_\alpha$; there is, however, no such function even in the real world by the assumption $\kappa\notin X$.

We can now conclude that

$$\kappa\notin\underline{X}^{\mathfrak{L}}. \tag{16}$$

In fact, according to (8), a sentence of $S_0(\subseteq S)$ says that

$$\forall \xi[\xi \in \underline{X} \Leftrightarrow (\xi \in AC \vee \exists \alpha < \underline{\kappa}(\alpha < \xi \,\&\, \xi \in M(\underline{Z}(\underline{\alpha}))))];$$

now the right-hand side of this equivalence fails for $\xi = \kappa$ according to (13) and (15).

It will be easy to see that (16) leads to a contradiction. At the beginning of the proof we picked a function f that is regressive and divergent on $\kappa \setminus X$. A sentence of $S_0 \subseteq S$ says that

$$\operatorname{dom}(\underline{f}) = \underline{\kappa} \setminus \underline{X},$$

and so by (12) and (16) we have

$$\kappa \in \operatorname{dom}(\underline{f}^{\mathfrak{L}}).$$

Now

$$\underline{f}^{\mathfrak{L}}(\kappa) = \delta < \kappa$$

for some δ, as a sentence of $S_0 \subseteq S$ says that $\underline{f}$ is regressive. This means that the sentences

$$\exists \xi \in \operatorname{dom}(\underline{f})[\underline{\alpha} \in \xi \,\&\, \underline{f}(\xi) = \underline{\delta}]$$

are valid in $\mathfrak{L}$ for any $\alpha < \kappa$. In fact, they are satisfied by $\xi = \kappa$ (cf. (11)). For a fixed α, either this sentence or its negation is true in $\langle V(\kappa'), =, \in \rangle$ (or, what is the same, in reality). So either this sentence or its negation must belong to $S_0 \subseteq S$. But the latter case would mean that this sentence is false in $\mathfrak{L}$, which is not the case. So each of these sentences is true in reality. That is, we have

$$\forall \alpha < \kappa \, \exists \xi \in \operatorname{dom}(f)[\alpha < \xi \,\&\, f(\xi) = \delta].$$

This contradicts the assumption that f is divergent. The proof is complete.

Next we consider a combinatorial property of weakly compact inaccessible cardinals, due to J. E. Baumgartner, that enables one to prove the above theorem in a direct mathematical way.

DEFINITION 30.6. The cardinal κ is said to satisfy *Baumgartner's principle* if, given arbitrary regressive functions $f_\alpha: \kappa \to \kappa$ $(\alpha < \kappa)$, there are ordinals $\eta_\alpha < \kappa$ such that

$$\forall X \subseteq \kappa[|X| < \kappa \Rightarrow \exists Y \subseteq \kappa[|Y| = \kappa \,\&\, \forall \alpha \in X \forall \xi \in Y\, f_\alpha(\xi) = \eta_\alpha]] \tag{17}$$

holds.

It is not difficult to prove that a cardinal κ satisfying Baumgartner's principle must be inaccessible, but this fact is irrelevant for purposes. The important fact is expressed by the following

THEOREM 30.7 (Baumgartner, unpublished). *An inaccessible cardinal satisfies Baumgartner's principle if and only if it is weakly compact.*

We shall give a proof of this theorem in the next section by using combinatorial properties equivalent to weak compactness. But it will be instructive to give a metamathematical proof of the 'if' part, since it lucidly tells where the ordinals η_α come from.

PROOF OF THE 'IF' PART. Assume that κ is weakly compact and inaccessible. As in the proof of the preceding theorem, let $\kappa' > \kappa$ be a cardinal such that $\langle V(\kappa'), =, \in\rangle \models \mathrm{ZFC}_n$ for a large enough integer n. We consider the language

$$\mathscr{L} = L_{\kappa\kappa}(\{\underline{\alpha} : \alpha \leq \kappa\} \cup \{\underline{f}_\alpha : \alpha < \kappa\} \cup \{\pi\}, \{=, \in\})$$

and, analogously as before, we construct a set of sentences $S = S_0 \cup S_1 \cup S_2$ of $\mathscr{L}$ as follows:

a) S_0 is the set of all sentences of $\mathscr{L}$ not containing π that are satisfied by $\langle V(\kappa'), =, \in\rangle$ if we interpret $\underline{\alpha}$ and $\underline{f}_\alpha$ as α and f_α, respectively.

b) S_1 consists of the single sentence $\pi \in \underline{\kappa}$.

c) We put

$$S_2 = \{\underline{\alpha} \in \pi : \alpha < \kappa\} .$$

As before, we have $|S| = \kappa$ by the inaccessibility of κ, and it is again easy to see that S is semantically consistent by noting that any subset S' of cardinality $< \kappa$ of S is satisfied by $\langle V(\kappa') =, \in\rangle$ if we interpret π as a large enough ordinal $< \kappa$. So S has a model $\mathfrak{A}$ by the weak compactness of κ. As the sentences under (9) and (10) belong to $S_0 \subseteq S$, by Mostowski's Collapsing Lemma used in the preceding proof, $\mathfrak{A}$ is again isomorphic to a transitive model

$$\mathfrak{B} = \langle B, =, \in\rangle;$$

this $\mathfrak{B}$ is of course also a model of S. By the same arguments as before, the relations in (11) and (12) are valid also for this model $\mathfrak{B}$. Put

$$\eta_\alpha = \underline{f}_\alpha^{\mathfrak{B}}(\kappa) \tag{18}$$

for any $\alpha < \kappa$. Note here that $\underline{f}_\alpha$ is defined on the whole of κ; hence, by a sentence of S_0, $\underline{f}_\alpha^{\mathfrak{B}}$ is defined on the whole of $\underline{\kappa}^{\mathfrak{B}}$; and by (12), κ belongs to this latter set. So the definition of η_α is meaningful. Furthermore, we have $\eta_\alpha < \kappa$ by the regressivity of $\underline{f}_\alpha^{\mathfrak{B}}$ (confirmed also by a sentence of S_0).

We only have to prove that (17) holds. To this end observe that, given a set $X \subseteq \kappa$ of cardinality $< \kappa$, the sentence

$$\exists\xi[\underline{\beta} < \xi < \underline{\kappa} \;\&\; \bigwedge_{\alpha \in X} \underline{f}_\alpha(\xi) = \underline{\eta_\alpha}]$$

holds in $\mathfrak{B}$ for every $\beta < \kappa$; in fact, it is satisfied by $\xi = \kappa$ (cf. (18), (11), and (12); the latter two are, as we pointed out above, also valid in the present model $\mathfrak{L}$). But this is a sentence of $\mathfrak{L}$ not containing π, so either itself or its negation must belong to $S_0 \subseteq S$; the latter case is impossible, since this sentence is true in $\mathfrak{L}$. So it belongs to S_0, i.e., it is true in $\langle V(\kappa'), =, \in\rangle$, i.e., it is true in reality. That is, we have

$$\exists\xi[\beta<\xi<\kappa \,\&\, \bigwedge_{\alpha\in X} f_\alpha(\xi)=\eta_\alpha]$$

for any $\beta<\kappa$. As X was chosen arbitrarily, this means that (17) is indeed satisfied, i.e., Baumgartner's principle holds for κ. This completes the proof of the 'if' part of the theorem. As mentioned above, the proof of the 'only if' part will be given in the next section (see Theorem 31.4).

31. BAUMGARTNER'S PRINCIPLE

In the preceding section we presented a metamathematical approach to the Hanf–Tarski result saying that $\kappa\to(\kappa)^2_2$ fails for many inaccessible cardinals. As this book is mainly about combinatorics, the reader might prefer to see a combinatorial proof of this result. This will be furnished in this and the next section. For the reader's convenience, we shall repeat some of what has been said in the preceding section; in particular, we shall give the definition of Baumgartner's principle again (Definition 31.3 below). But first we define the concept of normal ideals (we recall for this definition that a function f sending ordinals to ordinals is regressive if $f(\xi)<\xi$ holds for every nonzero ξ in its domain; cf. Section 5), and then prove a lengthy lemma about them. A similar, but conceptually simpler, result will be proved below in Section 34 (see Theorem 34.9), and the reader might benefit by studying that theorem as well as Definition 34.1 before reading the following definition and Lemma 31.2.

DEFINITION 31.1. Let F be a κ-complete field of sets with $\bigcup F=\kappa$, I an ideal over F, and, finally, H a set of functions mapping κ into κ. The ideal I is called *normal* with respect to H if it is κ-complete and proper, and, for every regressive function $f\in H$ and every set $X\in F\setminus I$, there is an ordinal $\alpha<\kappa$ such that

$$\{\xi\in X: f(\xi)=\alpha\}\notin I\,.$$

Although not required in the definition, we should stress that a normal ideal I as above is probably useless unless

$$\{\xi<\kappa: f(\xi)=\alpha\}\in F$$

holds for every regressive function $f\in H$ and for every $\alpha<\kappa$. Next we give a result asserting existence of normal prime ideals under certain conditions for cardinals κ satisfying $\kappa \to (\kappa)^2_2$:

LEMMA 31.2. *Let $\kappa > \omega$ be a cardinal satisfying the prime ideal property. Let H be a set of cardinality κ of functions mapping κ into κ that satisfies the following properties*:

(i) *H contains the identity function id_κ on κ and the constant functions $c_\alpha \equiv \alpha$ for all $\alpha < \kappa$;*

(ii) *if f, $g \in H$, then the characteristic function χ_X of the set $X = \{\xi < \kappa\colon f(\xi) < g(\xi)\}$ belongs to H ($\chi^X\colon \kappa \to 2$ is defined by stipulating that $\chi_X(\xi) = 1$ if and only if $\xi \in X$).*

(iii) *H is closed under composition, i.e., if $f, g \in H$ then $f \circ g \in H$.*

Let F be the κ-complete field of sets κ-generated by the sets

$$\{\xi < \kappa : f(\xi) = g(\xi)\}, \tag{1}$$

where f and g run over elements of H. Then there exists a prime ideal over F that is normal with respect to H.

PROOF. Note that the one element sets $\{\alpha\}$, $\alpha < \kappa$, belong to F. In fact, the functions id_κ and c_α belong to H by (i) and we have

$$\{\alpha\} = \{\xi < \kappa : \mathrm{id}_\kappa(\xi) = c_\alpha(\xi)\},$$

and this set belongs to F by (1). Write

$$I = \{X \subseteq \kappa : |X| < \kappa\},$$

and let J be a prime ideal over F that extends I; there is such a J since κ was assumed to satisfy the prime ideal property (note that F is κ-generated by κ sets, since $|H| = \kappa$, and so there are $\le \kappa$ sets listed in (1); in fact there are exactly κ of them since, as we have just seen, all the one element subsets of κ are among them).

Consider the following equivalence relation $\sim_J$ on H: given $f, g \in H$, write

$$f \sim_J g \Leftrightarrow \{\xi : f(\xi) = g(\xi)\} \notin J. \tag{2}$$

Note that the set on the right-hand side of this formula belongs to F, as the sets listed in (1) all belong to F. It is an easy exercise to verify that (2) indeed defines an equivalence relation. To show e.g. transitivity, assume that $f \sim_J g$ and $g \sim_J h$; then $f \sim_J h$ because we have

$$\{\xi < \kappa : f(\xi) \ne h(\xi)\} \subseteq \{\xi < \kappa : f(\xi) \ne g(\xi)\} \cup$$
$$\cup \{\xi < \kappa : g(\xi) \ne h(\xi)\} \in J$$

(in fact, both sets on the right-hand side belong to J, since their complements do not, and J is a prime ideal), and so the complement of the set on the left-hand side cannot belong to J. Let $[f]$ denote the equivalence class of the function $f \in H$. We define an ordering $<_J$ on the set

$$\bar{H} = \{[f] : f \in H\}$$

by putting

$$[f] <_J [g] \quad \text{iff} \quad \{\xi < \kappa : f(\xi) < g(\xi)\} \notin J \tag{3}$$

for any $f, g \in H$. Before discussing the soundness of this definition, note that

$$\{\xi < \kappa : f(\xi) < g(\xi)\} \in F \tag{4}$$

holds for any $f, g \in H$. Indeed, the characteristic function, say h, of the set on the left-hand side belongs to H by (ii); so does the constant function $c_1 \equiv 1$ by (i). Hence

$$\{\xi < \kappa : f(\xi) < g(\xi)\} = \{\xi < \kappa : h(\xi) = c_1(\xi)\} \in F$$

by (1).

Taking (4) into account, it is easy to verify that the definition in (3) is sound, i.e., that the truth value of the relation $[f] <_J [g]$ indeed depends only on the equivalence classes and not on the functions that were chosen to represent them. It is also easily seen that $<_J$ is a linear ordering. In fact, transitivity follows in the same way as it did for $\sim_J$; and to see that for any two $f, g \in H$ we have exactly one of the relations $[f] <_J [g]$, $[f] = [g]$, or $[g] <_J [f]$, we have only to point out that

$$\kappa = \{\xi < \kappa : f(\xi) < g(\xi)\} \cup \{\xi < \kappa : f(\xi) = g(\xi)\} \cup \{\xi < \kappa : g(\xi) < f(\xi)\}$$

obviously holds, and the sets on the right-hand side here belong to F by (1) and (4). Hence exactly one of them does not belong to J, since these sets are pairwise disjoint and J is a prime ideal over F.

We claim that $<_J$ is a wellordering. To see this, assume on the contrary that $\langle [f_n] : n < \omega \rangle$ is an infinite descending sequence of elements of $\bar{H}$ in the ordering $<_J$. Then each of the sets $\{\xi < \kappa : f_n(\xi) \le f_{n+1}(\xi)\}$ belongs to J (since its complement does not, in virtue of the relation $[f_{n+1}] <_J [f_n]$; the fact is used here that J is a prime ideal); hence we have

$$\{\xi < \kappa : \exists n[f_n(\xi) \le f_{n+1}(\xi)]\} = \bigcup_{n<\omega} \{\xi < \kappa : f_n(\xi) \le f_{n+1}(\xi)\} \in J,$$

as J is κ-complete and $\kappa > \omega$. So there is a $\xi < \kappa$ that does not belong to the set on the left-hand side. Then we have

$$f_0(\xi) > f_1(\xi) > f_2(\xi) > \dots;$$

this is a contradiction since there is no infinite descending sequence of ordinals, verifying our claim that $<_J$ is a wellordering.

Let now $h \in H$ be a function such that $[\bar{h}]$ is the least element of $\bar{H}$ in the wellordering $<_J$ that exceeds $[c_\alpha]$ for $\alpha < \kappa$, where c_α denotes the constant function $\equiv \alpha$ (h is often called a *minimal function* for J). Note that there is such an h, since, for the identity function id_κ on κ, $[\mathrm{id}_\kappa]$ exceeds all the $[c_\alpha]$. This remark also implies that $[h] \le_J [\mathrm{id}_\kappa]$; hence we may assume that

$$h(\xi) \le \xi \tag{5}$$

holds for any $\xi < \kappa$ (we can make this true by changing h on a set belonging to J). Put

$$J^* = \{X \in F : h^{-1}(X) \in J\}, \tag{6}$$

where

$$h^{-1}(X) = \{\xi < \kappa : h(\xi) \in X\}.$$

We claim that J^* is a prime ideal over F that is normal with respect to H. To show this, we first show that if $X \in F$, then

$$h^{-1}(X) \in F. \tag{7}$$

As the sets listed in (1) κ-generate F, it is enough to verify this in case X is one of the sets in (1), i.e., in case

$$X = \{\xi < \kappa : f(\xi) = g(\xi)\}$$

for some $f, g \in H$. We have

$$h^{-1}(X) = \{\xi < \kappa : h(\xi) \in X\} = (\xi < \kappa : f(h(\xi)) = g(h(\xi))\} =$$
$$= \{\xi < \kappa : (f \circ h)(\xi) = (g \circ h)(\xi)\},$$

and this last set also belongs to F, as $f \circ h, g \circ h \in H$ by (iii). This verifies (7). As J is a κ-complete prime ideal over F, the same is true for J^* by (6) and (7) (note here that the one element sets $\{\alpha\}$, $\alpha < \kappa$, belong to J^*, as $h^{-1}(\{\alpha\}) \in J$ holds because $[c_\alpha] \ne [h]$ by the choice of h; hence $\bigcup J^* = \kappa = \bigcup F$, and so J^* is indeed over F). We have yet to show that J^* is normal with respect to H. To this end, pick a regressive function $f \in H$. As $f(h(\xi)) < h(\xi)$ for all $\xi < \kappa$ unless $h(\xi) = 0$, we have

$$[f \circ h] <_J [h]; \tag{8}$$

in fact,

$$\{\xi < \kappa : (f \circ h)(\xi) \ge h(\xi)\} \subseteq h^{-1}(\{0\}) \in J$$

by the remark made just before in parentheses. (8) implies by the choice of h that

$$[f \circ h] = [c_\alpha]$$

holds for some $\alpha < \kappa$, i.e., that

$$\{\xi < \kappa : (f \circ h)(\xi) = \alpha\} \notin J. \tag{9}$$

It will now be easy to show that

$$\{\xi<\kappa: f(\xi)=\alpha\} \notin J^* . \tag{10}$$

Note first that the set on the left-hand side belongs to F by (1) (and (i)). We have

$$h^{-1}(\{\xi<\kappa: f(\xi)=\alpha\})=\{\xi<\kappa: f(h(\xi))=\alpha\} \notin J$$

by (9), and so (10) indeed holds in view of (6). (10) almost amounts to saying that J^* is normal with respect to H; but, in actual fact, we have to show somewhat more: given any $X \in F \setminus J^*$, the relation

$$\{\xi \in X: f(\xi)=\alpha\} \notin J^*$$

also holds. This is, however, obvious by (10), since $\kappa \setminus X \in J^*$, as J^* is a prime ideal. This verifies our claim that J^* is a prime ideal over F that is normal with respect to H. The proof of the lemma is complete.

We repeat the definition of Baumgartner's principle (with slight notational changes) given in the preceding section (cf. Definition 30.6).

DEFINITION 31.3. The cardinal κ is said to satisfy *Baumgartner's principle* if given an arbitrary set S of cardinality $\le\kappa$ of regressive functions on κ, there are ordinals η_f for each $f \in S$ such that

$$\forall B\subseteq S(|B|<\kappa\Rightarrow\exists Y\subseteq\kappa[|Y|=\kappa \,\&\, \forall f \in B\,\forall\xi \in Y\, f(\xi)=\eta_f]) . \tag{11}$$

The main result in this section is the following theorem (essentially identical to Theorem 30.7, but here we prove both implications).

THEOREM 31.4 (Baumgartner, unpublished). *An inaccessible cardinal κ satisfies Baumgartner's principle if and only if $\kappa\to(\kappa)^2_2$.*

It is not difficult to prove that Baumgartner's principle fails for all accessible cardinals; hence by Theorem 29.1, the above result holds for any $\kappa>\omega$ and not just for inaccessible κ. This fact is, however, of no interest to us here.

PROOF. 'If' part. Assume that $\kappa\to(\kappa)^2_2$ holds. Then, by Theorem 29.6, κ satisfies the prime ideal property. Given a set S of cardinality $\le\kappa$ of regressive functions on κ, let H be the smallest set of functions mapping κ into κ that includes S and satisfies (i)–(iii) of Lemma 31.2 (clearly, $|H|=\kappa$), and let F be the field of sets associated with this H as in that lemma. Then that lemma implies the existence of a prime ideal over F that is normal with respect to H; denote this by I. As the functions $f \in S$ are regressive and belong to H, we have

$$\{\xi<\kappa: f(\xi)=\eta_f\} \notin I$$

with some η_f for each $f \in S$. As I is a prime ideal, we have

$$\{\xi < \kappa : f(\xi) \neq \eta_f\} \in I .$$

Take a set $B \subseteq S$ of cardinality $< \kappa$. Then

$$\{\xi < \kappa : \exists f \in B\ f(\xi) \neq \eta_f\} = \bigcup_{f \in B} \{\xi < \kappa : f(\xi) \neq \eta_f\} \in I ,$$

because I is κ-complete (this is included in the definition of normality). Hence the complement of this set,

$$\{\xi < \kappa : \forall f \in B\ f(\xi) = \eta_f\} ,$$

does not belong to I; so it must have cardinality κ, as $\bigcup I = \bigcup F = \kappa$, and I is κ-complete. Taking this set as Y, we can see that (11) holds, which completes the proof of the 'if' part.

'Only if' part. Assume that κ satisfies Baumgartner's principle. We shall then show that κ satisfies the tree property; this will be enough in view of Theorem 29.6. To this end, let $\langle T, \prec \rangle$ be a κ-tree such that each of its levels has cardinality $< \kappa$. We have to prove that T has a branch of length κ. As $|T| = \kappa$, we may actually assume that $T = \kappa$. Define the set $S = \{f_\alpha : \alpha < \kappa\}$ by stipulating that $f_\alpha : \kappa \to 2$ is a function such that

$$f_\alpha(\xi) = 1 \quad \text{if} \quad \alpha \prec \xi \quad \text{and} \quad 0 \quad \text{otherwise.} \tag{12}$$

The functions f_α are regressive, except that we may have $f_\alpha(0) = 1$ or $f_\alpha(1) = 1$, i.e., $\alpha \prec 0$ or $\alpha \prec 1$, but it is harmless to assume that this is not the case (e.g. by supposing that the elements 0 and 1 were added to T as extra ones, incomparable to any other of its elements). According to Baumgartner's principle, there are ordinals $\eta_{f_\alpha} = \eta_\alpha < \kappa$ such that (11) holds (clearly $\eta_\alpha = 0$ or 1 in the present case). Put

$$b = \{\alpha < \kappa : \eta_\alpha = 1\} . \tag{13}$$

We claim that any two elements α and β of b are comparable in $\prec$. In fact, taking $B = \{f_\alpha, f_\beta\}$ in (11), there is a $\xi < \kappa$ such that $f_\alpha(\xi) = f_\beta(\xi) = 1$. We have $\alpha, \beta \prec \xi$ by (12), and so α and β are indeed comparable, since $\langle \kappa, \prec \rangle$ is a tree.

Next we claim that b has cardinality κ. To see this assume, on the contrary, that $|b| < \kappa$, and let X be a level of $\langle \kappa, \prec \rangle$ that is disjoint from b. Then $\eta_\alpha = 0$ for all $\alpha \in X$ by (13), and yet

$$\forall \alpha \in X\ f_\alpha(\xi) = 0$$

is possible by (12) only if ξ is in X or in a level below X; as the number of these ξ's is $< \kappa$, (11) must fail with $B = \{f_\alpha : \alpha \in X\}$ (note that $|B| \leq |X| < \kappa$, as each level of

$\langle\kappa, \prec\rangle$ has cardinality $<\kappa$). This is a contradiction, establishing our second claim. Our two claims show that $\langle\kappa, \prec\rangle$ has a branch of length κ; in fact, $b' = \{\xi : \exists\alpha\in b[\xi\prec\alpha]\}$ is such a branch. The proof is complete.

32. A COMBINATORIAL APPROACH TO THE HANF–TARSKI RESULT

In this section, using Baumgartner's principle, we shall show that $\kappa\to(\kappa)^2_2$ fails for many inaccessible cardinals. We start with a simple result which will characteristically show one of the ways in which Baumgartner's principle can be used. The result in itself is of not much interest to us here, because a far more comprehensive one will be proved below. Note that the result in (5) can be considered as a lemma for Theorem 32.4 below; it would, however, be uninspiring to formulate it as such.

THEOREM 32.1. *$\kappa\not\to(\kappa)^2_2$ holds for the first inaccessible cardinal κ.*

PROOF. Assume the contrary. Then Baumgartner's principle holds for κ by Theorem 31.4. Define κ regressive functions on κ as follows:

$$f_0(\xi) = \min\{\rho : 2^\rho \geq |\xi|\} \tag{1}$$

if this value is $<\xi$; otherwise put $f_0(\xi) = 0$; write

$$f_1(\xi) = \mathrm{cf}\,(\xi) \tag{2}$$

if ξ is a limit ordinal with $\mathrm{cf}\,(\xi) < \xi$; otherwise put $f_1(\xi) = 0$. If $\lambda<\kappa$ is a regular cardinal and $\xi<\kappa$ is an ordinal with $\mathrm{cf}\,(\xi) = \lambda$, then let

$$\langle f_{\lambda\alpha}(\xi) : \alpha<\lambda\rangle \tag{3}$$

be an increasing sequence of ordinals tending to ξ; if $\xi<\kappa$ and $\mathrm{cf}\,(\xi)\neq\lambda$, then put $f_{\lambda\alpha}(\xi) = 0$ for any $\alpha<\lambda$. This completes the definition of the set

$$S = \{f_0\}\cup\{f_1\}\cup\{f_{\lambda\alpha} : \lambda \text{ is regular } \& \alpha<\lambda<\kappa\} \tag{4}$$

of regressive functions. Let η_{f_0}, η_{f_1}, and $\eta_{f_{\lambda\alpha}}$ be the ordinals that correspond to these functions according to Baumgartner+s principle (see Definition 31.3).

Our first claim is that we must have $\eta_{f_0} = 0$. In fact, for any fixed η with $0<\eta<\kappa$ the relation $f_0(\xi) = \eta$ can hold only if $\eta = |\eta|$ and $2^{|\eta|}\geq|\xi|$, and this holds less than κ times, as $2^{|\eta|}<\kappa$. So (31.11) with $B = \{f_0\}$ implies our claim.

Our second claim is that $\eta_{f_1} = 0$. Assume the contrary; then we must have $\eta_{f_1} = \lambda$ for some regular cardinal $\lambda<\kappa$. Consider the set $B = \{f_1\}\cup\{f_{\lambda\alpha} : \alpha<\lambda\}$; (31.11) must hold with this set B. That is, there exists a set $Y\subseteq\kappa$ of cardinality κ such that

$$f_1(\xi) = \lambda \qquad \text{and} \qquad f_{\lambda\alpha}(\xi) = \eta_{f_{\lambda\alpha}}$$

for any $\xi \in Y$ Then, according to (2) and (3), the sequence

$$\langle \eta_{f_{\lambda\alpha}} : \alpha < \lambda \rangle$$

must tend to ξ for any $\xi \in Y$; this is a contradiction, since a sequence can tend only to at most one ξ. This establishes our claim.

We have therefore established that

$$\eta_{f_0} = \eta_{f_1} = 0, \tag{5}$$

and note for later reference that no other assumption was used about κ than that it satisfies Baumgartner's principle with the set S in (4) and that it is inaccessible. What we are now going to show is that (5) must fail for the first inaccessible cardinal. We shall do this by showing that if we take $B = \{f_0, f_1\}$ in (31.11), then (5) implies that for the set Y whose existence is claimed by (31.11) we must have that

$$\text{if} \quad \xi \in Y \text{ and } \xi > \omega, \text{ then } \xi \text{ is inaccessible.} \tag{6}$$

Hence $|Y| = \kappa$ clearly contradicts the assumption that κ is the first inaccessible cardinal.

To establish (6), choose a $\xi \in Y$ with $\xi > \omega$. Then (31.11) and (5) imply that

$$f_0(\xi) = f_1(\xi) = 0 \tag{7}$$

holds. We then have

$$2^{\rho} < |\xi| \tag{8}$$

for any cardinal $\rho < \xi$ by (1) and by $f_0(\xi) = 0$. Hence $\xi = |\xi|$ (because $2^{|\xi|} > \xi$ certainly holds), i.e., ξ is a cardinal and so, a fortiori, a limit ordinal; therefore by (2) (or, rather, by the clause after (2)) and the second equality in (7), we have

$$\xi = \mathrm{cf}\,(\xi)\,.$$

This and (8) imply that ξ is an inaccessible cardinal (note that we supposed $\xi > \omega$). Thus (6) is established. As mentioned, (6) leads to a contradiction, completing the proof of the theorem.

We now repeat the definition of the Mahlo operation given in Section 30 (we again stress that the operation defined here slightly differs from what Keisler and Tarski called Mahlo operation in [1964]).

DEFINITION 32.2. Let X be a class of ordinals. We define $M(X)$ by putting $\alpha \in M(X)$ iff either $\mathrm{cf}\,(\alpha) \leq \omega$ or $\alpha \setminus X$ is not stationary in α.

In other words, $M(X)$ contains the same limit ordinals as $\mathrm{nst}\,(On \setminus X)$, where On is the class of all ordinals, and the operation nst, the nonstationary points of a class of ordinals, was defined right after the announcement of Theorem 5.7. The proof of the next theorem, still not our main result, follows some ideas in the proof of Theorem 5.7. Note again that the implication (17) $\Rightarrow$ (19) below can be considered as a lemma for our main result, just as relation (5) above.

THEOREM 32.3. *Denote by AC the class of all accessible ordinals (i.e., of all ordinals that are not inaccessible cardinals). Let κ be an inaccessible cardinal, and assume that there is a set $Z \subseteq \kappa$ such that*

$$\kappa \notin AC \cup M(Z) \tag{9}$$

and

$$\kappa \in M(AC \cup M(Z)). \tag{10}$$

Then $\kappa \not\to (\kappa)^2_2$ holds.

PROOF. Assume that, on the contrary, $\kappa \to (\kappa)^2_2$ holds. Then Baumgartner's principle holds for κ by Theorem 31.4. For each $\xi \in M(Z) \setminus AC$, let f_ξ be a regressive function on $\xi \setminus Z$ that is divergent in ξ; as $\xi \setminus Z$ is not stationary in ξ, there is such a function by Neumer's theorem. (Theorem 5.3). The divergence of f_ξ means that, for each $\gamma < \xi$, there is a value $g_\gamma(\xi) < \xi$ such that

$$\forall \alpha \in \xi \setminus Z[\alpha > g_\gamma(\xi) \Rightarrow f_\xi(\alpha) > \gamma]. \tag{11}$$

This defines the regressive function $g_\gamma(\xi)$ for each $\gamma < \kappa$ and for each $\xi \in M(Z) \setminus AC$ with $\gamma < \xi < \kappa$; otherwise put $g_\gamma(\xi) = 0$. For each $\alpha \in \kappa \setminus Z$, write

$$h_\alpha(\xi) = f_\xi(\alpha) \tag{12}$$

if $\xi \in M(Z) \setminus AC$ and $\alpha < \xi < \kappa$ (i.e., when the right-hand side is defined), and write $h_\alpha(\xi) = 0$ otherwise. Finally, let h be a regressive function on κ such that

$$\begin{gathered} h \restriction (\kappa \setminus (AC \cup M(Z)) \quad \text{is divergent in } \kappa, \\ \text{and} \quad h(\xi) = 0 \quad \text{if} \quad \xi \in AC \cup M(Z) \quad \text{and} \quad \xi < \kappa. \end{gathered} \tag{13}$$

There is such an h by Neumer's theorem, since $\kappa \setminus (AC \cup M(Z))$ is not stationary in κ according to (10). Consider the set

$$\begin{gathered} S = \{f_0\} \cup \{f_1\} \cup \{f_{\lambda\alpha} : \lambda \quad \text{is regular } \& \ \alpha < \lambda < \kappa\} \cup \\ \cup \{g_\gamma : \gamma < \kappa\} \cup \{h_\alpha : \alpha \in \kappa \setminus Z\} \cup \{h\}, \end{gathered} \tag{14}$$

where the definitions of these functions were given in (1), (2), (3), (11), (12), and (13) in turn. According to Baumgartner's principle, for each $f \in S$ there is an ordinal $\eta_f < \kappa$ such that (31.11) holds.

First note that we must have

$$\eta_h = 0. \tag{15}$$

In fact, if we choose $B=\{h\}$ in (31.11), then this formula says that there is a set $Y\subseteq\kappa$ of cardinality κ such that $h(\xi)=\eta_h$ for any $\xi\in Y$. According to the divergence of the function in (13) we must have

$$|Y\cap(\kappa\setminus(AC\cup M(Z)))|<\kappa, \tag{16}$$

and so (15) indeed holds by the clause after (13). Note that the set S in (14) includes that in (4); hence (5) and its consequence (6) holds. Taking a set $B\subseteq S$ (= the S in (14)) containing f_0, f_1, and h, we must have for the Y in (31.11) by (6) and (16) that

$$|Y\setminus(M(Z)\setminus AC)|<\kappa.$$

As we can freely omit less than κ elements of Y, we may assume that

$$Y\subseteq(M(Z)\setminus AC)\cap\kappa. \tag{17}$$

It is this inclusion that we shall use in what follows, and it is of no relevance to us how we derived it. (This is important, because in the proof of the next theorem we shall obtain an analogous relation but in a slightly different way.) We shall prove that (17) and the fact that the set S includes the functions in

$$S'=\{g_\gamma:\gamma<\kappa\}\cup\{h_\alpha:\alpha\in\kappa\setminus Z\} \tag{18}$$

imply that

$$\kappa\in M(Z). \tag{19}$$

This of course contradicts our assumption (9), and this contradiction will complete the proof of the theorem.

To establish (19) we have to prove that $\kappa\setminus Z$ is not stationary in κ. To this end, by Neumer's theorem it will be enough to verify the following claim: The function

$$r(\alpha)=\eta_{h_\alpha}\qquad(\alpha\in\kappa\setminus Z)$$

is a regressive function on $\kappa\setminus Z$ that is divergent in κ. First note that r is regressive; in fact, $h_\alpha(\xi)<\alpha$ holds for any nonzero $\alpha\in\kappa\setminus Z$ by (12) (and the clause afterwards) and by the regressivity of f_ξ. Hence, taking $B=\{h_\alpha\}$ in (31.11), we can see that $r(\alpha)=\eta_{h_\alpha}<\alpha$ indeed holds for any nonzero $\alpha\in\kappa\setminus Z$. To see that r is divergent in κ, we shall show that

$$r(\alpha)>\gamma \tag{20}$$

holds for any $\gamma<\kappa$ whenever $\alpha\in\kappa\setminus Z$ and $\alpha>\eta_{g_\gamma}$. To this end, fix α and γ, and choose in (31.11) a set B containing h_α and g_γ. Then there is a set $Y\subseteq\kappa$ of cardinality κ satisfying that formula. By making sure that B contains certain other functions as well (namely f_0, f_1, and h), we can require, as we saw above, that (17) be also satisfied. Note now that (11) and (12) hold for all but less than κ elements ξ of the set $(M(Z)\setminus AC)\cap\kappa$; hence by (17) they hold for all but less than κ elements ξ of the set Y. Thus we have

$$\alpha>g_\gamma(\xi)\Rightarrow h_\alpha(\xi)>\gamma$$

for a large enough $\xi\in Y$ by (11) and (12). As $g_\gamma(\xi)=\eta_{g_n}$ and $h_\alpha(\xi)=\eta_{h_\alpha}$ hold here by (31.11) and the assumption $g_\gamma, h_\alpha\in B$, and $r(\alpha)=\eta_{h_\alpha}$ holds by the definition of r, this verifies (20). Hence, as claimed, r is a regressive function on $\kappa\setminus Z$ that is divergent in κ; so (19) follows, completing the proof of the theorem.

Our next result, the main one in the present section, is identical to Theorem 30.5, which we proved by metamathematical methods. All the ideas in the combinatorial proof below are essentially contained in the proofs of the preceding two theorems above. As before, AC denotes the class of all accessible ordinals.

THEOREM 32.4. *Let κ be an inaccessible cardinal, and assume that there are sets $Z_\mu\subseteq\kappa$ $(\mu<\kappa)$ such that, writing*

$$X=AC\cup\bigcup_{\mu<\kappa}(M(Z_\mu)\setminus(\mu\dot{+}1)),\tag{21}$$

we have $\kappa\notin X$ and $\kappa\in M(X)$. Then $\kappa\nrightarrow(\kappa)_2^2$.

REMARK. S. Shelah proved that if $V=L$ then this theorem gives an exact characterization of cardinals with $\kappa\nrightarrow(\kappa)_2^2$. In the other direction, M. Magidor constructed a model in which $\kappa\nrightarrow(\kappa)_2^2$ holds for the first inaccessible cardinal for which this is not confirmed by the above theorem.

PROOF. Assume, on the contrary, that $\kappa\rightarrow(\kappa)_2^2$ holds. Then Baumgartner's principle holds for κ by Theorem 31.4. Let h be a regressive function on κ such that

$$h\upharpoonright(k\setminus X)\quad\text{is divergent in }\kappa,\tag{22}$$

and let $h(\xi)=0$ for $\xi\in X\cap\kappa$; there is such an h, as $\kappa\in M(X)$. Let $k\colon\kappa\rightarrow\kappa$ be the function defined as follows:

$$k(\xi)=\min\{\mu\colon\xi\in M(Z_\mu)\setminus(\mu\dot{+}1)\}\tag{23}$$

if $\xi\in(X\cap\kappa)\setminus AC$ (in this case the set on the right-hand side is not empty), and put $k(\xi)=0$ otherwise. k is obviously regressive. Let f_0, f_1, and $f_{\lambda\alpha}$ be the functions defined in (1), (2), and (3), respectively, and let g_γ^μ and h_α^μ be the functions that are given by the same definitions as g_γ and h_α in (11) and (12), respectively, if the set Z there is replaced by Z_μ (note that f_ξ there also depends on

Z). Put

$$S=\{f_0\}\cup\{f_1\}\cup\{f_{\alpha\lambda}:\lambda \text{ is regular } \ \& \ \ \alpha<\lambda<\kappa\}\cup \\ \cup\bigcup_{\mu<\kappa}(\{g^\mu_\gamma:\gamma<\kappa\}\cup\{h^\mu_\alpha:\alpha\in\kappa\setminus Z_\mu\})\cup\{h\}\cup\{k\} \tag{24}$$

(the function h now differs from the one defined in (13)). We shall use Baumgartner's principle with this S. According to this, for each $f\in S$ there is an $\eta_f<\kappa$ such that (31.11) holds. Take a set $B\subseteq S$ that includes the functions f_0, f_1, h, and k, and consider the set Y whose existence is claimed in (31.11) for this B. Firstly, we must have

$$|Y\cap(\kappa\setminus X)|<\kappa$$

according to (22), and as there is no harm in omitting less than κ elements of Y, we may assume that

$$Y\subseteq X\cap\kappa\,.$$

Secondly, $\eta_{f_0}=\eta_{f_1}=0$ by (5), and so $|Y\cap AC|\leq\omega$ according to (6), i.e., we may assume $Y\subseteq\kappa\setminus AC$. Hence

$$Y\subseteq(X\cap\kappa)\setminus AC\,. \tag{25}$$

Now $k(\xi)=\eta_k$ holds for each $\xi\in Y$ by (31.11) because $k\in B$, which means, according to (23) and (25), that

$$Y\subseteq(M(Z_{\eta_k})\setminus AC)\cap\kappa\,.$$

This is the same as (17) above (expect that Z has a subscript here; but this subscript *does not* depend on the choice of B in (31.11), provided $k\in B$). Hence this relation implies the analogue of (19), i.e., that

$$\kappa\in M(Z_{\eta_k})\,.$$

This contradicts the assumption that $\kappa\notin X$, completing the proof of the theorem.

Our last theorem above shows that $\kappa\not\rightarrow(\kappa)^2_2$ holds for many inaccessible cardinals; in fact, it holds if $\kappa\in M(AC)$ (choose $Z_\mu=0$ for all $\mu<\kappa$ in (21)), or if $\kappa\in M(M(AC))$ (assuming that $\kappa\notin M(AC)$, choose $Z_\mu=M(AC)\cap\kappa$ for all $\mu<\kappa$), etc. The next section describes a large class of cardinals for which $\kappa\not\rightarrow(\kappa)^2_2$ holds by Theorem 32.4.

33. HANF'S ITERATION SCHEME

The aim of this short section is to describe a large class of inaccessible cardinals κ about which Theorem 32.4 confirms that $\kappa \not\to (\kappa)_2^2$. This class is obtained by an iteration scheme, due to Hanf, of the Mahlo operation. Hanf's scheme depends on an arbitrary wellordering R of κ. In discerning the meaning of the formulas below, it is perhaps best to take R first equal to the natural wellordering 'less than' of κ. But a short contemplation of these formulas will convince the reader that by taking R to be a longer wellordering, i.e., by taking $\mathrm{tp}\langle\kappa, R\rangle > \kappa$, we obtain a larger class M^R. (We shall make a further comment upon this below, after the definition of the iteration scheme.) This is why the parameter R is needed. To describe Hanf's iteration scheme, let R be an arbitrary wellordering of κ. Denote by α_γ^R the γth element in this wellordering ($\gamma < \mathrm{tp}\langle\kappa, R\rangle$). For a set $X \subseteq \kappa$ put

$$M^{(R,0)}(X) = X ;$$

if $\gamma < \mathrm{tp}\langle\kappa, R\rangle$ and $\gamma = \nu \dot{+} 1$ for some ordinal ν, then write

$$M^{(R,\gamma)}(X) = M^{(R,\nu)}(X) \cup [M(M^{(R,\nu)}(X)) \setminus (\alpha_\nu^R \dot{+} 1)] ,$$

and if γ is a limit ordinal $< \mathrm{tp}\langle\kappa, R\rangle$, then put

$$M^{(R,\gamma)}(X) = \bigcup_{\nu<\gamma} M^{(R,\nu)}(X) .$$

Finally set

$$M^R(X) = \bigcup_{\gamma < \mathrm{tp}\langle\kappa, R\rangle} M^{(R,\gamma)}(X) ,$$

and

$$M^*(X) = \bigcup_R M^R(X) ,$$

where, in the last formula, R runs over all wellorderings of κ.

To mention an example, if $<_\kappa$ denotes the natural wellordering of κ and R denotes the wellordering of type $\kappa \dot{+} \kappa$ of κ in which $\alpha_{\gamma \dot{+} n}^R = \gamma \dot{+} 2n$ and $\alpha_{\kappa \dot{+} \gamma \dot{+} n}^R = \gamma \dot{+} 2n \dot{+} 1$ hold for any integer n and for any limit ordinal $\gamma < \kappa$ (and for $\gamma = 0$), then it is easy to see that

$$M^R(X) = M^{<_\kappa}(M^{<_\kappa}(X))$$

holds for any $X \subseteq \kappa$ ($\kappa \geq \omega$). For further examples see Hanf's original paper [1964].

As a corollary of Theorem 32.4, we have

THEOREM 33.1 (Hanf [1964]). *If κ is an inaccessible cardinal and $\kappa \in M^*(AC \cap \kappa)$, then $\kappa \not\to (\kappa)_2^2$.*

In contrast to the remark after Theorem 32.4, one can show that $\kappa \nrightarrow (\kappa)^2_2$ holds e.g. for the first inaccessible cardinal with $\kappa \notin M^*(AC \cap \kappa)$, although we shall not do so. Again, in thinking about the meaning of this theorem, one should first consider the case $\kappa \in M^R(AC \cap \kappa)$ with R equal to the natural wellordering 'less than' of κ.

PROOF. According to the assumptions, we have $\kappa \in M^R(AC \cap \kappa)$ for some wellordering R of κ. This means that there is an ordinal γ with $\gamma \dotplus 1 < \text{tp}\,\langle \kappa, R\rangle$ such that

$$\kappa \notin M^{(R,\gamma)}(AC \cap \kappa) \tag{1}$$

and

$$\kappa \in M(M^{(R,\gamma)}(AC \cap \kappa)). \tag{2}$$

Define the function $f: \kappa \to \text{tp}\,\langle \kappa, R\rangle$ as follows:

$$f(\mu) = \nu \qquad \text{iff} \qquad \mu = \alpha^R_\nu .$$

For any $\mu < \kappa$ put

$$Z_\mu = (M^{(R, f(\mu))}(AC \cap \kappa)) \cap \kappa$$

if $f(\mu) < \gamma$, and $Z_\mu = 0$ otherwise. Defining X with these Z_μ's according to (32.21), it is easy to see that

$$X \cap (\kappa \dotplus 1) = M^{(R,\gamma)}(AC \cap \kappa) \cap (\kappa \dotplus 1)$$

holds; hence (1) and (2) imply that $\kappa \notin X$ and $\kappa \in M(X)$. So $\kappa \nrightarrow (\kappa)^2_2$ holds according to Theorem 32.4. This completes the proof.

34. SATURATED IDEALS, MEASURABLE CARDINALS, AND STRONG PARTITION RELATIONS

In this section we shall consider a large cardinal property stronger than weak compactness, called measurability. Measurability can be defined in terms of certain prime ideals; as these were already defined in Section 29, we could do without repeating the definition. The situation at present is, however, slightly simpler, since all the ideals considered here are ideals in the field $\mathscr{P}(\kappa)$ for some cardinal κ. So it would not be fair to recall here the more complicated concepts introduced in Section 28, and we therefore give the definitions here in complete detail; in the following definition κ is an arbitrary infinite cardinal.

DEFINITION 34.1. (i) A set I is called an *ideal* on κ if $\bigcup I = \kappa$ and, moreover, $X, Y \in I$ and $Z \subseteq X \cup Y$ imply $Z \in I$ for any X, Y, and Z. In this case it is customary to say that κ *carries* the ideal I. I is said to be *nontrivial* if $\kappa \notin I$. Given a cardinal λ, the ideal I on κ is said to be *λ-complete* if $\bigcup X \in I$ holds whenever $X \subseteq I$ and $|X| < \lambda$.

(ii) Given a cardinal λ and an ideal I on κ, I is said to be *λ-saturated* if there is no set $K \subseteq \mathscr{P}(\kappa) \setminus I$ of cardinality λ such that $X \cap Y \in I$ holds for any two distinct $X, Y \in K$. A 2-*saturated* ideal is called a *prime* ideal.

(iii) A nontrivial ideal I on κ is called *normal* if it is κ-complete and, given any set $X \subseteq \kappa$ with $X \notin I$ and any regressive function $f: X \to \kappa$, there is an ordinal $\xi < \kappa$ such that

$$\{\alpha \in X: f(\alpha) = \xi\} \notin I. \tag{1}$$

It is easy to see that an infinite cardinal κ carries a nontrivial κ-complete ideal if and only if κ is regular; the simplest such ideal is the set $[\kappa]^{<\kappa}$. ω does not carry a normal ideal, as one can easily ascertain; if $\kappa > \omega$ is regular, then the ideal of sets nonstationary in κ is a normal ideal on κ according to Fodor's theorem (Theorem 5.5). The notion of normal ideals is due to D. S. Scott, though Fodor proved his theorem several years prior to the birth of this notion. Just as we have two variants of Fodor's theorem for regular κ (Theorem 5.5 and Corollary 5.6), condition (1) in the definition of normality can be replaced by the following: given arbitrary sets $X_\xi \in I$, we have

$$X = \bigcup_{\xi < \kappa} (X_\xi \setminus (\xi \dotplus 1)) \in I. \tag{2}$$

In fact, the function $f: X \to \kappa$ defined by

$$f(\alpha) = \min\{\xi: \alpha \in X_\xi \setminus (\xi \dotplus 1)\} \qquad (\alpha \in X)$$

is regressive and so, assuming that (2) fails, i.e., that $X \notin I$, (1) will also fail with this f. Conversely, assuming that f is an arbitrary regressive function on a set $X \subseteq \kappa$, and putting

$$X_\xi = \{\alpha \in X: f(\alpha) = \xi\},$$

we have $X_\xi = X_\xi \setminus (\xi \dotplus 1)$ by the regressivity of f. If now (2) holds and $X_\xi \in I$ for all $\xi < \kappa$, then we obtain that $X = \bigcup_{\xi < \kappa} X_\xi \in I$, which means that (1) must hold if $X \notin I$. The question whether a cardinal κ carries a λ-saturated ideal is a difficult large cardinal problem in case $\lambda \leq 2^\kappa$; the answer is obviously yes in case $\lambda > 2^\kappa$. It is worth noting the following two results about saturated ideals. The first one is quite trivial:

LEMMA 34.2. *Let $\kappa \geq \omega$ and λ be cardinals, $\lambda \leq \kappa$, and let I be a λ-complete ideal on κ. Then I is λ-saturated if and only if there is no set $K \subseteq \mathscr{P}(\kappa) \setminus I$ of cardinality λ whose elements are pairwise disjoint.*

PROOF. The 'only if' part needs no comment. To see the 'if' part, assume that I is not λ-saturated, and let $M=\{X_\alpha: \alpha<\lambda\}\subseteq\mathscr{P}(\kappa)\setminus I$ be a set such that $X_\alpha\cap X_\beta\in I$ holds whenever $\alpha<\beta<\lambda$. Writing

$$X'_\alpha=X_\alpha\setminus\bigcup_{\beta<\alpha}X_\beta=X_\alpha\setminus\bigcup_{\beta<\alpha}(X_\alpha\cap X_\beta),$$

the set $K=\{X'_\alpha: \alpha<\lambda\}$ will consist of pairwise disjoint sets none of which belongs to I in view of the λ-completeness of this latter. The proof is complete.

The second lemma uses a minor trick:

LEMMA 34.3. *Let $\kappa>\omega$ be a regular cardinal and I a normal ideal on κ. Then I is κ^+-saturated if and only if there is no set $K\subseteq\mathscr{P}(\kappa)\setminus I$ of cardinality κ^+ such that $|X\cap Y|<\kappa$ holds for any two distinct $X, Y\in K$.*

PROOF. The 'only if' part follows from the observation that any subset of κ of cardinality $<\kappa$ belongs to I by the κ-completeness of I. To prove the 'if' part, assume that I is not κ^+-saturated, and let $M=\{X_\alpha: \alpha<\kappa^+\}\subseteq\mathscr{P}(\kappa)\setminus I$ be a set such that $X_\alpha\cap X_\beta\in I$ holds whenever $\alpha<\beta<\kappa^+$. For each $\beta<\kappa^+$, let $f_\beta: \beta\to\kappa$ be a 1-1 function (not necessarily *onto* κ), and put

$$X'_\beta=X_\beta\setminus Y_\beta,$$

where

$$Y_\beta=\bigcup_{\alpha<\beta}((X_\beta\cap X_\alpha)\setminus(f_\beta(\alpha)\dot{+}1)).$$

In view of the normality of I, (2) implies that $Y_\beta\in I$; hence $X'_\beta\notin I$. On the other hand, it is obvious that $X'_\alpha\cap X'_\beta\subseteq X_\alpha\cap X'_\beta\subseteq f_\beta(\alpha)\dot{+}1$ holds whenever $\alpha<\beta<\kappa^+$; hence we have $|X'_\alpha\cap X'_\beta|<\kappa$ in this case. This completes the proof of the lemma with $K=\{X'_\beta: \beta<\kappa^+\}$.

A classical result of S. Ulam [1930] says the following:

THEOREM 34.4. *Given any infinite cardinal κ, there is no nontrivial κ^+-complete and κ^+-saturated ideal on κ^+.*

For the proof we need a technical lemma, also due to Ulam. This lemma is very important in itself, because it has many other applications.

LEMMA 34.5 (Ulam [1930]). *Let κ be an infinite cardinal. Then there is a matrix $\langle U_{\alpha\beta}: \alpha<\kappa\ \&\ \beta<\kappa^+\rangle$ of subsets of κ^+ such that each row consists of pairwise disjoint sets and the union of each column contains all but $\leq\kappa$ elements of κ^+, i.e., such that $U_{\alpha\beta}$ and $U_{\alpha\beta'}$ are disjoint whenever $\alpha<\kappa$ and $\beta<\beta'<\kappa^+$, and we have*

$$|\kappa^+\setminus\bigcup_{\alpha<\kappa}U_{\alpha\beta}|\leq\kappa \tag{3}$$

for every $\beta<\kappa^+$.

The matrix whose existence is claimed in this lemma is called an *Ulam matrix* on κ^+.

Proof. For each $\xi<\kappa^+$, let $f_\xi:\xi\to\kappa$ be a 1-1 function (not necessarily onto κ), and write

$$U_{\alpha\beta}=\{\xi<\kappa^+:\beta<\xi\ \&\ f_\xi(\beta)=\alpha\}$$

for $\alpha<\kappa$ and $\beta<\kappa^+$. The sets $U_{\alpha\beta}$, $\beta<\kappa^+$, are pairwise disjoint for $\alpha<\kappa$ fixed by the one-to-oneness of f, and we clearly have

$$\bigcup_{\alpha<\kappa} U_{\alpha\beta}=\kappa^+\setminus(\beta\dot{+}1),$$

which verifies (3). The proof is complete.

Proof of Theorem 34.4. Assume I is a nontrivial κ^+-complete ideal on κ^+; we are going to show that I is not κ^+-saturated. To this end, consider an Ulam matrix $\langle U_{\alpha\beta}:\alpha<\kappa\ \&\ \beta<\kappa^+\rangle$ on κ^+, as described in the preceding lemma. By (3), there is an ordinal $f(\beta)<\kappa$ for each $\beta<\kappa^+$ such that $U_{f(\beta)\beta}\notin I$. There is a set $X\subseteq\kappa^+$ of cardinality κ^+ on which f is constant, say $f(\beta)=\alpha(<\kappa)$ for each $\beta\in X$. Then the set $\{U_{\alpha\beta}:\beta\in X\}\subseteq\mathscr{P}(\kappa^+)\setminus I$ consists of pairwise disjoint sets, showing that I is in fact not κ^+-saturated. The proof is complete.

It was proved by R. M. Solovay [1971], R. B. Jensen, and others that Theorem 34.4 can be extended with many regular limit or inaccessible cardinals replacing κ^+, e.g. for all those cardinals satisfying the assumptions of Theorem 32.4 (note that this theorem is identical to Theorem 30.5), and so, a fortiori, for all those satisfying the assumptions of Theorem 33.1. Their proofs, though often described in combinatorial terms, are essentially of metamathematical character. It is all the more interesting therefore that Ulam's original approach can be extended to work for some of these cardinals, especially because Ulam matrices can have various other applications. In order to describe the generalization of Ulam's result, we need the following concept:

Definition 34.6. Let $\kappa>\omega$ be a regular limit cardinal, and let $S\subseteq\kappa$ be a set. The matrix $\langle U_{\alpha\beta}:\beta\in S\ \&\ \alpha<\beta\rangle$ of subsets of κ is called a *triangular Ulam matrix on κ with support S* if S is stationary in κ, the sets $U_{\alpha\beta}$, $\alpha<\beta\in S$, are pairwise disjoing for any fixed $\alpha<\kappa$, and

$$|\kappa\setminus\bigcup_{\alpha<\beta} U_{\alpha\beta}|<\kappa \tag{4}$$

holds for each $\beta\in S$.

Hajnal's lemma on triangular Ulam matrices runs as follows (see Section 5 for the concepts used):

LEMMA 34.7 (Hajnal [1969]). *Let $\kappa > \omega$ be a regular limit cardinal such that there is a set S stationary in κ and a club C in κ, both consisting of uncountable cardinals, such that*

$$S \subseteq C \subseteq \{\lambda < \kappa\colon \text{ either } \lambda > \omega \text{ is regular and } S \cap \lambda \text{ is nonstationary in } \lambda \text{ or } \lambda \text{ is singular}\}, \tag{5}$$

where λ runs over cardinals. Then there is a triangular Ulam matrix on κ with support S.

PROOF. We first establish the following

CLAIM. *For each $\lambda \in C$, there is a 1-1 regressive function f_λ on $S \cap \lambda$.*

We construct the functions f_λ by recursion on $\lambda \in C$. Assume to this end that $f_{\lambda'}$ has already been constructed for each $\lambda' < \lambda$ with $\lambda' \in C$. If λ is the first element of C, then define f_λ as the empty function (i.e., the empty set). Suppose now that this is not the case, i.e., that $C \cap \lambda$ is not empty, and write

$$\rho = \sup(C \cap \lambda). \tag{6}$$

We have $\rho \in C$ by the definition of club. We distinguish three cases: a) $\rho < \lambda$, b) $\rho = \lambda$ and λ is regular, and c) $\rho = \lambda$ and λ is singular.

Ad a). If $\rho \notin S$, then we can simply put $f_\lambda = f_\rho$ in this case. If $\rho \in S$, then we can almost do the same, the only problem being that we also have to define $f_\lambda(\rho)$ in a way that f_λ be 1-1. But it is easy to make one value free for $f_\lambda(\rho)$ e.g. by putting

$$f'_\rho(\sigma) = 1 \dotplus f_\rho(\sigma) \tag{7}$$

for every $\sigma \in S \cap \rho$; then we can write $f_\lambda = f'_\rho \cup \{\langle \rho\, 0 \rangle\}$.

Ad b). Note in this case that there is a club $B' \subseteq \lambda \setminus S$ in λ according to (5). Write $B = B' \cap C$; B is also a club in λ because $\rho = \sup(C \cap \lambda) = \lambda$ in this case, and B consists of cardinals since it is a subset of C. Let τ_α, $\alpha < \lambda$, be an enumeration of the elements of B in increasing order; for each $\alpha < \tau$, f_{τ_α} is a 1-1 regressive function from $S \cap \tau_\alpha$ into τ_α. For each $\alpha < \lambda$, change the function $f_{\tau_{\alpha+1}} \upharpoonright (S \cap (\tau_\alpha, \tau_{\alpha+1}))$ into a 1-1 regressive function $f'_{\tau_{\alpha+1}}$ such that this latter function assumes its values in the interval $[\tau_\alpha, \tau_{\alpha+1})$. Noting that S consists of cardinals, and so the first element of $S \cap (\tau_\alpha, \tau_{\alpha+1})$ is $\geq \tau_\alpha^+$, it is easy to do this by putting

$$f'_{\tau_{\alpha+1}}(\sigma) = \tau_\alpha \dotplus f_{\tau_{\alpha+1}}(\sigma) \tag{8}$$

whenever $\sigma \in S \cap (\tau_\alpha, \tau_{\alpha+1})$. Put $f_\lambda = f_{\tau_0} \cup \bigcup_{\alpha < \lambda} f_{\tau_{\alpha+1}}$. Then f_λ is regressive and 1-1, and its domain is $S \cap \lambda$, as none of the τ_α's belong to S because $\tau_\alpha \in B \subseteq B' \subseteq \lambda \setminus S$.

Ad c). We use a similar argument as in the preceding case, although the situation now is slightly more complicated. Let $\langle \tau_\alpha : \alpha < \text{cf}(\lambda) \rangle$ be an increasing sequence of cardinals τ_α with $\text{cf}(\lambda) < \tau_\alpha \in C$ that tends to λ. Define the functions $f'_{\tau_{\alpha+1}}$ by formula (8) above for each ordinal α with $1 \leq \alpha < \text{cf}(\lambda)$; in case $\alpha = 0$, we shall have to make a slight change. Namely, we may have to define f_λ for certain τ_α's as well, as $\tau_\alpha \in S$ may well occur now. So let

$$f' : S \cap \{\tau_\alpha : \alpha < \lambda\} \to \{0\} \cup [\tau_0, \tau_0 \dot{+} \text{cf}(\lambda))$$

be a 1-1 function; f' is obviously regressive if we stipulate that $f'(\tau_0) = 0$ in case $\tau_0 \in S$, because we assumed $\tau_0 > \text{cf}(\lambda)$. So as to clear the range of f', define f'_{τ_1} by slightly changing (8) as follows:

$$f'_{\tau_1}(\sigma) = \tau_0 \dot{+} \text{cf}(\lambda) \dot{+} f_{\tau_1}(\sigma)$$

whenever $\sigma \in S \cap (\tau_0, \tau_1)$. So as to make the value 0 also free for f', define f'_{τ_0} according to (7) with τ_0 replacing ρ there. It is then clear that $f_\lambda = f' \cup f_{\tau_0} \cup \bigcup_{\alpha < \text{cf}(\lambda)} f_{\tau_{\alpha+1}}$ is a 1-1 regressive function on $S \cap \alpha$. Thus our claim is established.

It is now easy to define the matrix whose existence is claimed in the lemma to be proved: Let h be a 1-1 function from C onto κ, and for each α and β with $\alpha < \beta$ and $\beta \in S$ put

$$U_{\alpha\beta} = \{h(\lambda) : \beta < \lambda \in C \;\&\; f_\lambda(\beta) = \alpha\},$$

analogously as in the proof of Lemma 34.5. It is easy to show that $\langle U_{\alpha\beta} : \alpha < \beta \in S \rangle$ is a triangular Ulam matrix with support S; e.g., we have $\bigcup_{\alpha < \beta} U_{\alpha\beta} = h``(C \setminus (\beta \dot{+} 1))$, which entails (4) as h is onto κ. The proof is complete.

Let AC be the class of accessible ordinals, NRL be the class of those ordinals that are not uncountable regular limit cardinals, and, for a cardinal κ, denote by $<_\kappa$ the natural wellordering of κ. Then, using the notations of the preceding section, it is easy to check that the assumptions of the preceding lemma hold with some S and C whenever $\kappa \in M^{(\omega, <_\kappa)}(NRL) \setminus NRL$ or $\kappa \in M^{(<_\kappa, \omega)}(AC) \setminus AC$; i.e., the above lemma confirms the existence of a triangular Ulam matrix on these cardinals. It is not known, however, if there is a triangular Ulam matrix e.g. on the first cardinal not belonging to $M^{(\omega, <_\kappa)}(AC)$. The application of the above lemma that we aimed at is the following theorem of Hajnal [1969]:

THEOREM 34.8. *Let $\kappa > \omega$ be a regular limit cardinal, and assume that there is a triangular Ulam matrix on κ. Then κ does not carry any nontrivial κ-saturated κ-complete ideal.*

PROOF. Assume that I is a nontrivial κ-complete ideal on κ; we are going to show that I is not κ-saturated. To this end, consider a triangular Ulam matrix

$\langle U_{\alpha\beta}: \alpha < \beta \in S \rangle$ on κ; note here that S is stationary in κ according to Definition 34.6. By (4), there is an ordinal $f(\beta) < \beta$ for each $\beta \in S$ with $\beta \neq 0$ such that $U_{f(\beta)\beta} \notin I$. Neumer's or Fodor's theorem (the best reference here is Theorem 5.5) then implies that there is a set S' of cardinality κ such that f is constant on S', say $f(\beta) = \alpha$ for each $\beta \in S'$. Then the set $\{U_{\alpha\beta}: \beta \in S'\} \subseteq \mathscr{P}(\kappa) \setminus I$ consists of pairwise disjoint sets, showing that I is indeed not κ-saturated. The proof is complete.

According to the remark made before the theorem just proved, this theorem shows that there are many inaccessible or regular limit cardinals κ which do not carry nontrivial κ-saturated κ-complete ideals. It does not, however, give all the known results about κ-saturated κ-complete ideals on κ, since it does not confirm that the first cardinal κ not belonging to $M^{(\omega, <\kappa)}(AC)$ does not carry such a nontrivial ideal (cf. the remark made before the last theorem), although this is a well-known result (cf. our remark made after the proof of Theorem 34.4). Kunen [1970] proved that if $V = L$, then there is no κ^+-saturated κ-complete ideal on any uncountable cardinal κ. The assumption $V = L$ seems necessary here even in case of small κ, since Kunen also proved recently that a very strong large cardinal axiom ensures that it is consistent that $\aleph_1$ carries an $\aleph_2$-saturated $\aleph_1$-complete ideal.

Normal ideals are often much more convenient to handle than non-normal ones. Their applicability in the theory of saturated ideals is made possible by the following theorem of D. S. Scott:

THEOREM 34.9. *Let κ and λ be cardinals, κ regular, $3 \leq \lambda \leq \kappa^+$, and suppose that κ carries a nontrivial λ-saturated κ-complete ideal. Then κ also carries a λ-saturated normal ideal.*

The proof is analogous to that of Lemma 31.2, but the present situtaion is simplified by the fact that we here consider ideals in the complete field $\mathscr{P}(\kappa)$ of sets, while the fact that we consider λ-saturated ideals instead of prime ideals does not cause much complication.

PROOF. Let I be a nontrivial λ-saturated κ-complete ideal on κ, and define the relation $<_I$ on ${}^\kappa\kappa$ by putting

$$f <_I g \quad \text{iff} \quad \forall \xi < \kappa \; f(\xi) \leq g(\xi) \quad \text{and} \quad \{\xi < \kappa : f(\xi) = g(\xi)\} \in I$$

for any two functions $f, g \in {}^\kappa\kappa$. It is easy to check that $<_I$ is a partial ordering. There is no infinite sequence of functions $f_n \in {}^\kappa\kappa$, $n < \omega$, such that $f_{n+1} <_I f_n$ holds for each integer n. In fact, the contrary would mean in view of the κ-completeness of I that the set

$$\bigcup_{n<\omega} \{\xi < \kappa : f_{n+1}(\xi) = f_n(\xi)\}$$

does not equal κ. Picking a $\xi<\kappa$ not in this set, we would then have

$$f_0(\xi)>f_1(\xi)>\ldots,$$

which is impossible, as there is no infinite descending sequence of ordinals.

Let now f_0 be the identity function on κ (i.e., let $f_0(\xi)=\xi$ for each $\xi<\kappa$) and, having defined f_n for some integer n, choose, if possible, a function $f_{n+1}\in{}^{\kappa}\kappa$ with $f_{n+1}<_I f_n$ such that

$$\{\xi<\kappa: f_{n+1}(\xi)=\alpha\}\in I$$

for all $\alpha<\kappa$ (note that this clearly holds with f_0 replacing f_{n+1}). According to the remark just made, there is a least integer n for which it is impossible to choose such a function f_{n+1}. Write $h=f_n$. The function h obviously has the following important properties:

$$h(\xi)\leq\xi \tag{9}$$

holds for any $\xi<\kappa$, since $h\leq_I f_0$, and f_0 was chosen to be the identity function; for any ordinal $\alpha<\kappa$ we have

$$\{\xi<\kappa: h(\xi)=\alpha\}\in I, \tag{10}$$

and whenever f is a regressive function on κ, then we have

$$\{\xi<\kappa: f(h(\xi))=\alpha\}\notin I \tag{11}$$

for some $\alpha<\kappa$. To see this, one only has to observe that (10) with $\alpha=0$ implies $f\circ h<_I h$, as $f(h(\xi))<h(\xi)$ whenever $h(\xi)\neq 0$; if (11) were false, then we could choose $f_{n+1}=f\circ h$, a contradiction. (A function h with properties (9)–(11) is called a minimal function with respect to I.)

We are now in a position to define a λ-saturated normal ideal J on κ. For any $X\subseteq\kappa$ put $X\in J$ if and only if there is a regressive function f on X such that

$$\{\xi<\kappa: h(\xi)\in X\ \&\ f(h(\xi))=\alpha\}\in I \tag{12}$$

holds for any $\alpha<\kappa$. It is easy to see that J is a κ-complete ideal on κ; in fact, to see κ-completeness, assume that $X_\nu\in J$, $\nu<\rho$, are pairwise disjoint sets for some $\rho<\kappa$. Then there is a regressive function f_ν on X_ν for each $\nu<\rho$ such that (12) will be satisfied with X_ν and f_ν replacing X and f, respectively, and (12) with $f=\bigcup_{\nu<\rho} f_\nu\in J$ will confirm that $X=\bigcup_{\nu<\rho} X_\nu=J$. (11) shows that J is nontrivial.

J is also a normal ideal; this can be seen by arguments similar to those used in the proof of Fodor's theorem (Theorem 5.4) as follows: Assume that J is not normal. Then there is an $X\subseteq\kappa$ with $X\notin J$ and a regressive function g on X such that

$$X_\nu=\{\xi\in X: g(\xi)=\nu\}\in J$$

holds for any $\nu<\kappa$; i.e., by (12) there are regressive functions f_ν on X_ν, $\nu<\kappa$, such that

$$\{\eta\in\kappa: h(\eta)\in X \,\&\, g(h(\eta))=\nu \,\&\, f_\nu(h(\eta))=\alpha\}\in I \tag{13}$$

holds for any $\alpha<\kappa$. Define the regressive function f on $X=\bigcup_{\nu<\kappa} X_\nu$ by putting

$$f(\xi)=\max\{\nu, f_\nu(\xi)\}$$

whenever $\xi\in X_\nu$; then $f(\xi)\geq g(\xi)$ for all $\xi\in X$. Given any $\alpha<\kappa$, we have

$$\{\eta<\kappa: h(\eta)\in X \,\&\, f(h(\eta))\leq\alpha\}=$$

$$=\{\eta<\kappa: h(\eta)\in X \,\&\, g(h(\eta))\leq\alpha \,\&\, f(h(\eta))\leq\alpha\}=$$

$$=\bigcup_{\nu\leq\alpha}\{\eta<\kappa: h(\eta)\in X \,\&\, g(h(\eta))=\nu \,\&\, f(h(\eta))\leq\alpha\}=$$

$$=\bigcup_{\nu\leq\alpha}\{\eta<\kappa: h(\eta)\in X \,\&\, g(h(\eta))=\nu \,\&\, f_\nu(h(\eta))\leq\alpha\}=$$

$$=\bigcup_{\mu,\,\nu\leq\alpha}\{\eta<\kappa: h(\eta)\in X \,\&\, g(h(\eta))=\nu \,\&\, f_\nu(h(\eta))=\mu\}\in I,$$

since there are less than κ sets after the last union sign, and each of these belongs to I according to (13). This, however, means if we compare the two extreme sides of the above formula that $X\in J$. This is a contradiction, proving that J is normal.

Finally, we are going to show that the ideal J is λ-saturated. Assume the contrary; then Lemmas 34.2 and 34.3 imply in view of the assumption $\lambda\leq\kappa^+$ that there are sets $X_\mu\in\mathscr{P}(\kappa)\setminus J$; $\mu<\lambda$, such that

$$|X_\mu\cap X_\nu|<\kappa \tag{14}$$

holds whenever $\mu<\nu<\lambda$. Writing

$$Y_\mu=\{\xi<\kappa: h(\xi)\in X_\mu\},$$

we must have $Y_\mu\notin I$, since otherwise (12) with $f\equiv 0$ (the function constantly zero on κ) would imply $X_\mu\in J$. On the other hand, we have

$$Y_\mu\cap Y_\nu=\{\xi<\kappa: h(\xi)\in X_\mu\cap X_\nu\}\in I$$

whenever $\mu<\nu<\lambda$ in view of (10), (14), and the κ-completeness of I. This contradicts our assumption that I is λ-saturated. Thus the λ-saturatedness of J is also established. This completes the proof.

Our main concern in what follows is to prove a strong partition relation for measurable cardinals, to be defined below. We shall also briefly mention real-valued measurable cardinal since they are often related to combinatorial problems (in fact, they were mentioned in Section 11 while discussing ramifications of the Erdős–Dushnik–Miller theorem); the reader not interested in them may skip parts (ii)–(iv) of the following definition, as well as our short comments about real-valued measurable cardinals made afterwards.

DEFINITION 34.10. (i) An uncountable cardinal κ is called *measurable* if it carries a nontrivial κ-complete prime (i.e., 2-saturated) ideal.

(ii) If X is an indexed set of nonnegative real numbers, then their sum ΣX is defined as

$$\sup\{\Sigma Y: Y \subseteq X \ \&\ Y \text{ is finite}\},$$

where the sup here is either a real number or $+\infty$; the sum of finitely many real numbers above is to be understood in the usual sense.

(iii) Given an infinite cardinal κ, a *κ-additive real-valued probability measure on* κ is a function $m: \mathscr{P}(\kappa) \to [0, 1]$, this latter denoting an interval of real numbers, such that $m(\kappa) = 1$ and we have

$$m(\bigcup X) = \Sigma\langle m(x): x \in X\rangle$$

for any set X of cardinality $<\kappa$ of pairwise disjoint subsets of κ. m is called *nontrivial* if $m(\{\xi\}) = 0$ for any $\xi \in \kappa$.

(iv) An uncountable cardinal κ is called *real-valued measurable* if there is a nontrivial κ-additive real-valued probability measure on κ; in this case one also says that κ *carries* such a measure.

It is easy to see that a measurable cardinal is also real-valued measurable. In fact, if I is a nontrivial κ-complete prime ideal on $\kappa > \omega$, then the function m defined as $m(X) = 0$ whenever $X \in I$ and $m(X) = 1$ for any other $X \subseteq \kappa$ is a nontrivial real-valued probability measure on κ. On the other hand, if m is a nontrivial real-valued probability measure on κ, then the set

$$I = \{X \subseteq \kappa: m(X) = 0\}$$

is a nontrivial $\aleph_1$-saturated κ-complete ideal on κ. In fact, to see this, one only has to observe that the sum of $\aleph_1$ positive real numbers is $+\infty$; this latter assertion can easily be shown, since given $\aleph_1$ positive real numbers, there is a real number $\varepsilon > 0$ such that $\aleph_1$ among them are $> \varepsilon$. We shall soon see that a measurable cardinal must be inaccessible (see our remark immediately after the proof of Theorem 34.11). This is not so for real-valued measurable cardinals; in fact, a result of Solovay [1971] says that if ZFC + 'there is a measurable cardinal' is consistent, then ZFC + '$2^{\aleph_0}$ is real-valued measurable' is also consistent. A recent result of K. Prikry says that $2^{\aleph_0}$ is real-valued measurable, then $2\lambda = 2^{\aleph_0}$ for each λ with $\aleph_0 \leq \lambda < 2^{\aleph_0}$; Theorems 34.4 and 34.8 say that a real-valued measurable cardinal must be a 'very large' regular limit cardinal. A result of Erdős and Hajnal [1958] is the following:

THEOREM 34.11. *If* κ *is a measurable cardinal, then*

$$\kappa \to (\kappa)^{<\omega}_{\lambda} \tag{15}$$

holds for any cardinal $\lambda < \kappa$.

For the definition of the partition relation in (15) see Subsection 8.6. As pointed out by F. Rowbottom, if there is a measurable cardinal, then there must be an uncountable cardinal $\kappa <$ the first measurable cardinal satisfying the relation in (15); this can be seen by a simple argument involving indescribability. Hint: a measurable cardinal is Π^2_1-indescribable by a theorem of W. P. Hanf and D. S. Scott [1965] (see Silver [1971] for a proof), while saying that κ is the first cardinal $> \omega$ satisfying (15) with all $\lambda < \kappa$ can be expressed by a Π^1_2, or even by a Π^2_1, sentence. In order to prove the above theorem, we shall first prove the following lemma of F. Rowbottom [1971].

LEMMA 34.12. *Let* $\kappa > \omega$ *be measurable, I a normal prime ideal on* κ, *let n be an integer, and assume that* $f : [\kappa \setminus \{0\}]^n \to \kappa$ *is a function such that*

$$f(u) < \min(u) \tag{16}$$

holds for any $u \in [\kappa \setminus \{0\}]^n$. *Then there is an* $X \in \mathscr{P}(\kappa) \setminus I$ *such that f is constant on* $[X]^n$.

PROOF. We use induction on n; the assertion holds vacuously in case $n = 0$. Assume that $n > 0$ and the assertion holds with $n-1$ replacing n, and let $f : [\kappa \setminus \{0\}]^n \to \kappa$ be a function satisfying the requirements of the theorem. For each $\xi < \kappa$, define the function $f_\xi : [\kappa \setminus (\xi \dot{+} 1)]^{n-1} \to \kappa$ by putting

$$f_\xi(v) = f(\{\xi\} \cup v) \tag{17}$$

whenever $v \in [\kappa \setminus (\xi \dot{+} 1)]^{n-1}$; the induction hypothesis implies that for each $\xi < \kappa$ there is an $X_\xi \in \mathscr{P}(\kappa) \setminus I$ and a $g(\xi)$ such that

$$f_\xi``[X_\xi]^{n-1} = \{g(\xi)\} \tag{18}$$

holds; we have $g(\xi) < \xi$ according to (16) and (17). We obtain that $\kappa \setminus X_\xi \in I$ since I is a prime ideal, and so

$$\bigcup_{\xi < \kappa} (\kappa \setminus (X_\xi \cup (\xi \dot{+} 1)) \in I$$

by the normality of I (cf. (2)), i.e.,

$$X' = \bigcap_{\xi < \kappa} ((\xi \dot{+} 1) \cup X_\xi) \notin I .$$

Using the normality of I again (cf. (1)), there is an $\alpha<\kappa$ and an $X\subseteq X'$ with $X\notin I$ such that

$$g``X=\{\alpha\}. \tag{19}$$

We claim that

$$f(u)=\alpha \tag{20}$$

holds for any $u\in[X]^n$. In fact, let ξ be the least element of u, and write $v=$ $=u\setminus\{\xi\}$. Then we have $v\subseteq X\subseteq X'\subseteq(\xi\dot{+}1)\cup X_\xi$; the definition of v implies that $v\cap(\xi\dot{+}1)=0$; hence we actually have $v\subseteq X_\xi$. As $\xi\in X$, (17), (18), and (19) imply that

$$f(u)=f_\xi(v)=g(\xi)=\alpha,$$

which proves (20). This means that the conclusion of the theorem is valid, i.e., the proof is complete.

Proof of Theorem 34.11. There is a normal prime ideal I on κ in view of Theorem 34.9. Fix $\lambda<\kappa$ and let $f:[\kappa\setminus\lambda]^{<\omega}\to\lambda$ be a coloring; it is enough to prove that there is a set $X\in\mathscr{P}(\kappa)\setminus I$ such that the function $f\upharpoonright[X]^n$ is constant for each $n<\omega$. According to the theorem just proved, there is an $X_n\in\mathscr{P}(\kappa)\setminus I$ for each integer n such that $f\upharpoonright[X_n]^n$ is constant; then we have $X=\bigcap_{n<\omega}X_n\notin I$ because I is a κ-complete prime ideal. This X satisfies our requirements; the proof is complete.

Note that (15) with $\lambda=2$ obviously implies $\kappa\to(\kappa)^2_2$; hence a measurable cardinal is inaccessible in view of Theorem 29.1. Note also that we have

$$\omega\not\to(\omega)_2^{<\omega}. \tag{22}$$

In fact, put $f(x)=0$ for any finite set $x\subseteq\omega$ if $x_1-x_0\geq|x|$, where x_0 and x_1 are the first two elements of x, and put $f(x)=1$ otherwise. Then it is easy to check that f verifies the relation in (22). Observe that the Axiom of Choice is not used in establishing (22); hence it is interesting to compare this relation with the relation

$$\omega\to(\omega)_2^\omega, \tag{23}$$

which was shown by A. R. D. Mathias to be consistent with ZF + the Axiom of Dependent Choices (DC), which is weak form of the Axiom of Choice (see e.g. Jech [1971; p. 79]). This seems an anomaly, since one would expect the relation $\omega\to(\omega)_2^\omega$ to be stronger than $\omega\to(\omega)_2^{<\omega}$, but, obviously, the proof of (22) breaks down if one wants to use it to disprove (23). Of course, the relation in (23) is false in ZFC according to Theorem 12.1.

If we have $\kappa\not\to(\omega)_2^{<\omega}$ for an infinite cardinal κ, then the method of the proof of (24.2) with assumption d) in Lemma 24.1 implies that $2^\kappa\not\to(\omega)_2^{<\omega}$ holds, and if κ is singular and $\lambda \nrightarrow (\omega)_2^{<\omega}$ holds for all $\lambda < \kappa$, then similarly, the method of the proof of (22.5) in Theorem 22.1 shows that $\kappa \nrightarrow (\omega)_2^{<\omega}$ also holds. In this way it follows from (22) that

$$\lambda \nrightarrow (\omega)_2^{<\omega} \tag{24}$$

holds for every cardinal λ less than the first inaccessible cardinal κ_0. It is easy to derive from here that

$$\kappa_0 \nrightarrow (\omega \dot{+} 1)_2^{<\omega} \tag{25}$$

holds for this κ_0. In fact, according to (24) there is a coloring $f_\xi : [\xi]^{<\omega} \rightarrow 2$ for each *ordinal* $\xi < \kappa_0$ verifying the relation $\xi \nrightarrow (\omega)_2^{<\omega}$. Putting $f(x) = f_\xi(x \setminus \{\xi\})$ whenever $x \subseteq \kappa_0$ is a finite set with $\xi = \max x$, it is easy to see that f verifies the relation in (25). (25) implies in view of Theorem 34.11 that the first inaccessible cardinal is not measurable. One can push this argument further so as to prove that the first measurable cardinal κ must be the κth inaccessible cardinal. Of course, we knew this already e.g. in view of Theorem 34.8, or in view of the results on the relation $\kappa \rightarrow (\kappa)_2^2$ in Sections 30–33.

Finally we prove a simple lemma due to Silver [1966], which will be needed in Section 45:

LEMMA 34.13. *Let κ be an infinite cardinal and α be a limit ordinal. If*

$$\kappa \rightarrow (\alpha)_2^{<\omega} \tag{26}$$

holds, then we also have

$$\kappa \rightarrow (\alpha)_{2^{\aleph_0}}^{<\omega} \tag{27}$$

PROOF. Assume (26). Given a coloring

$$f : [\kappa]^{<\omega} \rightarrow {}^{\aleph_0}2,$$

we have to prove that there is a set X of order type α such that f is constant on $[X]^m$ for each integer m. For each integer m and every set $x \in [\kappa]^{\geq m}$, denote by $x|m$ the set formed by the first m elements of x. For any two integers m, n and for any set $x \in [\kappa]^{2^m \cdot 3^n}$ put

$$g(x) = (f(x|m))(n) \tag{28}$$

(note here that $f(x|m)$ is a function from $\aleph_0$ into 2, and so the value on the right-hand side here is 0 or 1); if this stipulation does not define $g(x)$, where $x \in [\kappa]^{<\omega}$, then put $g(x) = 0$ or 1 arbitrarily (e.g. put $g(x) = 0$ always). According to (26), there is a set $X \subseteq \kappa$ of order type α and there are integers $i_k < 2$ $(k < \omega)$ such that

$$g``[X]^k = \{i_k\} \tag{29}$$

holds for each $k<\omega$. It is easy to see then that f is constant on $[X]^m$ for each integer m. In fact, if x belongs to this set, then let $x' \in [X]^\omega$ be such that $x'|m=x$ (here we use the assumption that the order type α of x is a limit ordinal); then we have

$$f(x)=\langle g(x'|2^m \cdot 3^n): n<\omega\rangle=\langle i_{2^m \cdot 3^n}: n<\omega\rangle$$

(cf. (29)). The proof is complete.

CHAPTER VIII

DISCUSSION OF THE ORDINARY PARTITION RELATION WITH SUPERSCRIPT 2

In the first two sections of this chapter we collect the results obtained for the case $r=2$ of the ordinary partition relation, and an attempt is made to state them in a unified form. Understandably, this is a very unrewarding task, and the reader looking for interesting ideas might be disappointed by studying these two sections. Perhaps it is best for him to skip the detailed proofs here, or even to skip these sections entirely. We do not mean by saying this that these sections are superfluous. Their importance lies in the fact that they put the disarray of the large amount of information for the case $r=2$ contained in the previous three chapters in order. They give a practically usable test to see whether an ordinary partition relation with $r=2$ holds or fails if this can be decided with the aid of the known results; and, moreover, they help to locate the open problems, the most important of which is probably Problem 35.5. In Section 37, the third and last one in this chapter (which, by the way, the reader should not be afraid to consult, since it is not of the same character as the first two sections here) we shall show that this problem cannot be decided negatively with the aid of the methods used in this book. Since a positive solution also seems hopeless to us with present methods, it is quite possible that the solution of this problem will have a really stimulating effect on the development of combinatorial set theory.

35. DISCUSSION OF THE ORDINARY PARTITION SYMBOL IN CASE $r=2$

In this section we are going to collect our results and the open problems concerning the case $r=2$ of the ordinary partition symbol. So as to make the rather complicated discussion more intelligible, we are not going to formulate the theorems below in the most concise forms possible. We shall study the relation

$$\kappa \to (\lambda_\xi)^2_{\xi<\vartheta} \tag{1}$$

under the conditions

$$\kappa \geq \omega, \quad \vartheta \geq 2, \quad \text{and} \quad 3 \leq \lambda_\xi \leq \kappa \quad \text{for} \quad \xi < \vartheta, \tag{2}$$

which will be assumed throughout this section. To this end, we shall need the following

DEFINITION 35.1. Let $\vec{\lambda}=\langle\lambda_\xi:\xi<\vartheta\rangle$ and $\vec{\lambda}'=\langle\lambda'_\xi:\xi<\vartheta'\rangle$ be two sequences of cardinals. We write $\vec{\lambda}\ll\vec{\lambda}'$ (read: $\vec{\lambda}'$ dominates $\vec{\lambda}$) if there is a one-to-one function f from ϑ into ϑ' such that $\lambda_\xi\le\lambda'_{f(\xi)}$ holds for any $\xi<\vartheta$.

It is obvious that if $\vec{\lambda}'$ dominates $\vec{\lambda}$ and we have $\kappa\to(\lambda'_\xi)^2_{\xi<\vartheta'}$ then we also have $\kappa\to(\lambda_\xi)^2_{\xi<\vartheta}$. Our main goal would be to show that, whenever relation (1) holds, there is a sequence $\langle\lambda'_\xi:\xi<\vartheta'\rangle$ of a certain specified form dominating $\langle\lambda_\xi:\xi<\vartheta\rangle$ such that $\kappa\to(\lambda'_\xi)^2_{\xi<\vartheta'}$ holds. In this way, one would always be able to decide whether (1) holds or fails by simply checking whether or not such a dominating sequence exists. Unfortunately, we shall not be able to achieve this goal completely, though we shall come quite near to doing so. We start by a lemma listing the cases in which (1) fails by our earlier results. The use of the logarithm operation L defined in Section 7 will simplify our considerations.

LEMMA 35.2. (1) *fails provided at least one of the following conditions* (i)–(ix) *hold*:

(i) $\lambda_0,\lambda_1>L_{\lambda_0}(\kappa)$;

(ii) $\lambda_0=\kappa$ *and* $\lambda_1>\min(L_\kappa(\kappa), L_{\mathrm{cf}(\kappa)}(\mathrm{cf}(\kappa)))$;

(iii) $\lambda_0=\kappa$, $\lambda_1\ge\mathrm{cf}(\kappa)$, *and* $\mathrm{cf}(\kappa)\not\to(\mathrm{cf}(\kappa))^2_2$;

(iv) $\vartheta\ge L_3(\kappa)$;

(v) $|\{\xi<\vartheta:\lambda_\xi\ge\omega\}|\ge L_{\lambda_0}(\kappa)$;

(vi) $\lambda_0=\kappa$ *and* $\vartheta\ge L_3(\mathrm{cf}(\kappa))$;

(vii) $\lambda_0=\kappa$ *and* $|\{\xi:\lambda_\xi\ge\omega\}|\ge L_{\mathrm{cf}(\kappa)}(\mathrm{cf}(\kappa))$.

We can strengthen (v) *and* (vii) *in case* GCH *holds to*

(v′) $\vartheta\ge L_{\lambda_0}(\kappa)$ *and*

(vii′) $\lambda_0=\kappa$ and $\vartheta\ge L_{\mathrm{cf}(\mu)}(\mathrm{cf}(\kappa))$, *respectively*.

(viii) $\vartheta\ge\mathrm{cf}\,L_3(\kappa))$ *and there are cardinals* $\kappa_\xi<L_3(\kappa)$ *for* $\xi<\mathrm{cf}(L_3(\kappa))$ *such that their sum is* $L_3(\kappa)$ *and* $\lambda_\xi\ge\kappa_\xi$ *for every* $\xi<\mathrm{cf}(L_3(\kappa))$.

(ix) $2^{L_3(\kappa)}=\kappa$ *and at least one of the following conditions holds:*

a) $\lambda_0>L_3(\kappa)$ *and* $\lambda_1\ge L_3(\kappa)$;

b) $\lambda_0,\lambda_1\ge L_3(\kappa)$ *and* $\vartheta\ge L_3(\mathrm{cf}(L_3(\kappa)))$;

c) $\lambda_0,\lambda_1\ge L_3(\kappa)$ *and* $\mathrm{cf}(L_3(\kappa))\not\to(\mathrm{cf}(L_3(\kappa)))^2_2$;

d) $\lambda_0,\lambda_1\ge L_3(\kappa)$ *and* $\lambda_2>\mathrm{cf}(L_3(\kappa))$.

PROOF. We simply proceed by directly applying results established earlier. The monotonicity properties of the ordinary partition relation (see Subsection 9.6) will often be used without any explicit reference.

Ad (i). Writing $\rho=L_{\lambda_0}(\kappa)$, the definition of the logarithm function in Section 7 implies that there is a $\sigma<\lambda_0$ such that $\sigma^\rho\geq\kappa$. By a change of notation we obtain $\sigma^\rho\nrightarrow((\sigma\cdot\rho)^+,\rho^+)^2$ from (19.12). As $\sigma^\rho\geq\kappa$ and, moreover, $\sigma\cdot\rho<\lambda_0$ and $\rho<\lambda_1$ by our assumptions, we obtain $\kappa\nrightarrow(\lambda_0,\lambda_1)^2$. Hence (1) indeed fails.

Ad (ii). Suppose first that $\lambda_1>L_\kappa(\kappa)$. We may assume here (as we did in (2)) that $\lambda_1\leq\kappa$. But then we have $\lambda_0,\lambda_1>L_{\lambda_0}(\kappa)$ since $\lambda_0=\kappa$, which shows that this case is covered by (i).

Suppose now $\lambda_1>L_{\mathrm{cf}(\kappa)}(\mathrm{cf}(\kappa))$. Then the result established in the preceding paragraph before with $\mathrm{cf}(\kappa)$ replacing κ shows that $\mathrm{cf}(\kappa)\nrightarrow(\mathrm{cf}(\kappa),\lambda_1)$. Now Corollary 21.2 implies $\kappa\nrightarrow(\kappa,\lambda_1)^2$, which we wanted to show.

Ad (iii). We have $\mathrm{cf}(\kappa)\nrightarrow(\mathrm{cf}(\kappa),\lambda_1)^2$ by our assumptions; again, Corollary 21.2 implies the desired relation $\kappa\nrightarrow(\kappa,\lambda_1)^2$.

Ad (iv). We have $\kappa\nrightarrow(3)^2_{L_3(\kappa)}$ according to (19.18). Hence $\vartheta\geq L_3(\kappa)$ implies $\kappa\nrightarrow(\lambda_\xi)_{\xi<\vartheta}$, because we assumed in (2) that $\lambda_\xi\geq3$ holds for each $\xi<\vartheta$.

Ad (v). Writing $\rho=L_{\lambda_0}(\kappa)$, it is enough to show that $\kappa\nrightarrow(\lambda_0,(\omega)_\rho)^2$ holds (note that ρ is infinite by Theorem 7.2, and so $\rho-1=\rho$). To see this, observe that $\sigma^\rho\geq\kappa$ holds with some $\sigma<\lambda_0$ by the definition of the logarithm. Suppose first that $\rho\geq\lambda_0$. Then $\sigma<\rho$, and so $2^\rho=\sigma^\rho\geq\kappa$. We have $2^\rho\nrightarrow(3)^2_\rho$ according to (19.17), which implies that $\kappa\nrightarrow(\lambda_0,(\omega)_\rho)^2$.

Suppose now $\rho<\lambda_0$. By an alphabetic variant of (19.13) we have $\sigma^\rho\nrightarrow((\sigma\cdot\rho)^+,(\omega)_\rho)^2$. As $\sigma^\rho\geq\kappa$ and $\sigma,\rho<\lambda_0$, this means that $\kappa\nrightarrow(\lambda_0,(\omega)_\rho)^2$, which we wanted to show.

Ad (v′). We mentioned after (vii) that in case GCH holds we can strengthen (v) to (v′). To see this, we have to show that $\kappa\rightarrow(\lambda_0,(3)_\rho)^2$ holds under the assumption of GCH, where $\rho=L_{\lambda_0}(\kappa)$ again. First suppose $\rho\geq\lambda_0$. We mentioned above, in the proof of (v), that $2^\rho\geq\kappa$ and $2^\rho\nrightarrow(3)_\rho$ hold then, and these imply the desired result. Suppose now $\rho<\lambda_0$. By the definition of the logarithm, there is a $\sigma<\lambda_0$ such that $\sigma^\rho\geq\kappa$. On the other hand, GCH implies that $\sigma^\rho\leq\max(\sigma^+,\rho^+)\leq\lambda_0$; as $\lambda_0\leq\kappa$ by the assumptions made in (2), this means that $\lambda_0=\kappa$ and κ must be a successor cardinal. As σ can be chosen arbitrarily large as long as it is less than $\lambda_0=\kappa$, we may assume that $\kappa=\sigma^+$; but then, ρ being the least cardinal for which $\sigma^\rho\geq\sigma^+=\kappa$, GCH implies that $\rho=\mathrm{cf}(\sigma)$ (cf. e.g. Theorem 6.12). If σ is regular, then $\rho=\sigma$, and so $\kappa=\rho^+=2^\rho$ by GCH; we have $2^\rho\nrightarrow(3)^2_\rho$ according to (19.17), which implies the desired result. Assume now that σ is singular. According to Theorem 20.2, we have $\sigma^+\nrightarrow(\sigma^+,(3)_{\mathrm{cf}(\sigma)})^2$; note that GCH is again used here. Observing that $\kappa=\lambda_0=\sigma^+$ and $\rho=\mathrm{cf}(\sigma)$ in the present case, this establishes the desired result.

Ad (vi). The result in (iv) with $\mathrm{cf}(\kappa)$ replacing κ says that $\mathrm{cf}(\kappa)\not\to(3)^2_{L_3(\mathrm{cf}(\kappa))}$, and so we have a fortiori $\mathrm{cf}(\kappa)\not\to(\mathrm{cf}(\kappa), (3)_{L_3(\mathrm{cf}(\kappa))})^2$. Corollary 21.2 now implies $\kappa\not\to(\kappa, (3)_{L_3(\mathrm{cf}(\kappa))})$; since $\lambda_0=\kappa$ and each λ_ξ is ≥ 3 according to (2), this implies $\kappa\not\to(\lambda_\xi)^2_{\xi<\vartheta}$, which we wanted to show.

Ad (vii). Replacing κ with $\mathrm{cf}(\kappa)$ in (v) and putting $\lambda_0=\mathrm{cf}(\kappa)$ there, we obtain $\mathrm{cf}(\kappa)\not\to(\mathrm{cf}(\kappa), (\omega)_{L_{\mathrm{cf}(\kappa)}(\mathrm{cf}(\kappa))})^2$. Corollary 21.2 now entails $\kappa\not\to(\kappa, (\omega)_{L_{\mathrm{cf}(\kappa)}(\mathrm{cf}(\kappa))})^2$, which implies that (1) indeed fails.

Ad (vii′). We proceed similarly as before, but we use the stronger relation $\mathrm{cf}(\kappa)\not\to(\mathrm{cf}(\kappa), (3)_{L_{\mathrm{cf}(\kappa)}(\mathrm{cf}(\kappa))})^2$, which holds under the assumption of GCH according to (v′) if we make the same substitutions there as just before. We obtain $\kappa\not\to(\kappa,(3)_{L_{\mathrm{cf}(\kappa)}(\mathrm{cf}(\kappa))})^2$ by Corollary 21.2, which we wanted to show.

Ad (viii). In case $L_3(\kappa)$ is regular we have $\vartheta\geq L_3(\kappa)$, and so (1) fails by (iv). Assume therefore that $\rho=L_3(\kappa)$ is singular. We have $2^\rho\not\to(\kappa_\xi)^2_{\xi<\mathrm{cf}(\rho)}$ according to (19.15) in Corollary 19.8; as $2^\rho\geq\kappa$ by the definition of ρ and $\lambda_\xi\geq\kappa_\xi$ by our assumptions, this relation implies $\kappa\not\to(\lambda_\xi)^2_{\xi<\vartheta}$, which we wanted to show.

Ad (ix). Write $\rho=L_3(\kappa)$. Then $2^\varrho=\kappa$ by our assumption. We have $2^\varrho\not\to(\rho^+,\rho)^2$ according to (21.11), which implies a). To see b), observe that $2^\varrho\not\to(\rho,\rho,(3)_{L_3(\mathrm{cf}(\rho))})^2$ by (21.10). c) follows from Corollary 21.5(iii), which says that $\mathrm{cf}(\rho)\not\to(\mathrm{cf}(\rho))^2_2$ ensures $2^\varrho\not\to(\rho,\rho)^2$. d) is a consequence of (21.13), which says that $2^\varrho\not\to(\rho,\rho,(\mathrm{cf}(\rho))^+)^2$ holds. The proof of the lemma is complete.

Our next result analyses the situation when the conditions of the preceding lemma fail in terms of the relation $\lll$ given in Definition 35.1. This will help us to describe a condition which is as close to being necessary and sufficient for (1) to hold as is possible by our present knowledge.

LEMMA 35.3. *Assume that all the conditions* (i)–(ix) *of Lemma* 35.2 *fail with any rearrangement* $\langle\lambda'_\xi:\xi<\vartheta'\rangle$ *of* $\langle\lambda_\xi:\xi<\vartheta\rangle$ *replacing this latter sequence. Then there are cardinals* $\lambda,\rho,\sigma,\alpha,\beta,\tau$, *and* k_ν *for any* $\nu<\tau$ *satisfying the following conditions:*

$$\langle\lambda_\xi:\xi<\vartheta\rangle\lll\langle\lambda,(\rho)_\alpha,(\sigma)_\beta,(k_\nu)_{\nu<\tau}\rangle. \tag{3}$$

$$3\leq\lambda\leq\kappa \textit{ and } \rho=\begin{cases}L_\lambda(\kappa) \textit{ if } \lambda<\kappa,\\ \min\{L_\kappa(\kappa), L_{\mathrm{cf}(\kappa)}(\mathrm{cf}(\kappa))\} \textit{ if } \lambda=\kappa.\end{cases} \tag{4}$$

$$\begin{gathered}\sigma,\beta<\rho \textit{ and } \alpha<\mathrm{cf}(\rho), \textit{ and } k_\nu<\omega \textit{ for every } \nu<\tau;\\ \textit{furthermore, } \tau<L_3(\kappa), \textit{ and if } \lambda=\kappa, \textit{ then}\\ \tau<\min\{L_3(\kappa), L_3(\mathrm{cf}(\kappa))\}.\end{gathered} \tag{5}$$

$$\alpha=0 \textit{ provided either } (\mathrm{i})\ \lambda<\rho \textit{ or } (\mathrm{ii})\ \lambda=\kappa,\ \rho=\mathrm{cf}(\kappa), \textit{ and } \rho\not\to(\rho)^2_2. \tag{6}$$

$$\begin{gathered}\alpha=0 \quad \textit{provided} \quad \lambda\geq\rho, \quad \textit{and} \quad 2^{\varrho}=\kappa, \quad \textit{and} \\ \textit{at least one of the following conditions holds:} \\ \text{(i)} \quad \lambda>\rho, \quad \textit{or} \quad \text{(ii)} \quad \lambda=\rho \quad \textit{and} \quad \operatorname{cf}(\rho)\not\rightarrow(\operatorname{cf}(\rho))_2^2\,.\end{gathered} \tag{7}$$

$$\begin{gathered}\textit{If} \quad 2^{\varrho}=\kappa,\ \lambda=\rho, \quad \textit{and} \quad \rho>\operatorname{cf}(\rho), \quad \textit{then} \quad \alpha\leq 1; \\ \textit{if} \quad \alpha=1 \textit{ here, then we also have } \sigma\leq\operatorname{cf}(\rho) \quad \textit{and} \quad \beta,\ \tau<L_3(\operatorname{cf}(\rho))\,.\end{gathered} \tag{8}$$

$$\begin{gathered}\textit{If} \ \text{(v}')\ \textit{and}\ \text{(vii}')\ (\textit{see immediately after}\ \text{(vii)}\ \textit{of Lemma}\ 35.2)\ \textit{also fail,} \\ \textit{then}\ \tau=0.\end{gathered} \tag{9}$$

$$\textit{All these conditions are satisfied with } \lambda=\lambda_{\xi} \textit{ for some } \xi<\vartheta. \tag{10}$$

REMARK. (9) is important when GCH holds.

PROOF. For the sake of brevity, throughout the proof we shall refer to the assumption that the conditions (i), ..., (ix) of Lemma 35.2 fail with any rearrangement $\langle\lambda'_{\xi}\colon \xi<\vartheta'\rangle$ of $\langle\lambda_{\xi}\colon \xi<\vartheta\rangle$ replacing this latter sequence by using phrases such as "we obtain by $\neg$ (i),..., $\neg$ (ix)". For any ordinal η put

$$M_{\mapsto\eta}=\{\xi<\vartheta\colon \lambda_{\xi}\mapsto\eta\}\,,$$

where $\mapsto$ is any of the relation symbols $=$, $<$, $\leq$, $>$, and $\geq$.

Put

$$\rho_0=L_3(\kappa)\,.$$

By using $\neg$ (viii), it is easy to see that there exists a cardinal $\sigma_0<\rho_0$ such that

$$\text{(i)} \quad \vartheta=M_{\leq\sigma_0}\cup M_{\geq\rho_0} \quad \text{and (ii)} \quad |M_{\geq\rho_0}|<\operatorname{cf}(\rho_0)\,. \tag{11}$$

In fact, if the first relation here fails for each $\sigma_0<\rho_0$, then there are pairwise distinct $\nu_{\xi}<\vartheta$ for $\xi<\operatorname{cf}(\rho_0)$ such that the $\lambda_{\nu_{\xi}}$ are all $<\rho_0$ and their sequence tends to ρ_0; we can take $\lambda'_{\xi}=\kappa_{\xi}=\lambda_{\nu_{\xi}}$ for $\nu<\operatorname{cf}(\rho_0)$ to violate $\neg$ (viii) (i.e., to verify (viii) with λ'_{ξ} replacing λ_{ξ}). The second relation can also be checked easily by using $\neg$ (viii). Choose σ_0 the least possible in (11)(i), i.e., put

$$\sigma_0=\sup\{\lambda_{\xi}\colon \xi<\vartheta\ \&\ \lambda_{\xi}<\rho_0\}\,.$$

By $\neg$ (iv) we have

$$\vartheta<\rho_0\,. \tag{12}$$

$\neg$ (i) implies $|M_{>\rho_0}|\leq 1$; otherwise we could choose $\lambda'_0,\ \lambda'_1>\rho_0$, and we would have $\lambda'_0,\ \lambda'_1>L_{\lambda'_0}(\kappa)$ since $L_{\lambda'_0}(\kappa)\leq L_3(\kappa)=\rho_0$. It follows that there is a maximal λ_{ξ} with $\xi\in M_{\geq\rho_0}$, provided this latter set is not empty. Put $\lambda=\lambda_{\xi}$ with this ξ (so that (10) will hold), or, if $M_{\geq\rho_0}$ is empty, then put $\lambda=\lambda_{\xi}$ with an arbitrary ξ. Without loss of generality, we may assume $\xi=0$ in both cases; we shall then have $\lambda=\lambda_0$, which will make our notations below simpler. The choice of λ implies by (11)(i) that:

$$\begin{gathered}\text{Either (i) } \lambda=\lambda_0<\rho_0 \text{ and } \vartheta=M_{\leq\sigma_0}, \text{ or}\\ \text{(ii) } \lambda=\lambda_0\geq\rho_0 \text{ and } \lambda_\xi\leq\rho_0 \text{ for all } \xi \text{ with } 1\leq\xi<\vartheta, \text{ i.e.,}\\ \vartheta=\{0\}\cup M_{=\rho_0}\cup M_{\leq\sigma_0}.\end{gathered} \tag{13}$$

Note that the first relation in (4) is satisfied according to (2), since $\lambda=\lambda_0$. Define ρ by the second relation in (4). To verify (5)–(9), we distinguish two cases: (a) $\rho=\rho_0$ and (b) $\rho<\rho_0$.

Case (a). Putting $\alpha=|M_{=\rho_0}\setminus\{0\}|$, $\sigma=\sigma_0$, $\beta=|M_{\leq\sigma_0}|$, and $\tau=0$ (so that there are no k_ν's to be defined), we have $\langle\lambda_\xi:\xi<\vartheta\rangle\ll\langle\lambda,(\rho)_\alpha,(\sigma_\beta)\rangle$; hence (3) is satisfied. (9) holds even without the assumptions $\neg$ (v′) and $\neg$ (vii′). We have $\sigma=\sigma_0<\rho=\rho_0$ by the choice of σ_0, and $\alpha<\mathrm{cf}(\rho_0)$ and $\beta<\rho_0$ holds according to (11)(ii) and (12), respectively; hence (5) is satisfied. (6)(i) follows from (13)(i). To see (6)(ii), one only has to note that the assumptions here together with $\neg$ (iii) imply that we cannot have $\lambda_\xi\geq\mathrm{cf}(\kappa)=\rho$ for any ξ with $1\leq\xi<\vartheta$.

We have yet to verify (7) and (8) in the case considered. If either of these fail, then $\alpha\geq1$, and so we must have the alternative (ii) in (13), i.e.,

$$\lambda=\lambda_0\geq\rho_0=\rho\,.$$

We may also suppose that

$$2^\rho=\kappa\,,$$

since this is given as an assumption in both (7) and (8). As for (7)(i), if $\lambda=\lambda_0>\rho=\rho_0$, then $M_{=\rho_0}\setminus\{0\}=0$ by $\neg$ (ix)a), and so indeed $\alpha=0$. As for (7)(ii), the assumptions in this case are that $\lambda_0=\rho_0$ and $\mathrm{cf}(\rho_0)\not\to(\mathrm{cf}(\rho_0))^2_2$; thus $\neg$ (ix)c) implies that $M_{=\rho_0}\setminus\{0\}=0$ again, and so we have $\alpha=0$ in this case as well. To settle (8), suppose that $\lambda=\lambda_0=\rho=\rho_0>\mathrm{cf}(\rho_0)$. Then $|M_{=\rho_0}\setminus\{0\}|\leq1$ by $\neg$ (ix) d), and so $\alpha\leq1$. Assume $\alpha=1$. Noting that we put $\sigma_0=\sup\{\lambda_\xi:\xi<\vartheta\ \&\ \lambda_\xi<\rho_0\}$, as mentioned shortly after (11), we have $\sigma_0\leq\mathrm{cf}(\rho_0)$ again by $\neg$ (ix) d), and $\neg$ (ix) b) implies $\beta\leq\vartheta<L_3(\mathrm{cf}(\rho_0))$. As mentioned above, $\tau=0$ in Case (a), which is being considered. This establishes (8) as well.

Case (b). Put $\alpha=|M_{=\rho}|$, $\sigma=\sup\{\lambda_\xi:\xi\in M_{<\rho}\}$, $\beta=|M_{\geq\omega}\cap M_{<\rho}|$, $\tau=|M_{<\omega}|$, and let k_ν, $\nu<\tau$, enumerate all the finite ones among the λ_ξ's (that is, let h be a 1-1 map of τ onto $M_{<\omega}$, and put $k_\nu=\lambda_{h(\nu)}$). To verify (9), modify these definitions by putting $\beta=|M_{<\rho}|$ and $\tau=0$.

Note that $\lambda=\lambda_0\geq\rho_0$ must hold in the present case. In fact, if $\lambda=\kappa$, then this is true, since in this case we have $\rho_0=L_3(\kappa)\leq\kappa=\lambda$. If $\lambda<\kappa$, then $\rho=L_\lambda(\kappa)<\rho_0=L_3(\kappa)$ by the assumption made in Case (b), and so $\lambda>\rho_0$ by Theorem 7.10. (As we only need $\lambda\geq\rho_0$ here, one can spare the bother of looking up this reference: as $\rho=L_\lambda(\kappa)$, we certainly have $\lambda^\rho\geq\kappa$, and so $2^{\lambda\cdot\rho}\geq\kappa$; hence $\lambda\cdot\rho\geq L_3(\kappa)=\rho_0$, i.e., $\lambda\geq\rho_0$, since we assumed $\rho<\rho_0$.)

As $\lambda=\lambda_0\geq\rho_0>\rho$, the definition of ρ in (4) together with $\neg$ (i) and $\neg$ (ii) (for a permutation $\langle\lambda'_\xi\colon\xi<\vartheta\rangle$ of the λ_ξ's with $\lambda'_0=\lambda_0$) implies that $M_{>\rho}=\{0\}$, and so $\vartheta=\{0\}\cup M_{=\rho}\cup(M_{\geq\omega}\cap M_{<\rho})\cup M_{<\omega}$, which means means that (3) does indeed hold. By $\neg$ (iii) we have $\alpha=0$ in case $\lambda=\kappa$, $\rho=\mathrm{cf}(\kappa)$, and $\mathrm{cf}(\kappa)\nrightarrow(\mathrm{cf}(\kappa))^2_2$ hold; this establishes (6)(i). (6)(i) holds vacuously.

We claim that ρ is regular. If $\lambda<\kappa$, then this is true by the last sentence of Theorem 7.11, since by (4), $\rho=L_\lambda(\kappa)<L_3(\kappa)=\rho_0$ in this case. If $\lambda=\kappa$, then either $L_\kappa(\kappa)$ and $L_{\mathrm{cf}(\kappa)}(\mathrm{cf}(\kappa))$ are both regular, or $L_\kappa(\kappa)$ is singular; but then Theorem 7.15 implies that

$$L_\kappa(\kappa)\geq\mathrm{cf}(\kappa)\geq L_{\mathrm{cf}(\kappa)}(\mathrm{cf}(\kappa))=\rho$$

(cf. (4)), and the right-hand side here is regular again by Theorem 7.15, since $\mathrm{cf}(\kappa)$ is regular. Thus the regularity of ρ is established.

$\alpha+\beta<\rho$ follows from $\neg$ (v) and $\neg$ (vii) (only the former is needed when $\lambda=\lambda_0<\kappa$, but both are needed in case $\lambda=\lambda_0=\kappa$); to verify (9) (when β was defined differently), use $\neg$ (v′) and $\neg$ (vii′) instead ((v′) and (vii′) were mentioned immediately after (vii)). Now (5) easily follows: $\sigma<\rho$ by the definition of σ (cf. (11)(i)); $\beta<\rho$ since $\alpha+\beta<\rho$; $\alpha<\mathrm{cf}(\rho)$ since $\alpha+\beta<\rho$ and $\mathrm{cf}(\rho)=\rho$, as we saw this latter in the preceding paragraph. $k_\nu<\omega$ by definition; $\tau<\vartheta$, and, moreover, $\vartheta<L_3(\kappa)$, and if $\lambda=\kappa$, then $\vartheta<L_3(\mathrm{cf}(\kappa))$, according to $\neg$ (iv) and $\neg$ (vi), respectively; hence the inequalities for τ also follow.

We have yet to settle (7) and (8). But we have $2^\varrho\leq 2^\rho<\kappa$ because $\rho<\rho_0=L_3(\kappa)$ in the present case; hence (7) and (8) both hold vacuously. The proof of the lemma is complete.

By Lemmas 35.2 and 35.3, (1) can hold only if (3)–(8) hold for some sequence $\langle\lambda,(\rho)_\alpha,(\sigma)_\beta,(k_\nu)_{\nu<\tau}\rangle$. The best possible result would be that

$$\kappa\to(\lambda,(\rho)_\alpha,(\sigma)_\beta,(k_\nu)_{\nu<\tau})^2$$

holds whenever (4)–(8) are satisfied. Our next theorem, the main result of this section, collects the cases when we can really prove this.

THEOREM 35.4. *Let $\vartheta\geq 2$ be an ordinal, and let κ, $\lambda_\xi(\xi<\vartheta)$ be cardinals with $\kappa\geq\omega$ and $3\leq\lambda_\xi\leq\kappa$. Consider the following condition:*

There are cardinals λ, ρ, σ, α, β, τ, and k_ν $(\nu<\tau)$ satisfying (3)–(8). (14)

a) (14) *is necessary for*

$$\kappa\to(\lambda_\xi)^2_{\xi<\vartheta} \tag{15}$$

to hold.

b) (14) *is sufficient for* (15) *to hold provided* $\tau=0$ *and at least one of the following conditions hold*:

(i) κ *is regular*;

(ii) $\lambda<\kappa$;

(iii) κ *is strong limit, i.e.*, $\forall\kappa'<\kappa[2^{\kappa'}<\kappa]$;

(iv) $\rho=\omega$.

The difficulty in applying this result is to check whether (14) holds, especially because we are faced here with the problem of how to choose a λ for which (3)–(8) are satisfied. We claim that (3)–(8) are satisfied with some λ exactly if they are satisfied with

$$\lambda=\sup\{\lambda_\xi:\xi<\vartheta\}, \tag{16}$$

and if (3)–(8) are satisfied with $\tau=0$ and some λ, then they are also satisfied with $\tau=0$ and the above value of λ. After having chosen this value for λ, there should be no difficulty in checking (3)–(8) provided the cardinal exponentation function is known.

Before we go on to establish our claim, we point out that choosing λ as given in (16) does not adversely affect the applicability of Theorem 35.4. What we mean by this is the following: assume (3)–(8) are satisfied with two sets of parameters: $\lambda, \rho, \ldots$ and $\lambda', \rho', \ldots$ where λ is as given in (16), and $\lambda'\neq\lambda$. If our theorem is applicable with the second set of parameters, then it is also applicable with the first set of parameters. There are two reasons why this is not immediately clear: it may happen that α) $\lambda'<\kappa$ and $\lambda=\kappa$ or β) $\rho'=\omega$ and $\rho\neq\omega$. In case α), b(ii) of the theorem is applicable with the parameters $\lambda', \rho', \ldots$ while it is not applicable with the parameters $\lambda, \rho, \ldots$. We have

$$\kappa=(\lambda=)\sup\{\lambda_\xi:\xi<\vartheta\}$$

in this case; as we also have

$$\langle\lambda_\xi:\xi<\vartheta\rangle\ll\langle\lambda', (\rho')_{\alpha'}, (\sigma')_{\beta'}\rangle$$

according to (3) and $\tau'=0$ holds, we must have

$$\max\{\lambda', \rho', \sigma'\}=\kappa;$$

as $\lambda'<\kappa$ by our assumption and $\sigma'<\rho'$ by (5), $\rho'=\kappa$ must hold; i.e., we have $L_{\lambda'}(\kappa)=\kappa$ by (4), which means that κ is a strong limit cardinal. Hence b(iii) of the theorem is applicable with the parameters $\lambda, \rho, \ldots$. In case β), b(iv) of the theorem is applicable with the parameters $\lambda', \rho', \ldots$. If the theorem is not applicable with the parameters $\lambda, \rho, \ldots$, then we must have $\lambda=\kappa$ according to b(ii); $\rho=\min\{L_\kappa(\kappa), L_{\mathrm{cf}(\kappa)}(\mathrm{cf}(\kappa))\}$ by (4) in this case. As $\lambda'\leq\kappa$, we have $L_{\lambda'}(\kappa)\geq L_\kappa(\kappa)$ by Theorem 7.1.b), and so $\rho'\geq\min\{L_\kappa(\kappa), L_{\mathrm{cf}(\kappa)}\,\mathrm{cf}(\kappa))\}$ by (4); hence $\rho'=\omega<\rho$ cannot happen in the critical case $\lambda=\kappa$.

We now turn to establishing our claim that if (3)–(8) are satisfiable, then they are satisfiable with the value of λ given in (16). To this end assume that (3)–(8) are satisfied by $\lambda', \rho', \ldots$. We distinguish two cases: 1) $\lambda<\omega$ and 2) $\lambda\geq\omega$, where λ is given by (16). In Case 1) we have $\rho=L_\lambda(\kappa)=L_3(\kappa)\geq L_{\lambda'}(\kappa)=\rho'$. Putting $\alpha=0$, $\sigma=\lambda$, $\beta=\alpha'+\beta'$ $(<\rho'\leq\rho)$, $\tau=\tau'$, and $k_\nu=k'_\nu$, (3)–(8) are satisfied ((4)–(8) are rather easy to check; the reason why (3) holds is that $\sigma=\lambda=\sup\{\lambda_\xi: \xi<\vartheta\}$). In Case 2) we distinguish two subcases: 2a) $\lambda'>\lambda$ and 2b) $\lambda'<\lambda$. In Case 2a) we have $\rho=L_\lambda(\kappa)\geq L_{\lambda'}(\kappa)\geq\rho'$. If now $\rho>\min\{\rho',\lambda\}$, then put $\sigma=\min\{\rho',\lambda\}$, $\alpha=0$, $\beta=\alpha'+\beta'$, $\tau=\tau'$, and $k_\nu=k'_\nu$ $(\nu<\tau)$. It is easy to see that (3)–(8) will then hold. If $\rho\leq\min(\rho',\lambda)$ (i.e., if $\rho=\rho'\leq\lambda$), then put $\sigma=\sigma'$, $\alpha=\alpha'$, $\beta=\beta'$, ... Then (3)–(8) are again satisfied (the fine points to observe here are that (6)(i) is vacuous since $\rho\leq\lambda$ in this case, and so is (6)(ii), as $\lambda<\lambda'\leq\kappa$ now; and, moreover, if (7)(i), (ii) or (8) applies with λ, ρ, ..., then (7)(i) applies with λ', ρ', ..., and so $\alpha=\alpha'=0$). Consider Case 2b) now. As $\lambda\geq\omega$ and $\lambda'<\lambda=\sup\{\lambda_\xi: \xi<\vartheta\}$, we must have $\rho'\geq\lambda$ according to (3). In fact, this says that

$$\langle\lambda_\xi: \xi<\vartheta\rangle\ll\langle\lambda', (\rho')_{\alpha'}, (\sigma')_{\beta'}, (k'_\nu)_{\nu<\tau'}\rangle\,; \tag{17}$$

here the supremum of the sequence on the right-hand side must be greater than the supremum of the sequence on the left-hand side, which is λ. But $\lambda'<\lambda$ by our assumptions, and $k'_\nu<\omega$ $(\leq\lambda)$ and $\sigma'<\rho'$ according to (5). Hence we must indeed have $\rho'\geq\lambda$. We claim that this implies $\lambda\leq\sigma'$. In fact, we have $\lambda'<\lambda\leq\rho'$, and so $\alpha'=0$ according to (6)(i); hence (17) can hold only if $\lambda\leq\sigma'$ holds. As $\sigma'<\rho'$ $(\leq\kappa)$ according to (5), we have $\lambda<\rho'\leq\kappa$, and so $\rho=L_\lambda(\kappa)$ by (4). Now $\rho'=L_{\lambda'}(\kappa)>\lambda$ (note that $\lambda'\neq\kappa$ as $\lambda'<\lambda<\kappa$ at present), and so $\rho=L_\lambda(\kappa)=\rho'$ by an easy computation: in fact, if $\bar\lambda<\lambda$ and $\bar\rho<\rho'$, then $\bar\lambda^{\bar\rho}\leq 2^{\bar\lambda\cdot\bar\rho}<\kappa$ since $\lambda'\geq 3$ and $\bar\lambda\cdot\bar\rho<$ $<\lambda\cdot\rho'=\rho'=L_{\lambda'}(\kappa)$; hence $\bar\rho<L_\lambda(\kappa)$. This holds for any $\bar\rho<\rho'$, and so $\rho=$ $=L_\lambda(\kappa)\geq\rho'=L_{\lambda'}(\kappa)$; we have $\rho=L_\lambda(\kappa)\leq L_{\lambda'}(\kappa)=\rho'$ on the other hand, since $\lambda>\lambda'$. Thus $\rho=\rho'$ is established. Put $\sigma=\sigma'$, $\alpha=\alpha'=0$ (the latter equality holds by (6)(i), since $\lambda'<\lambda\leq\rho'$), $\beta=\beta'$, ... It is easy to see that (3)–(8) will hold then (note here that we saw $\lambda<\kappa$ above, and so only $\tau<L_3(\kappa)$ is required by (5), which holds as $\tau=\tau'<L_3(\kappa)$.This completes the proof of our claim that if (3)–(8) are satisfiable, then they are satisfiable with the λ given in (16).

Proof of Theorem 35.4. (14) is a necessary condition for (15) according to Lemmas 35.2 and 35.3. This settles a). We now turn to the verification of b). Assume therefore that $\tau=0$. We shall have to prove that

$$\kappa\rightarrow(\lambda, (\rho)_\alpha, (\sigma)_\beta)^2 \tag{18}$$

holds under the various conditions considered.

Ad (i). (4) and the regularity of κ implies that $\rho=L_\lambda(\kappa)$ holds in this case. If $\rho<\kappa$, then Corollary 17.2 implies that (18) holds. Assume therefore that $\rho=\kappa$. In this case κ is inaccessible; in fact, it is regular and $L_\lambda(\kappa)=\kappa$. If $\alpha=0$, then (18) follows from Corollary 17.3. If $\alpha>0$, then we cannot have $\lambda<\rho$ according to (6)(i), i.e., we must have $\lambda=\rho=\kappa=\text{cf}(\kappa)$; hence $\rho\to(\rho)^2_2$, i.e., $\kappa\to(\kappa)^2_2$, holds by (6)(ii). The equivalence of (iv) and (vi) in Theorem 29.6 then implies that $\kappa\to(\kappa)^2_{\kappa'}$ holds for any $\kappa'<\kappa$, and so (18) holds.

Ad (ii). Assume that (18) is false. Then κ is singular according to (i), which was established just before. As $\lambda<\kappa$ in this case, we have $\rho=L_\lambda(\kappa)$. If there is a regular cardinal κ_0 with $\lambda<\kappa_0<\kappa$ such that $2^{L_3(\kappa)}<\kappa_0$ (i.e., $L_3(\kappa_0)=L_3(\kappa)$) and $L_\lambda(\kappa_0)=L_\lambda(\kappa)$, then conditions (3)–(8) hold with κ_0 replacing κ, which implies according to (i) that (18) holds with κ_0 in place of κ; but then (18) also holds with κ. This contradicts our assumption; no such κ_0 can therefore exist, and so

$$\rho=L_\lambda(\kappa)=L_3(\kappa) \quad \text{and} \quad 2^\rho=\kappa \tag{19}$$

hold according to Theorem 7.16. We then have

$$\text{cf}(\rho)=\text{cf}(\kappa) \tag{20}$$

by Theorem 6.10.d (since the alternative c(i) of this theorem fails with $2^\rho=\kappa$ replacing κ^λ, as $2^{\rho'}<\kappa$ holds for any $\rho'<\rho=L_3(\kappa)$ by the definition of the logarithm operation).

If $\alpha=0$, then choose a $\rho_0<\rho$ such that $2^{\rho_0}\geq\lambda$ and $\rho_0\geq\sigma,\beta$ (this is possible in view of (19) and (5)); we have $(2^{\rho_0})^+\to((2^{\rho_0})^+,(\rho_0^+)_{\rho_0})^2$ according to Corollary 17.5, and this implies (18) under the assumption of $\alpha=0$, which is a contradiction. We must therefore have $\alpha>0$. Thus (6)(i), (19), and (7) imply that

$$\lambda=\rho \quad \text{and} \quad \text{cf}(\rho)\to(\text{cf}(\rho))^2_2\,; \tag{21}$$

hence cf (ρ) is inaccessible according to Theorem 29.1. Thus ρ cannot be regular; in fact, using (19), (20), and the singularity of κ, the regularity of ρ would entail that

$$\text{cf}(\rho)=2^{\text{cf}(\rho)}=2^\rho=\kappa>\text{cf}(\kappa)=\text{cf}(\rho)\,,$$

which is a contradiction. Thus we have cf $(\rho)<\rho$; therefore (8) implies with the aid of (19) and (21) that $\alpha\leq1$; as we saw $\alpha>0$ above, we must have $\alpha=1$; then (8) further implies that $\sigma\leq\text{cf}(\rho)$ and $\beta<(L_3(\text{cf}(\rho))\leq)\,\text{cf}(\rho)$; so we have

$$\kappa=2^\rho\to((\rho)_2,(\sigma)_\beta)^2$$

by Corollary 27.4 plus (19), (20) and (21). As we obtained $\alpha=1$ just before, and $\lambda=\rho$ according to (21), this means that (18) holds. This is again a contradiction, settling the case of condition (ii).

Ad (iii). By (i) and (ii) we may assume $\mathrm{cf}(\kappa)<\kappa$ and $\kappa=\lambda$. As κ is assumed to be strong limit in this case, we have $L_\kappa(\kappa)=\kappa$; hence $\rho=L_{\mathrm{cf}(\kappa)}(\mathrm{cf}(\kappa))$ according to (4) in this case. We claim that

$$\mathrm{cf}(\kappa)\to(\mathrm{cf}(\kappa),(\rho)_\alpha,(\sigma)_\beta)^2 \tag{22}$$

holds. To see this, it is enough to show according to (i) that (3)–(8) are satisfied with $\kappa'=\lambda'=\mathrm{cf}(\kappa)$ replacing κ and λ. This is obvious for (4) and (5) since $\rho=L_{\mathrm{cf}(\kappa)}(\mathrm{cf}(\kappa))$; (6)(i) is vacuous, and (6)(ii) is trivial since $\mathrm{cf}(\kappa')=$ $=\mathrm{cf}(\mathrm{cf}(\kappa))=\mathrm{cf}(\kappa)$, and so the assertion in (6)(ii) with the new parameters is identical to the one with the old parameters.

To check (7) and (8), assume now that $2^\varrho=\mathrm{cf}(\kappa)$ holds. Then $L_3(\mathrm{cf}(\kappa))\le\rho=$ $=L_{\mathrm{cf}(\kappa)}\,\mathrm{cf}(\kappa))$, and so $L_3(\mathrm{cf}(\kappa))=\rho$, as the reverse inequality also holds by the logarithm's being a monotonically nonincreasing function of its base (cf. Theorem 7.1.b). $\rho=L_{\mathrm{cf}(\kappa)}(\mathrm{cf}(\kappa))$ is regular according to Theorem 7.15. We have $\mathrm{cf}(\kappa)=\mathrm{cf}(2^\varrho)=\mathrm{cf}(\rho)$ in view of Theorem 6.10.d (as in the case of (20) above, the alternative c(i) in this theorem must fail since $\rho=L_3(2^\varrho)$ holds). Thus

$$\kappa'=\lambda'=\mathrm{cf}(\kappa)=\mathrm{cf}(\rho)=\rho,$$

and so (7)(i) holds vacuously. As for (7)(ii), we have to show that in case $\alpha>0$ we have $\mathrm{cf}(\rho)\to(\mathrm{cf}(\rho))^2_2$, which is the same as $\rho\to(\rho)^2_2$, as ρ is regular. This is indeed true, since in case $\rho\not\to(\rho)^2_2$ condition (6)(ii) involving the original parameters κ, λ, etc. implies $\alpha=0$. Finally, (8) holds vacuously because $\rho=\mathrm{cf}(\rho)$. Thus (22) is established. Using the assumption that κ is strong limit, (22) implies (18) via Corollary 27.5.

Ad (iv). α, β, and σ are finite in this case, hence the result follows from the Erdős–Dushnik–Miller theorem (Theorem 11.1; see (11.2) in particular). This completes the proof of the theorem.

Remarks. Lemma 35.2 is very likely not the best possible, and (v) and (vii) there can probably be replaced with (v′) and (vii′) even if GCH is not assumed. (This leads to Problem 20.1 stated above.) If this is indeed the case, then we can take $\tau=0$ in Lemma 35.3; this would mean that $\tau=0$ is also necessary for (1)$\equiv$(15) to hold.

The other question left open by Theorem 35.4 is what happens if $\kappa=\lambda>\mathrm{cf}(\kappa)$, κ is not a strong limit cardinal, and $\rho>\omega$. We formulate the simplest instance of this question.

Problem 35.5. Assume $2^{\aleph_0}=\aleph_1$ and $\aleph_\xi^{\aleph_0}<\aleph_{\omega_2}$ for every $\xi<\omega_2$. Does then

$$\aleph_{\omega_2}\to(\aleph_{\omega_2},\aleph_1)^2$$

hold?

Of course we have a positive answer here by Theorem 35.4.b(iii) unless $2^{\aleph_\xi} > \aleph_{\omega_2}$ for some $\xi < \omega_2$. We shall state this problem in a more general form and shall take a closer look at it in Section 37 (see Problem 37.1 below).

If we assume GCH, then our discussion is complete. We have

COROLLARY 35.6. *Assume* GCH *and* (2). *Then* (1) *holds if and only if there are cardinals satisfying conditions* (3)–(8) *of Lemma* 35.3 *with* $\tau = 0$.

PROOF. Lemmas 35.2, 35.3 (especially (9) there), and GCH entail that (1) implies the existence of cardinals satisfying (3)–(8) with $\tau = 0$; this settles the "only if" part. To see the "if" part, one only has to observe that if GCH holds, then κ is either regular or strong limit, and so the result follows from Theorem 35.4.b(i) and (iii). The proof is complete.

There is a more direct way of formulating this corollary. This will be shown in the next section.

36. DISCUSSION OF THE ORDINARY PARTITION RELATION IN CASE $r = 2$ UNDER THE ASSUMPTION OF GCH

The main result of the preceding section can be formulated in a relatively simple way if we assume GCH. To this end, for any cardinal κ put

$$\mathrm{cr}(\kappa) = \mathrm{cf}(\mathrm{cf}(\kappa)^-), \tag{1}$$

which is called the critical number of κ.

LEMMA 36.1. *If* GCH *holds, then*

$$\mathrm{cr}(\kappa) = \min\{L_\kappa(\kappa), L_{\mathrm{cf}(\kappa)}(\mathrm{cf}(\kappa))\}. \tag{2}$$

PROOF. We distinguish two cases: a) κ is a successor cardinal, and b) κ is a limit cardinal. In case a), $\kappa^- < \kappa$ and $\mathrm{cf}(\kappa) = \kappa$. Hence

$$\mathrm{cr}(\kappa) = \mathrm{cf}(\kappa^-) = L_\kappa(\kappa);$$

the second equality here follows from GCH and König's theorem (cf. Corollary 6.8 in particular). In case b),

$$L_\kappa(\kappa) = \kappa \geq \mathrm{cf}(\kappa) \geq L_{\mathrm{cf}(\kappa)}(\mathrm{cf}(\kappa));$$

here

$$L_{\mathrm{cf}(\kappa)}(\mathrm{cf}(\kappa)) = \mathrm{cf}(\kappa) = \mathrm{cr}(\kappa)$$

if cf (κ) is a limit cardinal, and, by König's theorem we obtain in the same way as above that

$$L_{\mathrm{cf}(\kappa)}(\mathrm{cf}(\kappa))=\mathrm{cf}(\mathrm{cf}(\kappa)^{-})=\mathrm{cr}(\kappa)$$

if cf (κ) is a successor cardinal. The proof is complete.

We shall also need the following observation, easily checked by the reader:

$$\mathrm{cr}(\kappa)=\mathrm{cf}(\kappa) \quad \text{iff} \quad \mathrm{cf}(\kappa)^{-}=\mathrm{cf}(\kappa). \tag{3}$$

We introduce two more operations: for any infinite cardinal κ we put

$$\mathrm{suc}(\kappa)=\begin{cases}1 & \text{if} \quad \kappa^{-}<\kappa \\ 0 & \text{if} \quad \kappa^{-}=\kappa,\end{cases} \tag{4}$$

where 'suc' is the first syllable of the word 'successor', and

$$\mathrm{com}(\kappa)=\begin{cases}1 & \text{if} \quad \mathrm{cf}(\kappa)\to(\mathrm{cf}(\kappa))^2_2 \quad \text{or} \quad \mathrm{cf}(\kappa)^{-}<\mathrm{cf}(\kappa) \\ 0 & \text{if} \quad \mathrm{cf}(\kappa)\not\to(\mathrm{cf}(\kappa))^2_2 \quad \text{and} \quad \mathrm{cf}(\kappa)^{-}=\mathrm{cf}(\kappa);\end{cases} \tag{5}$$

here the syllable 'com' is used on the pretext that the cardinals with $\kappa\to(\kappa)^2_2$ are exactly those which are inaccessible and weakly compact (see Theorem 30.3). We can now state the main result of the present section (recall the meaning of the relation $\ll$, given in Definition 35.1):

THEOREM 36.2. (Erdős–Hajnal–Rado [1965]) *Let* κ, ϑ, *and* λ_ξ $(\xi<\vartheta)$ *be cardinals such that* $\kappa\geq\omega$; $\vartheta\geq 2$, *and* $3\leq\lambda_\xi\leq\kappa$ *for any* $\xi<\vartheta$. *Assume* GCH. *Then*

$$\kappa\to(\lambda_\xi)^2_{\xi<\vartheta} \tag{6}$$

holds if and only if the sequence $\langle\lambda_\xi:\xi<\vartheta\rangle$ *is* $\ll$ *one of the sequences satisfying* (7) *or* (8):

$$\langle\kappa,(\mathrm{cr}(\kappa))_\alpha,(\sigma)_\beta\rangle, \quad \textit{where} \quad \sigma,\beta<\mathrm{cr}(\kappa) \quad \textit{and} \quad \alpha<\mathrm{cr}(\kappa)\cdot\mathrm{com}(\kappa)+1, \tag{7}$$

$$\langle(\kappa^{-})_\alpha,(\sigma)_\beta\rangle, \quad \textit{where} \quad \sigma,\beta<\kappa^{-} \quad \textit{and} \quad \alpha<\mathrm{cf}(\kappa^{-})\cdot\mathrm{suc}(\kappa)+1. \tag{8}$$

Here α, β, *and* σ *are cardinals, and* $+$ *denotes cardinal addition.*

PROOF. We assume GCH throughout the proof without necessarily mentioning this explicitly. According to Corollary 35.6, it is enough to show that a) any sequence satisfying (7) or (8) is dominated by (i.e., is $\ll$) some sequence $\langle\lambda,(\rho)_\alpha,(\sigma)_\beta\rangle$ satisfying (35.4)–(35.8), and b) any sequence $\langle\lambda,(\rho)_\alpha,(\sigma)_\beta\rangle$ satisfying (35.4)–(35.8) is dominated by some sequence satisfying (7) or (8).

Ad a). We deal separately with sequences satisfying (7) and (8). First we show:

Assume $\langle\kappa, (\mathrm{cr}(\kappa))_\alpha, (\sigma)_\beta\rangle$ satisfies (7). Then it satisfies (35.4)–(35.8) with $\lambda=\kappa$ and $\rho=\mathrm{cr}(\kappa)$. (9)

To see this, observe that (35.4) holds according to Lemma 36.1. $\mathrm{cr}(\kappa)$ is regular, and so (35.5) holds by (7). (35.6)(i) is vacuous. (35.6)(ii) can be verified as follows: suppose $\mathrm{cf}(\kappa)=\rho(=\mathrm{cr}(\kappa))$ and $\rho\nrightarrow(\rho)^2_2$. Then $\mathrm{cf}(\kappa)^-=\mathrm{cf}(\kappa)$ according to (3); as $\mathrm{cf}(\kappa)\nrightarrow(\mathrm{cf}(\kappa))^2_2$ also holds by our assumptions, we have $\mathrm{com}(\kappa)=0$ in view of (5). Hence $\alpha=0$ according to (7), which establishes (35.6)(ii). If $2^\rho=\kappa$, then $\kappa=\rho(=\mathrm{cr}(\kappa))$ by GCH, and so $\kappa=\mathrm{cf}(\kappa)=\mathrm{cf}(\kappa)^-$ by (3). Thus (35.7)(ii) holds; in fact, if $\mathrm{cf}(\rho)\nrightarrow(\mathrm{cf}(\rho))^2_2$, then $\mathrm{com}(\kappa)=0$ by (5), and so $\alpha=0$ by (7). (35.7)(i) and (35.8) are vacuous since we saw that $\rho=\mathrm{cf}(\rho)=\kappa$ under the assumption $2^\rho=\kappa$. This establishes (9).

Next we prove:

Any sequence satisfying (8) is dominated by the sequence $\langle\lambda, (\rho)_\alpha, (\sigma)_\beta\rangle$ satisfying (35.4)–(35.8) with $\rho=\kappa^-$, where (i) $\lambda=\kappa^-$ if $\kappa^-<\kappa$ and (ii) $\lambda=3$ if $\kappa^-=\kappa$. (10)

We only have to show that $\langle\lambda, (\rho)_\alpha, (\sigma)_\beta\rangle$ satisfies (35.4)–(35.8). We consider the cases (i) and (ii) separately.

Ad (i). $\lambda=\kappa^-<\kappa$, and $L_\lambda(\kappa)=\kappa^-=\rho$ by GCH, and so (35.4) follows. (35.5) follows from (8). (35.6) is vacuous. $2^\rho=\rho=\kappa^-<\kappa$ holds by GCH, and so (35.7) and (35.8) are also vacuous.

Ad (ii). $\lambda=3$ and $L_\lambda(\kappa)=L_3(\kappa)=\kappa=\kappa^-=\rho$ by GCH, and so (35.4) is true. By (8), we can see that (35.5) holds and $\alpha=0$, as $\mathrm{suc}(\kappa)=0$ in this case. Thus (35.6), (35.7), and (35.8) hold as well. (10) is established, and so part a) of the proof of the theorem is complete.

Ad b). Let $\langle\lambda, (\rho)_\alpha, (\sigma)_\beta\rangle$ be a sequence satisfying (35.4)–(35.8). We have to show that it is dominated by a sequence satisfying (7) or (8). We distinguish two cases: 1) $\lambda=\kappa$ and 2) $\lambda<\kappa$.

Ad 1). We prove that in this case

$$\langle\lambda, (\rho)_\alpha, (\sigma)_\beta\rangle \quad \text{is a sequence satisfying (7)}. \tag{11}$$

In fact, as $\lambda=\kappa$ in this case, we have $\rho=\min\{L_\kappa(\kappa), L_{\mathrm{cf}(\kappa)}(\mathrm{cf}(\kappa))\}$ by (35.4), i.e., $\rho=\mathrm{cr}(\kappa)$ according to Lemma 36.1. (35.5) implies $\alpha, \beta, \sigma<\mathrm{cr}(\kappa)$. Assume $\mathrm{com}(\kappa)=0$. Then $\mathrm{cf}(\kappa)=\mathrm{cr}(\kappa)=\rho$ and $\rho\nrightarrow(\rho)^2_2$ by (5) and (3); hence $\alpha=0$ in view of (35.6)(ii). Thus we always have $\alpha<\mathrm{cr}(\kappa)\cdot\mathrm{com}(\kappa)+1$, which settles (11).

Ad 2). We distinguish two cases: α) $\lambda<\kappa^-$ and β) $\lambda=\kappa^-$. First we show that

In case 2α), $\langle\lambda, (\rho)_\alpha, (\sigma)_\beta\rangle$ is dominated by $\langle(\kappa^-)_0, (\lambda+\sigma)_{\beta+1}\rangle$, and this latter sequence satisfies (8). (12)

In fact, by (35.4) and GCH we have $\rho=L_\lambda(\kappa)=\kappa^-$; as $\lambda<\kappa^-=\rho$ in this case, (35.6)(i) implies $\alpha=0$; hence

$$\langle\lambda, (\rho)_\alpha, (\sigma)_\beta\rangle \ll \langle(\kappa^-)_0, (\lambda+\sigma)_{\beta+1}\rangle.$$

Now (35.5) implies $\lambda+\sigma$, $\beta+1<\kappa^-$, establishing (12). It remains to show that

$$\text{In case } 2\beta),\ \langle\lambda, (\rho)_\alpha, (\sigma)_\beta\rangle \text{ is dominated by } \langle(\kappa^-)_{\alpha+1}, (\sigma)_\beta\rangle, \text{ and this latter sequence satisfies (8).} \tag{13}$$

As $\lambda<\kappa$ in case 2) and $\lambda=\kappa^-$ in case 2β), we have $\kappa^-<\kappa$. Hence suc $(\kappa)=1$. $\rho=L_\lambda(\kappa)=\kappa^-$ holds by (35.4) and GCH. Thus $\langle\lambda, (\rho)_\alpha, (\sigma)_\beta\rangle \ll \langle(\kappa^-)_{\alpha+1}, (\sigma)_\beta\rangle$. (35.5) now implies $\sigma, \beta<\rho=\kappa^-$ and $\alpha<\mathrm{cf}\,(\kappa^-)=\mathrm{cf}\,(\kappa^-)\cdot\mathrm{suc}\,(\kappa)+1$. This settles (13). The proof of the theorem is complete.

37. SIERPIŃSKI PARTITIONS

One of the main problems that remained open in the above discussion of the ordinary partition relation was the following

PROBLEM 37.1. Let κ be a singular cardinal and $\lambda>\omega$ a cardinal such that

$$\mathrm{cf}\,(\kappa)\to(\mathrm{cf}\,(\kappa), \lambda)^2 \tag{1}$$

holds and

$$\rho^\lambda<\kappa \tag{2}$$

holds for any $\rho<\kappa$. Does then

$$\kappa\to(\kappa, \lambda)^2 \tag{3}$$

hold?

(We stated this as Problem 35.5 above in the particular case when $\kappa=\aleph_{\omega_2}$, $\lambda=\aleph_1$, $2^{\aleph_0}=\aleph_1$, and $\aleph_\xi^{\aleph_0}<\aleph_{\omega_2}$ for every $\xi<\omega_2$.)

One has every reason to believe that the answer to this question is yes, though we have no idea how one could prove this. We can, however, show that one cannot prove for any choice of κ and λ that the answer is negative by using Sierpiński partitions only. We described Sierpiński partitions in Definition 19.5, but we give the definition here again, in a slightly modified form:

DEFINITION 37.2. Let κ, ρ, and σ be infinite cardinals. We say that

$$\kappa\overset{S}{\to}(\rho, \sigma)^2$$

(in words: κ S-arrows...) if any ordered set of cardinality κ includes a set having either order type ρ or order type σ^*. The negation of this relation is written as $\kappa \overset{\mathrm{S}}{\nrightarrow} (\rho, \sigma)^2$; in case this latter relation holds we may say that the relation $\kappa \nrightarrow (\rho, \sigma)^2$ is established by a *Sierpiński partition* (this agrees with the terminology introduced in Definition 19.5).

Lemma 19.4 above says that the relation $\kappa \overset{\mathrm{S}}{\nrightarrow} (\rho, \sigma)^2$ implies $\kappa \nrightarrow (\rho, \sigma)^2$. It is worth noting that, in each case above, when we proved a relation of the latter type, then we actually proved the stronger, i.e., the former relation instead. The main result of this section says that in Problem 37.1 at least $\kappa \overset{\mathrm{S}}{\rightarrow} (\kappa, \lambda)^2$ holds; hence, even if (3) might fail in certain cases, an entirely new idea is needed to prove this.

THEOREM 37.3. *Let κ be a singular cardinal and $\lambda \geq \omega$ a cardinal such that*

$$\mathrm{cf}(\kappa) \overset{\mathrm{S}}{\rightarrow} (\mathrm{cf}(\kappa), \lambda)^2 \tag{4}$$

and

$$\rho^\lambda < \kappa \tag{5}$$

for any $\rho < \kappa$. Then

$$\kappa \overset{\mathrm{S}}{\rightarrow} (\kappa, \lambda)^2 \tag{6}$$

holds.

PROOF. Write κ in the form

$$\kappa = \sum_{\alpha < \mathrm{cf}(\kappa)} \kappa_\alpha \tag{7}$$

where $\kappa_\alpha < \kappa$. Let $\langle X, \prec \rangle$ be an ordered set of cardinality κ. We have to prove that X includes a set Y having either order type κ or order type λ^*. Assuming that the second alternative fails, we shall prove the following

CLAIM. *There are pairwise disjoint sets $X_\alpha \subseteq X$, $\alpha < \mathrm{cf}(\kappa)$, such that X_α precedes X_β in the ordering $\prec$ for any $\alpha < \beta < \mathrm{cf}(\kappa)$ (i.e., $x \prec y$ whenever $x \in X_\alpha$ and $y \in X_\beta$), and*

$$|X_\alpha| > \kappa_\alpha^\lambda. \tag{8}$$

Note that $\kappa_\alpha^\lambda < \kappa$ according to (5); instead of (8) we might have required only $|X_\alpha| > \kappa_\alpha$, but there is a technical advantage in requiring the former. It is easy to see that the above claim implies that X contains a subset of order type κ, establishing (6). In fact, we have

$$(\kappa_\alpha^\lambda)^+ \rightarrow (\kappa_\alpha, \lambda)^2$$

according to Corollary 17.5 (we have even more, but this is all we need); hence, a fortiori,

$$(\kappa_\alpha^\lambda)^+ \overset{S}{\to} (\kappa_\alpha, \lambda)^2$$

holds by Lemma 19.4 (cf. the remark after Definition 37.2). Applying this to the set $\langle X_\alpha, \prec \rangle$, we can see by (8) that X_α includes a set Y_α having order type either κ_α or λ^*. The latter alternative fails according to our assumption; hence Y_α must have order type κ_α. Then $\bigcup_{\alpha < \mathrm{cf}(\kappa)} Y_\alpha$ has order type κ by (7). This verifies (6).

We have yet to prove the above Claim. To this end, define the equivalence relation $\sim$ on X as follows: for any two $x, y \in X$ with $x \preceq y$ put:

$$x \sim y \Leftrightarrow |\{z \in X : x \preceq z \preceq y\}| < \kappa .$$

We distinguish two cases: a) X consists of a single equivalence class, or b) X has at least two equivalence classes.

Ad a). In this case, we have

$$|\{z \in X : x \preceq z \preceq y\}| < \kappa \tag{9}$$

for any $x, y \in X$ with $x \prec y$. Take a set $Y \subseteq X$ that is wellordered in $\succ$ and cofinal in $\langle X, \succ \rangle$ (i.e., Y is cofinal in X downward). $|Y| < \lambda$ by our assumption; and so $|Y| < \mathrm{cf}(\kappa)$, as $\lambda \leq \mathrm{cf}(\kappa)$ according to (4). Hence, for any $y \in Y$ we have by (9) that

$$|\{z \in X : z \preceq y\}| = \bigcup_{x \in Y,\, x \prec y} |\{z \in Y : x \preceq z \preceq y\}| < \kappa .$$

Using (9) again, we can see that

$$|\{z \in X : z \preceq x\}| < \kappa$$

holds for any $x \in X$. As $|X| = \kappa$, it is easy to see from this that there is a sequence $\langle x_\alpha : \alpha < \mathrm{cf}(\kappa) \rangle$ increasing in $\prec$ of elements of X such that the set

$$X_\alpha = \{z \in X : x_\alpha \preceq z \prec x_{\alpha+1}\} \qquad (\alpha < \mathrm{cf}(\kappa))$$

has cardinality $> \kappa_\alpha^\lambda$ (note that $\kappa_\alpha^\lambda < \kappa$ by (5)). This establishes our Claim in case a).

Ad b). If there is an equivalence class X' of cardinality κ of X, then replacing X with X', our claim follows in exactly the same way as in case a); hence we may assume that each equivalence class of X has cardinality $< \kappa$. As X is the union of its equivalence classes, it follows that there are at least $\mathrm{cf}(\kappa)$ equivalence classes in X. Let $z_\alpha \in X$, $\alpha < \mathrm{cf}(\kappa)$, be representatives of distinct equivalence classes. As we assumed that X does not include a descending sequence of type λ, it follows from (4) that the set $\langle \{z_\alpha : \alpha < \mathrm{cf}(\kappa)\}, \prec \rangle$ includes an increasing sequence of type cf (κ), say $\langle x_\alpha : \alpha < \text{cf}(\kappa)\rangle$. Then the sets

$$X_\alpha = \{z \in X : x_\alpha \precsim z \prec x_{\alpha+1}\}$$

satisfy the requirements of our Claim; in fact, as x_α and $x_{\alpha+1}$ are not equivalent, we have $|X_\alpha| = \kappa > \kappa_\alpha^\lambda$. This proves the Claim also in case b), completing the proof of the theorem.

The discussion of the Sierpiński partition relation $\kappa \overset{\text{S}}{\rightarrow} (\rho, \sigma)^2$ presents no new problems. Namely, Theorem 35.4 is valid for this relation, as $\kappa \rightarrow (\rho, \sigma)^2$ implies $\kappa \overset{\text{S}}{\rightarrow} (\rho, \sigma)^2$ by Lemma 19.4, as we pointed out this in the remark after Definition 37.2, and, as mentioned in the same paragraph, whenever we proved $\kappa \nrightarrow (\rho, \sigma)^2$, we actually always proved the stronger relation $\kappa \overset{\text{S}}{\nrightarrow} (\rho, \sigma)^2$. As far as Sierpinski partition relations are concerned, Theorem 37.3 fills the only gap in Theorem 35.4; so, putting these two results together, one can give a complete discussion of the Sierpiński partition relation. We leave the details to the reader.

CHAPTER IX

DISCUSSION OF THE ORDINARY PARTITION RELATION WITH SUPERSCRIPT ≥ 3

In the next five sections we collect the results known to us concerning the ordinary partition relation for the case $r \geq 3$. Our discussion is based on the discussion of the case $r=2$ given in the preceding chapter, the Stepping-up Lemma, the special negative relations obtained in Section 25, the addition results for negative relations in Section 22, and on the Negative Stepping-up Lemma in Section 24. The main difficulty is that these two latter results are applicable only if certain inconvenient conditions are satisfied, and the main concern of Sections 38 and 39 below is to analyse these conditions so that we obtain results that can be efficiently used to derive a partition relation from an analogous one with a smaller superscript, and Sections 40–42 pick the fruits of these preparations. Even more than in the case of the first two sections of the preceding chapter, one might expect little enjoyment from reading these chapters, and so the reader will perhaps do well by consulting only the main results and problems obtained, and skipping the proofs of the former. As in the case of the preceding chapter, we feel it necessary to point out here that the importance of this chapter is, apart from giving a practically usable test to decide whether an ordinary partition relation with $r \geq 3$ holds or fails whenever this is known to us, that it locates the most important problems left open by our study. As interesting by-products of our discussion we obtain that there are no genuinely new problems for the case $r \geq 4$, i.e., all gaps in our knowledge in this case issue from gaps in the cases $r=2$ or 3; and, moreover, that the only problem remaining open with GCH is of the type given in Problem 25.8, which concerns the case $r=3$.

38. REDUCTION OF THE SUPERSCRIPT

The aim of this section is to show that the question whether or not the relation

$$\kappa \to (\lambda_\xi)^r_{\xi<\vartheta} \tag{1}$$

holds, where we shall usually assume that

$$r \geq 3, \quad \vartheta \geq 2, \quad \kappa \geq \omega, \quad \text{and} \quad \lambda_\xi > r \quad \text{for any} \quad \xi < \vartheta, \tag{2}$$

can be reduced to an analogous question involving a smaller r under fairly general circumstances. To this end, define the iterated logarithm $L_3^i(\kappa)$ by recursion on $i<\omega$ as follows:

$$L_3^0(\kappa)=\kappa \quad \text{and} \quad L_3^{i+1}(\kappa)=L_3(L_3^i(\kappa)). \tag{3}$$

It is obvious that

$$L_3^0(\kappa)\geq L_3^1(\kappa)\geq L_3^2(\kappa)\geq \ldots \geq \omega \tag{4}$$

(cf. Theorems 7.1b and 7.2). Moreover, one can easily check that

$$L_3^i(\kappa)=\min\{\rho: \exp_i(\rho)\geq\kappa\} \tag{5}$$

holds for any integer i, where the iterated exponential function is defined by

$$\exp_0(\kappa)=\kappa \quad \text{and} \quad \exp_{i+1}(\kappa)=\exp_i(2^\kappa) \tag{6}$$

for any $i<\omega$.

We shall also need the following two concepts.

DEFINITION 38.1. Let $r\geq 3$ be an integer and consider the sequence

$\vec{\lambda}=\langle\lambda_\xi: \xi<\vartheta\rangle$, where $\lambda_\xi\geq r+1$.

a) We say that $\mathrm{ns}(r,\vec{\lambda})$ holds if for any cardinal $\rho\geq\omega$, the relation

$$\rho\nrightarrow(\lambda_\xi-1)^{r-1}_{\xi<\vartheta}$$

implies

$$2^\rho\nrightarrow(\lambda_\xi)^r_{\xi<\vartheta}.$$

(Note that Lemma 24.1 says that $\mathrm{ns}(r,\vec{\lambda})$ often holds. The name Negative Stepping-up Lemma given to this lemma explains the meaning of the letters 'ns'.)

b) We say that $\mathrm{ad}(r,\vec{\lambda})$ holds if for every sequence $\langle\rho_\alpha: \alpha<\eta\rangle$ of infinite cardinals, the relations

$$\rho_\alpha\nrightarrow(\lambda_\xi)^r_{\xi<\vartheta} \quad \text{for all} \quad \alpha<\eta$$

and

$$|\eta|\nrightarrow(\lambda_\xi)^r_{\xi<\vartheta}$$

imply

$$\sum_{\alpha<\eta}\rho_\alpha\nrightarrow(\lambda_\xi)^r_{\xi<\vartheta}.$$

(Theorem 22.1 says that $\mathrm{ad}(r,\vec{\lambda})$ often holds; the letters 'ad' are justified by the fact that this theorem was obtained by addition of negative relations.)

We now prove a simple lemma that gives a sufficient condition for (1) to hold. It is a direct consequence of the Stepping-up Lemma (Lemma 16.1), but it will be useful to state it for further reference. The assumptions in (2) are not yet used here; in particular, the lemma is valid for $r=2$, and we shall exploit this fact below.

LEMMA 38.2. *Let i be an integer with $0<i\le r-1$, and assume that there is a cardinal ρ with $\exp_{i-1}(2^{\rho})<\kappa$ and*

$$\rho\to(\lambda_{\xi}-i)^{r-i}_{\xi<\vartheta} \tag{7}$$

Then (1) *holds.*

PROOF. Writing $\rho_0=\rho$ and $\rho_j=(\exp_{j-1}(2^{\rho}))^{+}$ for any integer $j>0$, we claim that (7) implies

$$\rho_j\to(\lambda_{\xi}+j-i)^{r+j-i}_{\xi<\vartheta}$$

for any integer j. In fact, this directly follows from the Stepping-up Lemma (Lemma 16.1) by induction on j; for the induction step one only has to observe that $\rho_{j+1}=(2^{\rho_j})^{+}$ holds for every j. As $\rho_i\le\kappa$ by our assumptions, the above relation with $j=i$ gives (1). The proof is complete.

Here is a partial converse to the lemma just proved:

LEMMA 38.3. *Let i be an integer with $0<i\le r-2$ and suppose that $\mathrm{cf}\,(L_3^i(\kappa))<L_3^{i-1}(\kappa)$ (i.e., either $L_3^i(\kappa)<L_3^{i-1}(\kappa)$ or $\mathrm{cf}\,(L_3^{i-1}(\kappa))<L_3^{i-1}(\kappa)$). Suppose, furthermore, that $\mathrm{ad}\,(r-i+1,\langle\lambda_{\xi}-i+1:\xi<\vartheta\rangle)$ and $\mathrm{ns}\,(j,\langle\lambda_{\xi}-r+j:\xi<\vartheta\rangle)$ hold for any j with $r-i<j\le r$ (see Definition 38.1 above). If*

$$\rho\not\to(\lambda_{\xi}-i)^{r-i}_{\xi<\vartheta} \tag{8}$$

holds whenever $\exp_{i-1}(2^{\rho})<\kappa$, then (1) *fails.*

PROOF. We distinguish two cases: a) $2^{L_3^i(\kappa)}<L_3^{i-1}(\kappa)$ and b) $2^{L_3^i(\kappa)}=L_3^{i-1}(\kappa)$ (note that $2^{L_3^i(\kappa)}>L_3^{i-1}(\kappa)$ cannot occur in view of the definition of the logarithm).

In case a) we clearly have $\exp_{i-1}(2^{L_3^i(\kappa)})<\kappa$ (cf. (5)), and so (8) holds with $\rho=L_3^i(\kappa)$. We claim that

$$L^{r-j}(\kappa)\not\to(\lambda_{\xi}-r+j)^{j}_{\xi<\vartheta} \tag{9}$$

holds for any j with $r-i\le j\le r$. In fact, for $j=r-i$ this is just (8) with $\rho=L_3^i(\kappa)$, and if $r-i<j\le r$ and (9) holds with $j-1$ replacing j, then $\mathrm{ns}\,(j,\langle\lambda_{\xi}-r+j:\xi<\vartheta\rangle)$ implies that (9) itself holds with j (note that we have $2^{L_3^{r-j+1}(\kappa)}\ge L_3^{r-j}(\kappa)$ by the definition of the logarithm). This verifies (9). Now (9) with $j=r$ says that

$$\kappa=L_3^0(\kappa)\not\to(\lambda_{\xi})^{r}_{\xi<\vartheta}$$

holds; hence (1) indeed fails.

In case b) we claim that

$$L_3^{i-1}(\kappa)\not\to(\lambda_{\xi}-i+1)^{r-i+1}_{\xi<\vartheta} \tag{10}$$

holds. To see this, observe first that

$$\rho\not\to(\lambda_{\xi}-i)^{r-i}_{\xi<\vartheta} \tag{11}$$

holds for any $\rho < L_3^i(\kappa)$ according to (8). In fact, we have

$$\exp_{i-1}(2^\rho) \le \exp_i(\rho) < \kappa$$

for any such ρ by (5). (11) and $\mathrm{ns}(r-i+1, \langle \lambda_\xi - i + 1 : \xi < \vartheta \rangle)$ imply that

$$2^\rho \nrightarrow (\lambda_\xi - i + 1)_{\xi<\vartheta}^{r-i+1} \tag{12}$$

holds for any ρ with $\omega \le \rho < L_3^i(\kappa)$. We can substitute

$$\rho = \min\{L_3^{i+1}(\kappa), \mathrm{cf}(L_3^i(\kappa))\}$$

here, because this ρ is $< L_3^i(\kappa)$. In fact, assuming $\rho \ge L_3^i(\kappa)$, we obtain in particular that $L_3^{i+1}(\kappa) = L_3^i(\kappa)$ holds; this, in conjunction with the equality $2^{\underset{\smile}{L_3^i(\kappa)}} = L_3^{i-1}(\kappa)$, which also holds in the case considered, implies that

$$L_3^i(\kappa) \le L_3^{i-1}(\kappa) = 2^{\underset{\smile}{L_3^i(\kappa)}} = 2^{\underset{\smile}{L_3^{i+1}(\kappa)}} \le L_3^i(\kappa),$$

i.e., $L_3^{i-1}(\kappa) = L_3^i(\kappa)$. Now we have $\mathrm{cf}(L_3^i(\kappa)) < L_3^{i-1}(\kappa)$ by the assumptions of the lemma to be proved, and so we have $\rho \le \mathrm{cf}(L_3^i(\kappa)) < L_3^{i-1}(\kappa) = L_3^i(\kappa)$, which contradicts the assumption $\rho \ge L_3^i(\kappa)$. Hence $\rho < L_3^i(\kappa)$, as claimed. (12) with this value of ρ gives

$$\mathrm{cf}(L_3^i(\kappa)) \nrightarrow (\lambda_\xi - i + 1)_{\xi<\vartheta}^{r-i+1}, \tag{13}$$

since clearly $2^\rho \ge \mathrm{cf}(L_3^i(\kappa))$ holds with this ρ. Noting that $L_3^i(\kappa) >$ the above value of ρ, which is $\ge \omega$, let now $\langle \rho_\alpha : \alpha < \mathrm{cf}(L_3^i(\kappa)) \rangle$ be an increasing sequence of infinite cardinals tending to $L_3^i(\kappa)$. Using $\mathrm{ad}(r-i+1, \langle \lambda_\xi - i + 1 : \xi < \vartheta \rangle)$, the relations in (12) and (13) imply

$$\sum_{\alpha < \mathrm{cf}(L_3^i(\kappa))} 2^{\rho_\alpha} \nrightarrow (\lambda_\xi - i + 1)_{\xi<\vartheta}^{r-i+1}.$$

As

$$L_3^{i-1}(\kappa) = 2^{\underset{\smile}{L_3^i(\kappa)}} = \sum_{\alpha < \mathrm{cf}(L_3^i(\kappa))} 2^{\rho_\alpha}$$

holds in case b), which is the case now being considered, this means that (10) does indeed hold. Using $\mathrm{ns}(j, \langle \lambda_\xi - r + j : \xi < \vartheta \rangle)$ for each j with $r - i + 1 \le j \le r$, we obtain from (10) by induction on j that

$$L_3^{r-j}(\kappa) \nrightarrow (\lambda_\xi - r + j)_{\xi<\vartheta}^{j}$$

holds for any such j (note that $2^{\underset{\smile}{L_3^{r-j+1}(\kappa)}} \ge L_3^{r-j}(\kappa)$). With $j = r$ this just means that $\kappa = L_3^0(\kappa) \nrightarrow (\lambda_\xi)_{\xi<\vartheta}^r$, which we wanted to show. The proof of the lemma is complete.

We next turn to the case $L_3^{i-1}(\kappa)=\mathrm{cf}\,(L_3^{i-1}(\kappa))=L_3^i(\kappa)$, which was left open by the lemma just proved. This case is relatively easy to handle, as $L_3^i(\kappa)=L_3^{i-1}(\kappa)$ is an inaccessible cardinal in this case.

LEMMA 38.4. *Let i be an integer with $0<i\leq r-2$, and assume that $\mathrm{cf}\,(L_3^i(\kappa))=L_3^{i-1}(\kappa)$. Assume, further, that* (2) *is satisfied. Then the following condition is sufficient for* (1) *to hold: either there is a $\rho<L_3^{i-1}(\kappa)$ such that*

$$\rho\rightarrow(\lambda_\xi-i)^{r-i}_{\xi<\vartheta}, \tag{14}$$

or we have

$$L_3^{i-1}(\kappa)\rightarrow(L_3^{i-1}(\kappa))^2_2 \tag{15}$$

and

$$\vartheta<L_3^{i-1}(\kappa) \quad \text{and} \quad \lambda_\xi\leq L_3^{i-1}(\kappa) \quad \text{for any} \quad \xi<\vartheta. \tag{16}$$

This condition is also **necessary** *for* (1) *provided* ns $(j,\langle\lambda_\xi+j-r\colon\xi<\vartheta\rangle)$ *holds for any j with $r-i+2\leq j\leq r$.*

PROOF. *Sufficiency.* $L_3^{i-1}(\kappa)$ is inaccessible in view of our assumptions. Therefore we have $\exp_{i-1}(2^\rho)<L_3^{i-1}(\kappa)\leq\kappa$ for any $\rho<L_3^{i-1}(\kappa)$, and so (14) implies (1) in view of Lemma 38.2. If (15) and (16) hold, then

$$L_3^{i-1}(\kappa)\rightarrow(L_3^{i-1}(\kappa))^r_\vartheta$$

also holds in virtue of the implication (vi)⇒(iv) in Theorem 29.6. As $L_3^{i-1}(\kappa)\leq\kappa$ and, moreover, $\lambda_\xi\leq L_3^{i-1}(\kappa)$ for any $\xi<\vartheta$ according to (16), we have $\kappa\rightarrow(\lambda_\xi)^r_{\xi<\vartheta}$ a fortiori.

Necessity. Assume that (14) fails for any $\rho<L_3^{i-1}(\kappa)$. We claim that then

$$\vartheta\geq L_3^{i-1}(\kappa) \quad \text{or} \quad \exists\xi<\vartheta[\lambda_\xi\geq L_3^{i-1}(\kappa)] \tag{17}$$

holds. In fact, in the contrary case there is a $\sigma<L_3^{i-1}(\kappa)$ such that $\vartheta\leq\sigma$ and $\lambda_\xi\leq\sigma$ for any $\xi<\vartheta$. Then

$$\sigma^+\rightarrow(\lambda_\xi)^1_{\xi<\vartheta}$$

holds. Using Lemma 38.2 with $r-i$ and $r-i-1$ replacing r and i, respectively, we obtain that

$$(\exp_{r-i-1}(\sigma))^+\rightarrow(\lambda_\xi+r-i-1)^{r-i}_{\xi<\vartheta}$$

holds. Putting $\rho=(\exp_{r-i-1}(\sigma))^+$, which is $<L_3^{i-1}(\kappa)$ by virtue of the inaccessibility of this latter cardinal, we obtain that (14) holds a fortiori with this ρ, which contradicts our assumption. This verifies (17).

Assume now that the conjunction of (15) and (16) also fails. We claim that then

$$L_3^{i-1}(\kappa)\not\rightarrow(\lambda_\xi-i+1)^{r-i+1}_{\xi<\vartheta} \tag{18}$$

holds. In fact, if (16) fails, then this is obvious. Assume therefore that (16) holds and (15) fails. The failure of (15) entails that

$$L_3^{i-1}(\kappa)\nrightarrow(L_3^{i-1}(\kappa),4)^3$$

by the implication (iv)⇒(v) in Theorem 29.6, since $L_3^{i-1}(\kappa)$ is inaccessible (actually, inaccessibility is not needed for this implication, as remarked at the beginning of the proof of Theorem 29.6). This relation, however, implies (18). In fact, $\lambda_\xi-i+1=\lambda_\xi\geq L_3^{i-1}(\kappa)$ for some $\xi<\vartheta$ according to the second relation in (17) (this must hold, as the first one fails in view of (16)); we have $r-i+1\geq 3$ by virtue of our assumptions on i; and, moreover, $\vartheta\geq 2$ and $\lambda_{\xi'}-i+1\geq(r+1)-$ $-i+1\geq 4$ holds for any $\xi'<\vartheta$ according to (2).

We claim that

$$L_3^{r-j}(\kappa)\nrightarrow(\lambda_\xi+j-r)^j_{\xi<\vartheta} \tag{19}$$

holds for any j with $r-i+1\leq j\leq r$. In fact, for $j=r-i+1$ this is just (18), and if (19) holds with $j-1$ replacing j, where $r-i+2\leq j\leq r$, then it also holds with j in view of ns (j, $\langle\lambda_\xi+j-r\colon\xi<\vartheta\rangle$), as $2^{L_3^{r-j+1}(\kappa)}\geq L_3^{r-j}(\kappa)$. With $j=r$ (19) just gives $\kappa\nrightarrow(\lambda_\xi)^r_{\xi<\vartheta}$, which we wanted to show. The proof is complete.

The next theorem sums up the results obtained.

Theorem 38.5 (Reduction Theorem). *Let $r\geq 3$, $\kappa\geq\omega$, $\vartheta\geq 2$, and $\lambda_\xi>r$ for every $\xi<\vartheta$. Let i be an integer with $0<i\leq r-2$. Consider the relation*

$$\kappa\rightarrow(\lambda_\xi)^r_{\xi<\vartheta} \tag{20}$$

and the following conditions:

(i) *Either* a) *there is an infinite cardinal ρ with* $\exp_{i-1}(2^\rho)<\kappa$ *and*

$$\rho\rightarrow(\lambda_\xi-i)^{r-i}_{\xi<\vartheta} \tag{21}$$

or b) *we have $\vartheta<L_3^{i-1}(\kappa)$, $\lambda_\xi\leq L_3^{i-1}(\kappa)$ for any $\xi<\vartheta$, and*

$$L_3^{i-1}(\kappa)\rightarrow(L_3^{i-1}(\kappa))^2_2\,. \tag{22}$$

(ii) *For every sequence $\langle\rho_\alpha\colon\alpha<\eta\rangle$ of infinite cardinals, the relations*

$$\rho_\alpha\nrightarrow(\lambda_\xi-i+1)^{r-i+1}_{\xi<\vartheta}\quad\text{for all}\quad\alpha<\eta$$

and

$$|\eta|\nrightarrow(\lambda_\xi-i+1)^{r-i+1}_{\xi<\vartheta}$$

imply

$$\sum_{\alpha<\eta}\rho_\alpha\nrightarrow(\lambda_\xi-i+1)^{r-i+1}_{\xi<\vartheta}\,.$$

(iii) *For every integer* j *with* $r-i<j\leq r$ *and every infinite cardinal* ρ, *the relation*

$$\rho \nrightarrow (\lambda_\xi - r + j - 1)^{j-1}_{\xi<\vartheta}$$

implies

$$2^\rho \nrightarrow (\lambda_\xi - r + j)^{j}_{\xi<\vartheta}.$$

(i) *is a sufficient condition for* (20) *to hold. If* (ii) *and* (iii) *are satisfied, then* (i) *is also a necessary condition for* (20).

A detailed discussion when (ii) and (iii) are satisfied is given in the next section. We note that the requirement in (i) on ρ being infinite is unimportant, and was included only in order to stress that we are considering partitions on infinite cardinals. In fact, if (21) is satisfied with a finite ρ, then it is also satisfied with $\rho=\omega$; the only trouble might be that $\exp_{i-1}(2^\omega)\geq\kappa$; in this case, however, $L_3^{i-1}(\kappa)=\omega$ by (5), and so (22) holds in view of Ramsey's theorem (Theorem 10.2), which means that (i) is satisfied.

Proof. *Sufficiency.* If (i)a) holds, then Lemma 38.2 implies (20), and if (i)b) holds, then the 'if' part of Lemma 38.4 implies (20); in fact, the assumption $\operatorname{cf}(L^i(\kappa))=L^{i-1}(\kappa)$ of that lemma holds since $L^{i-1}(\kappa)$ is inaccessible by (22) and Theorem 29.1, and, moreover, (15) and (16) hold in view of the assumptions in (i)b).

Necessity. Assume that (20) holds and (i)b) is false, and, moreover, that (ii) and (iii) are satisfied. If $L_3^{i-1}(\kappa)$ is not inaccessible, then (i)a) holds according to Lemma 38.3, and if $L_3^{i-1}(\kappa)$ is inaccessible, then it holds according to Lemma 38.4. In fact, to see the latter assertion, the conjunction of (15) and (16) in this lemma fails as (i)b) was assumed to be false; hence (14) must hold with some $\rho<L_3^{i-1}(\kappa)$; but then $\exp_{i-1}(2^\rho)<L_3^{i-1}(\kappa)\leq\kappa$, and so (i)a) indeed holds. The proof is complete.

It might be interesting to formulate the above result in the case $i=1$:

Corollary 38.6. *Let* $r\geq3$, $\kappa\geq\omega$, $\vartheta\geq2$, *and* $\lambda_\xi\geq r$ *for every* $\xi<\vartheta$. *Consider the relation*

$$\kappa\rightarrow(\lambda_\xi)^r_{\xi<\vartheta} \tag{23}$$

and the following conditions:

(i) *Either there is an infinite cardinal* ρ *with* $2^\rho<\kappa$ *and*

$$\rho\rightarrow(\lambda_\xi-1)^{r-1}_{\xi<\vartheta}$$

or we have $\vartheta<\kappa$ *and* $\lambda_\xi\leq\kappa$ *for every* $\xi<\vartheta$, *and*

$$\kappa\rightarrow(\kappa)^2_2.$$

(ii) *For every sequence* $\langle \rho_\alpha : \alpha < \eta \rangle$ *of infinite cardinals, the relations*

$$\rho_\alpha \nrightarrow (\lambda_\xi)^r_{\xi < \vartheta} \quad \textit{for all} \quad \alpha < \eta$$

and

$$|\eta| \nrightarrow (\lambda_\xi)^r_{\xi < \vartheta}$$

imply

$$\sum_{\alpha < \eta} \rho_\alpha \nrightarrow (\lambda_\xi)^r_{\xi < \vartheta} .$$

(iii) *For every infinite cardinal* ρ, *the relation*

$$\rho \nrightarrow (\lambda_\xi - 1)^{r-1}_{\xi < \vartheta}$$

implies

$$2^\rho \nrightarrow (\lambda_\xi)^r_{\xi < \vartheta} .$$

(i) *is a sufficient condition for* (23) *to hold. If* (ii) *and* (iii) *are satisfied, then* (i) *is also a necessary condition for* (23).

As we already mentioned in Definition 38.1, conditions (ii) and (iii) in this corollary as well as in the preceding theorem are often satisfied. In fact, it follows from Theorem 22.1 and Lemma 24.1 that it is enough to assume e.g. that $r - i \geq 4$ in Theorem 38.5, i.e., $r \geq 5$ in Corollary 38.6. For the remaining cases, some special assumptions are needed (e.g. cf $(\lambda_0) = \lambda_0 \geq \omega$ and $\lambda_1 \geq \omega$; cf. the results of the next section). For these cases, Theorem 38.5 does not contain all our results, since some gaps can be filled by using the special results obtained in Section 25. The filling of these gaps will be one of the aims of the next few sections. There remain however some problems which we are unable to settle. In case GCH is assumed, it will turn out that only one typical case cannot be covered by our results, and this case leads to Problem 25.8 stated above.

39. APPLICABILITY OF THE REDUCTION THEOREM

Our first aim now is to study the applicability of Theorem 38.5, i.e., to consider when conditions (ii) and (iii) of that theorem are satisfied. To this end we need the following simple lemma:

LEMMA 39.1. *For any cardinal* $\rho \geq \omega$ *and any integer* $r \geq 1$ *we have*

$$\exp_{r-1}(\rho) \nrightarrow (r+1)^r_\rho . \tag{1}$$

PROOF. The case $r = 1$ is trivial, and if $r = 2$, then the result is given in (19.17). Let $r = 3$. Then relation (24.5) in Lemma 24.1 with $\kappa = 2^\rho$, $\tau = \rho$ and $\lambda_\xi = 3$ implies

$$2^{2^\rho} \nrightarrow ((4)_\rho, (4)_\rho, 4, 4)^3$$

(here we used the fact that (1) holds with $r=2$); as $\rho\geq\omega$, this is just (1) with $r=3$. Let now $r\geq 4$, and assume that (1) holds with $r-1$ replacing r. Then, using (24.3) in Lemma 24.1 with $\kappa=\exp_{r-2}(\rho)$, $\tau=\rho$, and $\lambda_\xi=3$, we obtain that

$$\exp_{r-1}(\rho)\nrightarrow((r+1)_\rho, (r+1)_m)^r,$$

where m is finite; as $\rho\geq\omega$, (1) follows. The proof is complete.

The next lemma achieves a partial fulfilment of our first aim mentioned above:

LEMMA 39.2. *Let $r\geq 3$, $\kappa\geq\omega$, $\vartheta\geq 2$, $\lambda_\xi\geq r$ for every $\xi<\vartheta$, and let i be an integer with $0<i\leq r-2$. Condition* (ii) *of Theorem* 38.5 *holds except if $i=r-2$ and*

$$\begin{aligned}&\vartheta<\omega,\quad \lambda_\mu\geq\omega \quad \textit{for some} \quad \mu<\vartheta, \quad \textit{and}\\ &\lambda_\xi<\omega \quad \textit{for any} \quad \xi<\vartheta \quad \textit{different from } \mu.\end{aligned} \tag{2}$$

PROOF. Assume that condition (ii) of Theorem 38.5 fails, i.e., that there is a sequence $\langle\rho_\alpha:\alpha<\eta\rangle$ of infinite cardinals such that

$$\rho_\alpha\nrightarrow(\lambda_\xi-i+1)^{r-i+1}_{\xi<\vartheta} \tag{3}$$

for all $\alpha<\eta$, and

$$|\eta|\nrightarrow(\lambda_\xi-i+1)^{r-i+1}_{\xi<\vartheta}, \tag{4}$$

and yet

$$\sum_{\alpha<\eta}\rho_\alpha\rightarrow(\lambda_\xi-i+1)^{r-i+1}_{\xi<\vartheta}. \tag{5}$$

By this last relation and the preceding lemma we certainly have

$$\exp_{r-i}(|\vartheta|)<\sum_{\alpha<\eta}\rho_\alpha \tag{6}$$

here. We first show that, as claimed in (2), there is a $\mu<\vartheta$ with $\lambda_\mu\geq\omega$. By (6), either there is an $\alpha<\eta$ with $\exp_{r-i}(|\vartheta|)<\rho_\alpha$ or we have $\exp_{r-i}(|\vartheta|)<|\eta|$. Hence, if $\vartheta<\omega$ then by Ramsey's theorem (Theorem 10.2), and if $\vartheta\geq\omega$ then by (17.32), we obtain that either $\rho_\alpha\rightarrow(\omega)^{r-i+1}_\vartheta$ or $|\eta|\rightarrow(\omega)^{r-i+1}_\vartheta$ holds. So there must indeed be a $\mu<\vartheta$ with $\lambda_\mu\geq\omega$, since otherwise these relations would contradict (3) or (4). We can therefore assume that e.g.

$$\lambda_0\geq\omega. \tag{7}$$

Using this inequality, (22.5) in Theorem 22.1 implies that (3), (4), and (5) can hold simultaneously (i.e., (ii) of Theorem 38.5 can fail) only if $\lambda_\xi<\omega$ for any ξ with $1\leq\xi<\vartheta$ and $r-i+1=3$, i.e., $i=r-2$. We have yet to prove that $\vartheta<\omega$. But (3) and (4) with $i=r-2$ plus (7) imply via (22.6) of Theorem 22.1 that

$$\sum_{\alpha<\eta}\rho_\alpha\nrightarrow((\lambda_\xi-r+3)_{\xi<\vartheta}, 4)^3$$

holds, and in case $\vartheta \geq \omega$ this contradicts (5), since we then clearly have

$$\langle(\lambda_\xi - r+3)_{\xi<\vartheta}, 4\rangle \ll \langle \lambda_\xi - r+3 \colon \xi<\vartheta\rangle.$$

The proof is complete.

The following lemma gives the analogous result for condition (iii):

LEMMA 39.3. *Let $r \geq 3$, $\kappa \geq \omega$, $\vartheta \geq 2$, $\lambda_\xi > r$ for every $\xi<\vartheta$, and let i be an integer with $0<i\leq r-2$. Condition* (iii) *of Theorem* 38.5 *holds except if either $i=r-3$ and*

$$\begin{array}{l} \vartheta<\omega, \quad \omega \leq \operatorname{cf}(\lambda_\mu)<\lambda_\mu \quad \textit{for some} \quad \mu<\vartheta, \\ \textit{and} \quad \lambda_\xi<\omega \quad \textit{for any} \quad \xi<\vartheta \quad \textit{different from} \quad \mu, \end{array} \tag{8}$$

or $i=r-2$ and
we have

$$(*) \quad \langle(\lambda_\xi - r+3)_{\xi<\vartheta}, (\lambda_\xi - r+3)_{\xi<\vartheta}, 4, 4\rangle \not\ll \langle \lambda_\xi - r+3 \colon \xi<\vartheta\rangle; \tag{9}$$

moreover, $\lambda_\xi \geq \omega$ for some $\xi<\omega$; and either all $\lambda_\xi \geq \omega$ are singular, or there is exactly one $\mu<\vartheta$ with $\lambda_\mu \geq \omega$, and if this λ_μ is regular then $\vartheta<\omega$.

PROOF. Let $\rho \geq \omega$, $j>r-i$, and assume

$$\rho \nrightarrow (\lambda_\xi - r+j-1)^{j-1}_{\xi<\vartheta}. \tag{10}$$

We shall prove that then

$$2^\rho \nrightarrow (\lambda_\xi - r+j)^j_{\xi<\vartheta}$$

holds provided that the exceptions in the lemma fail (the additional assumption $j\leq r$ in (iii) of Theorem 38.5 will not be used). Assume that, on the contrary, we have

$$2^\rho \rightarrow (\lambda_\xi - r+j)^j_{\xi<\vartheta}. \tag{11}$$

By Lemma 39.1, this implies that $\exp_{j-1}(|\vartheta|)<2^\rho$, i.e., that

$$\exp_{j-2}(|\vartheta|)<\rho. \tag{12}$$

We may assume that

$$\lambda_0 \geq \omega; \tag{13}$$

in fact, if $\lambda_\xi<\omega$ for all $\xi<\vartheta$, then (10) fails in view of the relation

$$\rho \rightarrow (\omega)^{j-1}_{\xi<\vartheta},$$

which holds by Ramsey's theorem (Theorem 10.2) in case $\vartheta<\omega$, and by (12) and (17.32) in case $\vartheta \geq \omega$.

If $j \geq 5$, then (10) and (13) imply via (24.2) with assumption d) in Lemma 24.1 that (11) is false. Hence we must have $j \leq 4$. This means that $r-i$ equals 2 or 3, since $2 \leq r-i<j$. We consider two cases: 1) $r-i=3$ and 2) $r-i=2$.

Case 1). We shall prove that (8) holds in this case. Instead of assuming $r-i=3$ we shall only make the weaker assumption $j=4$. If λ_0, which is infinite according to (13), is regular, then (10) implies that (11) fails in view of (24.2) with assumption b) in Lemma 24.1; this is a contradiction, and so λ_0 must be singular. If there was a ξ with $1 \leq \xi < \vartheta$ such that $\lambda_\xi \geq \omega$, then we would obtain a contradiction again, as (24.2) with assumption c) in Lemma 24.1 entails that (10) and (11) cannot simultaneously hold. We have yet to prove that $\vartheta < \omega$ holds. Now (10) implies via (24.3) in Lemma 24.1 that

$$2^\rho \not\to ((\lambda_\xi - r + j)_{\xi<\vartheta}, (j+1)_m)^j$$

holds, where m is finite. If $\vartheta \geq \omega$ then this contradicts (11), since clearly

$$\langle(\lambda_\xi - r + j)_{\xi<\vartheta}, (j+1)_m\rangle \ll \langle\lambda_\xi - r + j \colon \xi < \vartheta\rangle$$

holds in this case. This shows that (8) is valid in case $j=4$, and so a fortiori in case $i=r-3$.

Case 2). We have to show that (9) holds in this case. As $r-i=2$ now, we have $j=3$ or 4. We saw above that (8) is valid in case $j=4$. Since (8) clearly implies (9), we may assume that $j=3$ holds. (10) implies

$$2^\rho \not\to ((\lambda_\xi - r + 3)_{\xi<\vartheta}, (\lambda_\xi - r + 3)_{\xi<\vartheta}, 4, 4)^3$$

via (24.5) in Lemma 24.1; this contradicts (11) unless inequality (*) in (9) holds. If there are two distinct $\xi, \xi' < \vartheta$ such that $\lambda_\xi, \lambda_{\xi'} \geq \omega$ and λ_ξ is regular, then (10) and (11) again contradict each other in view of (24.2) with assumption a) in Lemma 24.1. We have yet to prove that $\vartheta < \omega$ holds in case λ_0 is regular (cf. (13)). Now (10) implies

$$2^\rho \not\to ((\lambda_\xi - r + 3)_{\xi<\vartheta}, 4)^3$$

in this case in view of (24.4) in Lemma 24.1. In case $\vartheta \geq \omega$ this contradicts (11), since clearly

$$\langle(\lambda_\xi - r + 3)_{\xi<\vartheta}, 4\rangle \ll \langle\lambda_\xi - r + 3 \colon \xi < \vartheta\rangle$$

holds then. The proof of the lemma is complete.

40. CONSEQUENCES OF THE REDUCTION THEOREM

The lemmas of the preceding section supply us with an explicit way of checking whether Theorem 38.5 is applicable, i.e., whether its conditions (ii) and (iii) are satisfied. With this tool in our possession, we can proceed to establish the analogue of Lemma 35.2. Throughout our discussion we shall assume that

$$\kappa \geq \omega, \quad r \geq 3, \quad \vartheta \geq 2, \quad \text{and} \quad \lambda_\xi > r \quad \text{for every} \quad \xi < \vartheta. \tag{1}$$

First we establish a simple corollary of the results of the preceding sections that will be needed in the next two lemmas.

Corollary 40.1. *Assume* (1). *Let* i *be an integer with* $0 \leq i \leq r-2$, *and suppose that*

$$L_3^i(\kappa) \nrightarrow (\lambda_\xi - i)_{\xi<\vartheta}^{r-i}. \tag{2}$$

Then

$$\kappa \nrightarrow (\lambda_\xi)_{\xi<\vartheta}^r \tag{3}$$

holds except if $i = r-3$ *and* (39.8) *holds or* $i = r-2$ *and* (39.9) *holds.*

Proof. If $i=0$, then (2) and (3) are identical; so we may assume that $i>0$. Lemmas 39.2 and 39.3 imply that (ii) and (iii) in Theorem 38.5 hold (to see this, observe that (39.2) implies (39.9)). (i) a) in Theorem 38.5 fails in view of (2), since $\exp_{i-1}(2^\rho) < \kappa$ there implies $\rho \leq L_3^i(\kappa)$ by (38.5). So does (i) b) in that theorem, as (38.22) implies $L_3^i(\kappa) = L_3^{i-1}(\kappa)$ via Theorem 29.1, and so (38.22) contradicts (2). Hence this theorem implies (3), which we wanted to prove.

Using this, we can get an orientation as to when we can expect a positive arrow relation:

Lemma 40.2. *Assume* (1). *If*

$$\kappa \rightarrow (\lambda_\xi)_{\xi<\vartheta}^r,$$

then $\lambda_\xi \leq L_3^{r-2}(\kappa)$ *for all* $\xi < \vartheta$.

Proof. It is enough to show that

$$\kappa \rightarrow ((L_3^{r-2}(\kappa)), r+1)^r. \tag{4}$$

We have

$$L_3^{r-3}(\kappa) \nrightarrow ((L_3^{r-2}(\kappa))^+, 4)^3 \tag{5}$$

in view of Corollary 25.2, as $2^{L_3^{r-2}(\kappa)} \geq L_3^{r-3}(\kappa)$ holds. Using the preceding corollary with $i = r-3$, we can see that (5) implies (4), since the sequence $\langle (L_3^{r-2}(\kappa))^+, r+1 \rangle$ does not satisfy (39.8). This completes the proof.

Next we give an analogue of Lemma 35.2, which will be derived from that lemma with the aid of Corollary 40.1. This will not cover all the cases in which we know that the relation $\kappa \nrightarrow (\lambda_\xi)_{\xi<\vartheta}^r$ holds; the remaining cases, which involve some special assumptions about the iterated logarithms of κ, will be dealt with later, at the end of this section.

LEMMA 40.3. *Assume (1). Write* $\kappa_0 = L_3^{r-2}(\kappa)$. *The relation*

$$\kappa \nrightarrow (\lambda_\xi)^r_{\xi<\vartheta} \tag{6}$$

holds provided at least one of the following conditions hold:

(i) $\lambda_0, \lambda_1 > L_{\lambda_0}(\kappa_0)$;

(ii) $\lambda_0 \geq \kappa_0$ *and* $\lambda_1 > \min\{L_{\kappa_0}(\kappa_0), L_{\mathrm{cf}(\kappa_0)}(\mathrm{cf}(\kappa_0))\}$;

(iii) $\lambda_0 \geq \kappa_0$, $\lambda_1 \geq \mathrm{cf}(\kappa)$, *and* $\mathrm{cf}(\kappa_0) \nrightarrow (\mathrm{cf}(\kappa_0))^2_2$;

(iv) $\vartheta \geq L_3(\kappa_0)$;

(v) $|\{\xi<\vartheta: \lambda_\xi \geq \omega\}| \geq L_{\lambda_0}(\kappa_0)$;

(vi) $\lambda_0 \geq \kappa_0$ *and* $\vartheta \geq L_3(\mathrm{cf}(\kappa_0))$;

(vii) $\lambda_0 \geq \kappa_0$ *and* $|\{\xi<\vartheta: \lambda_\xi \geq \omega\}| \geq L_{\mathrm{cf}(\kappa_0)}(\mathrm{cf}(\kappa_0))$;

(viii) $\vartheta \geq \mathrm{cf}(L_3(\kappa_0))$ *and there are cardinals* $\kappa_\xi < L_3(\kappa_0)$ *such that their sum is* $L_3(\kappa_0)$ *and* $\lambda_\xi \geq \kappa_\xi$ *for every* $\xi < \mathrm{cf}(L_3(\kappa_0))$;

(ix) $\kappa_0 = 2^{\underset{\smile}{L_3(\kappa_0)}} (= 2^{\underset{\smile}{L_3^{r-1}(\kappa)}})$ *and at least one of the following conditions holds:*

a) $\lambda_0 > L_3(\kappa_0)$ *and* $\lambda_1 \geq L_3(\kappa_0)$;

b) $\lambda_0, \lambda_1 \geq L_3(\kappa_0)$ *and* $\vartheta \geq L_3(\mathrm{cf}(L_3(\kappa_0)))$;

c*) $\lambda_0, \lambda_1 \geq L_3(\kappa_0)$, $\lambda_2 \geq \omega$, *and* $\mathrm{cf}(L_3(\kappa_0)) \nrightarrow (\mathrm{cf}(L_3(\kappa_0)))^2_2$;

c**) $\lambda_0, \lambda_1 \geq L_3(\kappa_0)$, $L_3(\kappa_0) \nrightarrow (L_3(\kappa_0))^2_2$, *and either* $\mathrm{cf}(L_3(\kappa_0)) = L_3(\kappa_0)$ *or* $L_3(\kappa_0) = \kappa_0$;

d) $\lambda_0, \lambda_1 \geq L_3(\kappa_0)$ *and* $\lambda_2 > \mathrm{cf}(L_3(\kappa_0))$.

As in Lemma 35.2, (vii) can be strengthened if GCH holds, though we cannot quite get here the analogue of (vii′) in that lemma, mentioned immediately after (vii). This strengthening will be given in Lemma 40.8 below. We cannot strengthen (v) even under the assumption of GCH. The closest we can get to (v′) (mentioned after (vii)) in Lemma 35.2 is (iv). (ix) c in Lemma 35.2 has no exact analogue; (ix) c* and c** here cover somewhat less ground. A typical problem that remains open is the following:

PROBLEM 40.4. Assume $2^{\underset{\smile}{\aleph_{\omega_1}}} > \aleph_{\omega_1}$ and $2^{2^{\underset{\smile}{\aleph_{\omega_1}}}} > 2^{2^{\aleph_\alpha}}$ for every $\alpha < \omega_1$. Does then the relation

$$2^{2^{\underset{\smile}{\aleph_{\omega_1}}}} \nrightarrow (\aleph_{\omega_1}, \aleph_{\omega_1})^3 \tag{7}$$

hold?

As S. Shelah pointed out, the answer to a similar problem with $\aleph_\omega$ replacing $\aleph_{\omega_1}$ easily follows from his result Corollary 27.4. That is, we have the following: if $2^{\aleph_\omega} > \aleph_\omega$ and $2^{2^{\aleph_\omega}} > 2^{2^{\aleph_n}}$ for every $n < \omega$, then

$$2^{2^{\aleph_\omega}} \to (\aleph_\omega, \aleph_\omega)^3 \tag{8}$$

holds. In fact, we have $2^{\aleph_\omega} \to (\aleph_\omega, \aleph_\omega)^2$ according to the result of Shelah just mentioned, and so

$$(2^{2^{\aleph_\omega}})^+ \to (\aleph_\omega, \aleph_\omega)^3$$

follows by Corollary 38.6. As we have

$$2^{2^{\aleph_\omega}} = \sum_{n<\omega} 2^{2^{\aleph_n}} < \prod_{n<\omega} 2^{2^{\aleph_n}} = 2^{2^{\aleph_\omega}}$$

here by König's theorem (cf. Corollary 6.2), the above relation implies (8). The essential difference between (7) and (8) is that $\mathrm{cf}\,(L_3^2(2^{2^{\aleph_\omega}})) = \aleph_0$ and $\mathrm{cf}\,(L_3^2(2^{2^{\aleph_1}}) = \aleph_1$, and while $\aleph_0 \to (\aleph_0)_2^2$ holds, we have $\aleph_1 \nrightarrow (\aleph_1)_2^2$ (cf. Ramsey's theorem and Theorem 29.1). Note that we have

$$2^{2^{\aleph_\omega}} \nrightarrow (\aleph_\omega, \aleph_\omega, (4)_\omega)^3$$

and

$$2^{2^{\aleph_{\omega_1}}} \nrightarrow (\aleph_{\omega_1}, \aleph_{\omega_1}, (4)_{\omega_1})^3$$

by (ix) b in Lemma 40.3 and

$$2^{2^{\aleph_{\omega_1}}} \nrightarrow (\aleph_{\omega_1}, \aleph_{\omega_1}, \aleph_0)^3$$

by (ix) c* in that lemma under the given conditions, while (ix) c** is not applicable here.

PROOF OF LEMMA 40.3. We usually proceed by combining Lemma 35.2 with Corollary 40.1, although occasionally we shall use other results as well. That is, where the Negative Stepping-up Lemma (or, rather, its consequence as formulated in Corollary 40.1) does not suffice, we shall use what might be called the Pseudo Negative Stepping-up Lemma, i.e., Theorem 25.6.

Ad (i)–(v) *and* (vii). We have to prove the relations

(i′) $\kappa \nrightarrow (\lambda_0, (L_{\lambda_0}(\kappa_0)^+)^r$ if $\lambda_0 > L_{\lambda_0}(\kappa_0)$;

(ii′) $\kappa \nrightarrow (\kappa_0, (\min\{L_{\kappa_0}(\kappa_0), L_{\mathrm{cf}(\kappa_0)}(\mathrm{cf}(\kappa_0))\})^+)^r$

(iii′) $\kappa \nrightarrow (\kappa_0, (\mathrm{cf}(\kappa_0))^+)^r$ if $\mathrm{cf}(\kappa_0) \nrightarrow (\mathrm{cf}(\kappa_0))_2^2$

(iv′) $\kappa \nrightarrow (r+1)^r_{L_3(\kappa_0)}$;

(v′) $\kappa\nrightarrow(\lambda_0,\ (\omega)_{L_{\lambda_0}(\kappa_0)})^r$;

(vii′) $\kappa\nrightarrow(\kappa_0,\ (\omega)_{L_{\mathrm{cf}(\kappa_0)}(\mathrm{cf}(\kappa_0))})^r$.

According to (i)–(v) and (vii), respectively, in Lemma 35.2 with κ_0 replacing κ, we have the following analogous relations (these can be formally obtained from the above relations by replacing κ by κ_0 and r by 2):

(i″) $\kappa_0\nrightarrow(\lambda_0, (L_{\lambda_0}(\kappa_0))^+)^2$ if $\lambda_0>L_{\lambda_0}(\kappa_0)$;

(ii″) $\kappa_0\nrightarrow(\kappa_0, (\min\{L_{\kappa_0}(\kappa_0), L_{\mathrm{cf}(\kappa_0)}(\mathrm{cf}(\kappa_0))\})^+)^2$;

(iii″) $\kappa_0\nrightarrow(\kappa_0, (\mathrm{cf}(\kappa_0))^+)^2$ if $\mathrm{cf}(\kappa_0)\nrightarrow(\mathrm{cf}(\kappa_0))^2_2$;

(iv″) $\kappa_0\nrightarrow(3)^2_{L_3(\kappa_0)}$;

(v″) $\kappa_0\nrightarrow(\lambda_0,\ (\omega)_{L_{\lambda_0}(\kappa_0)})^2$;

(vii″) $\kappa_0\nrightarrow(\kappa_0,\ (\omega)_{L_{\mathrm{cf}(\kappa_0)}(\mathrm{cf}(\kappa_0))})^2$.

It is easy to check that each sequence on the right-hand side here violates (39.9). Hence relations (i″)–(v″) and (vii″) imply (i′)–(v′) and (vii′), respectively, via Corollary 40.1 with $i=r-2$.

Ad (vi). It is sufficient to show that

(vi′) $\kappa\nrightarrow(\kappa_0,\ (r+1)_{L_3(\mathrm{cf}(\kappa_0))})^r$

holds. The trouble here is that the sequence on the right-hand side here might satisfy (39.9) and so (vi) in Lemma 35.2 is not applicable. But noting that it violates (39.8), we can use Corollary 40.1 with $i=r-3$, provided we can show that

$$L_3^{r-3}(\kappa)\nrightarrow(\kappa_0,\ (4)_{L_3(\mathrm{cf}(\kappa_0))})^3 \tag{9}$$

holds. We have

$$2^{\kappa_0}\nrightarrow(\kappa_0,\ (4)_{L_3(\mathrm{cf}(\kappa_0))})^3$$

according to Corollary 25.7. As $2^{\kappa_0}=2^{L_3^{r-2}(\kappa)}\geq L_3^{r-3}(\kappa)$, the relation in (9) follows from here. Hence (vi) is established.

Ad (viii). If $L_3(\kappa_0)$ is regular, then the result follows from (iv), which has already been established. Hence we may assume that $L_3(\kappa_0)$ is singular. In this case it means no restriction of generality to assume that the κ_ξ's are regular, and then we have to prove that

$$\kappa\nrightarrow(\kappa_\xi)^r_{\xi<\mathrm{cf}(L_3(\kappa_0))}$$

holds. We have

$$\kappa_0\nrightarrow(\kappa_\xi)^2_{\xi<\mathrm{cf}(L_3(\kappa_0))}$$

according to (viii) in Lemma 35.2 with κ_0 replacing κ. As the sequence on the right-hand side violates (39.9) in view of the regularity of the κ_ξ's, this relation implies the former one via Corollary 40.1 with $i=r-2$.

Ad (ix) a, c*, *and* d. It is sufficient to show that

a′) $\kappa \nrightarrow ((L_3(\kappa))^+, L_3(\kappa_0))^r$;

c*′) $\kappa \nrightarrow (L_3(\kappa_0), L_3(\kappa_0), \omega)^r$ if $\operatorname{cf}(L_3(\kappa_0)) \nrightarrow (\operatorname{cf}(L_3(\kappa_0)))^2_2$;

and

d′) $\kappa \nrightarrow (L_3(\kappa_0), L_3(\kappa_0), \operatorname{cf}(L_3(\kappa_0)))^r$.

According to (ix)a, c, and d, respectively, in Lemma 35.2 with κ_0 replacing κ, we have

a″) $\kappa_0 \nrightarrow ((L_3(\kappa_0))^+, L_3(\kappa_0))^2$;

c″) $\kappa_0 \nrightarrow (L_3(\kappa_0), L_3(\kappa_0))^2$ if $\operatorname{cf}(L_3(\kappa_0)) \nrightarrow (\operatorname{cf}(L_3(\kappa_0)))^2_2$;

d″) $\kappa_0 \nrightarrow (L_3(\kappa_0), L_3(\kappa_0), \operatorname{cf}(L_3(\kappa_0)))^2$.

The sequences on the right-hand sides of a″) and d″) violate (39.9). This is not necessarily true in the case of c″), but it is true for its following weakened version:

c*″) $\kappa_0 \nrightarrow (L_3(\kappa_0), L_3(\kappa_0), \omega)^2$ if $\operatorname{cf}(L_3(\kappa_0)) \nrightarrow (\operatorname{cf}(L_3(\kappa_0)))^2_2$.

a″), c*″), and d″) imply a′), c*′), and d′), respectively, via Corollary 40.1 with $i=r-2$.

Ad (ix) c**. We have to show that

c**′) $\kappa \nrightarrow (L_3(\kappa_0), L_3(\kappa_0))^r$

holds, provided $L_3(\kappa_0) \nrightarrow (L_3(\kappa_0))^2_2$ and either $L_3(\kappa_0)$ is regular or *l* n the former case, the sequence on the right-hand side violates (39.9), and so the result follows from c″) above (in the proof of (ix)a, c*, and d) via Corollary 40.1 with $i=r-2$. So we may suppose that $\operatorname{cf}(\kappa_0) < \kappa_0 = L_3(\kappa_0)$, i.e., that κ_0 is a singular strong limit cardinal. We are then going to prove the stronger version

$$\kappa \nrightarrow ((\operatorname{cf}(\kappa_0))^+, \kappa_0)^r$$

instead of c**′). The right-hand side here violates (39.9), and so it is sufficient to prove

$$\kappa_0 \nrightarrow ((\operatorname{cf}(\kappa_0))^+, \kappa_0)^2$$

according to Corollary 40.1 with $i=r-2$. This is, however, confirmed by Corollary 21.3.

Ad (ix) b. We have to prove that

b′) $\kappa \nrightarrow (L_3(\kappa_0), L_3(\kappa_0), (r+1)_{L_3(\operatorname{cf}(L_3(\kappa_0)))})^r$

holds. As the sequence on the right-hand side violates (39.8), it is sufficient to show according to Corollary 40.1 with $i=r-3$ that

$$L_3^{r-3}(\kappa) \nrightarrow (L_3(\kappa_0), L_3(\kappa_0), (4)_{L_3(\mathrm{cf}\,(L_3(\kappa_0)))})^3 \tag{10}$$

holds. We are going to use Corollary 25.9 with $\lambda = L_3(\kappa_0)$ replacing κ. This says that

$$2^{2^{\underset{\smile}{\lambda}}} \nrightarrow (\lambda, \lambda, (4)_\rho)^3 \tag{11}$$

holds provided (α) $2^\rho \geq \mathrm{cf}\,(\lambda)$ and (β) $2^\sigma < 2^{\underset{\smile}{\lambda}}$ for any cardinal $\sigma < \lambda$. Clearly, (α) holds with

$$\rho = L_3(\mathrm{cf}\,(\lambda)) = L_3\,(\mathrm{cf}\,(L_3\,(\kappa_0)))\,,$$

and (β) holds in view of the definition of λ, since $2^{\underset{\smile}{L_3(\kappa_0)}} = \kappa_0$ holds by our assumptions in (ix) of the lemma to be proved. Hence (11) is valid with the above value of ρ; and since we have

$$2^{2^{\underset{\smile}{\lambda}}} = 2^{2^{\underset{\smile}{L_3(\kappa_0)}}} = 2^{\kappa_0} = 2^{L_3^{r-2}(\kappa)} \geq L_3^{r-3}(\kappa)\,,$$

the relation in (10) is also valid, and this is what we wanted to show. The proof of the lemma is complete.

So far we made use only of Corollary 40.1 instead of the stronger Theorem 38.5. The following result strengthens Lemma 40.2 provided certain special assumptions on the iterated logarithms of κ are satisfied.

LEMMA 40.5. *Assume* (1). *Write $L^i = L_3^i(\kappa)$ for any integeri. Assume, further, that*

$$2^{\underset{\smile}{L^{r-2}}} = L^{r-3} \qquad \textit{and} \qquad L^{r-2} \nrightarrow (L^{r-2})_2^2\,. \tag{12}$$

Suppose that $\lambda_0 \geq L^{r-2}$. Then

$$\kappa \nrightarrow (\lambda_\xi)_{\xi<\vartheta}^r \tag{13}$$

holds provided at least one of the following conditions is satisfied:

(i) $\lambda_1 \geq r+2\,;$

(ii) $L^{r-2} = L^{r-3}\,;$

(iii) $\mathrm{cf}\,(L^{r-2}) = L^{r-2}\,;$

(iv) $r=3$ *and* $\mathrm{cf}\,(L^{r-2}) \nrightarrow (\mathrm{cf}\,(L^{r-2}))_2^2\,;$

(v) $\vartheta \geq 3\,.$

The ideal result would be obtained if we could prove (13) without any of the assumptions (i)–(v); the relevant open problems will be mentioned after the proof. The rationale behind the assumption in (12) is that, as the reader can easily check, it guarantees that, in case $i=r-2$, (2) in Corollary 40.1 is an essentially stronger requirement than the failure of (i) in Theorem 38.5; hence we can expect to prove more by applying this latter. We shall nonetheless get along by using only the former result and not the latter, plus some special tricks.

PROOF. It is sufficient to prove that, under the assumption of (12), we have

(i′) $\kappa \nrightarrow (L^{r-2}, r+2)^r$;

(ii′) $\kappa \nrightarrow (L^{r-2}, r+1)^r$ if $L^{r-2}=L^{r-3}$.

(iii′) $\kappa \nrightarrow (L^{r-2}, r+1)^r$ if $\mathrm{cf}\,(L^{r-2})=L^{r-2}$;

(iv′) $\kappa \nrightarrow (L^1, 4)^3$ if $\mathrm{cf}\,(L^1) \nrightarrow (\mathrm{cf}\,(L^1))^2_2$ $(r=3)$;

(v′) $\kappa \nrightarrow (L^{r-2}, r+1, r+1)^r$.

To this end, we first show that

$$\kappa \nrightarrow (L^{r-2})^{r-1}_3 \tag{14}$$

and

$$\kappa \nrightarrow (L^{r-2})^{r-1}_2 \quad \text{provided either} \quad L^{r-2}=L^{r-3} \quad \text{or} \quad \mathrm{cf}\,(L^{r-2})=L^{r-2} \quad \text{or} \quad \mathrm{cf}\,(L^1) \nrightarrow (\mathrm{cf}\,(L^1))^2_2 \quad \text{and} \quad r=3. \tag{15}$$

To see these, observe that we have

$$\kappa \nrightarrow (L^{r-2}, L^{r-2}, (\mathrm{cf}\,(L^{r-2}))^+)^{r-1}$$

by (ix)d in Lemma 35.2 or 40.3 according as $r=3$ or $r>3$. This proves (14) in case L^{r-2} is singular. Assume therefore that L^{r-2} is regular. Then $\mathrm{cf}\,(L^{r-2}) \nrightarrow (\mathrm{cf}\,(L^{r-2}))^2_2$ in view of (12), and so we even obtain (15) (which is clearly stronger than (14)) in this case by (ix) c in Lemma 35.2 or (ix) c** in Lemma 40.3 according as $r=3$ or $r>3$. It remains to see that (15) holds if either $r=3$ and $\mathrm{cf}(L^1) \nrightarrow (\mathrm{cf}(L^1))^2_2$ or $L^{r-2}=L^{r-3}$, but this follows from (ix) c in Lemma 35.2 and (ix) c** in Lemma 40.3, respectively (note that in the latter case there is nothing to prove if $r=3$, since then we have $L^{r-2}=L^{r-3}=L^0=\kappa$, and so (15) simply means that $L^{r-2} \nrightarrow (L^{r-2})^2_2$, which is assumed in (12)). So (14) and (15) are established.

Using (25.3) in Theorem 25.1 with $\lambda=L^{r-2}$, $\tau=3$ or 2, and with $r-1$ replacing r, we obtain that

$$\kappa \nrightarrow (L^{r-2}, r+2)^r \quad \text{if (14) holds, and}$$

$$\kappa \nrightarrow (L^{r-2}, r+1)^r \quad \text{if (15) holds.}$$

Thus (i′)–(iv′) are satisfied. (v′) follows from (14) by (25.2) in Theorem 25.1 with $\lambda = L^{r-2}$, $\tau = 3$, and with $r-1$ replacing r. The proof is complete.

We now formulate two typical problems that remain open:

PROBLEM 40.6. (i) Assume that $2^{\aleph_\omega} > \aleph_\omega$ and $2^{2^{\aleph_\omega}} > 2^{2^{\aleph_n}}$ for all $n < \omega$. Is it true that

$$\text{(a)}\quad 2^{\aleph_\omega} \to (\aleph_\omega, 4)^3 \quad \text{or (b)} \quad 2^{2^{\aleph_\omega}} \to (\aleph_\omega, 5)^4\ ? \tag{16}$$

(ii) Assume $2^{\aleph_{\omega_1}} > \aleph_{\omega_1}$ and $2^{2^{\aleph_{\omega_1}}} > 2^{2^{\aleph_\alpha}}$ for all $\alpha < \omega_1$. Does then

$$2^{2^{\aleph_{\omega_1}}} \nrightarrow (\aleph_{\omega_1}, 5)^4 \tag{17}$$

hold?

For an appreciation of the difference between (i) and (ii) here see the remarks after Problem 40.4. The question whether (16)(a) holds was already asked in Problem 22.2. Note also that Problems 40.4 and 40.6 are closely related. In fact, (7) implies (17) in view of (25.3) in Theorem 25.1. By Lemma 40.5 we know only that the following relations hold under the assumptions of (i) and (ii), respectively:

$$2^{\aleph_\omega} \nrightarrow (\aleph_\omega, 5)^3;\ 2^{\aleph_\omega} \nrightarrow (\aleph_\omega, 4, 4)^3;\ 2^{\aleph_{\omega_1}} \nrightarrow (\aleph_{\omega_1}, 4)^3$$

$$2^{2^{\aleph_\omega}} \nrightarrow (\aleph_\omega, 6)^4;\ 2^{2^{\aleph_\omega}} \nrightarrow (\aleph_\omega, 5, 5)^4$$

$$2^{2^{\aleph_{\omega_1}}} \nrightarrow (\aleph_{\omega_1}, 6)^4,\ 2^{2^{\aleph_{\omega_1}}} \nrightarrow (\aleph_{\omega_1}, 5, 5)^4 .$$

(17) remains an open problem despite the relation $2^{\aleph_{\omega_1}} \nrightarrow (\aleph_{\omega_1}, 4)^3$, as the sequence $\langle \aleph_{\omega_1}, 5 \rangle$ satisfies (39.8), and so Lemma 40.1 with $i = r-3$ is not applicable.

The next result answers a slightly simpler question than that asked in Problem 40.4:

LEMMA 40.7. *Assume* (1). *Write* $L^i = L_3^i(\kappa)$ *for any integer* i. *Assume, further, that*

$$2^{L^{r-2}} = L^{r-3} \quad \textit{and} \quad L^{r-2} \nrightarrow (L^{r-2})_2^2, \tag{18$\equiv$(12)}$$

and

$$2^{L^{r-1}} = L^{r-2} . \tag{19}$$

If $\lambda_0, \lambda_1 \geq L^{r-1}$ *then*

$$\kappa \nrightarrow (\lambda_\xi)^r_{\xi < \vartheta}$$

holds.

This says, in particular, that under the assumptions of Problem 40.4 we have

$$2^{2^{\aleph_{\omega_1}}} \nrightarrow (\aleph_{\omega_1}, \aleph_{\omega_1})^3$$

and under the assumptions of (8) we have

$$2^{2^{\aleph_\omega}} \nrightarrow (\aleph_\omega, \aleph_\omega)^3 .$$

PROOF. As the sequence $\langle L^{r-1}, L^{r-1}\rangle$ violates (39.8), it is enough to prove

$$L^{r-3} \nrightarrow (L^{r-1}, L^{r-1})^3 \tag{20}$$

in view of Corollary 40.1 with $i=r-3$. Let $\langle \rho_\alpha : \alpha < \mathrm{cf}\,(L^{r-1})\rangle$ be an increasing sequence of cardinals tending to L^{r-1}. Then, clearly, we have

$$L^{r-3} = \sum_{\alpha < \mathrm{cf}\,(L^{r-1})} 2^{2^{\rho_\alpha}} \tag{21}$$

by (19) and the first relation in (18). We have $2^{\rho_\alpha} \nrightarrow (\rho_\alpha^+)^2_2$ by (19.14), and so $2^{2^{\rho_\alpha}} \nrightarrow (\rho_\alpha^+)^3_2$ follows by (24.2) with assumption a) in Lemma 24.1; hence we have a fortiori

$$2^{2^{\rho_\alpha}} \nrightarrow (L^{r-1}, L^{r-1})^3 \tag{22}$$

for any $\alpha < \mathrm{cf}\,(L^{r-1})$. We claim that we also have

$$\mathrm{cf}\,(L^{r-1}) \nrightarrow (L^{r-1}, L^{r-1})^3 . \tag{23}$$

In fact, in the contrary case we must have $\mathrm{cf}\,(L^{r-1}) = L^{r-1}$, and we would have a fortiori

$$L^{r-1} \to (L^{r-1}, L^{r-1})^2 .$$

As $L^{r-2} = 2^{L^{r-1}_{\smile}} = L^{r-1}$ in this case by (19) and Theorem 29.1, this contradicts the second relation in (18). Thus (23) is established. Now (21), (22), and (23) imply (20) via (22.5) in Theorem 22.1. The proof is complete.

We now turn to the strengthenings of (vii) in Lemma 40.3 in case GCH holds (cf. (vii′), mentioned immediately after (vii), in Lemma 35.2).

LEMMA 40.8. *Assume* GCH. *Assume* (1), *and write* $\kappa_0 = L_3^{r-2}(\kappa)$. *Suppose, further, that*

$$\lambda_0 = \kappa_0 \qquad \textit{and} \qquad \vartheta \geq L_{\mathrm{cf}\,(\kappa_0)}(\mathrm{cf}\,(\kappa_0)) . \tag{24}$$

Then

$$\kappa \nrightarrow (\lambda_\xi)^r_{\xi<\vartheta}$$

holds provided either (i) κ_0 *is regular or* (ii) $\lambda_1 \geq \omega$.

PROOF. We have to prove the following relations:

$$\kappa \nrightarrow (\kappa_0, (r+1)_{L_{\mathrm{cf}(\kappa_0)}(\mathrm{cf}(\kappa_0))})^r \text{ if } \mathrm{cf}\,(\kappa_0) = \kappa_0$$

and

$$\kappa \nrightarrow (\kappa_0, \omega, (r+1)_{L_{\mathrm{cf}(\kappa_0)}(\mathrm{cf}\,(\kappa_0))})^r .$$

The sequences on the right-hand side here do not satisfy (39.9); hence it follows from Corollary 40.1 with $i = r-2$ that it is sufficient to prove that we have

$$\kappa_0 \nrightarrow (\kappa_0, (3)_{L_{\mathrm{cf}(\kappa_0)}(\mathrm{cf}(\kappa_0))})^2 \quad \text{if} \quad \mathrm{cf}(\kappa_0) = \kappa_0$$

and

$$\kappa_0 \nrightarrow (\kappa_0, \omega, (3)_{L_{\mathrm{cf}(\kappa_0)}(\mathrm{cf}(\kappa_0))})^2 .$$

Both of these relations hold by (vii′) (mentioned after (vii)) of Lemma 35.2. The proof is complete.

41. THE MAIN RESULT FOR THE CASE $r \geq 3$

The following rather lengthy theorem does not contain any new results. It is just a systematic recapitulation of results already proved.

THEOREM 41.1. *Assume $\kappa \geq \omega$, $\vartheta \geq 2$, $r \geq 3$, and $\lambda_\xi > r$ for $\xi < \vartheta$. Put $L^i = L_3^i(\kappa)$ for any integer i. Consider the following conditions*:

(○) $\quad 2^{\underset{\smile}{L^{r-2}}} < L^{r-2} \quad$ *or* $\quad L^{r-2} \to (L^{r-2})^2_2$,

(○○) $\quad 2^{\underset{\smile}{L^{r-2}}} = L^{r-3} \quad$ *and* $\quad L^{r-2} \nrightarrow (L^{r-2})^2_2$,

(+) $\quad 2^{\underset{\smile}{L^{r-1}}} < L^{r-2}$,

(++) $\quad 2^{\underset{\smile}{L^{r-1}}} = L^{r-2}$.

Then exactly one of (○) *and* (○○), *and exactly one of* (+) *and* (++) *holds.*

Consider the relation

(□) $$\kappa \to (\lambda_\xi)^r_{\xi < \vartheta} .$$

We have the following results:

(i) $\lambda_\xi \leq L^{r-2}$ *for all $\xi < \vartheta$ is always necessary for* (□).

(ii) *Each of the following conditions implies that* (□) *is false*:

(ii.1) $\langle \lambda, L_\lambda(L^{r-2})^+ \rangle \ll \langle \lambda_\xi : \xi < \vartheta \rangle$ *for some* $\lambda > L_\lambda(L^{r-2})$;

(ii.2) $\langle L^{r-2}, L_{\mathrm{cf}(L^{r-2})}(\mathrm{cf}(L^{r-2}))^+ \rangle \ll \langle \lambda_\xi : \xi < \vartheta \rangle$;

(ii.3) $\langle L^{r-2}, L_{L^{r-2}}(L^{r-2})^+ \rangle \ll \langle \lambda_\xi : \xi < \vartheta \rangle$;

(ii.4) $\langle L^{r-2}, \mathrm{cf}(L^{r-2}) \rangle \ll \langle \lambda_\xi : \xi < \vartheta \rangle$ *and* $\mathrm{cf}(L^{r-2}) \nrightarrow (\mathrm{cf}(L^{r-2}))^2_2$;

(ii.5) $\vartheta \geq L^{r-1}$, *i.e.*, $(r+1)_{L^{r-1}} \ll \langle \lambda_\xi : \xi < \vartheta \rangle$;

(ii.6) $\langle \lambda, (\omega)_{L_\lambda(L^{r-2})} \rangle \ll \langle \lambda_\xi : \xi < \vartheta \rangle$ *for some λ with $r < \lambda \leq L^{r-2}$*;

(ii.7) $\langle L^{r-2}, (r+1)_{L_3(\mathrm{cf}(L^{r-2}))}\rangle \ll \langle \lambda_\xi : \xi < \vartheta\rangle$;

(ii.8) $\langle L^{r-2}, (\omega)_{L_{\mathrm{cf}(L^{r-2})}(\mathrm{cf}(L^{r-2}))}\rangle \ll \langle \lambda_\xi : \xi < \vartheta\rangle$;

(ii.9) *There are* $\rho_\alpha < L^{r-1}$ *for* $\alpha < \mathrm{cf}(L^{r-1})$ *such that* $\sum_{\alpha<\mathrm{cf}(L^{r-1})} \rho_\alpha = L^{r-1}$ *and* $\langle \rho_\alpha : \alpha < \mathrm{cf}(L^{r-1})\rangle \ll \langle \lambda_\xi : \xi < \vartheta\rangle$.

$(++)$ *is assumed for* (ii.10)–(ii.15):

(ii.10) $\langle (L^{r-1})^+, L^{r-1}\rangle \ll \langle \lambda_\xi : \xi < \vartheta\rangle$;

(ii.11) $\langle L^{r-1}, L^{r-1}, \mathrm{cf}(L^{r-1})^+\rangle \ll \langle \lambda_\xi : \xi < \vartheta\rangle$;

(ii.12) $\langle L^{r-1}, L^{r-1}, \omega\rangle \ll \langle \lambda_\xi : \xi < \vartheta\rangle$ *and* $\mathrm{cf}(L^{r-1}) \nrightarrow (\mathrm{cf}(L^{r-1}))^2_2$;

(ii.13) $\langle L^{r-1}, L^{r-1}, (r+1)_{L_3(\mathrm{cf}(L^{r-1}))}\rangle \ll \langle \lambda_\xi : \xi < \vartheta\rangle$;

(ii.14) $\langle L^{r-1}, L^{r-1}\rangle \ll \langle \lambda_\xi : \xi < \vartheta\rangle$, $L^{r-1} \nrightarrow (L^{r-1})^2_2$, *and either* $\mathrm{cf}(L^{r-1}) = L^{r-1}$ *or* $L^{r-1} = L^{r-2}$;

(ii.15) $\langle L^{r-1}, L^{r-1}\rangle \ll \langle \lambda_\xi : \xi < \vartheta\rangle$ *and* (○○) *holds.*

GCH *is assumed for* (ii.16) *and* (ii.17):

(ii.16) $\langle L^{r-2}, \omega, (r+1)_{L_{\mathrm{cf}(L^{r-2})}(\mathrm{cf}(L^{r-2}))}\rangle \ll \langle \lambda_\xi : \xi < \vartheta\rangle$;

(ii.17) $\langle L^{r-2}, (r+1)_{L_{L^{r-2}}(L^{r-2})}\rangle \ll \langle \lambda_\xi : \xi < \vartheta\rangle$ *and* $\mathrm{cf}(L^{r-2}) = L^{r-2}$.

Put

(Exception 1) $\langle L^{r-1}, L^{r-1}\rangle \ll \langle \lambda_\xi : \xi < \vartheta\rangle \ll \langle L^{r-1}, L^{r-1}, (k'_\nu)_{\nu<\tau'}\rangle$ *for some* $\tau' < L_3(\mathrm{cf}(L^{r-1}))$ *and* $k'_\nu < \omega$ *for every* $\nu < \tau'$, *and, moreover,* $\mathrm{cf}(L^{r-1}) \nrightarrow (\mathrm{cf}(L^{r-1}))^2_2$, $\mathrm{cf}(L^{r-1}) < L^{r-1} < L^{r-2}$, *and* (○) *and* $(++)$ *hold;*

and consider the following condition

(∗) *There is a sequence* $\langle \lambda, (\rho)_\alpha, (\sigma)_\beta, (k_\nu)_{\nu<\tau}\rangle$ *satisfying conditions* (35.4)–(35.8) *of Lemma* 35.3 *with* L^{r-1} *replacing* κ *such that*

$$\langle \lambda_\xi : \xi < \vartheta\rangle \ll \langle \lambda, (\rho)_\alpha, (\sigma)_\beta, (k_\nu)_{\nu<\tau}\rangle .$$

(iii) (□) *implies* (∗) *with* $\lambda = \lambda_\xi$ *for some* $\xi < \vartheta$ *provided* (Exception 1) *is false. If* GCH *holds then* (□) *implies that* (∗) *holds with* $\lambda = L^{r-2}$ *and* $\tau = 0$ *provided there is a* $\mu < \vartheta$ *such that* $\lambda_\mu = L^{r-2}$ *and either* $\mathrm{cf}(\lambda_\mu) = \lambda_\mu$ *or there is a* $\nu < \vartheta$ *with* $\nu \neq \mu$ *and* $\lambda_\nu \geq \omega$.

(iv) *Assume* (○) *holds. Then* (∗) *is a sufficient condition for* (□) *provided* $\tau = 0$ *and at least one of the following conditions holds:*

(iv.1) *L^{r-2} is regular;*

(iv.2) $\lambda < L^{r-2}$;

(iv.3) *L^{r-2} is a strong limit cardinal;*

(iv.4) $\rho = \omega$.

As a corollary of these results we have

(v) *If* (o) *holds then* (□) *holds provided Theorem* 35.4 *confirms*

$$L^{r-2} \to (\lambda_\xi - r + 2)^2_{\xi < \vartheta},$$

and if (Exception 1) *is false then* (□) *fails provided Lemma* 35.2 *confirms*

$$L^{r-2} \nrightarrow (\lambda_\xi - r + 2)^2_{\xi < \vartheta}.$$

Write

(Exception 2) $\vartheta = 2$, $\{L^{r-2}, r+1\} = \{\lambda_0, \lambda_1\}$, $L^{r-2} < L^{r-3}$, $\mathrm{cf}(L^{r-2}) < L^{r-2}$, *and either* $r > 3$ *or* $\mathrm{cf}(L^{r-2}) \to (\mathrm{cf}(L^{r-2}))^2_2$.

We have

(vi) *If* (oo) *holds and* (Exception 2) *is false, then* $\lambda_\xi < L^{r-2}$ *for every* $\xi < \vartheta$ *is necessary for* (□).

(vii) *Assume* (oo) *and* (+) *hold. The following condition is necessary for* (□) *provided* (Exception 2) *is false*:

(**) $$\langle \lambda_\xi : \xi < \vartheta \rangle \ll \langle \lambda, (L_\lambda(L^{r-2}))_\alpha, (\sigma)_\beta, (k_\nu)_{\nu < \tau} \rangle,$$

where $\lambda < L^{r-2}$, $\alpha < \mathrm{cf}(L_\lambda(L^{r-2}))$, $\sigma, \beta < L_\lambda(L^{r-2})$, $\tau < L^{r-1}$, *and* $k_\nu < \omega$ *for every* $\nu < \tau$; *and if* $\lambda < L_\lambda(L^{r-2})$, *then* $\alpha = 0$.

Condition (**) *is sufficient provided either* $\tau = 0$ *or* $L_\lambda(L^{r-2}) = L^{r-1}$.

(viii) *If* (oo) *and* (+ +) *hold and* (Exception 2) *is false, then the following condition is necessary and sufficient for* (□):

(***) $\langle \lambda_\xi : \xi < \vartheta \rangle \ll \langle \lambda, (\sigma)_\beta \rangle$, *where* $\lambda < L^{r-2}$ *and* $\sigma, \beta < L^{r-1}$.

Remarks. (Exception 1) leads to Problem 40.4. Make the assumption now that (o) holds (which assumption in a sense covers the nondegenerate cases), and suppose also that (Exception 1) is false. Then (v) above says that the general problem can be reduced to the case $r = 2$ whenever we *know* the answer in this latter case. That is why this result, though formally incomparable with that given by the Reduction Theorem (Theorem 38.5), seems to be stronger and more effective than the latter.

(Exception 2) leads to Problem 40.6. We should like to point out that if GCH is assumed then (vii) is vacuous; in fact $2^{\underset{\smile}{L^{r-2}}} = L^{r-3}$ holds according to (oo), and so $L^{r-2} = L^{r-3}$ is a strong limit by GCH, i.e., $L^{r-1} = L_3(L^{r-2}) = L^{r-2}$, which contradicts (+). The problem left open in (vii) is not a genuinely new one, and it corresponds to the analogous problem for $r=2$, which leads to Problem 20.1.

The reason that in case (oo) and (+ +) hold we have a simple result according to (viii) is that this is the case if GCH is assumed and L^{r-2} is a strong limit cardinal with $L^{r-2} \not\to (L^{r-2})^2_2$; cf. (36.7) in Theorem 36.2.

Proof. To see that exactly one of (o) and (oo), and exactly one of (+) and (+ +) holds, one has to observe that $2^{\underset{\smile}{L^{r-2}}} \leq L^{r-3}$ and $2^{\underset{\smile}{L}} \leq L^{r-2}$, respectively, hold by the definition of the (iterated) logarithm, and the rest follows by a simple exercise in propositional calculus. (i) holds by Lemma 40.2; (ii) under conditions (ii.1), (ii.2) $\vee$ (ii.3), (ii.4)–(ii.9) holds in turn by (i), (ii), (iii)–(viii) in Lemma 40.3; (ii) holds with (ii.10)–(ii.14) by Lemma 40.9(ix)a, d, c*, b, and c**, respectively; (ii) holds with (ii.15) by Lemma 40.7, and with (ii.16) and (ii.17) by Lemma 40.8.

Ad (iii). Assume (□) holds and yet (*) fails with $\lambda = \lambda_\xi$ for any $\xi < \vartheta$. This means in view of Lemma 35.3 (cf. (35.10) especially) that at least one of conditions (i)–(ix) of Lemma 35.2 is fulfilled with a suitable rearrangement $\langle \lambda'_\xi : \xi < \vartheta' \rangle$ of the sequence $\langle \lambda_\xi : \xi < \vartheta \rangle$ and with L^{r-2} replacing κ. Note that these conditions are identical to conditions (i)–(ix) in Lemma 40.3, except that (ix)c is replaced by (ix)c* and c** in the latter. Yet none of the conditions (i)–(ix) in Lemma 40.3 holds for $\langle \lambda'_\xi : \xi < \vartheta' \rangle$, since the contrary would mean by this lemma that (□) fails. Hence (ix)c of Lemma 35.2 with L^{r-2} replacing κ must hold for $\langle \lambda'_\xi : \xi < \vartheta' \rangle$. That is, we have

$$\langle L^{r-1}, L^{r-1} \rangle \ll \langle \lambda_\xi : \xi < \vartheta \rangle, \quad (++) \text{ holds, and } \operatorname{cf}(L^{r-1}) \not\to \operatorname{cf}(L^{r-1})^2_2. \quad (1)$$

We may assume here that $\lambda_0, \lambda_1 \geq L^{r-1}$. As (□) holds, we also have $\lambda_0, \lambda_1 \leq L^{r-1}$ by (ii.10) in the theorem being proved; $\lambda_\xi < \omega$ for any ξ with $2 \leq \xi < \vartheta$ by (ii.12), $\vartheta < L_3(\operatorname{cf}(L^{r-1}))$ by (ii.13), and we have $\operatorname{cf}(L^{r-1}) < L^{r-1} < L^{r-2}$ by (ii.14); finally, (oo) fails by (ii.15), i.e., (o) is true. All these mean that (Exception 1) holds. This settles the first claim in (iii).

We now turn to the second claim, in which GCH is assumed. Assume also that (□) holds and (*) fails. Then (1) holds, as we saw just before; but then (□) must fail in view of (ii.14) in the theorem being proved, as GCH and (+ +), the latter of which holds by (1), imply that $L^{r-1} = 2^{\underset{\smile}{L^{r-1}}} = L^{r-2}$. This is a contradiction showing that (*) holds. Suppose now that (*) fails with $\tau = 0$. By Lemma 35.3 (especially by (35.9)), we can then see that one of conditions (v′) and (vii′) (see after (vii)) of Lemma 35.2 holds with L^{r-2} replacing κ. That is, we have either

$$\langle \lambda, (r+1)_{L_\lambda(\kappa)} \rangle \ll \langle \lambda_\xi : \xi < \vartheta \rangle$$

for some λ with $r+1 \leq \lambda < L^{r-2}$, or

$$\langle L^{r-2}, (r+1)_{L_{\mathrm{cf}(L^{r-2})}(\mathrm{cf}(L^{r-2}))} \rangle \ll \langle \lambda_\xi : \xi < \vartheta \rangle .$$

If GCH holds, then $L_\lambda(L^{r-2}) = L_3(L^{r-2}) = L^{r-1}$ for any λ with $3 \leq \lambda < L^{r-2}$, i.e., the first assertion is false by (ii.5) above. If the second assertion holds, then by the assumptions in the clause being proved of (iii) we have either

$$\langle L^{r-2}, \omega, (r+1)_{L_{\mathrm{cf}(L^{r-2})}(\mathrm{cf}(L^{r-2}))} \rangle \ll \langle \lambda_\xi : \xi < \vartheta \rangle$$

or L^{r-2} is regular, which is impossible by (ii.16) or (ii.17) above, respectively. To see that we can take $\lambda = L^{r-2}$ in $(*)$ one needs only to consider the remark made immediately after (35.16). The proof of (iii) is complete.

Ad (iv). $(*)$ and $\tau = 0$ imply

$$L^{r-2} \to (\lambda_\xi)^2_{\xi < \vartheta}$$

via Theorem 35.4, and so

$$L^{r-2} \to (\lambda_\xi - r + 2)^2_{\xi < \vartheta}$$

holds a fortiori. Observe that we have either $\exp_{r-3}(2^{L^{r-2}}) < \kappa$ or $L^{r-2} \to (L^{r-2})^2_2$ by (o) (cf. (38.5)). If the first alternative holds, then (i)a of Theorem 38.5 is true with $\rho = L^{r-2}$ and $i = r-2$. If the first alternative fails and the second one holds, then $2^{L^{r-2}} = L^{r-3}$ and $L^{r-2} \to (L^{r-2})^2_2$; thus L^{r-2} is inaccessible by Theorem 29.1, i.e., $L^{r-2} = L^{r-3}$, and so $L^{r-3} \to (L^{r-3})^2_2$. Thus (i)b of Theorem 38.5 holds now with $i = r-2$. Hence it follows by this theorem with $i = r-2$ from the last centered line that

$$\kappa \to (\lambda_\xi)^r_{\xi < \vartheta}$$

holds, which we wanted to show.

(v) immediately follows from (iii) and (iv).

Ad (vi). Assume that (oo) and ($\square$) hold, and yet we have $\lambda_\xi \geq L^{r-2}$ for some $\xi < \vartheta$; say $\lambda_0 \geq L^{r-2}$. By ($\square$) and Lemma 40.2 we then have $\lambda_0 = L^{r-2}$, and Lemma 40.5(i), (ii), (iii), (v), and (iv) imply $\lambda_1 = r+1$, $L^{r-2} < L^{r-3}$, $\mathrm{cf}(L^{r-2}) < L^{r-2}$, $\vartheta = 2$, and either $r > 3$ or $\mathrm{cf}(L^{r-2}) \to (\mathrm{cf}(L^{r-2}))^2_2$, respectively. Thus (Exception 2) holds; this is what we wanted to prove.

Ad (vii). We begin by discussing some of the consequences of (oo) and (+):

If (oo) holds, then L^{r-2} is a limit cardinal and $\mathrm{cf}(L^{r-2}) = \mathrm{cf}(L^{r-3})$. (2)

This follows by Theorem 6.10 d.

If (oo) holds, then (Exception 1) is false. (3)

Indeed, (Exception 1) requires that (o) holds, and this contradicts (oo).

If $(+)$ holds, then the sequence $\langle\lambda, (\rho)_\alpha, (\sigma)_\beta, (k_\nu)_{\nu<\tau}\rangle$ satisfies conditions (35.4)–(35.8) of Lemma 35.3 with $\lambda<L^{r-2}$ and with L^{r-2} replacing κ if and only if $\rho=L_\lambda(L^{r-2})$, $\alpha<\mathrm{cf}\,(L_\lambda(L^{r-2}))$, $\sigma, \beta<L_\lambda(L^{r-2})$, $\tau<L^{r-1}$, and $k_\nu<\omega$ for every $\nu<\tau$, and $\alpha=0$ in case $\lambda<L_\lambda(L^{r-2})$. (4)

To see this, one only has to observe that (35.6)(ii) is vacuous because $\lambda<L^{r-2}$ and (35.7)–(35.9) are vacuous since $2^\rho\leq 2^{L^{r-1}\smile}<L^{r-2}$ by $(+)$. Finally we need

Assume $(\circ\circ)$ and $(+)$ hold and $\lambda<L^{r-2}$. Then there is a regular κ' such that $\lambda<\kappa'<L^{r-2}$, $L_\lambda(L^{r-2})=L_\lambda(\kappa')$, and $2^{L^{r-2}\smile}<\kappa'$. (5)

This follows from Theorem 7.16, as L^{r-2} is a limit cardinal by (2).

We now embark upon the proof of the first part of (vii). Suppose to this end that $(\circ\circ)$, $(+)$, and $(\square)$ hold and (Exception 2) fails. As (Exception 1) is then false by (3), we can conclude by the first sentence of (iii) that $(*)$ holds for a sequence $\langle\lambda, (\rho)_\alpha, (\sigma)_\beta, (k_\nu)_{\nu<\tau}\rangle$ with $\lambda=\lambda_\xi$ for some $\xi<\vartheta$. We have $\lambda=\lambda_\xi<L^{r-2}$ by (vi) above. $\lambda<L^{r-2}$ implies via (4) that $(**)$ holds; this proves the first assertion.

We now turn to the second assertion in (vii). Assume that $(\circ\circ)$, $(+)$, and $(**)$ hold and either a) $L_\lambda(L^{r-2})=L^{r-1}$ or b) $\tau=0$. To see that $(\square)$ holds, we have only to prove in view of Theorem 38.5 with $i=r-2$ that there is a $\kappa'<L^{r-2}$ such that

$$\kappa'\to(\lambda, (L_\lambda(L^{r-2}))_\alpha, (\sigma)_\beta, (k_\nu)_{\nu<\tau})^2$$

holds. Choose a regular cardinal κ' satisfying (5). Then by Theorem 35.4.b(i) we have

$$\kappa'\to(\lambda, (L_\lambda(\kappa'))_\alpha, (\sigma)_\beta)^2$$

in any case, and

$$\kappa'\to(\lambda, (L_\lambda(\kappa'))_\alpha, (\sigma)_\beta, (k_\nu)_{\nu<\tau})^2$$

in case $L_\lambda(\kappa')=L_\lambda(L^{r-2})=L^{r-1}$ (in this case we can use that theorem with $\beta'=$ $=\beta+\tau$, $\tau'=0$), since (35.4), (35.5), and (35.6)(i) hold with κ' and $L_\lambda(\kappa')$ replacing κ and ρ in view of the equality $L_\lambda(L^{r-2})=L_\lambda(\kappa')$ (cf. (5)) and $(**)$, and, finally, (35.6)(ii), (35.7), and (35.8) are vacuous with these substitutions since $\lambda<\kappa'$ and $2^{L^{r-2}\smile}<\kappa'$ hold according to (5). This settles (vii).

Ad (viii). Assume that $(\circ\circ)$ and $(++)$ hold, and (Exception 2) is false. To prove necessity, suppose that $(\square)$ is valid. We have $\lambda_\xi<L^{r-2}$ for any $\xi<\vartheta$ by (vi) above, and $\langle L^{r-1}, L^{r-1}\rangle\not\leq\langle\lambda_\xi\colon \xi<\vartheta\rangle$ by (ii.15). As (ii.9) also fails, it is easy to conclude that $(***)$ holds, proving necessity. We now turn to the proof of sufficiency. Assume to this end that $(***)$ holds. We nave to prove that $\kappa\to(\lambda, (\sigma)_\beta)^2$ is true, and by Theorem 38.5 with $i=r-2$ it is enough to prove that there is a $\rho < L^{r-2}$ such that

$$\rho \to (\lambda, (\sigma)_\beta)^2 \tag{6}$$

holds. We have $2^{L^{r-1}} = L^{r-2}$ according to $(++)$. Hence there is an infinite $\rho_0 < L^{r-1}$ such that $\sigma, \beta \leq \rho_0$ and $\lambda \leq 2^{\rho_0}$ $(< L^{r-2})$. We have $(2^{\rho_0})^+ \to ((2^{\rho_0})^+, (\rho_0^+)_{\rho_0})^2$ according to Corollary 17.5. Hence (6) holds with $\rho = (2^{\rho_0})^+$, and $\rho < L^{r-2}$ since L^{r-2} is a limit cardinal according to (2). This completes the proof of Theorem 41.1.

42. THE MAIN RESULT FOR THE CASE $r \geq 3$ WITH GCH

Let κ be an infinite cardinal. For any integer i, define κ^{-i} by recursion as follows:

$$\kappa^{-0} = \kappa, \ \kappa^{-(i+1)} = (\kappa^{-i})^-, \tag{1}$$

where we recall that λ^- is the immediate cardinal predecessor of the cardinal λ if this exists, and $\lambda^- = \lambda$ otherwise. Clearly, if $\kappa = \aleph_\alpha$ then $\kappa^{-i} = \aleph_{\alpha - i}$. It is easy to see that, under the assumption of GCH, we have

$$\kappa^{-i} = L_3^i(\kappa) \tag{2}$$

for any integer i. Put

$$h(\kappa) = \min\{i < \omega \colon \kappa^{-i} \text{ is a limit cardinal}\}. \tag{3}$$

Note that it is clear by (2) that if GCH holds then

$$h(\kappa) = \min\{i < \omega \colon L_3^{i+1}(\kappa) = L_3^i(\kappa)\}. \tag{4}$$

Finally, we remind the reader that the critical number cr (κ) of κ was defined by formula (36.1), that is, we have

$$\operatorname{cr}(\kappa) = \operatorname{cf}(\operatorname{cf}(\kappa)^-). \tag{5}$$

We now turn to the corollary of the theorem in the preceding section for the case when GCH holds:

THEOREM 42.1 (Erdős–Hajnal–Rado [1965]). *Assume* GCH, *and let* $\kappa \geq 2$, $\vartheta \geq 2$, $3 \leq r < \omega$, *and* $\lambda_\xi > r$ *for any* $\xi < \vartheta$. *Consider the following conditions:*

(o′) $\quad h(\kappa) \geq r-2$ *or* $\kappa^{-(r-2)} \to (\kappa^{-(r-2)})_2^2$,

(oo′) $\quad h(\kappa) < r-2$ *and* $\kappa^{-(r-2)} \not\to (\kappa^{-(r-2)})_2^2$.

Then exactly one of (o′) *and* (oo′) *holds. Consider the relation*

(□) $$\kappa \to (\lambda_\xi)_{\xi < \vartheta}^r$$

and the following conditions (for the definition of the operations com *and* suc *see* (36.4) *and* (36.5)):

there are cardinals α, β, *and* σ *with*

$$\langle \lambda_\xi : \xi < \vartheta \rangle \ll \langle \kappa^{-(r-2)}, (\mathrm{cr}\,(\kappa^{-(r-2)}))_\alpha, (\sigma)_\beta \rangle \tag{6}$$

such that $\alpha < \mathrm{cr}\,(\kappa^{-(r-2)}) \cdot \mathrm{com}\,(\kappa^{-(r-2)}) + 1$ *and* $\beta, \sigma < \mathrm{cr}\,(\kappa^{-(r-2)})$

(*note that* + *denotes cardinal addition*), *and*

there are cardinals α, β, *and* σ *with*

$$\langle \lambda_\xi : \xi < \vartheta \rangle \ll \langle (\kappa^{-(r-1)})_\alpha, (\sigma)_\beta \rangle \tag{7}$$

such that $\alpha < \mathrm{cf}\,(\kappa^{-(r-1)}) \cdot \mathrm{suc}\,(\kappa^{-(r-2)}) + 1$ *and* $\beta, \sigma < \kappa^{-(r-1)}$.

Put

(Exception 3) $\kappa^{-(r-2)} > \mathrm{cf}\,(\kappa^{-(r-2)}) > \mathrm{cf}\,(\kappa^{-(r-2)})^- > \mathrm{cr}\,(\kappa^{-(r-2)})$, *and*
$\langle \kappa^{-(r-2)}, (r+1)_{\mathrm{cr}\,(\kappa^{-(r-2)})} \rangle \ll \langle \lambda_\xi : \xi < \vartheta \rangle \ll \langle \kappa^{-(r-2)}, (k'_\nu)_{\nu < \tau'} \rangle$,
where $\tau' < \mathrm{cf}\,(\kappa^{-(r-2)})^-$ *and* $k'_\nu < \omega$ *for any* $\nu < \tau'$.

We have the following results:

a) $\lambda_\xi \leq \kappa^{-(r-2)}$ *for any* $\xi < \vartheta$ *is necessary for* ($\square$).

b) *Assume* (o′) *holds. Then* (6) *or* (7) *is a sufficient condition for* ($\square$), *and if* (Exception 3) *fails then* (6) *or* (7) *is a necessary condition as well.*

c) *If* (oo′) *holds, then it is necessary and sufficient for* ($\square$) *that*

$$\langle \lambda_\xi : \xi < \vartheta \rangle \ll (\sigma)_\beta$$

holds with some $\beta, \sigma < \kappa^{-(r-2)}$.

Remark. If (Exception 3) holds then we do not know whether or not ($\square$) holds even in case $r=3$. This leads to Problem 25.8 discussed earlier.

Proof. It is easy to see by elementary propositional calculus that exactly one of (o′) and (oo′) holds; it is also important to note the fact used below often, without explicitly mentioning it, that under the assumption of GCH, (o′) and (oo′) are equivalent to conditions (o) and (oo), respectively, of the preceding section; this easily follows from (3) and from the equality $2^\lambda = \lambda$, true for any infinite cardinal λ if GCH is assumed.

a) follows from Theorem 41.1(i).

Ad b). Assume that (o′) holds and either (6) or (7) is satisfied. Then (*) of Theorem 41.1 holds with $\tau = 0$ by (36.9) and (36.10). Taking into consideration that, under the assumption of GCH, every infinite cardinal is either regular or strong limit, ($\square$) follows by Theorem 41.1(iv.1) or (iv.3). This proves the sufficiency of (6) or (7).

We now turn to the proof of necessity. Assume to this end that (o′) and ($\square$) hold, and (Exception 3) fails. Noting that (Exception 1) always fails under GCH (in fact, (Exception 1) requires in particular that $(++)$ hold, which simply becomes $L^{r-1}=L^{r-2}$ under the assumption of GCH, and it also requires the inequality $L^{r-1}<L^{r-2}$, which contradicts the former equality), we can conclude by the first sentence of (iii) in Theorem 41.1 that $(*)$ holds there with $\lambda=\lambda_\xi$ for some $\xi<\vartheta$ (we may assume $\lambda=\lambda_0$ without any loss of generality), i.e., that we have

$$\langle \lambda_\xi : \xi<\vartheta\rangle \ll \langle \lambda, (\rho)_\alpha, (\sigma)_\beta, (k_\nu)_{\nu<\tau}\rangle, \tag{8}$$

and the sequence on the right-hand side satisfies conditions (35.4)–(35.8) of Lemma 35.3 with $\kappa^{-(r-2)}$ replacing κ. Here we have

$$\tau<L_3(\kappa^{-(r-2)})=\kappa^{-(r-1)} \quad \text{and} \quad k_\nu<\omega \quad \text{for any} \quad \nu<\tau \tag{9}$$

according to (35.5). We distinguish two cases: 1) $\lambda=\lambda_0<\kappa^{-(r-2)}$ and 2) $\lambda=\lambda_0=\kappa^{-(r-2)}$.

Case 1. By (36.12) and (36.13) with $\kappa^{-(r-2)}$ replacing κ we have

$$\langle \lambda, (\rho)_\alpha, (\sigma)_\beta\rangle \ll \langle (\kappa^{-(r-1)})_{\alpha'}, (\sigma')_{\beta'}\rangle$$

with some α', β', and σ' satisfying

$$\alpha'<\operatorname{cf}(\kappa^{-(r-1)})\cdot \operatorname{suc}(\kappa^{-(r-2)})+1 \quad \text{and} \quad \beta', \sigma'<\kappa^{-(r-1)}. \tag{10}$$

By (8) we then have

$$\langle \lambda_\xi : \xi<\vartheta\rangle \ll \langle (\kappa^{-(r-1)})_{\alpha'}, (\sigma')_{\beta'}, (k_\nu)_{\nu<\tau}\rangle.$$

So we can conclude by (9) and (10) that in case $\kappa^{-(r-1)}=\omega$ (7) is confirmed by the inequality

$$\langle \lambda_\xi : \xi<\vartheta\rangle \ll \langle (\kappa^{-(r-1)})_{\alpha'}, (\sigma'+k)_{\beta'+\tau}\rangle,$$

where $k=\sup\{k_\nu : \nu<\tau\}$, and in case $\kappa^{-(r-1)}>\omega$ (7) is confirmed by the inequality

$$\langle \lambda_\xi : \xi<\vartheta\rangle \ll \langle (\kappa^{-(r-1)})_{\alpha'}, (\sigma'+\aleph_0)_{\beta'+\tau}\rangle.$$

Thus (7) holds in Case 1.

Case 2. We shall prove that (6) holds in this case; assume, on the contrary, that (6) fails. We have $\lambda=\lambda_0=\kappa^{-(r-2)}$ in the case considered and, by the second clause of (iii) in Theorem 41.1, we may assume that

$$\begin{array}{l}\text{either } \operatorname{cf}(\kappa^{-(r-2)})<\kappa^{-(r-2)} \text{ and } \lambda_\xi<\omega \\ \text{for any } \xi \text{ with } 1\leq\xi<\vartheta \text{ or } \tau=0 \text{ in (8).}\end{array} \tag{11}$$

By (36.11) we obtain that

$$\rho = \text{cr}\,(\kappa^{-(r-2)}),\quad \sigma, \beta < \text{cr}\,(\kappa^{-(r-2)}),\quad \text{and} \\ \alpha < \text{cr}\,(\kappa^{-(r-2)}) \cdot \text{com}\,(\kappa^{-(r-2)}) + 1. \tag{12}$$

If $\tau < \text{cr}\,(\kappa^{-(r-2)})$, then we can conclude from (8) and (12) and the equality $\lambda = \kappa^{-(r-2)}$, valid in the present case, that (6) holds with $\beta' = \beta + \tau$ replacing β (σ should also be replaced by $\sigma' = \sigma + \sup\{k_\nu : \nu < \tau\}$ or $\sigma' = \sigma + \aleph_0$ according as $\text{cr}\,(\kappa^{-(r-2)})$ equals or is greater than $\aleph_0$). This contradicts our assumption that (6) fails; hence we must have

$$\tau \geq \text{cr}\,(\kappa^{-(r-2)}) \tag{13}$$

that is, we cannot have $\tau = 0$, and so

$$\text{cf}\,(\kappa^{-(r-2)}) < \kappa^{-(r-2)} \quad \text{and} \quad \lambda_\xi < \omega \tag{14}$$

for any ξ with $1 \leq \xi < \vartheta$ by (11). As $\lambda = \kappa^{-(r-2)}$ and (35.5) holds with $\kappa^{-(r-2)}$ replacing κ according to what was said after (8), we have

$$\tau < L_3(\text{cf}\,(\kappa^{-(r-2)})) = \text{cf}\,(\kappa^{-(r-2)})^-. \tag{15}$$

Comparing this with (13), we obtain that

$$\text{cf}\,(\kappa^{-(r-2)})^- > \text{cr}\,(\kappa^{-(r-2)}). \tag{16}$$

This implies via (36.3) that

$$\text{cf}\,(\kappa^{-(r-2)}) > \text{cf}\,(\kappa^{-(r-2)})^-. \tag{17}$$

Now we have by the equality $\lambda = \lambda_0 = \kappa^{-(r-2)}$, (13), and the assumption $\lambda_\xi > r$ of the theorem being proved that

$$\langle \kappa^{-(r-2)}, (r+1)_{\text{cr}\,(\kappa^{-(r-2)})} \rangle \ll \langle \lambda_\xi : \xi < \vartheta \rangle. \tag{18}$$

On the other hand, we have

$$|\vartheta| \leq 1 + \alpha + \beta + \tau < \text{cf}\,(\kappa^{-(r-2)})^-,$$

where the first inequality holds by (8), and the second one by (12), (15), and (16). Hence, by (14) and (8) we can conclude that

$$\langle \lambda_\xi : \xi < \vartheta \rangle \ll \langle \kappa^{-(r-2)}, (k'_\nu)_{\nu < \tau'} \rangle, \tag{19}$$

where $\tau' = |\vartheta| - 1 < \text{cf}\,(\kappa^{-(r-2)})^-$ and $k'_\nu < \omega$ for any $\nu < \tau'$. The first inequality in (14), and (16), (17), (18), and (19) confirm that (Exception 3) holds. This is a contradiction, completing the proof of assertion b).

Ad c). Assume (oo′) holds. Then by (4) we have that $L_3^{r-3}(\kappa) = L_3^{r-2}(\kappa) = L_3^{r-1}(\kappa)$, and so (Exception 2) is false, and (++) holds by GCH. The assertion now follows from (viii) of Theorem 41.1 by observing that $\kappa^{-(r-2)} = L_3^{r-2}(\kappa) = L_3^{r-1}(\kappa)$. The proof of the theorem is complete.

CHAPTER X

SOME APPLICATIONS OF COMBINATORIAL METHODS

The next few sections contain some applications of partition relations and combinatorial methods connected with them in topology and in the theory of set mappings, etc., and the culmination of this chapter is the proof of a recent inequality on cardinal exponentiation due to F. Galvin and A. Hajnal. The chapter ends with a discussion of results of J. Ketonen, T. J. Jech and K. Prikry on the relationship between cardinal exponentiation and saturated ideals.

43. APPLICATIONS IN TOPOLOGY

Here we give some applications in topology of the results and methods relating to the ordinary partition relation. We do not strive at obtaining the most general results; our aim is, rather, to show the usefulness of partition relations in topology by picking characteristic examples. We first give a few basic concepts. A *topological space* is a pair $\langle E, \mathcal{O}\rangle$ such that E is a set (the underlying set) and $\mathcal{O}\subseteq\mathcal{P}(E)$ is such that $0\in\mathcal{O}$, $E\in\mathcal{O}$, and if $H\subseteq\mathcal{O}$ then $\bigcup H\in\mathcal{O}$ and, finally, if $H\subseteq\mathcal{O}$, $H\neq 0$ is finite then $\bigcap H\in\mathcal{O}$. The elements of E are called points, $\mathcal{O}$ is said to be the *topology* on E, and the elements of $\mathcal{O}$ are called *open sets*. $X\subseteq E$ is *closed* if $E\setminus X$ is open. The closure cl (X) of a set $X\subseteq E$ is the least closed set including X, i.e.,

$$\mathrm{cl}\,(X)=\bigcap\{Y\subseteq E\colon X\subseteq Y\,\&\,E\setminus Y\in\mathcal{O}\}.$$

If $x\in E$ and $x\in X\in\mathcal{O}$, then X is called a neighborhood of the point x. More generally, if $X\subseteq E$ and $X\subseteq X'\in\mathcal{O}$, then X' is called a *neighborhood* of the set X. One often writes briefly E instead of $\langle E, \mathcal{O}\rangle$. E is a *Hausdorff space* if any two distinct points in E have disjoint neighborhoods. All spaces considered below will be assumed to be Hausdorff. Given a topological space $\langle E, \mathcal{O}\rangle$ and a set $E'\subseteq E$, the pair

$$\langle E', \{X\cap E'\colon X\in\mathcal{O}\}\rangle$$

is also a topological space. This space is called a *subspace* of E; the topology on E' is called its *relative* (or *subspace*) topology. Given a point x of a topological space $\langle E, \mathcal{O}\rangle$, a set $B\subseteq\mathcal{O}$ is called a *neighborhood base* of (or at) x if for every $X\in\mathcal{O}$ with

$x \in X$ there is a $Y \in B$ such that $x \in Y \subseteq X$. B is called a *base* for the space $\langle E, \mathcal{O} \rangle$ if it is a neighborhood base at each point x of E.

This concludes our review of the main concepts of topology. What comes next is a series of characteristic applications of combinatorial methods in topology. When necessary, a short paragraph preceding a result will explain the topological concepts not explained so far.

Our first result, due to A. Hajnal and I. Juhász [1967], is obtained by an application of the partition relation

$$(2^{\aleph_0})^+ \to (\aleph_1)^2_\omega \tag{1}$$

(see Corollary 17.5). To explain the topological concepts needed, a topological space is *first countable* if each point has a countable neighborhood base; it has the *Suslin property* if it has no $\aleph_1$ pairwise disjoint open sets.

THEOREM 43.1. *A first countable Hausdorff space with Suslin property has cardinality* $\leq 2^{\aleph_0}$.

PROOF. Let $\langle E, \mathcal{O} \rangle$ be a topological space with the required properties and assume, on the contrary, that

$$|E| > 2^{\aleph_0}. \tag{2}$$

For each point $x \in E$ let $\{U_x^n : n < \omega\}$ be a countable neighborhood base of x; we may assume without loss of generality that $x \in U_x^n$. Put

$$V_x^n = \bigcap_{i<n} U_x^i;$$

then, clearly,

$$\{V_x^n : 1 \leq n < \omega\}$$

is also a neighborhood base of x. As E is a Hausdorff space, any two distinct points x and y have disjoint neighborhoods X and Y, respectively. We have

$$V_x^n \subseteq X \qquad \text{and} \qquad V_y^n \subseteq Y$$

for any large enough integer n. Then, obviously,

$$V_x^n \cap V_y^n = 0$$

holds. Let $f(\{xy\})$ be the least integer n for which this relation holds. Then

$$f: [E]^2 \to \omega$$

is a coloring, and according to (1) and (2), there is a homogeneous set $H \subseteq E$ of cardinality $\aleph_1$; let m be the color of this set (i.e., $f``[H]^2 = \{m\}$). Then we have

$$V_x^m \cap V_y^m = 0$$

for any two distinct $x, y \in H$ by the definition of the coloring f. Hence

$$\{V_x^m : x \in H\}$$

is a set of cardinality $\aleph_1$ of pairwise disjoint open sets, contradicting our assumption that the space $\langle E, \mathcal{O}\rangle$ has the Suslin property.

Our next result, also due to Hajnal and Juhász [1967], gives a more sophisticated application of the partition relation

$$(2^{2^\kappa})^+ \to (\kappa)^3_4, \tag{3}$$

which holds for any infinite κ (cf. (17.30) with $\lambda = 3$ and $\rho = \kappa$). It uses the notion of discrete topological space: $\langle E, \mathcal{O}\rangle$ is *discrete* if $\mathcal{O} = \mathcal{P}(E)$; for this, it is obviously enough to require that each one-element set be open.

THEOREM 43.2. *If κ is infinite, then a Hausdorff space of cardinality $> 2^{2^\kappa}$ has a discrete subspace of cardinality κ.*

This result can be sharpened as follows: For any infinite cardinal κ, a Hausdorff space of cardinality $> \sum_{\lambda < \kappa} 2^{2^\lambda}$ has a discrete subspace of cardinality κ; i.e., one can "usually" lower the bound for the cardinality of the space in case κ is not a successor cardinal. This can be proved by a minor modification of the proof of Theorem 43.4; see our remark right after that proof. By improving the general Canonization Lemma, Shelah [1981] can prove this result even if κ is singular and the cardinality of the space is $> 2^{2^\lambda}$ for all $\lambda < \kappa$ (cf. Theorem 43.3 for the role of the General Canonization Lemma).

PROOF. Let $\langle E, \mathcal{O}\rangle$ be a Hausdorff space such that $|E| > 2^{2^\kappa}$, and let $\prec$ be a wellordering of E. For any $x, y \in E$ with $x \prec y$ let U^0_{xy} and U^1_{xy} be neighborhoods of x and y, respectively, such that

$$U^0_{xy} \cap U^1_{xy} = 0. \tag{4}$$

For any $x_0, x_1, x_2 \in E$ with $x_0 \prec x_1 \prec x_2$ and for $i, j = 0$ or 1 put

$$\{x_0 x_1 x_2\} \in I_i \quad \text{if} \quad x_0 \notin U^i_{x_1 x_2} \tag{5}$$

and

$$\{x_0 x_1 x_2\} \in I^j \quad \text{if} \quad x_2 \notin U^j_{x_0 x_1}. \tag{6}$$

Writing $I_{ij} = I_i \cap I^j$, it is easy to see by (4) that $I = \langle I_{00}, I_{01}, I_{10}, I_{11}\rangle$ is a partition of $[E]^3$ into four (not necessarily disjoint) parts. According to (3), there is a homogeneous set X of cardinality κ; we may assume that X has order type κ in the wellordering $\prec$. Let x_α, $\alpha < \kappa$, be the enumeration of the elements of X in increasing order and let

$$Y=\{x_\alpha\colon \alpha \text{ is an odd ordinal } <\kappa\};$$

here an odd ordinal is a limit ordinal (or 0) $\dot{+}$ an odd integer.

We claim that Y is a discrete subspace of E. To this end we have to show that any $x_\alpha \in Y$ has a neighborhood U_α such that

$$U_\alpha \cap Y=\{x_\alpha\}. \tag{7}$$

Let $\beta=\alpha \dot{-} 1$ and $\gamma=\alpha \dot{+} 1$. We claim that

$$U_\alpha=U^1_{x_\beta x_\alpha}\cap U^0_{x_\alpha x_\gamma} \tag{8}$$

satisfies (7). Note that U_α is a neighborhood of x_α since it is the intersection of two of its neighborhoods. To see that (7) holds, let $x_\xi \in Y$ be a point distinct from x_α; we shall show that

$$x_\xi \notin U_\alpha. \tag{9}$$

To this end, we distinguish four cases: 1) $\xi<\alpha$ and $[X]^3\subseteq I_0$, 2) $\xi<\alpha$ and $[X]^3\subseteq I_1$, 3) $\xi>\alpha$ and $[X]^3\subseteq I^0$, and 4) $\xi>\alpha$ and $[X]^3\subseteq I^1$. (Note that in cases 1) and 2) $\xi<\beta$ and in cases 3) and 4) $\xi>\gamma$ holds since ξ is an odd ordinal, as $x_\xi\in Y$). In case 1), we have $x_\xi\notin U^0_{x_\alpha x_\gamma}$ by (5), in case 2) $x_\xi\notin U^1_{x_\beta x_\alpha}$ by (5), in case 3) $x_\xi\notin U^0_{x_\alpha x_\gamma}$ by (6), and in case 4) $x_\xi\notin U^1_{x_\beta x_\alpha}$ by (6). So (9) follows from (8). Hence (7) holds; i.e., Y is a discrete subspace of cardinality κ of E. This completes the proof.

Our next result exploits the same idea plus the General Canonization Lemma (Lemma 28.1). This is the first instance in this book that we canonize triples; so far, the Canonization Lemma was only applied in the case of pairs.

THEOREM 43.3. *If κ is a singular strong limit cardinal, then a Hausdorff space of cardinality $\geq\kappa$ has a discrete subspace of cardinality κ.*

PROOF. Let $\langle\kappa_\xi\colon \xi<\mathrm{cf}(\kappa)\rangle$ be a sequence of infinite cardinals tending to κ such that (28.2) and (28.3) are satisfied with $r=2$, $\vartheta=\mathrm{cf}(\kappa)$, and $\tau=4$. Let $\langle E, \mathcal{O}\rangle$ be a Hausdorff space with $|E|=\kappa$ (this is a harmless assumption, since if $|E|>\kappa$, then we may work with a subspace of cardinality κ of E). Let $E=\bigcup_{\xi<\mathrm{cf}(\kappa)} A_\xi$ be a decomposition of E into pairwise disjoint sets such that (28.4) holds (i.e., $|A_\xi|\geq(\exp_3(\kappa_\xi))^+$). Let $\prec_\xi$ be a wellordering of A_ξ and, for any $x, y\in E$ put

$$\begin{aligned} x\prec y \text{ if } x\in A_\xi,\ y\in A_\eta \text{ and } \xi<\eta \text{ for some } \xi,\ \eta<\mathrm{cf}(\kappa)\\ \text{or } x,\ y\in A_\xi \text{ and } x\prec_\xi y \text{ for some } \xi<\mathrm{cf}(\kappa). \end{aligned} \tag{10}$$

As in the preceding proof, for any two points $x, y\in E$ with $x\prec y$ pick neighborhoods U^0_{xy} and U^1_{xy} of x and y, respectively, so that (4) hold, and define I_i

and I^j by (5) and (6). We again obtain a partition $I=\langle I_{00}, I_{01}, I_{10}, I_{11}\rangle$ of $[E]^3$ as above. Write

$$f(\{x_0x_1x_2\})=k-1 \tag{11}$$

if the kth class of the above partition is the first one containing $\{x_0x_1x_2\}$ ($k=1, 2, 3, 4$). By Lemma 28.1, there are sets $B_\xi \subseteq A_\xi$, $\xi<\mathrm{cf}(\kappa)$, with $|B_\xi|\geq\kappa_\xi^+$ such that the sequence $\langle B_\xi: \xi<\mathrm{cf}(\kappa)\rangle$ is canonical with respect to the coloring f.

Making the harmless assumption that the order type of B_ξ is κ_ξ^+ in the wellordering $\prec$, let x_ξ^α, $\alpha<\kappa_\xi^+$, be an enumeration of its elements in increasing order in $\prec$. Put

$$Y=\{x_\xi^\alpha: \xi<\mathrm{cf}(\kappa),\quad \alpha<\kappa_\xi^+,\quad \text{and } \alpha \text{ is odd}\}.$$

We claim that Y is a discrete subspace of E. To this end, we have to show that any $x_\xi^\alpha\in Y$ has a neighborhood U_ξ^α such that

$$U_\xi^\alpha\cap Y=\{x_\xi^\alpha\} \tag{12}$$

holds. We claim that the neighborhood

$$U_\xi^\alpha=U^1_{x_\xi^{\alpha-1}x_\xi^\alpha}\cap U^0_{x_\xi^\alpha x_\xi^{\alpha+1}} \tag{13}$$

of x_ξ^α satisfies (12). To see (12), let $x_\eta^\delta\in Y$ be arbitrary; we shall show that

$$x_\eta^\delta\notin U_\xi^\alpha. \tag{14}$$

In case $\eta=\xi$ the same argument works as in the preceding proof, since the set $Y\cap B_\xi$ is homogeneous with respect to the coloring f (and so with respect to the partition I defined after (6)) by the definition of canonicity. If e.g. $\eta<\xi$ and

$$f(\{x_\eta^\delta x_\xi^{\alpha-1}x_\xi^\alpha\})=f(\{x_\eta^\delta x_\xi^\alpha x_\xi^{\alpha+1}\})=k$$

holds for some $k<4$ (the first equality holds by canonicity; this is the point where canonicity is used), then in case $k=0$ or 1 we have $\{x_\eta^\delta x_\xi^\alpha x_\xi^{\alpha+1}\}\in I_0$ by (11), and so $x_\eta^\delta\notin U^0_{x_\xi^\alpha x_\xi^{\alpha+1}}$ by (5), and in case $k=2$ or 3 we have $\{x_\eta^\delta x_\xi^{\alpha-1}x_\xi^\alpha\}\in I_1$, and so $x_\eta^\delta\notin U^1_{x_\xi^{\alpha-1}x_\xi^\alpha}$ by (5), verifying (13) in case $\eta<\xi$. The case $\eta>\xi$ can be dealt with similarly, using (6) instead of (5). The proof of the theorem is complete.

Call a topological space *right*(*left*)-*separated* if it has a wellordering under which each initial (final) segment is open. The next result, due to J. de Groot [1965], applies a simple tree argument. As pointed out by Shelah, a proof using the relation $(2^\kappa)^+\to\langle\kappa\rangle_2$ (see (15.8)) and ideas similar to those used in the proof of Theorem 43.2 also works (see below for hints).

THEOREM 43.4. *If κ is an infinite cardinal, then a Hausdorff space of cardinality $>2^\kappa$ has a subspace of cardinality κ that is right-separated.*

HINTS for Shelah' proof. Let $\prec$ be a wellordering of the space $\langle E, \mathcal{O}\rangle$, consider the partition $I=\langle I^0, I^1\rangle$ of $[E]^3$ defined by (6), and assuming $|E|>2^\kappa$, let $X=\{x_\alpha: \alpha<\kappa\}$ be end-homogeneous with respect to this partition. Given $\alpha<\kappa$ odd, either $x_{\alpha+2}\notin U^0_{x_\alpha x_{\alpha+1}}$ or $x_{\alpha+2}\notin U^1_{x_\alpha x_{\alpha+1}}$. In case the former alternative happens throw away $x_{\alpha+1}$ and otherwise throw away x_α. The remaining set is right-separated by end-homogeneity.

PROOF. Let $\langle E, \mathcal{O}\rangle$ be a Hausdorff space with $|E|>2^\kappa$. We are going to construct a partition tree $\langle E, S, R, T\rangle$ by transfinite recursion as follows (cf. Section 14). We put

$$\operatorname{dom}(S)=\bigcup_{\alpha\leq|S|^+} {}^\alpha 2; \tag{15}$$

for any function $f\in\operatorname{dom}(S)$ the set $S(f)$ will be a closed subset of E. Assume that $S(f)$ has already been defined and write $\operatorname{dom}(f)=\alpha$. In case $\alpha<|S|^+$ we have to define $S(f\cup\{\langle\alpha 0\rangle\})$ and $S(f\cup\{\langle\alpha 1\rangle\})$. If $|S(f)|\leq 1$, define both sets as empty (so $|R(f)|=1$ by (14.3)); if $|S(f)|\geq 2$, then let $s(f\cup\{\langle\alpha 0\rangle\})$ and $s(f\cup\{\langle\alpha 1\rangle\})$ be two distinct elements of $S(f)$, and let U_0 and U_1, respectively, be disjoint neighborhoods of them. Put

$$S(f\cup\{\langle\alpha 0\rangle\})=S(f)\setminus U_0 \tag{16}$$

and

$$S(f\cup\{\langle\alpha 1\rangle\})=S(f)\setminus U_1. \tag{17}$$

Note that in this case we have

$$S(f\cup\{\langle\alpha 0\rangle\})\cup S(f\cup\{\langle\alpha 1\rangle\})=S(f)\setminus(U_0\cap U_1)=S(f),$$

i.e., (14.3) implies that

$$R(f)=0. \tag{18}$$

This completes the definition of S, and with it that of the whole partition tree $\langle E, S, R, T\rangle$. Observe that $S(f)$ is always a closed set. In fact, $S(0)=E$ is closed. If $g\in\operatorname{dom}(S)$ and $S(g\upharpoonright\xi)$ is closed for all $\xi<\operatorname{dom}(g)$, then we have

$$S(g)=\bigcap_{\xi<\operatorname{dom}(g)} S(g\upharpoonright\xi)$$

in case $\operatorname{dom}(g)$ is a limit ordinal, and so $S(g)$ is also closed in this case. Finally, if $\operatorname{dom}(g)=\alpha+1$ for some α, then g has either the form $f\cup\{\langle\alpha 0\rangle\}$ or $f\cup\{\langle\alpha 1\rangle\}$, and so $S(g)$ is either empty (when $|S(f)|\leq 1$), or equals one of the sets in (16) and (17); in both cases, $S(g)$ is closed provided that $S(f)=S(g\upharpoonright\alpha)$ is closed. Note that we always have

$$|R(f)|\leq 1;$$

in fact $|R(f)|=1$ occurs in case $|S(f)|=1$ and $|R(f)|=0$ in all other cases (cf. (18)). Taking, as usual, $T=\{f\in \mathrm{dom}(S)\colon S(f)\neq 0\}$, we obtain by Theorem 14.3.b with $(2^\kappa)^+$ and κ replacing κ and λ that T has a path of length $\kappa\dot{+}1$ (note for this that each element of T can have at most two immediate successors by (15)), i.e., there is an $f\in T$ with $\mathrm{dom}(f)=\kappa$. As $S(f)\neq 0$ in view of $f\in T$, the point $s(f\upharpoonright\alpha)$ is defined for every successor ordinal $\alpha<\kappa$; put

$$X=\{s(f\upharpoonright(\alpha\dot{+}1))\colon \alpha<\kappa\}.$$

The subspace X of E is right-separated; in fact, for each $\alpha<\kappa$, the set $E\setminus S(f\upharpoonright\alpha)$ is a neighborhood of $\{s(f\upharpoonright(\xi\dot{+}1))\colon \xi<\alpha\}$ that does not intersect the set $\{s(f\upharpoonright(\xi\dot{+}1))\colon \alpha\leq\xi<\kappa\}$. One can also see from here that the points $s(f\upharpoonright(\alpha\dot{+}1))$ are pairwise distinct, and so X has cardinality κ. The proof is complete.

Remark. If we want to prove the sharpening of Theorem 43.2 mentioned immediately after that theorem, then we have to construct almost the same tree as in the preceding proof, where now we have $|E|>\sum_{\lambda<\kappa} 2^{2^\lambda}$ instead of $|E|>2^\kappa$. The modification that we have to make is the following: having defined $S(f)$, put

$$S'(f)=S(f)\setminus \mathrm{cl}\,(\{s(f\upharpoonright(\xi\dot{+}1))\colon \xi\dot{+}1\leq\alpha\}),$$

where $\alpha=\mathrm{dom}(f)$, and replace $S(f)$ with $S'(f)$ in the definitions of $s(f\cup\{\langle\alpha 0\rangle\})$ and $s(f\cup\{\langle\alpha 1\rangle\})$, and on the right-hand sides of (16) and (17) above. Noting that the closure of a set of cardinality λ in a Hausdorff space has cardinality $\leq 2^{2^\lambda}$, it is easy to conclude that there is an $f\in T$ with $\mathrm{dom}(f)=\kappa$. The sequence $\langle s(f\upharpoonright(\alpha\dot{+}1))\colon \alpha<\kappa\rangle$ is left-separated in view of the definition of S', and it is right-separated by the argument used in the preceding proof. Hence its elements form a discrete subspace of cardinality κ.

Our next aim is to establish A.V. Arhangel'skiĭ's [1969] famous result saying that every first countable compact Hausdorff space has cardinality $\leq 2^{\aleph_0}$, which settled a nearly fifty year old conjecture of P. S. Alexandrov. For this, we need the following concept: a sequence $\langle x_\xi\colon \xi<\alpha\rangle$ of points of a topological space is *free* if for each $\eta<\alpha$ the closures $\mathrm{cl}\,(\{x_\xi\colon \xi<\eta\})$ and $\mathrm{cl}\,(\{x_\xi\colon \eta\leq\xi<\alpha\})$ are disjoint (it is easy to see that in this case $\{x_\xi\colon \xi<\alpha\}$ is discrete in the subspace topology). The main part of Arhangel'skiĭ's result is contained in the following

Lemma 43.5. *Let $\kappa>\omega$ be a regular cardinal, $\langle E, \mathcal{O}\rangle$ a topological space of cardinality $>2^\kappa$, and assume that for every set $X\subseteq E$ of cardinality $<\kappa$ we have*

$$|\mathrm{cl}(X)|\leq 2^\kappa \tag{19}$$

and

$$\mathrm{cl}\,(X) \text{ is the intersection of at most } 2^\kappa \text{ open sets.} \tag{20}$$

Then there is a free sequence of order type κ in E.

PROOF. We shall define a partition tree $\langle E, S, R, T\rangle$ such that

$$\operatorname{dom}(S)=\bigcup_{\alpha\leq\kappa}{}^{\alpha}(2^{\kappa});$$

we shall take care that $S(f)$ is a closed set for every $f\in\operatorname{dom}(S)$. For each $g\in T$ with $\operatorname{dom}(g)$ a successor ordinal we shall also define a point $s(g)\in S(g)$. Let $\alpha<\kappa$ be an ordinal and $f\in\operatorname{dom}(S)$ a function such that $S(f)$ has already been defined and is a nonempty closed set; suppose that the set

$$A_f=\{s(f\restriction(\xi\dot{+}1)):\xi<\alpha\}$$

has also been defined already. Let

$$\operatorname{cl}(A_f)=\bigcap_{\nu<2^{\kappa}}A^{\nu} \tag{21}$$

be a representation of $\operatorname{cl}(A_f)$ as an intersection of 2^{κ} open sets; such a representation exists according to (20). Write

$$S(f\cup\{\langle\alpha\nu\rangle\})=S(f)\setminus A_f^{\nu}, \tag{22}$$

and if $S(f\cup\{\langle\alpha\nu\rangle\})$ is not empty, then select an element of this set as $s(f\cup\{\langle\alpha\nu\rangle\})$ arbitrarily. This completes the definition of the partition tree $\langle E, S, R, T\rangle$.

Observe that the set $S(g)$ is closed for every $g\in T$. Indeed, if $\operatorname{dom}(g)$ is a limit ordinal, then $S(g)=\bigcap_{\xi<\operatorname{dom}(g)}S(g\restriction\xi)$, and so the closedness of $S(g)$ follows from that of the $S(g\restriction\xi)$'s; and if g has the form $f\cup\{\langle\alpha\nu\rangle\}$, where $\alpha=\operatorname{dom}(f)$ and $\nu<2^{\kappa}$, then $S(g)$ is closed in view of (22), since $S(f)$ is closed (by the induction hypothesis) and A_f^{ν} is open. Note also that we have

$$\operatorname{cl}(A_{g\restriction\xi})\cap S(g\restriction(\xi\dot{+}1))=0 \tag{23}$$

for every $g\in T$ and every $\xi<\operatorname{dom}(g)$; indeed, this follows simply from (21) and (22) by taking $f=g\restriction\xi$ and $\nu=g(\xi)$ there. This means that the sequence

$$\langle g(\xi\dot{+}1):\xi\dot{+}1<\operatorname{dom}(g)\rangle$$

is a free sequence. Hence we only have to show that there is a $g\in T$ with $\operatorname{dom}(g)=\kappa$. Assume that this is not the case; then for any $f\in T$ there is an $\alpha<\kappa$ such that $\operatorname{dom}(f)=\alpha$; hence

$$R(f)=S(f)\setminus\bigcap_{\nu<2^{\kappa}}S(f\cup\{\langle\alpha\nu\rangle\})\subseteq\bigcap_{\nu<2^{\kappa}}A_f^{\nu}=\operatorname{cl}(A_f)$$

by (21) and (22), and so

$$|R(f)|\leq 2^{\kappa}$$

according to (19). Observing that $(2^\kappa)^\rho = 2^\kappa$ holds for any $\rho < \kappa$ in view of the regularity of κ by Theorem 6.10.e, we can use Theorem 14.3.a with $(2^\kappa)^+$ and κ replacing κ and λ, respectively. We obtain that T has a path of length $\kappa \dotplus 1$; this contradicts our assumption that dom $(g) < \kappa$ for every $g \in T$, completing the proof.

Here the combinatorial part of this section is essentially over. As a final act, we show how to derive Arhangel'skiĭ's result from the above lemma:

COROLLARY 43.6. *Every first countable compact Hausdorff space has cardinality* $\leq 2^{\aleph_0}$.

By an argument similar to that given in the proof below, one can establish the stronger results in which compact' is replaced by 'Lindelöf' (a space is *Lindelöf* if every open cover has a countable subcover); for the proof see the remark at the end of this subsection.

PROOF. Let $\langle E, \mathcal{O} \rangle$ be a first countable compact Hausdorff space. First observe that (19) and (20) are satisfied with $\kappa = \aleph_1$ by $\langle E, \mathcal{O} \rangle$. In fact, as for (19), let $X \subseteq E$ be an arbitrary set and let $x \in \mathrm{cl}\,(X)$. Then, by first countability, it is easy to find a sequence $\langle x_n : n < \omega \rangle$ of elements of X that converges to x (in fact, pick $x_n \in X$ from the intersection of the first n neighborhoods of x in a given countable neighborhood base). There are two important consequences of this: firstly,

$$\mathrm{cl}\,(X) = \bigcup \{\mathrm{cl}\,(Y) : Y \subseteq X \;\&\; |Y| \leq \omega\}, \tag{24}$$

and secondly,

$$|\mathrm{cl}\,(X)| \leq |X|^{\aleph_0}; \tag{25}$$

this latter holds because in a Hausdorff space a convergent sequence can have only one limit point, and the cardinal on the right-hand side is an upper bound of the number of convergent sequences that can be formed by elements of X. If we assume $|X| \leq \aleph_0$, then (19) (with $\kappa = \aleph_1$) follows from (25).

To see (20) with $\kappa = \aleph_1$, we show that any closed set $Y \subseteq E$ of cardinality $\leq 2^{\aleph_0}$ is the intersection of $\leq 2^{\aleph_0}$ open sets. To this end, for each $y \in Y$ let $\{U_y^n : n < \omega\}$ be a countable neighborhood base of y (assume $y \in U_y^n$ for $n < \omega$). Then, by E being a Hausdorff space, for every $x \in E \setminus Y$ and every $y \in Y$ there is an integer $n(x, y)$ such that $x \notin U_y^{n(x,y)}$. Since

$$\{U_y^{n(x,y)} : y \in Y\}$$

is an open cover of the closed set Y, by the compactness of E there is a finite set $Y(x) \subseteq Y$ such that

$$Y \subseteq \bigcup_{y \in Y(x)} U_y^{n(x,y)}.$$

As x does not belong to the set on the right-hand side, we have

$$Y=\bigcap\{\bigcup_{y\in Y(x)} U_y^{n(x,y)}: x\in E\setminus Y\}.$$

The sets on the right-hand side are open, and there are only $\le 2^{\aleph_0}$ different ones among them. This establishes (20) with $\kappa=\aleph_1$.

Hence $|E|\le 2^{\aleph_0}$ will follow from Lemma 43.5 with $\kappa=\aleph_1$ if we show that E cannot have a free sequence of order type $\aleph_1$. Assume, on the contrary, that $\langle x_\xi:\xi<\aleph_1\rangle$ is such a sequence. Writing $X=\{x_\xi:\xi<\aleph_1\}$, we have

$$\mathrm{cl}\,(X)=\bigcup_{\alpha<\aleph_1}\mathrm{cl}\,(\{x_\xi:\xi<\alpha\})$$

by (24). As

$$\mathrm{cl}\,(\{x_\xi:\xi<\alpha\})\cap\mathrm{cl}\,(\{x_\xi:\alpha\le\xi<\aleph_1\})=0$$

holds for any $\alpha<\aleph_1$ since the the sequence $\langle x_\xi:\xi<\aleph_1\rangle$ is free, we can see from here that

$$\bigcap_{\alpha<\aleph_1}\mathrm{cl}\,(\{x_\xi:\alpha\le\xi<\aleph_1\})\subseteq\mathrm{cl}\,(X)\setminus\bigcup_{\alpha<\aleph_1}\mathrm{cl}(\{x_\xi:\xi<\alpha\})=0\,;$$

no intersection of countably many sets on the left-hand side is empty, however. This contradicts the compactness of the space E (it would contradict even Lindelöfness). Hence E cannot indeed have a free sequence of order type $\aleph_1$, which completes the proof.

44. FODOR'S AND HAJNAL'S SET-MAPPING THEOREMS

Given a set E, a *set mapping on* E is a function $f: E\to\mathscr{P}(E)$ such that $x\notin f(x)$ holds for every $x\in E$. (Set mappings *over* E will be considered in the next section.) A subset H of E is called *free with respect to* f if $H\cap f(x)=0$ holds for every $x\in H$. Note that set mappings were briefly considered earlier (cf. Lemma 20.3).

In this section we shall consider set mappings where a cardinality restriction is imposed upon the images, and we shall consider the question whether large free sets exist. If f is a set mapping on κ, then we might first ask how large a free set must exist if we assume only $|f(\alpha)|<\kappa$ for every $\alpha\in\kappa$. A brief reflection shows, however, that this assumption does not even ensure the existence of a two element free set (clearly, a one element set is always free); indeed, take $f(\alpha)=$ $=\alpha(=\{\beta:\beta<\alpha\})$ for every $\alpha\in\kappa$. One might expect more if we assume that $|f(\alpha)|<\lambda$ holds with some cardinal $\lambda<\kappa$ for all $\alpha\in\kappa$. Ruziewicz [1936] conjectured in 1936 that this assumption ensures the existence of a free set of cardinality κ; this conjecture was settled affirmatively by Hajnal in 1960 (see Theorem 44.3 below).

The following theorem, which answers Ruziewicz's conjecture for regular κ (see Corollary 44.2 below), is due to Fodor [1952]. The case $\lambda=\omega$ of this theorem was already known to de Bruijn and Erdős [1951].

THEOREM 44.1. *Let κ and λ be infinite cardinals, and let f be a set mapping on κ such that $|f(x)|<\lambda$ holds for every $x\in\kappa$. Then κ can be represented as the sum of λ sets free with respect to f.*

PROOF. We may assume $\lambda<\kappa$. We define a matrix $\langle x_{\alpha\beta}:\alpha<\mu\,\&\,\beta<\lambda\rangle$ of elements of κ such that the sets $\{x_{\alpha\beta}:\alpha<\mu\}$, $\beta<\lambda$, are pairwise disjoint free sets and $\kappa=\{x_{\alpha\beta}:\alpha<\mu\,\&\,\beta<\lambda\}$ (there may be identical ones among the $x_{\alpha\beta}$'s). Assuming that $\langle x_{\alpha\beta}:\alpha<\xi\,\&\,\beta<\lambda\rangle$ has already been defined for some ordinal ξ, put $E_\xi=\{x_{\alpha\beta}:\alpha<\xi\,\&\,\beta<\lambda\}$, and assume that E_ξ is closed under f (i.e., $f(x)\subseteq E_\xi$ holds whenever $x\in E_\xi$). If $E_\xi=\kappa$, then we put $\mu=\xi$; in this case the definition of the matrix is completed. Otherwise let $F_\xi=\{y_{\xi\gamma}:\gamma<\lambda\}\subseteq\kappa\setminus E_\xi$ be a set of cardinality $\le\lambda$ such that $E_\xi\cup F_\xi$ is closed under f. Such an F_ξ clearly exists. Define $x_{\xi\beta}$ by transfinite recursion as $y_{\xi\gamma}$ for the least γ such that $y_{\xi\gamma}\in F_\xi\setminus\{x_{\xi\delta}:\delta<\beta\}$ and such that the set $\{y_{\xi\gamma}\}\cup\{x_{\alpha\beta}:\alpha<\xi\}$ is free with respect to f; if there is no such $y_{\xi\gamma}$ (this can happen only in case $\xi>0$ provided we make sure that $|F_0|=\lambda$), then put $x_{\xi\gamma}=x_{0\beta}$.

We claim that $F_\xi\subseteq\{x_{\xi\beta}:\beta<\lambda\}$ ($\subseteq E_\xi\cup F_\xi$) holds; this will imply that $E_{\xi+1}$ is also closed under f. In fact, assume, on the contrary, that $\gamma<\lambda$ is the least ordinal such that $y_{\xi\gamma}\notin\{x_{\xi\beta}:\beta<\lambda\}$. Then for every $\beta<\lambda$, either the set $\{y_{\xi\gamma}\}\cup\{x_{\alpha\beta}:\alpha<\xi\}$ is not free or we must have $x_{\xi\beta}=y_{\xi\delta}$ for some $\delta<\gamma$. We show that this is absurd by showing that the former possibility can occur only for $<\lambda$ of β's (this is obviously true for the latter possibility). Indeed, the set E_ξ is closed under f, and so $y_{\xi\gamma}\notin f(x_{\alpha\beta})$ holds for any $\alpha<\xi$ and $\beta<\lambda$; and, on the other hand, as $|f(y_{\xi\gamma})|<\lambda$ by our assumptions, $f(y_{\xi\gamma})\cap\{x_{\alpha\beta}:\alpha<\xi\}\neq 0$ can hold only for less than λ of β's, because these sets are pairwise disjoint. This establishes our claim.

Eventually, we shall have $\kappa=\{x_{\alpha\beta}:\alpha<\mu\,\&\,\beta<\lambda\}$ for some ordinal μ, and the matrix $\langle x_{\alpha\beta}:\alpha<\mu\,\&\,\beta<\lambda\rangle$ will be defined according to our requirements. The proof is complete.

A corollary of this result, proved by D. Lázár [1936] long before the preceding theorem was established, is the following:

COROLLARY 44.2. *Let κ and λ be cardinals, κ regular and $\lambda<\kappa$, and let f be a set mapping on κ such that $|f(x)|<\lambda$ holds for every $x\in\kappa$. Then there exists a set of cardinality κ that is free with respect to f.*

PROOF. If $\lambda\ge\omega$, then the assertion follows immediately from the preceding theorem. If $\kappa>\omega$, then we may assume $\lambda\ge\omega$; so the only case that remains to be settled is when $\kappa=\omega$ and $\lambda<\omega$. In this case Theorem 44.1 is not applicable, since $\lambda\geq\omega$ is assumed there (though an extension of this theorem to fintie λ's, due to de Bruijn and Erdős [1951], which says that, under the same assumptions, κ splits up into $2\lambda-1$ free sets, would be applicable). But a simple argument invoking Ramsey's theorem settles this case as well. To see this, define a coloring $g:[\omega]^2\to 3$ as follows: for each $m, n<\omega$ with $m<n$ we put $g(\{mn\})=0$ if $m\in f(n)$, and $g(\{mn\})=1$ if this is not the case but $n\in f(m)$; finally write $g(\{mn\})=2$ if $\{mn\}$ is a free set with respect to f. Ramsey's theorem says that there is an infinite set $X\subseteq\omega$ that is homogeneous with respect to this coloring. We show that this X must be a set free with respect to f. To this end, let $x_0, x_1, \ldots$ be an enumeration of the elements of X in increasing order (in the ordering 'less than' of integers). If the color of X were 0, then we would have $\{x_i: i<\lambda\}\subseteq f(x_\lambda)$, and if it were 1, then we would have $\{x_i: 1\leq i<\omega\}\subseteq f(x_0)$; either of these contradicts the assumption that $|f(x)|<\lambda$ for all $x\in\omega$. Hence X must have color 2; this means that X is a set free with respect to f.

Finally, Hajnal [1961] proved the following theorem, answering Ruziewicz's conjecture affirmatively:

THEOREM 44.3. *Let κ and λ be cardinals, κ infinite and $\lambda<\kappa$, and let f be a set mapping on κ such that $|f(x)|<\lambda$ holds for all $x\in\kappa$. Then there is a set of cardinality κ that is free with respect to f.*

PROOF. By the preceding result we may assume that κ is singular; hence we may also assume that λ is regular and $\lambda>\mathrm{cf}(\kappa)$ (indeed, if this is not the case already, then replace λ with a larger $\lambda'<\kappa$ that satisfies these requirements). Let $\langle\kappa_\xi:\xi<\mathrm{cf}(\kappa)\rangle$ be an increasing sequence of regular cardinals tending to κ such that $\kappa_0>\lambda$. It is easy to see then that there are free sets $A_\xi\subseteq\kappa$ of cardinality κ_ξ for each $\xi<\kappa$; indeed, one only has to apply Corollary 44.2 with κ_ξ replacing κ. Write

$$B_\xi=A_\xi\setminus\bigcup_{\alpha<\xi}(A_\alpha\cup\bigcup\{f(x):x\in A_\alpha\}); \tag{1}$$

it is clear that $|B_\xi|=\kappa_\xi$ holds, as the set subtracted from A_ξ has smaller cardinality. We shall define a matrix $\langle C_{\xi\mu}:\xi<\mathrm{cf}(\kappa)\,\&\,\mu<\lambda\rangle$ and a sequence $\langle\eta_\xi:\xi<\mathrm{cf}(\kappa)\rangle$ by transfinite recursion such that, for any $\xi<\mathrm{cf}(\kappa)$,

$$C_{\xi\mu}\subseteq B_\xi,\ |C_{\xi\mu}|=\kappa_\xi, \tag{2}$$

$$C_{\xi\mu}\cap C_{\xi\mu'}=0 \tag{3}$$

hold for all $\mu, \mu'<\lambda$ with $\mu\neq\mu'$, and

$$f(x)\cap\bigcup\{C_{\alpha\nu}:\eta_\xi\leq\nu<\lambda\}=0 \tag{4}$$

holds for all $x \in \bigcup_{\mu<\mathrm{cf}(\kappa)} C_{\xi\mu}$ and for all $\alpha<\xi$. Assume that for some $\xi<\lambda$ the matrix $\langle C_{\xi'\mu}: \xi'<\xi \,\&\, \mu<\lambda\rangle$ and the sequence $\langle \eta_{\xi'}: \xi'<\xi\rangle$ have already been constructed so that (2)–(4) hold for all $\xi'<\xi$ replacing ξ; we are going to define $C_{\xi\mu}$, $\mu<\lambda$, and η_ξ. To this end, write

$$B_\xi^\eta = \{x \in B_\xi : f(x) \cap \bigcup_{\eta \le \nu < \lambda} \bigcup_{\alpha<\xi} C_{\alpha\nu} = 0\}$$

for each $\eta<\lambda$. Noting that λ is regular and that the sets $\bigcup_{\alpha<\xi} C_{\alpha\nu}$, $\nu<\lambda$, are pairwise disjoint, the assumption $|f(x)|<\lambda$ implies that each $x \in B_\xi$ must belong to some B_ξ^η. Hence, as $|B_\xi|=\kappa_\xi$ is regular and $\kappa_\xi>\lambda$, it follows that $|B_\xi^\eta|=\kappa_\xi$ for some $\eta<\lambda$. Write $\eta_\xi=\eta$ for this η, and let $C_{\xi\mu}$, $\mu<\lambda$, be a decomposition into λ pairwise disjoint sets, each of cardinality κ_ξ, of $B_\xi^{\eta_\xi}$. It is obvious then that (2)–(4) hold for this ξ. This completes the definition of the matrix

$$\langle C_{\xi\mu}: \xi<\mathrm{cf}(\kappa) \,\&\, \mu<\lambda\rangle \text{ and the sequence } \langle \eta_\xi : \xi<\mathrm{cf}(\kappa)\rangle.$$

It is easy now to finish the proof of the theorem: As $\lambda>\mathrm{cf}(\kappa)$ and λ is regular, there must be an $\eta<\lambda$ such that $\eta_\xi \le \eta$ for all $\xi<\mathrm{cf}(\kappa)$. Then $C=\bigcup_{\xi<\mathrm{cf}(\kappa)} C_{\xi\eta}$ is a free set by (1) and (4) and because each $C_{\xi\eta}$ is free; this latter is the case since $C_{\xi\eta} \subseteq B_\xi \subseteq A_\xi$, and A_ξ itself is free. C has cardinality $\sum_{\xi<\mathrm{cf}(\kappa)} \kappa_\xi = \kappa$ by (2). The proof is complete.

45. SET MAPPINGS OF TYPE >1

In the preceding section we discussed questions involving set mappings *on* a set. In this section we turn to the usually more intricate questions involving set mappings over a set: given a set E, a *set mapping over* E is a function f from a subset of $\mathscr{P}(E)$ into $\mathscr{P}(E)$ such that $X \cap f(X)=0$ for every X in the domain of f. A subset H of E is called *free with respect to* f if $H \cap f(X)=0$ holds for every $X \in \mathscr{P}(H) \cap \mathrm{dom}(f)$. Given cardinals κ, λ, ρ, and σ, we shall study the following question: for an arbitrary set mapping $f: [\kappa]^\lambda \to [\kappa]^{<\rho}$, does there exist a free set of cardinality σ? The string of symbols

$$(\kappa, \lambda, \rho) \to \sigma \tag{1}$$

will mean that the answer to this question is affirmative for all such f, and $(\kappa, \lambda, \rho) \not\to \sigma$ will mean that this is not the case. For such set mappings f, λ is called the *type* and ρ, the *order*, of f.

The case $\lambda=1$ was covered in the preceding section: in fact, a set mapping $f: \kappa \to \mathscr{P}(\kappa)$ on κ can be identified with the set mapping $f': [\kappa]^1 \to \mathscr{P}(\kappa)$ over κ

by putting $f'(\{x\})=f(x)$ for all $x\in\kappa$. (For example, Theorem 44.3 can be expressed in symbols by saying that $(\kappa, 1, \lambda)\to\kappa$ holds if $\kappa>\lambda$ and $\kappa\geq\omega$ hold.) The case $\lambda\geq\omega$ is entirely settled by the following theorem:

THEOREM 45.1 (Erdős–Hajnal [1958]). *If κ and λ are infinite cardinals, then*

$$(\kappa, \lambda, 2)\not\to\lambda \tag{2}$$

holds.

Note that every set of cardinality $<\lambda$ here is free by definition, and so $(\kappa, \lambda, 2)\to\sigma$ trivially holds for any $\sigma<\lambda$. The proof of this result shows some similarities to that of Theorem 12.1.

PROOF. First we prove the relation

$$(\lambda, \lambda, 2)\not\to\lambda\,. \tag{3}$$

To this end, let $h\colon [\lambda]^\lambda\to[\lambda]^\lambda$ be a 1-1 mapping such that $h(X)\subset X$ for every $X\in[\lambda]^\lambda$ ($\subset$ means proper inclusion). It is easy to construct such an h by transfinite recursion: in fact, let $\langle A_\xi:\xi<2^\lambda\rangle$ be a wellordering of $[\lambda]^\lambda$, and choose $h(A_\xi)$ as the first $A_\eta\subset A_\xi$ $(\eta<2^\lambda)$ that has not been chosen as $h(A_\alpha)$ for any $\alpha<\xi$. For any $X\in[\lambda]^\lambda$, let $f'(h(X))$ be an arbitrary element of $X\setminus h(X)$, and for any $Y\in[\lambda]^\lambda\setminus\operatorname{ra}(h)$ define $f'(Y)$ arbitrarily. Put $f(Y)=\{f'(Y)\}$ for every $Y\in[\lambda]^\lambda$. It is easy to see that no subset of λ of cardinality is free with respect to the set mapping f; this verifies relation (3).

We now turn to the proof of (2). To this end, let $S\subseteq[\kappa]^\lambda$ be a maximal system of almost disjoint sets; i.e., let S be such that

$$|X\cap Y|<\lambda \text{ for any two distinct } X,\ Y\in S\,, \tag{4}$$

and

$$\text{if}\quad X\in[\kappa]^\lambda,\quad\text{then}\quad |X\cap Y|=\lambda\quad\text{for some}\quad Y\in S\,. \tag{5}$$

For each $X\in S$, let $f_X\colon[X]^\lambda\to[X]^1$ be a set mapping such that no subset of X of cardinality λ is free with respect to f_X; there is such a set mapping according to (3). Given a set $Z\in[\kappa]^\lambda$, put

$$f(Z)=f_X(Z) \tag{6}$$

if $Z\subseteq X$ for some $X\in S$ (note that if there is such an X, then it is uniquely determined by Z in view of (4)); if there is no such $X\in S$, then put e.g. $f(Z)=0$. We claim that f verifies relation (2). In fact, assume on the contrary that the set $X\in[\kappa]^\lambda$ is free with respect to f. We have $|X\cap Y|=\lambda$ for some $Y\in S$ by (5), and it is easy to see by (6) that the set $X\cap Y$ must be free with respect to f_Y; this, however, contradicts the definition of f_Y. The proof is complete.

An interesting intermediate case between $\lambda<\omega$ and $\lambda\geq\omega$ in (1) is the case of the set mappings $f:[\kappa]^{<\omega}\to[\kappa]^{\rho}$. Write

$$(\kappa, <\omega, \rho)\to\sigma \tag{7}$$

in case there is a free set of cardinality σ for all such f, and write $\not\to$ instead of $\to$ if this is not the case. Baumgartner, and later Devlin [1973a], proved that $V=L$ implies $(\kappa, <\omega, 2)\not\to\omega$ provided $\kappa\not\to(\omega)_2^{<\omega}$. Devlin and Paris [1973] proved that $(\kappa, <\omega, 2)\to\aleph_1$ contradicts $V=L$; in fact, it implies that $0^\#$ exists (for $0^\#$, see Solovay [1967] or e.g. Devlin [1973; p. 197]). On the other hand, we have:

THEOREM 45.2. *Let κ and λ be infinite cardinals such that $\kappa\to(\lambda)_2^{<\omega}$ holds. Then we have*

$$(\kappa, <\omega, \sigma)\to\lambda$$

for every cardinal $\sigma<\lambda$.

As we have $\kappa\to(\kappa)_2^{<\omega}$ for a measurable cardinal according to Theorem 34.11, this result implies that $(\kappa, <\omega, \sigma)\to\kappa$ holds for every $\sigma<\kappa$ for a measurable cardinal κ. A theorem of Solovay says that this conclusion holds even if we only assume that $\kappa>\omega$ is a cardinal such that there is a κ-saturated ideal on κ (see Definition 34.1(ii) for the definition of saturated ideals).

In the proofs of the above theorem and of several results below, we shall repeat in a refined form the trick that we used to prove Corollary 44.2 in case $\kappa=\omega$ and $\lambda<\kappa$. For this, given a set mapping f over κ such that dom $(f)\subseteq[\kappa]^{<\omega}$, we shall define a coloring g mapping (a subset of) $[\kappa]^{<\omega}$ into ω as follows: For a finite subset $X=\{x_0, \ldots, x_{k-1}\}$ of κ, where $x_0<\ldots<x_{k-1}$, put

$$g_f(X)=i \tag{8}$$

if $i<k$ is the least integer such that $x_i\in f(X\setminus\{x_i\})$ holds; if no such i exists, then we put $g_f(X)=k$. If dom $(f)=[\kappa]^{<\omega}$, then we shall define g_f on the whole of $[\kappa]^{<\omega}$, and if dom $(f)=[\kappa]^n$ for some integer n, then we shall define g_f on $[\kappa]^{n+1}$. It is obvious in both cases that a set $H\subseteq\kappa$ is free with respect to f if and only if $g_f(X)=|X|$ holds for every set $X\subset H$, $X\in$ dom (g_f); hence in particular, a set free with respect to f is homogeneous with respect to g_f. Note that if $X\in$ dom (g_f) and $|X|\leq 1$ then $g_f(X)=|X|$ holds trivially; hence this equality has to be checked only in case $|X|\geq 2$.

PROOF OF THEOREM 45.2. Given a set mapping $f:[\kappa]^{<\omega}\to[\kappa]^{<\sigma}$, define the coloring $g_f:[\kappa]^{<\omega}\to\omega$ according to (8). We assumed that $\kappa\to(\lambda)_2^{<\omega}$ holds; by Lemma 34.13 the relation $\kappa\to(\lambda)_\omega^{<\omega}$ also holds; hence there is a set H of order type λ that is homogeneous with respect to g_f; i.e., there are integers i_n, $n<\omega$, such that $g_f(X)=i_n$ hold whenever $X\subseteq H$ and $|X|=n$. We claim that the set H is free. Indeed, assume, on the contrary, that $i_n<n$, holds for some n with $2\leq n<\omega$.

Denote by h_ξ the ξth element of H for any $\xi<\lambda$, and put

$$Y=\{h_j:j<i_n\}\cup\{h_{i_n+\sigma+k}:k<n-i_n-1\}.$$

For each ordinal ξ with $i_n\leq\xi<\sigma$, the set $Y\cup\{h_\xi\}$ is an n-element subset of H the i_nth element of which is h_ξ. So we have $g_f(Y\cup\{h_\xi\})=i_n$, i.e.,

$$h_\xi\in f(Y)$$

holds by the definition of g_f. As ξ is an arbitrary ordinal here with $i_n\leq\xi<\sigma\dot{+}k$, this means that $|f(Y)|\geq\sigma$, which contradicts our assumptions. The proof is complete.

We now turn to the case of set mappings of a finite type. The proof of the next theorem applies the same trick as that of the preceding one.

THEOREM 45.3. *Let $r\geq 1$ and n be integers. Then*

$$(\omega, r, n)\rightarrow\omega$$

holds.

PROOF. Let $f:[\omega]^r\rightarrow[\omega]^{<n}$ be a set mapping; we have to show that there is an infinite free set. Define the coloring $g_f:[\omega]^{r+1}\rightarrow r+2$ as described in (8). According to Ramsey's theorem, there exists an infinite homogeneous set H with respect to this coloring. Let i be the color of H; this means that $g(X)=i$ holds for all $X\in[H]^{r+1}$. We claim that H is a free set with respect to f, i.e., that $i=r+1$ holds. In fact, assume $i\leq r$, and put

$$Y=\{h_j:j<i\}\cup\{h_k:i+n\leq k<n+r-1\},$$

where h_j denotes the jth element of H. As $g(Y\cup\{h_l\})=i$ holds for each l with $i\leq l<i+n$, we have $h_l\in f(Y)$ for these l according to the definition of g. This, however, contradicts the assumption that $|f(Y)|<n$. The proof is complete.

The tricks used above exploited the existence of a large homogeneous set. The existence of a large canonical set (see definition 27.1) can also be exploited in a similar way. This is illustrated by the following result; it should be noted, however, that this result will be strengthened below (cf. Theorem 45.6).

THEOREM 45.4. *Assume κ is a singular strong limit cardinal and $\lambda<\kappa$. Then*

$$(\kappa, r, \lambda)\rightarrow\kappa$$

holds for every integer r.

PROOF. Given a set mapping $f:[\kappa]^r\rightarrow[\kappa]^{<\lambda}$, define the coloring

$$g_f:[\kappa]^{r+1}\rightarrow r+2$$

as given by (8). According to Corollary 28.2, there are pairwise disjoint sets B_ξ, $\xi<\mathrm{cf}(\kappa)$, of cardinality $<\kappa$ such that $\bigcup_{\xi<\mathrm{cf}(\kappa)} B_\xi$ has cardinality κ and the sequence $\langle B_\xi : \xi<\mathrm{cf}(\kappa)\rangle$ is canonical with respect to g_f. Clearly, we may assume that $|B_\xi|>\lambda$ holds and the order type of B_ξ is an infinite cardinal for each $\xi<\mathrm{cf}(\kappa)$, and every element of B_ξ precedes every element of B_η whenever $\xi<\eta<\mathrm{cf}(\kappa)$, (even if the B_ξ's do not originally satisfy those requirements, they possess subsets that do). We claim that $B=\bigcup_{\xi<\mathrm{cf}(\kappa)} B_\xi$ is a free set. Assume, on the contrary, that $g_f(X)=i$ holds with some $i\leq r$ for an $(r+1)$-element subset X of B. Let $x_0,\ldots,x_r$ be the elements of X in increasing order, and define ξ_k by $x_k\in B_{\xi_k}$ $(k\leq r)$; clearly, $\xi_0\leq\ldots\leq\xi_r$. Define i' as the largest integer $\leq r$ for which $\xi_{i'}=\xi_i$. Denote by b_ξ^α the αth element of B_ξ; let η be the ordinal for which $x_i=b_{\xi_i}^\eta$. Put

$$Y=(X\setminus\{x_j : i\leq j\leq i'\})\cup\{b_{\xi_i}^{\eta\dot{+}\lambda\dot{+}j} : i<j\leq i'\}.$$

Note that all the b_{ξ_i}'s are defined because the order type of B_{ξ_i} is an infinite cardinal $>\lambda$ according to our assumptions. By canonicity, we have

$$i=g_f(X)=g_f(Y\cup\{b_i^{\eta\dot{+}\alpha}\})$$

for each $\alpha<\lambda(\dot{+}i\dot{+}1)$; hence $b_{\xi_i}^{\eta\dot{+}\alpha}\in f(Y)$ holds for all $\alpha<\lambda$ by the definition of g_f in (8). This contradicts our assumption that $|f(Y)|<\lambda$, completing the proof.

Obtaining large free sets via homogeneous (or canonical) sets is not a precise enough method. The next theorem illustrates that the use of end-homogeneous sets can lead to better results.

THEOREM 45.5. *Let κ and λ be cardinals such that $\kappa>\lambda$ and $\kappa\geq\omega$, and let $r\geq 2$ be an integer. Then*

$$((\exp_{r-2}(2^\kappa))^+, r, \lambda)\to\kappa$$

holds.

PROOF. Write $\kappa_1=\kappa$ and $\kappa_{i+1}=(2^{\kappa_i})^+$ for each integer $i\geq 1$. Then $\kappa_r = \exp_{r-2}(2^\kappa)^+$ holds for each integer $r\geq 2$. It is enough to show that

$$(\kappa_r, r, \lambda)\to\kappa \tag{9}$$

holds for each $r\geq 1$; we shall use induction on r. The case $r=1$ follows from Hajnal's Theorem 44.3. Assume now that $r\geq 2$ and (9) holds with $r-1$ replacing r. Let $f:[\kappa_r]^r\to[\kappa_r]^{<\lambda}$ be a set mapping; we want to define a coloring analogously to the definition of g_f under (8), but now we need a sligthly more complicated definition. We shall specify a mapping $g'_f:[\kappa_r]^{r+1}\to{}^{r+1}2$ as follows: given a set $X=\{x_0,\ldots,x_r\}\subseteq\kappa_r$, where $x_0<x_1<\ldots<x_r$, put

$$g'_f(X)=\langle g'_f(X)(0), g'_f(X)(1), \ldots, g'_f(X)(r)\rangle \tag{10}$$

where $g'_f(X)(i)=1$ if and only if $x_i\in f(X\setminus\{x_i\})$, and it is 0 otherwise $(0\leq i\leq r)$. As

$\kappa_r = (2^{\kappa}\cup)^+$, it follows from the relation $(2^{\kappa}\cup)^+ \to \langle \kappa_{r-1} \dot{+} 1 \rangle_{2^{r+1}}$ (cf. (15.18) in Corollary 15.3) that there is a set $H \subseteq \kappa_r$ of order type $\kappa_{r-1} \dot{+} 1$ that is end-homogeneous with respect to g'_f. Let ζ be the maximal element of H, and write $H' = H \setminus \{\zeta\}$. Let $X \subseteq H'$ be an arbitrary set of cardinality $r-1$, and let $\xi \in H'$ be an ordinal exceeding $\max X$. We claim that

$$f(X \cup \{\xi\}) \cap H' \subseteq f(X \cup \{\zeta\}) \tag{11}$$

holds. To see this, we first show that if $\alpha < \xi$ and $\alpha \in H' \setminus X$, then

$$\alpha \in f(X \cup \{\xi\}) \tag{12}$$

implies

$$\alpha \in f(X \cup \{\zeta\}). \tag{13}$$

Indeed, assume (12) and let $i < r+1$ be the integer such that α is the $(i+1)$st element of the set $X \cup \{\xi\} \cup \{\alpha\}$. Then $g'_f(X \cup \{\xi\} \cup \{\alpha\})(i) = 1$ follows from (12) according to (10); hence we have $g'_f(X \cup \{\zeta\} \cup \{\alpha\})(i) = 1$ by end-homogeneity, and so (13) holds again by (10). So it is enough to show in order to complete the proof of (11) that (12) fails for the remaining values of α, i.e., that

$$\alpha \notin f(X \cup \{\xi\}) \tag{14}$$

whenever either $\alpha \in H'$ and $\alpha > \xi$ or $\alpha \in X \cup \{\xi\}$. In the latter case this holds by the stipulation that the image under a set mapping of a set is disjoint from this set, made in the definition of set mappings; hence we only have to deal with the case when $\alpha \in H'$ and $\alpha > \xi$. In this case α is the last (i.e., the $(r+1)$-st) element of the set $X \cup \{\xi\} \cup \{\alpha\}$, and so, assuming that (14) fails, we have $g'_f(X \cup \{\xi\} \cup \{\alpha\})(r) = 1$. By end-homogeneity, $g'_f(X \cup \{\xi\} \cup \{\alpha'\})(r) = 1$ holds then for any $\alpha' \in H'$ with $\alpha' > \xi$; i.e., $\alpha' \in f(X \cup \{\xi\})$ holds for any such α'. This, however, contradicts our assumption that $|f(X \cup \{\xi\})| < \lambda < \kappa_{r-1}$, verifying (14). Thus (11) is established.

Define now the set mapping $f' \colon [H']^{r-1} \to [H']^{<\lambda}$ by putting

$$f'(X) = f(X \cup \{\zeta\}) \cap H'$$

for every $X \in [H']^{r-1}$. As $|H'| = \kappa_{r-1}$, the induction hypothesis implies that there exists a set $F \subseteq H'$ of cardinality κ that is free with respect to f'. It follows from (11) that F is free also with respect to f. The proof of the theorem is complete.

Later on we shall comment on the strength of the above theorem, but at present we give another, more sophisticated, application of the method used here in order to give a generalization of Theorem 45.4. Instead of obtaining a large end-homogeneous set now, we shall obtain a set that might be called 'end-canonical'.

THEOREM 45.6. *Let $r \geq 1$ be an integer and $\tau \geq \omega$ a cardinal, and let $\langle \kappa_\xi : \xi < \tau \rangle$ be a sequence of cardinals such that $\tau < \kappa_0$ and*

$$\exp_{r-1}(\kappa_\xi) < \exp_{r-1}(\kappa_\eta) \tag{15}$$

holds whenever $\xi < \eta < \tau$, and let λ be a cardinal $< \kappa_0$. Then we have

$$(\sum_{\xi<\tau} \exp_{r-1}(\kappa_\xi), r, \lambda) \to \sum_{\xi<\tau} \kappa_\xi . \tag{16}$$

PROOF. Instead of (16), we shall prove the apparently (but not in fact) stronger statement

$$(\sum_{\xi<\tau} \exp_{r-1}(\kappa_\xi), r, \lambda^+) \to \sum_{\xi<\tau} \kappa_\xi . \tag{16'}$$

We shall use induction on r. (16′) holds in case $r=1$ by Hajnal's Theorem 44.3. Let now $r>1$ and assume that (16′) holds with $r-1$ replacing r and with any $\lambda' < \kappa_0$ replacing λ (note that (15) obviously remains valid if we replace r with $r-1$ there). Writing

$$\kappa^{(i)} = \sum_{\xi<\tau} \exp_{i-1}(\kappa_\xi)$$

for any integer $i \geq 1$, assume that we are given a set mapping $f: [\kappa^{(r)}]^r \to [\kappa^{(r)}]^{\leq \lambda}$. We have to prove that there is a set of cardinality $\kappa^{(1)}$ that is free with respect to f. To this end, choose sets $A_\alpha \subseteq \kappa^{(r)}$, $\alpha < \tau$, such that $|A_\alpha| = (\exp_{r-1}(\kappa_\alpha))^+$ and such that each element of A_α precedes any element of A_β whenever $\alpha < \beta < \tau$. Define the coloring $g'_f : [\kappa^{(r)}]^{r+1} \to {}^{r+1}2$ as was done in (10) (replace κ_r there with $\kappa^{(r)}$). For each $\alpha < \tau$, we shall define a set $B_\alpha \subseteq A_\alpha$ such that B_α has order type $\exp_{r-2}(\kappa_\alpha) \dot{+} 1$ and, denoting by ζ_α the last element of B_α, we have

$$g'_f(X \cup \{\xi\}) = g'_f(X \cup \{\zeta_\alpha\}) \tag{17}$$

whenever $X \in [\bigcup_{\nu<\tau} B_\nu]^r$, $X \subseteq \xi$, and $\xi \in B_\alpha$ (the sequence $\langle B_\nu : \nu < \tau \rangle$ might be called end-canonical with respect to the coloring g_f; the arguments used here do in fact form a part of the proof of the General Canonization Lemma [Lemma 28.1]). Fix $\alpha < \tau$, and assume that the sets B_β, $\beta < \alpha$ have already been defined. For every $n \leq r+1$, call two sets $X, Y \in [A_\alpha]^n$ equivalent if

$$g'_f(Z \cup X) = g'_f(Z \cup Y)$$

holds for every $Z \in [\bigcup_{\beta<\alpha} B_\beta]^{r+1-n}$; it is easy to see that, for each n, there are at most $2^{(r+1)\cdot(1+\Sigma_{\beta<\alpha}|B_\beta|)} \leq 2^{(r+1)\cdot \exp_{r-2}(\kappa_\alpha)} = \exp_{r-1}(\kappa_\alpha)$ equivalence classes. Let $g''_f : [A_\alpha]^{\leq r+1} \to \exp_{r-1}(\kappa_\alpha)$ be a coloring such that two sets have the same color exactly if they have the same cardinality and they are equivalent in the sense described just before. It follows from the relation $(2^\kappa)^+ \to \langle \kappa^+ \dot{+} 1 \rangle_{2^\kappa}$ (cf. (15.17) in Corollary 15.3) with $\kappa = \exp_{r-2}(\kappa_\alpha)$ that there is a set of order type $\exp_{r-2}(\kappa_\alpha) \dot{+} 1$ that is end-homogeneous with respect to g''_f. Choose this set as B_α. It is easy to see that (17) holds for this B_α.

Writing $B = \bigcup_{\nu < \tau} B_\nu$, we claim that we have

$$f(Y \cup \{\xi\}) \cap B \subseteq f(Y \cup \{\zeta_\alpha\}) \cap B \tag{18}$$

whenever $Y \in [B]^{r-1}$ and $Y \subseteq \xi \in B_\alpha$ holds ($\alpha < \tau$). In fact, if $\gamma \in B \setminus Y$ and $\gamma < \xi$, then (17) with $X = Y \cup \{\gamma\}$ implies that $\gamma \in f(Y \cup \{\xi\})$ holds exactly if $\gamma \in f(Y \cup \{\zeta_\alpha\})$ holds. If $\gamma \in Y$ or $\gamma = \xi$, then $\gamma \notin f(Y \cup \{\xi\})$ by the definition of set mappings. Finally, if $\gamma \in B_\beta$ with $\alpha \leq \beta < \tau$ and $\gamma > \xi$, then $\gamma \notin f(Y \cup \{\xi\})$ also holds. In fact, (17) with γ replacing ξ and with $X = Y \cup \{\xi\}$ implies that $\gamma \in f(Y \cup \{\xi\})$ holds if and only if $\zeta_\beta \in f(Y \cup \{\xi\})$. So, if the former relation holds with one γ, then it must hold with all γ satisfying $\gamma \in B_\beta$ and $\gamma > \xi$; this is, however, impossible, as B_β has order type $\exp_{r-2}(\kappa_\beta) \dot{+} 1$, and $|f(Y \cup \{\xi\})| \leq \lambda$ according to our assumption. Therefore $\gamma \notin f(Y \cup \{\xi\})$ indeed holds in this case. Thus (18) is completely verified.

Consider now the set mapping $f' : [B]^{r-1} \to [B]^{\leq \lambda \cdot \tau}$ defined as follows: given any $Y \in [B]^{r-1}$ let $\alpha_Y < \tau$ be the least ordinal such that $Y \subseteq \zeta_{\alpha_Y}$, and put

$$f'(Y) = \bigcup_{\alpha_Y \leq \alpha < \tau} (f(Y \cup \{\zeta_\alpha\}) \cap B).$$

As the cardinality of B equals $\sum_{\xi < \tau} \exp_{r-2}(\kappa_\xi)$ and $\lambda \cdot \tau < \kappa_0$, there is a set F free with respect to f' by the induction hypothesis. It follows from (18) that F is also free with respect to f. The proof is complete.

Some remarks about the strength of the above theorems, especially of Theorem 45.5, seem appropriate here. We first mention the following simple result of Kuratowski and Sierpiński (see Kuratowski [1951]):

THEOREM 45.7. *For any integer k and any ordinal α we have*

$$(\aleph_{\alpha+k}, k+1, \aleph_\alpha) \nrightarrow k+2. \tag{19}$$

PROOF. We use induction on k. The case $k=0$ is simple and has been mentioned in the introduction of the preceding section (the following set mapping $f_0 : [\aleph_\alpha]^1 \to [\aleph_\alpha]^{<\aleph_\alpha}$ has no two-element free set: for each $\xi \in \aleph_\alpha$, let $f(\{\xi\}) = \xi [= \{\gamma : \gamma < \xi\}]$). Let now $k > 0$ and assume that the assertion is valid with $k-1$ replacing k; let $f_{k-1} : [\aleph_{\alpha+(k-1)}]^{k-1} \to [\aleph_{\alpha+(k-1)}]^{<\aleph_\alpha}$ be a set mapping which has no free set of $k+1$ elements. For each $\xi<\aleph_{\alpha+k}$ let $h_\xi:\xi\to\aleph_{\alpha+(k-1)}$ be a 1-1 function, and for any set $X\subseteq\aleph_{\alpha+k}$ of $k-1$ elements and any ordinal ξ with $X\subseteq\xi<\aleph_{\alpha+k}$ put

$$f_k(X\cup\{\xi\})=\{\eta<\xi: h_\xi(\eta)\in f_{k-1}(h_\xi``X)\}.$$

Then f_k is a set mapping from $[\aleph_{\alpha+k}]^k$ into $[\aleph_{\alpha+k}]^{<\aleph_\alpha}$. If $Y\subseteq\aleph_{\alpha+k}$ is a free set of $k+2$ elements with respect to f_k, and ξ is the maximal element of Y, then $h_\xi``(Y\setminus\{\xi\})$ must be free with respect to f_{k-1}; this latter is impossible by the induction hypothesis, showing that f_k verifies (19). The proof is complete.

According to this result, no large free sets can be expected to exist for a set mapping f: $[\aleph_1]^2\to[\aleph_1]^{<\omega}$; in fact, it follows that if GCH holds, then the cardinal $(\exp_{r-2}(2^\kappa))^+$ in Theorem 45.5 cannot be replaced with a smaller one (see below). Moreover, a forcing argument shows that if $\aleph_\alpha$ is regular and GCH holds, then for each integer $r\geq 2$ there is a cardinal preserving Cohen extension in which $2^{\aleph_\alpha}=\aleph_{\alpha+r}$ and

$$(\aleph_{\alpha+r}, r, 2)\nrightarrow\aleph_{\alpha+1}$$

holds (see Hajnal–Máté [1975], Theorem 3.5). This shows that, even without GCH, the cardinal $(\exp_{r-2}2^\kappa)^+$ in Theorem 45.5 cannot be replaced with a smaller one in case $r=2$ and κ is a successor cardinal. If $2^{\aleph_0}=\aleph_1$, then Theorem 45.5 with $\kappa=\aleph_1$ implies $(\aleph_2, 2, \omega)\to\aleph_1$. Our next theorem, due to Erdős and Hajnal [1958], says with $\kappa=\aleph_1$ that this cannot be improved if $2^{\aleph_1}=\aleph_2$.

THEOREM 45.8. *Let κ be an infinite cardinal and assume that $2^\kappa=\kappa^+$. Then*

$$(\kappa^+, 2, 2)\nrightarrow\kappa^+.$$

PROOF. Let A_α, $\alpha<\kappa^+$, be an enumeration of the elements of $[\kappa^+]^\kappa$, and for each $\xi<\kappa^+$ define the set mapping g_ξ: $\xi\to[\xi]^1$ so that no set $A_\alpha\subseteq\xi$ with $\alpha<\xi$ be free with respect to g_ξ. This can easily be done as follows. Rearrange the set $\{A_\alpha$: $\alpha<\xi\ \&\ A_\alpha\subseteq\xi\}$ into order type $\nu_\xi\leq\kappa$: $\{B^\xi_\eta$: $\eta<\nu_\xi\}$. Using transfinite recursion, pick two distinct elements γ_η, $\delta_\eta\in B^\xi_\eta\setminus\{\gamma_\zeta, \delta_\zeta:\zeta<\gamma\}$, and put $g_\xi(\gamma_\eta)=$ $=\{\delta_\eta\}$ $(\eta<\nu_\xi)$.

Having defined the set mappings g_ξ, put $f(\{\eta\xi\})=g_\xi(\eta)$ for any η, ξ with $\eta<\xi<\kappa^+$. What we have to show is that there is no set of cardinality κ^+ that is free with respect to f. Assume, on the contrary, that $x\subseteq\kappa^+$ is such a set. Then $A_\alpha\subset X$ holds for some $\alpha<\kappa^+$; let $\xi\in X$ be such that $\alpha<\xi$ and $a_\alpha\leq\xi$. Then, by the above construction, A_α is not free with respect to g_ξ; hence $A_\alpha\cup\{\xi\}\subset X$ is not free with respect to f. This is a contradiction, completing the proof.

This theorem shows that Theorem 45.5 is best possible in case $r=2$, $2^\kappa=\kappa$, and κ is a successor cardinal. It is also best possible in case GCH holds, $r=3$, and κ is a successor cardinal, since if GCH holds and $\aleph_\alpha$ is regular, then

$$\mathrm{GCH}+(\aleph_{\alpha+2}, 3, 2)\nrightarrow\aleph_{\alpha+1}$$

holds in a cardinal preserving Cohen extension (Hajnal–Máté [1975]). Recently, J. Burgess [1979] and K. J. Devlin showed that this relation also holds at least in case $\alpha=0$ if $V=L$. There are no good results in case $r\geq 4$; one would expect that

$$\mathrm{GCH}+(\aleph_3, 4, 2)\nrightarrow\aleph_1$$

is consistent, but we cannot prove this.

We shall further study set mappings in the next section; to conclude this one we should like to indicate an alternative approach to Theorem 45.5. It works only in case κ is regular, since it makes use of the partition relation $(\exp_{r-2}(2^{\kappa}))^{+}\to(\kappa)^{r}_{\lambda}$, which was established only for $\lambda<\mathrm{cf}(\kappa)$ (see (17.30)). For the sake of simplicity, we shall confine ourselves only to the instructive particular case of Theorem 45.5 when $r=2$, $\kappa=\aleph_1$ and $\lambda=\aleph_0$; that is we shall give an alternative proof of the relation

$$((2^{\aleph_0})^{+}, 2, \omega)\to\aleph_1\,. \tag{20}$$

PROOF OF (20). Write $\kappa=(2^{\aleph_0})^{+}$. Given a set mapping $f:[\kappa]^2\to[\kappa]^{<\omega}$, we have to prove that there exists a free set of cardinality $\aleph_1$. For $\alpha<\kappa$ let $g_\alpha:\kappa\to[\kappa]^{<\omega}$ be the set mapping defined by

$$g_\alpha(\beta)=f(\{\alpha\beta\})$$

($\beta<\kappa$ and $\beta\neq\alpha$; for $\beta=\alpha$ put $g_\alpha(\beta)=0$). Theorem 44.1 of Fodor with $\lambda=\omega$ implies that κ can be represented as the sum of pairwise disjoint sets $E_{\alpha n}$, $n<\omega$, each of which is free with respect to g_α. Define a coloring h: $[\kappa]^2\to\omega\times\omega$ as follows: for any α, β with $\alpha<\beta<\kappa$ put $h(\{\alpha\beta\})=\langle mn\rangle$ if $\alpha\in E_{\beta m}$ and $\beta\in E_{\alpha n}$ ($m, n<\omega$). As $\kappa=(2^{\aleph_0})^{+}$, it follows from the partition relation $(2^{\aleph_0})^{+}\to(\aleph_1)_{\aleph_0}$ (cf. Corollary 17.5) that there is a set $H\subseteq\kappa$ of cardinality $\aleph_1$ that is homogeneous with respect to h; i.e., there are integers m, n such that $h(\{\alpha\beta\})=\langle m, n\rangle$ holds for any two distinct $\alpha, \beta\in H$. We claim that the set H is free with respect to f. In fact, let $\alpha, \beta, \gamma\in H$ be ordinals such that $\alpha<\beta<\gamma$. Then $\alpha\notin f(\{\beta\gamma\})$, as $\alpha, \beta\in E_{\gamma m}$, and we have $\alpha\notin g_\gamma(\beta)=f(\{\beta\gamma\})$ since $E_{\gamma m}$ is free with respect to g_γ. We can establish $\beta\notin f(\{\alpha\gamma\})$ and $\gamma\notin f(\{\alpha\beta\})$ similarly. The proof is complete.

This proof of course works for any regular κ. As for the case $r>2$, given a set mapping $f:[\kappa]^r\to[\kappa]^{<\lambda}$, we have to define the set mappings $g_X:\kappa\to[\kappa]^{<\lambda}$ for any $X\in[\kappa]^{r-1}$ by putting $g_X(\alpha)=f(X\cup\{\alpha\})$ for any $\alpha\in\kappa\setminus X$ ($g_X(\alpha)=0$ if $\alpha\in X$), and then the coloring h: $[\kappa]^r\to{}^r\lambda$ can be defined analogously as we did above, with appropriate changes. The details are left to the reader (see also Erdős–Hajnal [1958]).

46. FINITE FREE SETS WITH RESPECT TO SET MAPPINGS OF TYPE >1

We start out with the counterpart of Theorem 45.7, due to Kuratowski and Sierpiński (see Kuratowski [1951]).

THEOREM 46.1. *For any integer k and any ordinal α we have*

$$(\aleph_{\alpha+k}, k, \aleph_\alpha) \to k+1. \tag{1}$$

PROOF. We use induction on k. In case $k=0$ relation (1) holds vacuously (in fact, the case $k=1$ is also easy, and is settled by Theorem 44.3). Let f be a set mapping, f: $[\aleph_{\alpha+k}]^k \to [\aleph_{\alpha+k}]^{<\aleph_\alpha}$. Assume now that $k>1$, and define a set mapping g on the $(k-1)$-element subsets of $\aleph_{\alpha+k} \setminus \aleph_\alpha = \{\xi\colon \aleph_\alpha \le \xi < \aleph_{\alpha+k}\}$ as follows: if $X \in [\aleph_{\alpha+k} \setminus \aleph_\alpha]^{k-1}$, then put

$$g(X) = \bigcup_{\xi < \aleph_\alpha} f(X \cup \{\xi\}).$$

Then $|g(X)| < \aleph_{\alpha+1}$, and so the relation

$$(\aleph_{\alpha+k}, k-1, \aleph_{\alpha+1}) \to k,$$

which holds in view of the induction hypothesis, implies that there exists a set $Y \subseteq \aleph_{\alpha+k} \setminus \aleph_\alpha$ of k elements that is free with respect to g. Let η be an arbitrary element of the set $\aleph_\alpha \setminus f(Y)$ (this set is nonempty since $f(Y)$ has cardinality $<\aleph_\alpha$). It is easy to see that $Y \cup \{\eta\}$ is a set of $k+1$ elements that is free with respect to f. The proof is complete.

The above theorem can be improved in the case $k=2$ or 3 (and, of course, in case $k=1$), but nothing is known in case $k \ge 4$; for instance, it is not known whether or not the relation

$$(\aleph_4, 4, \aleph_0) \to 6 \tag{2}$$

can be proved in ZFC. If GCH holds, then of course we have $(\aleph_4, 4, \aleph_0) \to \aleph_1$ in view of Theorem 45.5. The result that improves Theorem 46.1 in case $k=2$ and 3 is the following:

THEOREM 46.2. *For any integer n and any ordinal α we have*

$$(\aleph_{\alpha+2}, 2, \aleph_\alpha) \to n \tag{3}$$

(Hajnal–Máté [1975]) *and*

$$(\aleph_{\alpha+3}, 3, \aleph_\alpha) \to n \tag{4}$$

(Hajnal).

PROOF OF (3). Let f: $[\aleph_{\alpha+2}]^2 \to [\aleph_{\alpha+2}]^{<\aleph_\alpha}$ be a set mapping. We construct a matrix $\langle x_{m\xi}\colon m<\omega \,\&\, \xi<\aleph_\alpha\rangle$ of pairwise distinct elements of $\aleph_{\alpha+2}$ and a sequence $H_0 \supseteq H_1 \supseteq \ldots \supseteq H_m \supseteq \ldots (m < \omega)$ of subsets of $\aleph_{\alpha+2}$ with the following properties:

(i) $|H_m| = \aleph_{\alpha+2}$ and $\{x_{m\xi}: \xi < \aleph_\alpha\} \subseteq H_m$ holds for every $m < \omega$,

(ii) $u \notin f(\{x_{m\xi}v\})$ holds for every $m < \omega$, $\xi < \aleph_\alpha$, and $u, v \in H_{m+1}$.

The construction proceeds by finite recursion as follows. Take $H_0 = \aleph_{\alpha+2}$, and, having defined H_m for $m \le r$ and $x_{m\xi}$ for $m < r$ and $\xi < \aleph_\alpha$, where $r < \omega$, choose pairwise distinct elements of H_r as $x_{r\xi}$, $\xi < \aleph_\alpha$. Define the set mapping g_r on $H'_r = H_r \setminus \{x_{r\xi}: \xi < \aleph_\alpha\}$ by putting

$$g_r(u) = \bigcup_{\xi < \aleph_\alpha} f(\{x_{r\xi}u\})$$

for any $u \in H'_r$. As $|H'_r| = \aleph_{\alpha+2}$ and $|g_r(u)| < \aleph_{\alpha+1}$, by Theorem 44.3 there exists a set $H''_r \subseteq H'_r$ of cardinality $\aleph_{\alpha+2}$ that is free with respect to g_r. Put $H_{r+1} = H''_r$. This finishes the construction of the matrix $\langle x_{m\xi}: m < \omega \;\&\; \xi < \aleph_\alpha \rangle$ and the sets H_m, $m < \omega$.

It is now easy to construct an n-element set $\{x_i: i < n\}$ that is free with respect to f. Choose $x_i \in \{x_{i\xi}: \xi < \aleph_\alpha\}$ arbitrarily for $i = n-1$ and $i = n-2$. If $i < n-2$ and x_j has already been defined for any j with $i < j \le n-1$, then choose x_i such that

$$x_i \in \{x_{i\xi}: \xi < \aleph_\alpha\} \setminus \bigcup_{i < j < j' \le n-1} f(\{x_j x_{j'}\}). \tag{5}$$

It is easy to see that the set $\{x_i: i < n\}$ is free with respect to f. In fact, if $i < j < k \le n-1$ then $x_k \notin f(\{x_i x_j\})$ and $x_j \notin f(\{x_i x_k\})$ hold in view of (ii) as $x_j, x_k \in H_{i+1}$, and, moreover, $x_i \notin f(\{x_j x_k\})$ holds in view of (5). The proof of (3) is complete.

For the proof of (4) we need a lemma on set mappings of type 1. This lemma should be compared to Lemma 20.3; the difference here is that we do not assume the regularity of λ, and so we are able to derive only a weaker result.

LEMMA 46.3. *Let λ be an infinite cardinal, k an integer, and let $E = \bigcup_{i<k} E_i$, where E_i is a set of cardinality $> \lambda$ for every $i < k$. Let $f: E \to \mathscr{P}(E)$ be a set mapping such that $|f(x)| < \lambda$ holds for any $x \in E$. Then there is a set X free with respect to f such that $|X \cap E_i| > \lambda$ holds for every $i < k$.*

PROOF. First we prove the weaker assertion that there is a free transversal with respect to f; more precisely, we prove that the following Claim holds:

CLAIM. *Under the assumptions of the above lemma, there is a free set Y such that $Y \cap E_i$ is nonempty for any $i < k$.*

Proof of Claim. Define the sets $F_i \subseteq E_i$, $i<k$, by recursion as follows. Given $i<k$, if F_j has already been defined for all $j<i$, then choose F_i as an arbitrary subset of cardinality λ of the set

$$E_i \setminus \bigcup \{f(x): x \in \bigcup_{j<i} F_j\}.$$

Now define the elements $y_i \in F_i$, $i<k$, by recursion as follows. Given $i<k$, if y_j has already been defined for all j with $i<j<k$, then choose y_i as an arbitrary element of the set

$$F_i \setminus \bigcup_{i<j<k} f(y_j).$$

It is easy to see that $Y=\{y_i: i<k\}$ is a free set, completing the proof of the claim.

We now return to the proof of the lemma. Define the pairwise disjoint finite sets Y_ξ, $\xi<\lambda^+$, with the aid of the above Claim by transfinite recursion as follows: Having defined the sets Y_η for all $\eta<\xi$, where $\xi<\lambda^+$, let Y_ξ be a finite free subset of $E \setminus \bigcup_{\eta<\xi} Y_\eta$ such that $Y_\xi \cap E_i$ is nonempty for each $i<k$.

Define now the set mapping $g: \lambda^+ \to P(\lambda^+)$ as follows: for each $\xi<\lambda^+$ put

$$g(\xi)=\{\eta<\lambda^+: Y_\eta \cap \bigcup_{x \in Y_\xi} f(x) \neq 0\}.$$

Clearly, $\xi \notin g(\xi)$ (since Y_ξ is free) and $|g(\xi)|<\lambda$ for all $\xi<\lambda^+$. According to e.g. Theorem 44.3, there is a set $H \subseteq \lambda^+$ of cardinality λ^+ that is free with respect to g. It is easy to see that the set

$$X=\bigcup\{Y_\xi: \xi \in H\}$$

satisfies the requirements of the lemma, completing the proof.

Proof of (4). Let $f: [\aleph_{\alpha+3}]^3 \to [\aleph_{\alpha+3}]^{<\aleph_\alpha}$ be a set mapping. We start out in a way analogous to the proof of (3) and construct a matrix $\langle x_{m\xi}: m<\omega \,\&\, \xi<\aleph_{\alpha+1}\rangle$ of pairwise distinct elements of $\aleph_{\alpha+3}$ and a sequence $H_0 \supseteq H_1 \supseteq \ldots \supseteq H_m \supseteq \ldots$ $(m<\omega)$ of subsets of $\aleph_{\alpha+3}$ with the following properties:

(i) $|H_m|=\aleph_{\alpha+3}$ and $\{x_{m\xi}: \xi<\aleph_{\alpha+1}\} \subseteq H_m$ for every $m<\omega$.

(ii). $u \notin f(\{x_{l\xi} x_{m\eta} v\})$ holds whenever $l<m<\omega$, $\xi, \eta<\aleph_{\alpha+1}$ and $u, v \in H_{m+1}$.

The construction proceeds by finite recursion as follows. Take $H_0=\aleph_{\alpha+3}$, $x_{0\xi}=\xi$ $(\xi<\aleph_{\alpha+1})$, $H_1=\aleph_{\alpha+3} \setminus \aleph_{\alpha+1}$ $(=\{\gamma: \aleph_{\alpha+1} \leq \gamma<\aleph_{\alpha+3}\})$. Having defined H_m for $m \leq r$ and $x_{m\xi}$ for $m<r$ and $\xi<\aleph_{\alpha+1}$, where $1 \leq r<\omega$, choose pairwise distinct elements of H_r as $x_{r\xi}$, $\xi<\aleph_{\alpha+1}$. Define the set mapping g_r on $H_r'=$ $=H_r \setminus \{x_{r\xi}: \xi<\aleph_{\alpha+1}\}$ by putting

$$g_r(u)=\bigcup\{f(\{x_{m\xi} x_{r\eta} u\}): m<r \,\&\, \xi, \eta<\aleph_{\alpha+1}\}$$

for any $u \in H'_r$. As $|H'_r| = \aleph_{\alpha+3}$ and $|g_r(u)| < \aleph_{\alpha+2}$, by Theorem 44.3 there exists a set $H''_r \subseteq H'_r$ of cardinality $\aleph_{\alpha+3}$ that is free with respect to g_r. Put $H_{r+1} = H''_r$. This finishes the construction of the matrix $\langle x_{m\xi} : m < \omega \;\&\; \xi < \aleph_{\alpha+1} \rangle$ and of the sets H_m, $m < \omega$.

We are now going to construct an n-element set $\{x_i : i < n\}$ that is free with respect to f such that $x_i \in \{x_{i\xi} : \xi < \aleph_{\alpha+1}\}$. Simultaneously with the construction of this set, we also construct sequences of sets $E^i_j \subseteq \{x_{j\xi} : \xi < \aleph_{\alpha+3}\}$ for $i \leq n-2$ and $j \leq i$, and we will have $x_i \in E^i_i$ for $i \leq n-2$. For $i = n-1$ and $i = n-2$ pick $x_i \in \{x_{i\xi} : \xi < \aleph_{\alpha+1}\}$ arbitrarily, and write $E^{n-2}_j = \{x_{j\xi} : \xi < \aleph_{\alpha+1}\}$ for $j \leq n-2$. Given $m < n-2$, if x_i and E^i_j has already been defined in case $m < i < n$ and $j \leq i$, then define the set mapping h_m on $E^{m+1} = \bigcup_{j \leq m} E^{m+1}_j$ by putting

$$h_m(u) = \bigcup_{m+1 < i < n} f(\{u x_{m+1} x_i\}) .$$

Clearly $|h_m(u)| < \aleph_\alpha$, and so, by the lemma just proved (Lemma 46.3), there is a set X_m free with respect to h such that $X_m \cap E^{m+1}_j$ has cardinality $\aleph_{\alpha+1}$ for each $j \leq m$. Put

$$E^m_j = X_m \cap E^{m+1}$$

for $j \leq m$. Pick an arbitrary element of E^m_m as x_m. This finishes the construction of the set $\{x_i : i < n\}$.

It is easy to see that this set is free with respect to f. In fact, let $i_0 < i_1 < < i_2 < i_3 < n$. Then $x_{i_3} \notin f(\{x_{i_0} x_{i_1} x_{i_2}\})$ and $x_{i_2} \notin f(\{x_{i_0} x_{i_1} x_{i_3}\})$ follow from (i) and (ii) above, similarly to the proof of (3), since $x_{i_2}, x_{i_3} \in H_{i_1+1}$; and, moreover, $x_{i_1} \notin f(\{x_{i_0} x_{i_2} x_{i_3}\})$ and $x_{i_0} \notin f(\{x_{i_1} x_{i_2} x_{i_3}\})$ also follow, since $x_{i_0}, x_{i_1} \in E^{i_2-1}$, and this latter set is free with respect to h_{i_2}. The proof of (4), and so that of Theorem 46.2 is complete.

47. INEQUALITIES FOR POWERS OF SINGULAR CARDINALS

After Cohen's result on the independence of the continuum hypothesis, it was soon proved by W.B. Easton [1970] that powers of regular cardinals can be simultaneously given 'arbitrary reasonable' values (that is, König's theorem and the monotonicity of cardinal exponentiation has to be obeyed, and the values have to be prescribed by an 'absolute' operation); for about a decade, however, nothing was known about how powers of singular cardinals depend on powers of smaller cardinals except for the relatively simple Bukovsky–Hechler result (see

Corollary 6.11), and Solovay's [1974] result that the continuum hypothesis holds for every singular strong limit cardinal exceeding a strongly compact cardinal (see also Drake's book [1974]). Silver's theorem that the continuum hypothesis cannot first fail at a singular cardinal of uncountable cofinality (see below) was the first real breakthrough. Silver's results have been further improved by the results of Galvin and Hajnal. Important contributions were made by Baumgartner, Benda, Jech, Magidor, Prikry, Shelah, and Solovay. In another direction, Jensen [1974b] showed that the failure of the GCH at a singular strong limit cardinal implies that $0^{\#}$ exists (see e.g. Devlin–Jensen [1974] and Jech [1978], p. 356). Later Dodd and Jensen [1981] also proved that the same implies the existence of a measurable cardinal in an inner model.

Silver's original proof used model-theoretic methods; Baumgartner, Prikry, and Jensen observed that this proof can also be formulated in terms of elementary combinatorial set theory, using almost disjoint transversals. The results of Galvin and Hajnal are obtained in this latter way.

DEFINITION 47.1. Let κ be an uncountable cardinal, and let $\langle A_\alpha : \alpha<\kappa\rangle$ be a sequence of sets. A set $F \subseteq \bigtimes_{\alpha<\kappa} A_\alpha$ is called an *a. d. t.* (*almost disjoint transversal system*) for $\langle A_\alpha : \alpha<\kappa\rangle$ if

$$|\{\alpha<\kappa : f(\alpha)=g(\alpha)\}|<\kappa$$

holds for any two distinct $f, g \in F$.

The usefulness of a. d. t.'s is illustrated by the following simple

LEMMA 47.2. *Let κ be a cardinal, $\langle \kappa_\alpha : \alpha<\kappa\rangle$ a sequence of cardinals, and let $A=\langle A_\alpha : \alpha<\kappa\rangle$, where $A_\alpha = \bigtimes_{\beta<\alpha} \kappa_\beta$. Then there is an a. d. t. F for A with $|F| = \prod_{\alpha<\kappa} \kappa_\alpha$.*

PROOF. For each $h \in \bigtimes_{\alpha<\kappa} \kappa_\alpha$ and $\alpha<\kappa$ put

$$f_h(\alpha) = h \upharpoonright \alpha .$$

Then $F=\{f_h : h \in \bigtimes_{\alpha<\kappa} \kappa_\alpha\}$ is an a. d. t. for A and $|F| = |\bigtimes_{\alpha<\kappa} \kappa_\alpha| = \prod_{\alpha<\kappa} \kappa_\alpha$. The proof is complete.

From now on in this section κ will denote a fixed regular cardinal $>\omega$. For any sequence $A=\langle A_\alpha : \alpha<\kappa\rangle$ put

$$T(A) = \sup\{|F| : F \text{ is an a. d. t. for } A\}. \tag{1}$$

We make a trivial but important remark: if $A=\langle A_\alpha\colon \alpha<\kappa\rangle$ and $B=\langle B_\alpha\colon \alpha<\kappa\rangle$ are such that $|A_\alpha|\leq|B_\alpha|$ for all $\alpha<\kappa$, then

$$T(A)\leq T(B). \tag{2}$$

In particular,

$$T(\langle A_\alpha\colon \alpha<\kappa\rangle)=T(\langle |A_\alpha|\colon \alpha<\kappa\rangle).$$

Note that if φ is a function in ${}^\kappa On$, then the notation $T(\varphi)$ makes good sense, since φ is the same as the sequence $\langle\varphi(\alpha)\colon \alpha<\kappa\rangle$. If $\varphi\equiv\delta$ for some ordinal δ, i.e. if $\varphi(\alpha)=\delta$ for all $\alpha<\kappa$, then put

$$T(\varphi)=T(\kappa,\delta). \tag{3}$$

In order to obtain good estimates for $T(\varphi)$, where φ is an arbitrary function in ${}^\kappa On$, we are going to define a partial ordering $<_{I,X}$ on ${}^\kappa On$, depending on a normal ideal I on κ (see Definition 34.1) and on a set $X\in\mathscr{P}(\kappa)\setminus I$ as follows: given $\varphi,\psi\in{}^\kappa On$, we put $\varphi<_{I,X}\psi$ if and only if

$$\{\alpha\in X\colon \varphi(\alpha)\geq\psi(\alpha)\}\in I.$$

Note that $<_{I,X}$ is well-founded, i.e., there is no infinite $<_{I,X}$-descending sequence (cf. the proofs of Theorem 34.9 and Lemma 31.2). For $\varphi\in{}^\kappa On$, we define the ***I, X-rank*** $\|\varphi\|_{I,X}$ of φ by generalized transfinite recursion as follows: $\|\varphi\|_{I,X}$ is the least ordinal exceeding $\|\psi\|_{I,X}$ for all ψ with $\psi<_{I,X}\varphi$, i.e.,

$$\|\varphi\|_{I,X}=\sup\{\|\psi\|_{I,X}\dot{+}1\colon \psi<_{I,X}\varphi\}. \tag{4}$$

If $X\supseteq Y\in\mathscr{P}(\kappa)\setminus I$, then clearly

$$\|\varphi\|_{I,X}\leq\|\varphi\|_{I,Y} \tag{5}$$

holds since the partial order $<_{I,Y}$ is finer than $<_{I,X}$. Define the *I-rank* $\|\varphi\|_I$ of φ by

$$\|\varphi\|_I=\|\varphi\|_{I,\kappa}.$$

Also, write $<_I$ for $<_{I,\kappa}$.

From now on in this section, I will be a fixed normal ideal on κ, which, as remarked above, is a fixed uncountable cardinal. For the purposes of this section, the concept of I-rank will be important only if I is the ideal of nonstationary sets in κ.

We shall need the following lemma, originally due to Galvin and Hajnal; part (e) is, however, an improvement on their original result by Shelah, and part (f) is an addition to it obtained by J. E. Baumgartner and M. Benda simultaneously.

LEMMA 47.3. *Let* $\varphi \in {}^\kappa On$.

(a) *If* $\mu < \kappa^+$, *then there is a function* $\varphi_\mu \in {}^\kappa\kappa$ *not depending on* I *such that* $\|\varphi_\mu\|_I = \mu$ *and, for all* $\psi \in {}^\kappa On$, $\|\psi\|_I > \mu$ *if and only if* $\psi >_I \varphi_\mu$. *If* φ'_μ *is a function satisfying the conditions imposed on* φ_μ, *then* $\{\alpha < \kappa: \varphi_\mu(\alpha) \neq \varphi'_\mu(\alpha)\} \in I$.

(b) *If* $\mu < \kappa$, *then we can put* $\varphi_\mu \equiv \mu$; *i.e.,* $\|\varphi\|_I > \mu$ *if and only if* $\{a < \kappa: \varphi(\alpha) \leq \mu\} \in I$.

(c) *We can put* $\varphi_\kappa = \mathrm{id}_\kappa$ *(the identity on* κ*); i.e.,* $\|\varphi\|_I > \kappa$ *if and only if* $\{\alpha < \kappa: \varphi(\alpha) \leq \alpha\} \in I$.

(d) $\|\varphi\|_I < (\prod \langle |\varphi(\alpha)|: \alpha < \kappa \,\&\, \varphi(\alpha) \neq 0\rangle)^+$.

(e) *Let* $\lambda > \kappa$ *be a regular cardinal such that we have* $\tau^\kappa < \lambda$ *for all cardinals* $\tau < \lambda$ *with* $\mathrm{cf}(\tau) = \kappa$. *If* $\varphi \in {}^\kappa\lambda$, *then* $\|\varphi\|_I < \lambda$.

(f) *Let* $\kappa = \aleph_1$ *and let* I *be the ideal of nonstationary sets in* $\aleph_1$. *If Chang's conjecture (see Subsection 8.5) holds and* $\varphi \in {}^{\aleph_1}\aleph_1$, *then* $\|\varphi\|_I < \aleph_2$.

The assertion in (f) can be stated more generally for arbitrary uncountable κ as follows: (f′) *If* I *is the ideal of nonstationary sets in* κ, then we have $\|\varphi\|_I < \kappa^+$ *for all* $\varphi \in {}^\kappa\kappa$ *if and only if for every mapping* $h: [\kappa^+]^{<\omega} \to \kappa$ *there is a* $\xi < \kappa$ *such that for all* $\alpha < \kappa$ *there is a set* $A \subset \kappa^+$ *with* $\mathrm{tp}\, A \geq \alpha$ *and*

$$\sup \{h(X): X \in [A]^{<\omega}\} = \xi.$$

This generalization is due to Galvin. To see that the 'if' part here includes (f) of the above lemma, one only has to note that the well-known fact that Chang's conjecture is equivalent to the partition relation $\aleph_2 \to [\aleph_1]^{<\omega}_{\aleph_1, \aleph_0}$; the proof of the 'if' part of (f′) is essentially the same as the proof of (f), given below, in the above lemma. The 'only if' part of (f′) sheds additional light on the situation.

PROOF OF LEMMA 47.3. *Ad* (a). Put $\varphi_0(\alpha) = 0$ for all $\alpha < \kappa$. Let $0 < \mu < \kappa^+$, and assume that φ_ν has already been defined for all $\nu < \mu$. As $\mathrm{cf}(\mu) \leq \kappa$, let $\langle \mu_\xi: \xi < \kappa\rangle$ a sequence of ordinals such that $\mu = \sup \{\mu_\xi \dot{+} 1: \xi < \kappa\}$ and put

$$\varphi_\mu(\alpha) = \sup \{\varphi_{\mu_\xi}(\alpha) \dot{+} 1: \xi < \alpha\}. \tag{6}$$

This defines the functions φ_μ, $\mu < \kappa^+$. One can easily prove by induction on μ that $\varphi_\nu <_I \varphi_\mu$ holds whenever $\nu < \mu < \kappa^+$ no matter what the ideal I is, and so $\|\varphi_\mu\|_I \geq \mu$ easily follows from (4).

Next we prove that, for all $\mu < \kappa^+$ and for all $\psi \in {}^\kappa On$, $\|\psi\| > \mu$ holds if and only if $\psi >_I \varphi_\mu$. The 'if' part is obvious, since $\psi >_I \varphi_\mu$ implies $\|\psi\|_I > \|\varphi_\mu\|_I \geq \mu$ by the definition of I-rank in (4). To prove the 'only if' part, we use induction on μ. Assume that $\|\psi\|_I > \mu$ and yet $\psi \not>_I \varphi_\mu$. The former implies by (4) that there is a $\psi' \in {}^\kappa On$ with $\psi' <_I \psi$ and $\|\psi'\|_I \geq \mu$, and the latter that

$$\{\alpha<\kappa: \psi(\alpha)\leq\varphi_\mu(\alpha)\}\notin I,$$

i.e., that

$$X=\{\alpha<\kappa: \psi'(\alpha)<\varphi_\mu(\alpha)\}\notin I. \tag{7}$$

For any $\xi<\kappa$ put

$$X_\xi=\{\alpha<\kappa: \psi'(\alpha)\leq\varphi_{\mu_\xi}(\alpha)\}$$

(see (6) for μ_ξ). As $\mu_\xi<\mu$, we have $\|\psi'\|_I>\mu_\xi$, and so $\psi'>_I\varphi_\xi$ by the induction hypothesis; hence $X_\xi\in I$. As $X=\bigcup_{\xi<\kappa}(X_\xi\setminus(\xi\dot{+}1))$ by the definition of φ_μ in (6), we have $X\in I$ by the normality of I, which contradicts (7). This contradiction proves the 'only'if'part of the assertion in the first sentence of this paragraph.

To show that $\|\varphi_\mu\|_I=\mu$ it remains to be verified that $\|\varphi_\mu\|_I\leq\mu$. But $\|\varphi_\mu\|_I>\mu$ would imply $\varphi_\mu>_I\varphi_\mu$ by what we have just proved, and this latter is absurd.

Finally, we are going to establish the assertion in the second sentence of (a); writing $\varphi'_\mu\dot{+}1$ for the function defined by $(\varphi'_\mu\dot{+}1)(\alpha)=\varphi'_\mu(\alpha)\dot{+}1$ for $\alpha<\kappa$, we have $\|\varphi'_\mu\dot{+}1\|_I>\|\varphi'_\mu\|_I=\mu$, and so $\varphi'_\mu\dot{+}1>_I\varphi_\mu$. Similarly, $\varphi_\mu\dot{+}1>_I\varphi'_\mu$. Hence $\{\alpha<\kappa: \varphi_\mu(\alpha)\neq\varphi'_\mu(\alpha)\}\in I$, which we wanted to show.

Ad (b). Given $\mu<\kappa$, put $\mu_\xi=\xi$ if $\xi<\mu$ and $\mu_\xi=0$ if $\mu\leq\xi<\kappa$ in the definition of φ_μ in (6). Assuming that $\varphi_{\mu'}\equiv\mu'$ holds for all $\mu'<\kappa$, we get $\varphi_\mu(\alpha)=\mu$ for all α with $\mu\leq\alpha<\kappa$ according to (6). Changing the values of φ_μ at fewer than κ places does not affect the claims made about it in (a); hence we may assume $\varphi_\mu(\alpha)=\mu$ for all $\alpha<\mu$. The assertion in the second clause follows from (a).

Ad (c). Let $\mu=\kappa$, and put $\mu_\xi=\xi$ for all $\xi<\kappa$ in (6). Then (6) with $\varphi_{\mu'}\equiv\mu'$ for $\mu'<\kappa$ (cf. (b)) gives $\varphi_\kappa=\mathrm{id}_\kappa$. The assertion in the second clause again follows from (a).

Ad (d). For any ψ, $\psi'\in{}^\kappa On$ put

$$\psi\equiv_I\psi' \quad \text{iff} \quad \{\alpha<\kappa: \psi(\alpha)\neq\psi'(\alpha)\}\in I.$$

Clearly, if $\psi\equiv_I\psi'$, then $\|\psi\|_I=\|\psi'\|_I$. Note that if $\psi<_I\varphi$, then there is a ψ' with $\psi'\equiv_I\psi$ such that $\psi'(\alpha)<\varphi(\alpha)$ or $\psi'(\alpha)=\varphi(\alpha)=0$ holds for every $\alpha<\kappa$, and such that the number of these ordinals ψ' is $\leq\tau$, where we put

$$\tau=\prod\langle|\varphi(\alpha)|: \alpha<\kappa\ \&\ \varphi(\alpha)\neq0\rangle,$$

hence there are $\leq\tau$ I-ranks below $\|\varphi\|_I$. Since all ordinals $<\|\varphi\|_I$ must be represented among these ranks, we can conclude that $\|\varphi\|_I<\tau^+$, which we wanted to prove.

Ad (e). Assume the contrary, i.e., that $\|\varphi\|_I=\|\varphi\|_{I,\kappa}\geq\lambda$ holds for some $\varphi\in{}^\kappa\lambda$. Let $\nu\geq\lambda$ be the least ordinal for which there are $X\in\mathscr{P}(\kappa)\setminus I$ and $\varphi\in{}^\kappa\lambda$ such that $\|\varphi\|_{I,X}=\nu$ (it is easy to show by (4) that $\nu=\lambda$, but we will not need this). Fix X and φ with this property. Write

$$J=\{Y\subseteq\kappa: X\cap Y\notin I\Rightarrow\|\varphi\|_{I,X\cap Y}>\nu\}.$$

We claim that J is a nontrivial κ-complete ideal on κ; in fact, it is also normal, but we shall not show this since we shall not need it. J is clearly nontrivial, as $I \subseteq J$ and $X \notin J$; the latter holds as $\|\varphi\|_{I,X} = \nu$. It is also easy to see by (5) that if $Z \subseteq Y$ and $Y \in J$ then also $Z \in J$. For κ-completeness, we only have to show that

$$\bigcup_{\alpha<\gamma} Y_\alpha \in J$$

whenever $\gamma < \kappa$ and $Y_\alpha \in J$ for all $\alpha < \gamma$; here we may assume without loss of generality that $Y_\alpha \subseteq X$ and $Y_\alpha \cap Y_\beta = 0$ whenever $\alpha, \beta < \gamma$ are distinct. We may also assume that $Y_\alpha \notin I$ for all $\alpha < \gamma$, since those Y_α's which belong to I will not make any difference (namely, if $Y, Y' \subseteq \kappa$, $Y \notin I$, and $Y' \in I$ then clearly $\|\varphi\|_{I,Y} = \|\varphi\|_{I,Y\cup Y'}$). Now $Y_\alpha \in J$ under these assumptions means that $\|\varphi\|_{I,Y_\alpha} > \nu$, and we shall have to show that $Y = \bigcup_{\alpha<\gamma} Y_\alpha \in J$, i.e., that $\|\varphi\|_{I,Y} > \nu$. This will be easy; in fact, we are going to show by induction on μ that, for any ordinal μ and any function $\psi \in {}^\kappa On$, if $\|\psi\|_{I,Y_\alpha} \geq \mu$ holds for all $\alpha < \gamma$, then $\|\psi\|_{I,Y} \geq \mu$ also holds. To this end, assume that this is true for all $\mu' < \mu$. $\|\psi\|_{I,Y_\alpha} \geq \mu$ means that for all $\mu' < \mu$ there is a ψ'_α with $\psi'_\alpha <_{I,Y_\alpha} \psi$ and $\|\psi'_\alpha\| \geq \mu'$. Defining ψ' by stipulating $\psi' \restriction Y_\alpha = \psi'_\alpha \restriction Y_\alpha$ (ψ' may be defined arbitrarily outside Y), we have $\psi' <_{I,Y} \psi$ and, by the induction hypothesis, $\|\psi'\|_{I,Y} \geq \mu'$. Since $\mu' < \mu$ was arbitrary, this proves that $\|\psi\|_{I,Y} \geq \mu$. Putting $\mu = \nu + 1$ and $\psi = \varphi$, this shows that $\|\psi\|_{I,Y} > \nu$, i.e., that $Y = \bigcup_{\alpha<\gamma} Y_\alpha \in J$, which we wanted to show.

Next we establish the following

CLAIM. *There is a set $Y \in \mathcal{P}(X) \setminus J$ and a set $F \subseteq {}^\kappa\lambda$ of cardinality $< \lambda$ such that F is cofinal to φ in the partial ordering $<_{I,Y}$, i.e., for every $\psi \in {}^\kappa On$ such that $\psi <_{I,Y} \varphi$ there is an $f \in F$ such that $\psi <_{I,Y} f <_{I,Y} \varphi$.*

Note that this claim immediately leads to a contradiction, proving part (e) of the lemma to be proved. In fact, by the cofinality of F we have

$$\|\varphi\|_{I,Y} = \sup\{\|f\|_{I,Y} \dot{+} 1 : f \in F \,\&\, f <_{I,Y} \varphi\}.$$

The left-hand side here equals ν; in fact, it is $\geq \nu$ by (5), as $Y \subseteq X$, but it cannot be $> \nu$ as $Y \notin I$. So $\|f\|_{I,Y} < \nu$ for the f's considered on the right-hand side, and then $\|f\|_{I,Y} < \lambda$ as $\nu \geq \lambda$ was chosen minimal. As $|F| < \lambda$ and λ is regular, this means that the supremum on the right-hand side is less than λ. Thus the above equality cannot hold; this is the contradiction we wanted to derive.

Next we turn to the proof of the above Claim. To this end, consider the function h on κ defined by $h(\xi) = \mathrm{cf}(\varphi(\xi))$ $(\xi < \kappa)$. We distinguish two cases: 1) there is an $Y \in \mathcal{P}(X) \setminus J$ such that h is constant on Y, and 2) there is no such Y.

Ad 1). Let the constant value of h on Y be $\rho (< \lambda)$. Clearly, ρ is a regular cardinal. For each $\xi \in Y$ let $\langle h_\alpha(\xi) : \alpha < \rho \rangle$ be an increasing sequence tending to

$\varphi(\xi)$ (outside Y define h_α arbitrarily). If $\rho>\kappa$, then put $F=\{h_\alpha: \alpha<\rho\}$. Then $|F|=\rho<\lambda$ and F is obviously cofinal to φ in $<_{I,Y}$. If $\rho\leq\kappa$, then let F be the set of all functions $f\in{}^\kappa On$ such that $f(\xi)\in\{h_\alpha(\xi): \alpha<\rho\}$ for all $\xi<\kappa$. Then $|F|==2^\rho\leq 2^\kappa=\kappa^\kappa<\lambda$, and, again, F is obviously cofinal to φ in $<_{I,Y}$.

Ad 2). Let η be the least ordinal such that

$$Y=Y_\eta=\{\xi\in X: h(\xi)<\eta\}\notin J.$$

As $\{\xi\in X: h(\xi)=\alpha\}\in J$ for all ordinals α in this case, it is clear by the κ-completeness of J that η must be the supremum of an increasing sequence of type κ of values assumed by h; hence cf $(\eta)=\kappa$, and since h only assumes cardinals as values (cofinalities are cardinals), η must be a cardinal. Also, $\eta<\lambda$, as $\lambda>\kappa$ is regular and $\eta\leq\lambda$. Noting that $h(\xi)=\text{cf}\,(\varphi(\xi))$, for each $\xi\in Y$ let $F_\xi\subseteq\varphi(\xi)$ be a set of cardinality $h(\xi)<\eta$ that is cofinal to $\varphi(\xi)$; let $F_\xi=\{0\}$ if $\xi\in\kappa\setminus Y$. Putting $F=\bigtimes_{\xi<\kappa}F_\xi$, F is clearly cofinal to φ in $<_{I,Y}$, and $|F|\leq\eta^\kappa<\lambda$ by the assumption on λ. This completes the proof of the Claim, and hence of part (e) of the lemma.

Ad (f). Assume Chang's conjecture (see Subsection 8.5). It is easy to see from the original, model-theoretic formulation of Chang's conjecture that the following partition relation holds then:

$$\aleph_2\to[\aleph_1]^{<\omega}_{\aleph_1,\aleph_0}, \tag{8}$$

which means that given any coloring h: $[\aleph_2]^{<\omega}\to\aleph_1$, there is a set $A\subseteq\aleph_2$ of cardinality $\aleph_1$ such that $\sup\{h(X): X\in[A]^{<\omega}\}<\aleph_1$.

Let $\varphi\in{}^{\aleph_1}On$ be such that $\|\varphi\|_I\geq\aleph_2$, where I is the ideal of sets nonstationary in $\aleph_1$. We have to show that $\varphi(\xi)\geq\aleph_1$ for some $\xi<\aleph_1$. Using part (a) of the lemma being proved (with $\kappa=\aleph_1$), we have $\varphi_\mu<_I\varphi_\nu<_I\varphi$ whenever $\mu<\nu<\aleph_1$, so there is a set $C_{\mu\nu}$ closed unbounded in $\aleph_1$ such that $\varphi_\mu(\xi)<\varphi_\nu(\xi)<\varphi(\xi)$ holds for all $\xi\in C_{\mu\nu}$ $(\mu<\nu<\aleph_1)$. For any finite $X\subseteq\aleph_2$ put

$$h(X)=\min\bigcap\{C_{\mu\nu}: \mu,\nu\in X\ \&\ \mu<\nu\}.$$

Then h is a mapping from $[\aleph_2]^{<\omega}$ into $\aleph_1$, so by (8) (or, rather, by the clause after (8)) there is a set $A\subseteq\aleph_2$ of cardinality $\aleph_1$ such that

$$\xi=\sup\{h(X): X\in[A]^{<\omega}\}<\omega_1. \tag{9}$$

It is easy to see that $\xi\in C_{\mu\nu}$ whenever $\mu,\nu\in A$ and $\mu<\nu$. In fact, assuming this is not the case for some μ, ν as above, we have $\xi'=\sup\,(C_{\mu\nu}\cap(\xi+1))<\xi$. Namely, we cannot have $\xi'=\xi$, as $\xi'\in C_{\mu\nu}$ by the closedness of the latter. By (9), there is $X\in[A]^{<\omega}$ such that $h(X)>\xi'$. Then $h(X\cup\{\mu\nu\})>\xi$ by the definition of h, but this contradicts (9); so indeed $\xi\in C_{\mu\nu}$ as claimed. But then $\varphi_\mu(\xi)<\varphi_\nu(\xi)<\varphi(\xi)$ whenever $\mu,\nu\in A$ and $\mu<\nu$. As $|A|=\aleph_1$, this means that $\varphi(\xi)\geq\aleph_1$, which we wanted to show. The proof of the lemma is complete.

Given a cardinal ρ and an ordinal γ, we recall that $\rho^{+\gamma}$ denotes the γth successor of ρ; i.e. $\rho^{+0}=\rho$, $\rho^{+(\gamma+1)}=(\rho^{+\gamma})^{+}$, and, for limit γ, $\rho^{+\gamma}=\sup_{\beta<\gamma}\rho^{+\beta}$.

We are now in a position to formulate the Galvin–Hajnal [1975] result that, together with the last lemma above, gives most of the results included in this book concerning powers of singular cardinals. In the applications I will be the ideal of nonstationary subsets of κ.

THEOREM 47.4. *Let κ be an uncountable regular cardinal, and let I be a normal ideal on κ. Let $\langle\sigma_\alpha\colon \alpha<\kappa\rangle$ be a nondecreasing continuous sequence of infinite cardinals (i.e., $\sigma_\alpha=\bigcup_{\beta<\alpha}\sigma_{\beta+1}$ for $\alpha>0$), and let $\varphi\in{}^{\kappa}On$. Put $\psi(\alpha)=\sigma_\alpha^{+\varphi(\alpha)}$ for $\alpha<\kappa$ and let*

$$\Delta=2^{\kappa}\cdot\sum_{\alpha<\kappa}T(\kappa,\sigma_\alpha). \tag{10}$$

(see (3)). Then

$$T(\psi)\leq\Delta^{+\|\varphi\|_I}. \tag{11}$$

PROOF. We use induction on $\|\varphi\|_I$. Assume first that $\|\varphi\|_I=0$ (the assertion with this assumption is a theorem of Erdős, Hajnal, and Milner [1968] in case the σ_α's are strictly increasing). According to Lemma 47.3(b), this assumption implies that $\{\alpha<\kappa\colon \varphi(\alpha)=0\}\notin I$, and therefore we also have

$$X_0=\{\alpha<\kappa\colon \varphi(\alpha)=0\;\&\;\alpha \text{ is a limit ordinal}\}\notin I,$$

as I is a normal ideal, which implies that the set of successor ordinals $<\kappa$ belongs to I. Let F be an a. d. t. for ψ. For $f\in F$ and $\alpha\in X_0$ we have $f(\alpha)<\sigma_\alpha$; by the continuity of the sequence σ_α, there is an $h_f(\alpha)<\alpha$ such that $f(\alpha)<\sigma_{h_f(\alpha)}$. The function h_f assumes a constant value $\beta_f<\kappa$ on a set $X_f\in\mathscr{P}(X_0)\setminus I$ in view of the normality of I. For fixed $X\subseteq\kappa$ and $\beta<\kappa$, let

$$F_{X,\beta}=\{f\in F\colon X_f=X\;\&\;\beta_f=\beta\}.$$

It is easy to see that $|F_{X,\beta}|\leq T(\kappa,\sigma_\beta)\leq\Delta$. Since there are only 2^{κ} pairs $\langle X,\beta\rangle$, it follows that $|F|\leq 2^{\kappa}\cdot\Delta=\Delta$.

Assume now $\|\varphi\|_I=\nu>0$, and suppose that the assertion holds for any φ' with $\|\varphi'\|_I<\nu$ replacing φ. Let F be an a. d. t. for ψ. For each $f\in F$, define $\varphi_f,\psi_f\in{}^{\kappa}On$ by putting $\varphi_f(\alpha)=\min\{\beta\colon |f(\alpha)|\leq\sigma_\alpha^{+\beta}\}$ and $\psi_f(\alpha)=\sigma_\alpha^{+\varphi_f(\alpha)}$ for all $\alpha<\kappa$. Note that we have either $\varphi_f(\alpha)<\varphi(\alpha)$ or $\varphi_f(\alpha)=\varphi(\alpha)=0$ for $\alpha<\kappa$; as $\|\varphi\|_I=\nu>0$ according to our assumptions, it follows that $\|\varphi_f\|_I\sim\nu$ for all $f\in F$. That is, we have $F=\bigcup_{\mu<\nu}F_\mu$, where $F_\mu=\{f\in F\colon \|\varphi_f\|_I=\mu\}$. As $\nu\leq\Delta^{+\nu}$, in order to see that $|F|\leq\Delta^{+\|\varphi\|_I}=\Delta^{+\nu}$ it will suffice to show that $|F_\mu|\leq\Delta^{+\nu}$ for each $\mu<\nu$. In fact, we shall show that $|F_\mu|\leq\Delta^{+\mu+1}$.

To this end, let $\mu<\nu$ be fixed. We define a set mapping H on F_μ as follows: we put

$$H(f)=\{g\in F_\mu\colon \forall\alpha<\kappa[g(\alpha)\leq f(\alpha)]\}\setminus\{f\}$$

for every $f\in F_\mu$. Then $H(f)$ is an a. d. t. for $f\dot{+}1$. Since $|f(\alpha)\dot{+}1|<\sigma_\alpha^{+\varphi_f(\alpha)}=\psi_f(\alpha)$ for $\alpha<\kappa$, it follows that $|H(f)|\leq T(\psi_f)$. But $\psi_f(\alpha)=\sigma_\alpha^{+\varphi_f(\alpha)}$ and $\|\varphi_f\|=\mu<\nu$, so $T(\psi_f)\leq\Delta^{+\mu}$ by the induction hypothesis. That is, $|H(f)|\leq\Delta^{+\mu}$ for each $f\in F_\mu$.

Now suppose that, contrary to our claim, $|F_\mu|\geq\Delta^{+\mu\dot{+}2}$. Then by Hajnal's Theorem 44.3 it follows that there is a set $F'\subseteq F_\mu$ of cardinality $\Delta^{+\mu\dot{+}2}$ that is free with respect to H. In particular, since $\Delta^{+\mu\dot{+}2}>2^\kappa$, there is a sequence $\langle f_\xi\colon \xi<(2^\kappa)^+\rangle$ with $f_\xi\in F_\mu$ and $f_\xi\notin H(f_\eta)$ for $\xi<\eta<(2^\kappa)^+$. That is, for $\xi<\eta<$ $<(2^\kappa)^+$, there is $\alpha=\alpha(\xi,\eta)<\kappa$ such that $f_\xi(\alpha)>f_\eta(\alpha)$. Now by the Erdős–Rado theorem (cf. (17.32) in Corollary 17.10) we have $(2^\kappa)^+\rightarrow(\kappa^+)^2_\kappa$ and so in particular $(2\kappa)^+\rightarrow(\omega)^2_\kappa$. Hence there is an infinite sequence $\xi_0<\xi_1<\ldots<\xi_n<$ $<\ldots<(2^\kappa)^+$ $(n<\omega)$ and a fixed $\alpha<\kappa$ such that $\alpha(\xi_m,\xi_n)=\alpha$ for $m<n<\omega$. That is, $f_{\xi_0}(\alpha)>f_{\xi_1}(\alpha)>\ldots$, a contradiction. So we must have $|F_\mu|\leq\Delta^{+\mu+1}\leq\Delta^{+\nu}$ as claimed. This completes the proof of the theorem.

We are now turning to the corollaries of the theorem just proved. The most general among these that we shall need is the following:

THEOREM 47.5 (Galvin–Hajnal [1975]). *Let κ be an uncountable regular cardinal, I a normal ideal on κ, $\langle\sigma_\alpha\colon\alpha<\kappa\rangle$ a nondecreasing continuous sequence of infinite cardinals, and define Δ according to* (10).

Let $\varphi\in{}^\kappa On$, and let $\langle\kappa_\alpha\colon\alpha<\kappa\rangle$ be a sequence of cardinals such that

$$\prod_{\beta<\alpha}\kappa_\beta\leq\sigma_\alpha^{+\varphi(\alpha)} \tag{12}$$

holds for every $\alpha<\kappa$. Then

$$\prod_{\alpha<\kappa}\kappa_\alpha\leq\Delta^{+\|\varphi\|_I}. \tag{13}$$

PROOF. Define $\psi\in{}^\kappa On$ by putting $\psi(\alpha)=\sigma_\alpha^{+\varphi(\alpha)}$ for any $\alpha<\kappa$. According to Lemma 47.2, it is enough to show that $T(\psi)\leq\Delta^{+\|\varphi\|_I}$ (cf. also (2) above). This is, however, confirmed by Theorem 47.4. The proof is complete.

We are going to derive various corollaries of this result. Corollaries 47.6–47.13 are improvements by Shelah of the original Galvin–Hajnal results, brought about by his improvement of Lemma 47.3(e).

COROLLARY 47.6 (Galvin–Hajnal–Shelah). *Let κ and λ be uncountable regular cardinals such that $\lambda>\kappa$ and $\tau^\kappa<\lambda$ holds for all cardinals $\tau<\lambda$ with $cf(\tau)=\kappa$. Let, further, σ be an infinite cardinal such that $\sigma^\kappa<\sigma^{+\lambda}$, and $\langle\kappa_\alpha\colon\alpha<\kappa\rangle$ a sequence of* *cardinals such that*

$$\prod_{\beta<\alpha} \kappa_\beta < \sigma^{+\lambda} \tag{14}$$

holds for every $\alpha<\kappa$. *Then*

$$\prod_{\alpha<\kappa} \kappa_\alpha < \sigma^{+\lambda}. \tag{15}$$

Proof. Choose $\sigma_\alpha = \sigma$ for all $\alpha<\kappa$ in the preceding theorem, and define $\varphi \in {}^\kappa On$ by putting

$$\varphi(\alpha) = \min\{\xi : \prod_{\beta<\alpha} \kappa_\beta \le \sigma^{+\xi}\}$$

for all $\alpha<\kappa$ (i.e., either $\varphi(\alpha)=0$ or $\prod_{\beta<\alpha} \kappa_\beta = \sigma^{+\varphi(\alpha)}$). Then we have $\|\varphi\|_I < \lambda$ for any normal ideal I on κ (the simplest is to choose I as the ideal of nonstationary sets in κ) according to Lemma 47.3(e). It is easy to see that with the Δ in (10) we have $\Delta \le 2^\kappa \cdot \sigma^\kappa < \sigma^{+\lambda}$, where the second inequality holds by our assumption $\sigma^\kappa < \sigma^{+\lambda}$. So, using the preceding theorem, we get

$$\prod_{\alpha<\kappa} \kappa_\alpha \le \Delta^{+\|\varphi\|_I} < \Delta^{+\lambda} \le \sigma^{+\lambda}$$

(as $\Delta \ge \sigma$ obviously, we actually have equality instead of the last inequality), which completes the proof.

Corollary 47.7 (Galvin–Hajnal–Shelah). *Let* κ, λ, *and* σ *be as in Corollary* 47.6. *If* $\rho^\tau < \sigma^{+\lambda}$ *for all* $\tau<\kappa$, *then* $\rho^\kappa < \sigma^{+\lambda}$.

Proof. Put $\kappa_\alpha = \rho$ in the preceding result.

Corollary 47.8 (Galvin–Hajnal–Shelah). *Let* κ, λ, *and* σ *be as in Corollary* 47.6, *and let* ρ *be a cardinal such that* $\mathrm{cf}(\rho)=\kappa$. *If* $2^\tau < \sigma^{+\lambda}$ *for all* $\tau<\rho$, *then* $2^\rho < \sigma^{+\lambda}$.

Proof. Write $\rho = \sum_{\alpha<\kappa} \rho_\alpha$, $\rho_\alpha < \rho$, and put $\kappa_\alpha = 2^{\rho_\alpha}$. Then, for any $\alpha<\kappa$, $\prod_{\beta<\alpha} \kappa_\beta = 2^\tau < \sigma^{+\lambda}$, where $\tau = \sum_{\beta<\alpha} \rho_\alpha < \rho$. Hence $2^\rho = \prod_{\alpha<\kappa} \kappa_\alpha < \sigma^{+\lambda}$ according to Corollary 47.6.

Corollary 47.9 (Galvin–Hajnal–Shelah). *Let* κ *be an uncountable regular cardinal* ρ *and* τ *cardinals* $\tau\ge 2$, *and* ξ *and* η *ordinals*, $\eta<\kappa$. *If* $\rho^\vartheta < \aleph_{\xi \dot{+} (\tau^\kappa)^{+\eta \dot{+} 1}}$ *for all cardinals* $\vartheta<\kappa$, *and* $\aleph_\xi^\kappa < \aleph_{\xi \dot{+} (\tau^\kappa)^{+\eta \dot{+} 1}}$ *then* $\rho^\kappa < \aleph_{\xi \dot{+} (\tau^\kappa)^{+\eta \dot{+} 1}}$.

Proof. Put $\sigma = \aleph_\xi$ and $\lambda = (\tau^\kappa)^{+\eta \dot{+} 1}$ in Corollary 47.7.
Using the same substitutions, Corollary 47.8 gives the following:

COROLLARY 47.10 (Galvin–Hajnal–Shelah). *Let κ be an uncountable regular cardinal, ρ and τ cardinals, assume $\tau \geq 2$ and* $\mathrm{cf}(\rho)=\kappa$, *and let ξ and η be ordinals, $\eta<\kappa$. If $2^{\vartheta}<\aleph_{\xi\dot{+}(\tau^{\kappa})^{+}\eta\dot{+}1}$ for all cardinals $\vartheta<\rho$ and $\aleph_{\xi}^{\kappa}<\aleph_{\xi\dot{+}(\tau^{\kappa})^{+}\eta\dot{+}1}$, then $2^{\rho}<\aleph_{\xi\dot{+}(\tau^{\kappa})^{+}\eta\dot{+}1}$.*

We now mention characteristic particular cases of the above result:

COROLLARY 47.11. *Let ξ and η be ordinals with* $\mathrm{cf}(\xi)>\eta,\omega$. *If $2^{\aleph_\alpha}<\aleph_{(|\xi|^{\mathrm{cf}(\xi)})^{+}\eta\dot{+}1}$ for all $\alpha<\xi$, then $2^{\aleph_\xi}<\aleph_{(|\xi|^{\mathrm{cf}(\xi)})^{+}\eta\dot{+}1}$.*

PROOF. Replace ξ, η, τ, ρ, and κ with 0, η, $|\xi|$, $\aleph_\xi$, and $\mathrm{cf}(\xi)$, respectively, in the preceding corollary.

COROLLARY 47.12. *Let ξ and η be ordinals with* $\mathrm{cf}(\xi)>\eta,\ \omega$. *Assume $\aleph_\alpha^{\sigma}<\aleph_{(|\xi|^{\mathrm{cf}(\xi)})^{+}\eta\dot{+}1}$ for all cardinals $\sigma<\mathrm{cf}(\xi)$ and for all ordinals $\alpha<\xi$. Then $\aleph_\xi^{\mathrm{cf}(\xi)}<\aleph_{(|\xi|^{\mathrm{cf}(\xi)})^{+}\eta\dot{+}1}$.*

PROOF. The result follows from Corollary 47.9 with 0, η, $|\xi|$, $\mathrm{cf}(\xi)$, and $\aleph_\xi$ replacing ξ, η, τ, κ, and ρ, since we have

$$\aleph_\xi^{\sigma}\leq \sum_{\alpha<\xi}\aleph_\alpha^{\sigma}<\aleph_{(|\xi|^{\mathrm{cf}(\xi)})^{+}\eta\dot{+}1}$$

for any cardinal $\sigma<\mathrm{cf}(\xi)$, where the first inequality holds in view of Theorem 6.3.

The last two results with $\xi=\omega_1$ give:

COROLLARY 47.13. *Let η be an ordinal with $\eta<\omega_1$.*

(i) *If* $2^{\aleph_\alpha}<\aleph_{(2^{\aleph_1})^{+}\eta\dot{+}1}$ *for all* $\alpha<\omega_1$, *then* $2^{\aleph_{\omega_1}}<\aleph_{(2^{\aleph_1})^{+}\eta\dot{+}1}$.
(ii) *If* $\aleph_\alpha^{\aleph_0}<\aleph_{(2^{\aleph_1})^{+}\eta\dot{+}1}$ *for all* $\alpha<\omega_1$, *then* $\aleph_{\omega_1}^{\aleph_1}<\aleph_{(2^{\aleph_1})^{+}\eta\dot{+}1}$.

The last result in this series (there are more to come later) involves Chang's conjecture. Its original proof involved more complicated ideas than used here; the Baumgartner–Benda result (Lemma 47.3(f)) made it possible to treat this result in the present framework.

COROLLARY 47.14 (M. Magidor [1977]). *Assume Chang's conjecture (see Subsection* 8.5), *and assume that $2^{\aleph_\alpha}<\aleph_{\omega_1}$ holds for all $\alpha<\omega_1$. Then $2^{\aleph_{\omega_1}}<\aleph_{\omega_2}$.*

PROOF. We are going to apply Theorem 47.5. To this end, write $\kappa=\aleph_1$, $\sigma_\alpha=\aleph_0$, $\kappa_\alpha=\aleph_\alpha$ $(\alpha<\omega_1)$ and define $\varphi\in{}^{\aleph_1}On$ by stipulating that

$$\aleph_{\varphi(\alpha)}=\prod_{\beta<\alpha}\aleph_\beta(\leq 2^{\aleph_\alpha}<\aleph_{\omega_1})$$

holds for $\alpha<\omega_1$. Then $\varphi(\alpha)<\aleph_1$ always holds, and so, choosing I as the ideal of nonstationary sets on $\aleph_1$, we obtain $\|\varphi\|_I<\aleph_2$ by Lemma 47.3(f). It is easy to see that $\Delta = 2^{\aleph_1} < \aleph_{\omega_1}$ in the present case (Δ was defined in (10)), and so

$$\prod_{\alpha<\aleph_1} \aleph_\alpha < (\aleph_{\omega_1})^{+\aleph_2} = \aleph_{\omega_2}$$

holds. Here the left-hand side equals

$$\prod_{\alpha<\aleph_1} \aleph_\alpha = \prod_{\alpha<\aleph_1} 2^{\aleph_\alpha} = 2^{\Sigma_{\alpha<\aleph_1}\aleph_\alpha} = 2^{\aleph_{\omega_1}},$$

which completes the proof.

These results leave the following problems open:

Problem 47.15. (i) If $2^{\aleph_n} < \aleph_\omega$ for all $n < \omega$, is there a bound for $2^{\aleph_\omega}$?

(ii) Let α be the least ordinal such that $\aleph_\alpha = \alpha$ and $\mathrm{cf}(\alpha) > \omega$, and assume that $2^{\aleph_\beta} < \aleph_\alpha$ for all $\beta < \alpha$. Is there a bound for $2^{\aleph_\alpha}$?

Added in proof: S. Shelah [1980] proved that the answer to problem 47.15(ii) is affirmative. We only state the simplest instance of his general result.

Let $\aleph^1_\nu$: $\nu \in On$ be an enumeration of the fixed points of the $\aleph$ operation. Assume that $\aleph^1_{\omega_1}$ is strong limit. Then $2^{\aleph^1_{\omega_1}} < \aleph^1_{(2^{2^{\aleph_1}})^+}$.

The problem remains open whether the same hypothesis implies $2^{\aleph^1_{\omega_1}} < \aleph^1_{(2^{\aleph_1})^+}$. Shelah's proof of the above-mentioned result is no longer purely combinatorial. It uses a mixture of the methods of Silver's original metamathematical proof and the methods described above. It also makes use of a result of A. Dodd and R. B. Jensen that if GCH fails at a singular strong-limit cardinal, then there is a measurable cardinal in an inner model.

Recently, Shelah [1982, p. 444] also settled Problem 47.15(i) by proving that $\aleph_\xi^{\mathrm{cf}(\xi)} < \aleph_{(|\xi|^{\mathrm{cf}(\xi)})^+}$ holds for every limit ordinal ξ. This improves Corollary 47.12 even in case $\mathrm{cf}(\xi) > \omega$.

Results of Jensen [1974], [1974a], [1974b]; Jech and Prikry [1976], and Magidor [1977a], [1977b] show that these problems are related to large cardinal hypotheses.

Before the above results were obtained. Silver established several results connecting cardinal exponentiation at a singular cardinal of cofinality $> \omega$ with cardinal exponentiation at ordinals below, his main result essentially saying that the GCH cannot first fail at such a cardinal. Prior to that, nothing like this was thought to be provable according to the general concensus of set theorists.

Before discussing Silver's results, we need the following easy consequence of Theorem 47.5.

Corollary 47.16. *Let κ and λ be infinite cardinals with $\kappa = \mathrm{cf}(\lambda) > \omega$ such that $\rho^\kappa < \lambda$ for all $\rho < \lambda$. Let $\langle \lambda_\alpha : \alpha < \kappa \rangle$ be a strictly increasing continuous sequence of infinite cardinals converging to λ, and let $\varphi \in {}^\kappa On$. Let I be a normal* *ideal on κ. If we have*

$$\prod_{\beta<\alpha} \lambda_\beta \leq \lambda_\alpha^{+\varphi(\alpha)} \tag{16}$$

for every $\alpha<\kappa$, then

$$\lambda^\kappa \leq \lambda^{+\|\varphi\|_I}. \tag{17}$$

What we really need instead of the assumption that $\langle\lambda_\alpha\colon \alpha<\kappa\rangle$ is strictly increasing is that $\lambda_\alpha<\lambda$ for all $\alpha<\kappa$. Allowing nonstrictly increasing sequences would not, however, make the assertion stronger.

Proof. Take $\sigma_\alpha=\kappa_\alpha=\lambda_\alpha$ in Theorem 47.5. It is then easy to see that for the cardinal Δ in (10) we have

$$\Delta \leq \sum_{\alpha<\kappa} \lambda_\alpha^\kappa \leq \lambda,$$

and so (13) in Theorem 47.5 implies (17), completing the proof.

Corollary 47.17 (Silver [1975]). *Let κ and λ be infinite cardinals with $\kappa=\mathrm{cf}(\lambda)>\omega$ such that $\rho^\kappa<\lambda$ for all $\rho<\lambda$. Let $\langle\lambda_\alpha\colon \alpha<\kappa\rangle$ be a strictly increasing continuous sequence of infinite cardinals converging to λ, and suppose that the set $\{\alpha<\kappa\colon \prod_{\beta<\alpha}\lambda_\beta\leq\lambda_\alpha^{+\mu}\}$ is stationary in κ for some ordinal $\mu<\kappa$. Then $\lambda^\kappa\leq\lambda^{+\mu}$. If also $2^\tau\leq\lambda^\kappa$ for all $\tau<\lambda$ then $2\lambda\leq\lambda^{+\mu}$.*

Proof. The first assertion follows from the preceding corollary with I chosen as the ideal of nonstationary sets in view of Lemma 47.3(b). As for the assertion in the last sentence, this follows from the first assertion by noting that under the additional assumption $\forall\tau<\lambda[2^\tau\leq\lambda^\kappa]$ we have

$$2^\lambda=2^{\Sigma_{\alpha<\kappa}\lambda_\alpha}=\prod_{\alpha<\kappa}2^{\lambda_\alpha}\leq\prod_{\alpha<\kappa}\lambda^\kappa=\lambda^\kappa.$$

The proof is complete.

Corollary 47.18 (Silver [1975]). *Let λ be a singular strong limit cardinal with $\mathrm{cf}(\lambda)=\kappa>\omega$, let $\langle\lambda_\alpha\colon\alpha<\kappa\rangle$ be a strictly increasing sequence of infinite cardinals converging to λ, and let $\mu<\kappa$ be an ordinal. If $\{\alpha<\kappa\colon \prod_{\beta<\alpha}\lambda_\beta\leq\lambda_\alpha^{+\mu}\}$ is stationary in κ, then $2^\lambda\leq\lambda^{+\mu}$.*

Proof. The assertion follows from the last sentence of the preceding corollary.

Corollary 47.19 (Silver [1975]). *Let λ be a singular cardinal with $\mathrm{cf}(\lambda)>\omega$, and let $\mu<\mathrm{cf}(\lambda)$ be an ordinal. If $\{\sigma<\lambda\colon 2^\sigma\leq\sigma^{+\mu}\}$ is stationary in λ, then $2^\lambda\leq\lambda^{+\mu}$.*

Proof. Note that λ is a strong limit cardinal. Let $\langle\lambda_\alpha\colon\alpha<\kappa\rangle$ be a strictly increasing continuous sequence of cardinals tending to λ. Then $\{\alpha<\kappa\colon 2^{\lambda_\alpha}\leq\lambda_\alpha^{+\mu}\}$

is stationary in κ (see Theorem 5.2). Since $\prod_{\beta<\alpha} \lambda_\beta \leq \lambda_\alpha^{\lambda_\alpha} = 2^{\lambda_\alpha}$, the result follows from the preceding corollary.

Corollary 47.20 (Silver [1975]). *Let $\mu < \omega_1$. If $2^{\aleph_1} < \aleph_{\omega_1}$ and $\{\alpha < \omega_1 : \aleph_\alpha^{\aleph_0} \leq \leq \aleph_{\alpha+\mu}\}$ is stationary in $\aleph_1$, then $\aleph_{\omega_1}^{\aleph_1} \leq \aleph_{\omega_1+\mu}$.*

Proof. By Corollary 47.17, since we have $\aleph_\alpha^{\aleph_1} = 2^{\aleph_1} \cdot \aleph_\alpha^{\aleph_0} < \aleph_{\omega_1}$ for any $\alpha < \omega_1$, as is easily seen. In fact, the equality above is trivial in case $\alpha \leq 1$; if $\alpha > 1$, then, using induction on $\alpha < \omega_1$ and applying Corollary 6.4, we obtain

$$\aleph_\alpha^{\aleph_1} \leq \left(\prod_{1 \leq \xi < \alpha} \aleph_{\xi+1} \right)^{\aleph_1} = \prod_{1 \leq \xi < \alpha} \aleph_{\xi+1}^{\aleph_1} = \prod_{1 \leq \xi < \alpha} (\aleph_\xi^{\aleph_1} \cdot \aleph_{\xi+1}) =$$

$$= \prod_{1 \leq \xi < \alpha} (2^{\aleph_1} \cdot \aleph_\xi^{\aleph_0} \cdot \aleph_{\xi+1}) \leq \prod_{1 \leq \xi < \alpha} (2^{\aleph_1} \cdot \aleph_\alpha^{\aleph_0} \cdot \aleph_\alpha) = 2^{\aleph_1} \cdot \aleph_\alpha^{\aleph_0} \leq \aleph_\alpha^{\aleph_1},$$

which we wanted to show. The proof is complete.

Corollary 47.21 (Silver [1975]). *Let $\mu < \omega_1$. If $2^{\aleph_\alpha} < \aleph_{\omega_1}$ for all $\alpha < \omega_1$, and $\{\alpha < \omega_1 : \aleph_\alpha^{\aleph_0} \leq \aleph_{\alpha+\mu}\}$ is stationary in $\aleph_1$, then $2^{\aleph_{\omega_1}} < \aleph_{\omega_1+\mu}$.*

Proof. By Corollary 47.18.

Corollary 47.22 (Silver [1975]). *Let $\mu < \omega_1$. If $\{\alpha < \omega_1 : 2^{\aleph_\alpha} < \aleph_{\alpha+\mu}\}$ is stationary in ω_1, then $2^{\aleph_{\omega_1}} < \aleph_{\omega_1+\mu}$.*

Proof. By Corollary 47.19.

The next result is meant only as an illustration of further consequences of Corollary 47.16.

Corollary 47.23 (Silver [1975]). *If $2^{\aleph_\alpha} < \aleph_{\omega_1}$ for all $\alpha < \omega_1$, and $\{\alpha < \omega_1 : \aleph_\alpha^{\aleph_0} \leq \leq \aleph_{\alpha \cdot 2}\}$ is stationary in $\aleph_1$, then $2^{\aleph_{\omega_1}} < \aleph_{\omega_1 \cdot 2}$.*

Proof. Corollary 47.16 with $\kappa = \aleph_1$, $\lambda = \aleph_{\omega_1}$, and I being the ideal of nonstationary sets on $\aleph_1$ gives $2^{\aleph_{\omega_1}} \leq \aleph_{\omega_1 \cdot 2}$ in view of Lemma 47.3(c). Equality here is impossible in view of König's theorem (Theorem 6.9), completing the proof.

48. CARDINAL EXPONENTIATION AND SATURATED IDEALS

The moral of the results of the preceding section is that cardinal powers of singular cardinals (of cofinality $>\omega$, at least) are usually affected by cardinal powers of smaller cardinals. Something similar happens at regular cardinals carrying certain saturated ideals (for saturated ideals, see Section 34). The first result in this direction is Scott's result (see below) saying that the GCH cannot first fail at a measurable cardinal; this result was recently generalized by Ketonen. That something similar can happen for small cardinals was observed by Baumgartner, who found a connection between the values of $2^{\aleph_0}$ and $2^{\aleph_1}$, and the maximal number of almost disjoint subsets of $\aleph_1$ (i.e., sets such that any two of them have countable intersection). This result has been generalized by Jech and Prikry to the case in which the ideal of countable sets on $\aleph_1$ is replaced by an arbitrary $\aleph_1$-complete ideal.

Before we can discuss the result outlined above, we need some preliminaries about saturated ideals. To this end, let κ be an infinite cardinal and let I be a (3-complete) ideal on κ. The saturation sat (I) of I is a cardinal defined as follows:

$$\text{sat}\,(I)=\min\,\{\lambda\colon I \text{ is } \lambda\text{-saturated}\}. \tag{1}$$

We need the following simple lemma:

LEMMA 48.1 (Tarski). *Let $\kappa\geq\omega$ be a cardinal, and let I be a (3-complete) ideal on κ. Then sat (I) is either finite or regular.*

PROOF. Assume that $\lambda=\text{sat}\,(I)$ is singular, and let $\langle\lambda_\xi\colon\ \xi<\text{cf}\,(\lambda)\rangle$ be an increasing sequence of regular cardinals tending to λ; assume that $\lambda_0>\text{cf}\,(\lambda)$. As I is not λ_ξ-saturated, for each $\xi<\text{cf}\,(\lambda)$ there are sets $X^\xi_\alpha\in\mathscr{P}(\kappa)\setminus I$, $\alpha<\lambda_\xi$, such that $X^\xi_\alpha\cap X^\xi_\beta\in I$ whenever $\alpha<\beta<\lambda_\xi$. We may assume that for each $\xi<\text{cf}\,(\lambda)$ there is an $\alpha_\xi<\lambda_\xi$ and an $\eta_\xi<\text{cf}\,(\lambda)$ such that I is λ_{η_ξ}-saturated on X^ξ_α (i.e., such that $I|X^\xi_\alpha=\{Y\subseteq\kappa\colon\ Y\cap X^\xi_\alpha\in I\}$ is λ_{η_ξ}-saturated) for each α with $\alpha_\xi\leq\alpha<\lambda_\xi$. In fact, if this is not the case, then we can take an increasing sequence $\langle\gamma_\nu\colon\ \nu<\text{cf}\,(\lambda)\rangle$ of ordinals $<\lambda_\xi$ such that I is not λ_ν-saturated on $x^\xi_{\gamma_\nu}$, i.e., there are sets $Y^{\gamma_\nu}_\beta\in\mathscr{P}(X^\xi_{\gamma_\nu})\setminus I$, $\beta<\lambda_\nu$, such that $Y^{\gamma_\nu}_\beta\cap Y^{\gamma_\nu}_{\beta'}\in I$ whenever $\beta<\beta'<\lambda_\nu$; then the set $\{Y^{\gamma_\nu}_\beta\colon\ \nu<\text{cf}\,(\lambda)\,\&\,\beta<\lambda_\nu\}$, which consists of sets that are almost disjoint with respect to I, shows that I is not λ-saturated. This is a contradiction.

So make the assumption mentioned above and choose an increasing sequence ξ_ν, $\nu<\text{cf}\,(\lambda)$, of cardinals such that $\xi_\nu>\eta_{\xi_\mu}$ for any $\mu<\nu$. This assumption means that for any $\mu<\nu$ and any α with $\alpha_{\xi_\mu}\leq\alpha<\lambda_{\xi_\mu}$ the set

$$\{\beta<\lambda_{\xi_\nu}\colon X^{\xi_\mu}_\alpha\cap X^{\xi_\nu}_\beta\notin I\}$$

has cardinality $<\lambda_{\xi_\nu}$ (since the sets after the colon are almost disjoint with respect to I). Let β_ν be the supremum of the union of all these sets as μ runs over all ordinals $<\nu$ and α runs over the indicated range; $\beta_\nu<\lambda_{\xi_\nu}$ by the regularity of λ_{ξ_ν}. Then the set

$$\{X^{\xi_\nu}_\alpha\colon\nu<\text{cf}\,(\lambda)\,\&\,\max\,\{\alpha_{\xi_\nu},\beta_{\xi_\nu}\}<\alpha<\lambda_{\xi_\nu}\}$$

has cardinality λ and consists of sets almost disjoint with respect to I. This is again a contradiction, showing that λ must be regular. The proof is complete.

Again, as in the preceding section, the results here below can be approached either by elementary methods or by model-theoretic methods using generic ultrapowers. Unlike the situation in the preceding section where the elementary method gave more general results than can at present be obtained by model-theoretic methods, here the model-theoretic approach does not seem to be inferior to the elementary one, and all the results discussed below could just as well be obtained by using the former. In fact, the former approach here seems to have some distinct advantages. True to the spirit of this book, we shall nonetheless apply combinatorial methods to derive the results below. The framework here slightly differs from the one used in the preceding section; instead of full transversals, we shall have to use partial transversals here, as described in the following definition:

DEFINITION 48.2. Let κ be an uncountable regular cardinal, I a κ-complete nontrivial ideal on κ (see Definition 34.1(i)), $x \in \mathscr{P}(\kappa) \setminus I$ a set, and $A = \langle A_\alpha : \alpha \in X \rangle$ a sequence of sets. A set F is called a *sad pat* (*s*et of *a*lmost *d*isjoint *pa*rtial *t*ransversals) for A and I if the following hold: (i) each $f \in F$ is a function with $\operatorname{dom}(f) \in \mathscr{P}(X) \setminus I$ such that if $\alpha \in \operatorname{dom}(f)$ then $f(\alpha) \in a_\alpha$ and (ii) if $f, g \in F$ and $f \neq g$ then $\operatorname{dom}(f \cap g) \in I$ (this latter is the same as saying that

$$\{\alpha \in \operatorname{dom}(f) \cap \operatorname{dom}(g) : f(\alpha) = g(\alpha)\} \in I).$$

From now on to the end of this section, κ will denote an uncountable regular cardinal, and I a κ-complete nontrivial ideal on κ. The relevance of sad pats for cardinal exponentiation is illustrated by the following lemma, which is an analogue of Lemma 47.2.

LEMMA 48.3. *Let $\langle \kappa_\alpha : \alpha < \kappa \rangle$ be a sequence of cardinals and write $A = \langle A_\alpha : \alpha < \kappa \rangle$, where $A_\alpha = \bigtimes_{\beta < \alpha} \kappa_\beta$. Then there is a sad pat F for A and I with $|F| = \prod_{\alpha < \kappa} \kappa_\alpha$.*

PROOF. The result follows directly from Lemma 47.2, as an a. d. t. for A is also a sad pat for A and I no matter what the ideal I is. (The result is also true in case $\kappa = \omega$, but this case is of no interest to us here.) —

For any sequence $A = \langle A_\alpha : \alpha \in X \rangle$, where $X \notin I$, put

$$PT(A, I) = \sup \{|F|^+ : F \text{ is a sad pat for } A \text{ and } I\}. \tag{2}$$

It is worth comparing this to (47.1). We make similar remarks as we made after (47.1). If $A = \langle A_\alpha : \alpha \in X \rangle$ and $B = \langle B_\alpha : \alpha \in X' \rangle$ are such that $X \subseteq X'$ and $|A_\alpha| \leq |B_\alpha|$ for all $\alpha \in X$, then

$$PT(A, I) \leq PT(B, I). \tag{3}$$

In particular

$$PT(\langle A_\alpha : \alpha \in X\rangle, I) = PT(\langle |A_\alpha| : \alpha \in X\rangle, I).$$

Note that if φ is a function in ${}^\kappa On$, then the notation $PT(\varphi, I)$ makes good sense, since φ is the same as the sequence $\langle \varphi(\alpha) : \alpha < \kappa\rangle$. If $\varphi \equiv \delta$ for some ordinal δ, i.e., if $\varphi(\alpha) = \delta$ for all $\alpha < \kappa$, then put

$$PT(\varphi, I) = PT(\kappa, \delta, I). \tag{4}$$

The following result, due to Jech and Prikry, will enable us to derive our main result concerning the relationship between cardinal exponentiation and saturated ideals on κ in case κ is a successor cardinal.

THEOREM 48.4 (Jech–Prikry [1976]). *Let κ and ρ be infinite cardinals such that κ is a successor cardinal and $\kappa \le \rho < \kappa^{+\kappa}$. Then*

$$PT(\kappa, \kappa^-, I) \le \text{sat}\,(I) \tag{5}$$

and

$$PT(\kappa, \rho, I) \le (\rho \cdot \text{sat}\,(I))^+ \tag{6}$$

(sat (I) *was defined in* (1)).

PROOF. Write $\lambda = \text{sat}\,(I)$. Then λ is regular by Lemma 48.1 and

$$\lambda > \kappa \tag{7}$$

in view of Theorem 34.4. For an ordinal μ write

$$c_\mu = \langle \mu : \alpha < \kappa\rangle; \tag{8}$$

i.e., let c_μ denote the constant function on κ with value μ.

We prove (5) first. To this end, let F be a sad pat for c_{κ^-}, and assume, contrary to what we want to prove, that $|F| \ge \lambda$. For $f \in F$ and $\xi < \kappa^-$ put

$$X_\xi^f = \{\alpha \in \text{dom}\,(f) : f(\alpha) = \xi\}.$$

Clearly, $\bigcup_{\xi < \kappa^-} X_\xi^f = \text{dom}\,(\text{f}) \notin I$; so, by the κ-completeness of I we obtain that for each f there is a $\xi_f < \kappa^-$ such that $X_{\xi_f}^f \notin I$. By (7) and the regularity of λ, there is a $\xi < \kappa^-$ such that the set

$$F_\xi = \{f \in F : \xi_f = \xi\}$$

has cardinality $\ge \lambda$. Then the set $\{X_\xi^f : f \in F_\xi\}$ refutes the λ-saturatedness of I; in fact if $f \ne f'$ and $f, f' \in F_\xi$, then $X_\xi^f, X_\xi^{f'} \notin I$;and $X_\xi^f \cap X_\xi^{f'} \in I$ (this latter holds because f and f' coincide on $X_\xi^f \cap X_\xi^{f'}$). This is a contradiction, establishing (5).

We are now going to prove (6). We distinguish three cases: 1) $\rho = \kappa$, 2) $\rho > \kappa$ is regular, and 3) $\rho > \kappa$ is singular.

Ad 1). Assume, on the contrary, that (6) fails with $\rho=\kappa$, and let F be a sad pat of cardinality $\lambda^+=(\mathrm{sat}\,(I))^+=(\rho\cdot\mathrm{sat}\,(I))^+$ (the latter equality here holds by (7)) for the constant function c_κ on κ. Define the set mapping H on F as follows: For any $f\in F$ put

$$H(f)=\{g\in F\colon \{\alpha\in\mathrm{dom}\,(f)\cap\mathrm{dom}\,(g)\colon g(\alpha)<f(\alpha)\}\notin I\}.$$

For a fixed f, we want to estimate the cardinality of $H(f)$. To this end, write

$$\bar{g}=g\upharpoonright\{\alpha\in\mathrm{dom}\,(f)\cap\mathrm{dom}\,(g)\colon g(\alpha)<f(\alpha)\}$$

for any $g\in H(f)$. Note that if $g_1, g_2\in H(f)$ are distinct, then so are $\bar{g}_1$ and $\bar{g}_2$, as g_1 and g_2 are almost disjoint partial transversals; and, moreover, the set of $\bar{g}$'s $(g\in H(f))$ is a sad pat for f. Hence, using some earlier observations (see below), we obtain that

$$\begin{aligned}&H(f)\leq \text{ the number of } \bar{g}\text{'s}<PT(f, I)\leq\\&\leq PT(c_{\kappa^-}\upharpoonright\mathrm{dom}\,(f), I)\leq PT(\kappa, \kappa^-, I)\leq\\&\leq \mathrm{sat}\,(I)=\lambda;\end{aligned}$$

here the third and fourth inequalities follow from (3), and the fifth holds in view of (5). Thus $H(f)<\lambda$ for any $f\in F$; as $|F|=\lambda^+$, it follows from Hajnal's Theorem 44.3 that there is a set $F'\subseteq F$ of cardinality λ^+ that is free with respect to H. We claim that the elements of the set

$$X=\{\mathrm{dom}\,(f)\colon f\in F'\}$$

are almost disjoint with respect to I, i.e., that $\mathrm{dom}\,(f_1)\cap\mathrm{dom}\,(f_2)\in I$ for any two distinct f_1, $f_2\in F'$. In fact,

$$\begin{aligned}&\mathrm{dom}\,(f_1)\cap\mathrm{dom}\,(f_2)=\\&=\{\alpha<\kappa\colon f_1(\alpha)<f_2(\alpha)\}\cup\{\alpha<\kappa\colon f_1(\alpha)=f_2(\alpha)\}\cup\{\alpha<\kappa\colon f_1(\alpha)>f_2(\alpha)\},\end{aligned}$$

and the first set on the right-hand side here belongs to I because $f_1\notin H(f_2)$ (this latter holds since F' is a free set), the third, because $f_2\notin H(f_1)$, and the second, because f_1 and f_2 are almost disjoint partial transversals. The set X above has the same cardinality as F', i.e., $|X|=\lambda^+$. This, however contradicts the assumption that I is λ-saturated.

In the remaining two cases we use induction on ρ, that is, we assume that (6) holds if we replace ρ with any ρ' satisfying $\kappa\leq\rho'<\rho$.

Ad 2). Let F be a sad pat for the function c_ρ (cf. (8)), and put

$$F_\xi=\{f\in F\colon \forall\alpha\in\mathrm{dom}\,(f)\,[f(\alpha)<\xi]\}$$

for any ordinal $\xi<\rho$. We have

$$F=\bigcup_{\xi<\rho} F_\xi,$$

as $\rho>\kappa$ is regular in the case considered. So, using the induction hypothesis, we obtain that

$$|F|\leq \sum_{\xi<\rho}|F_\xi|\leq \sum_{\xi<\rho}|\xi|\cdot \text{sat}(I)\leq \rho\cdot \text{sat}(I)<(\rho\cdot \text{sat}(I))^+,$$

showing that (6) holds in the present case.

Ad 3). As before, let F be a sad pat for c_ρ. Let $\langle\rho_\xi\colon \xi<\text{cf}(\rho)\rangle$ be an increasing sequence of cardinals $\geq\kappa$ tending to ρ, and put

$$F_\xi=\{f\in F\colon \{\alpha\in\text{dom}(f)\colon f(\alpha)<\rho_\xi\}\notin I\}$$

for any ordinal $\xi<\text{cf}(\rho)$. Noting that $\text{cf}(\rho)<\kappa$ in view of the assumption $\rho<\kappa^{+\kappa}$, the κ-completeness of I implies that

$$F=\bigcup_{\xi<\text{cf}(\lambda)} F_\xi.$$

We claim that

$$|F_\xi|<PT(\kappa,\rho_\xi,I).$$

In fact, for $f\in F_\xi$ put

$$\bar{f}=f\upharpoonright\{\alpha\in\text{dom}(f)\colon f(\alpha)<\rho_\xi\};$$

then $\{\bar{f}\colon f\in F_\xi\}$ is a sad pat for c_{ρ_ξ}, and so

$$|F_\xi|=|\{\bar{f}\colon f\in F_\xi\}|<PT(\kappa,\rho_\xi,I),$$

as was claimed. Hence, using the induction hypothesis, we obtain that

$$|F|\leq \sum_{\xi<\text{cf}(\lambda)}|F_\xi|\leq \sum_{\xi<\text{cf}(\lambda)}\rho_\xi\cdot \text{sat}(I)\leq \rho\cdot \text{sat}(I)<(\rho\cdot \text{sat}(I))^+,$$

showing that (6) holds. The proof is complete.

The above theorem and Lemma 48.3 give the following result due to Jech and Prikry [1976]. In case I is the set of all subsets of cardinality $<\kappa$ of κ, this was obtained by Baumgartner [1970] earlier.

THEOREM 48.5 (Jech and Prikry [1976]). *Let τ be an infinite cardinal, $\kappa=\tau^+$, and assume that $2^\tau<\kappa^{+\kappa}$. Let I be a κ-complete nontrivial ideal on κ. Then*

$$2^\kappa\leq 2^\tau\cdot \text{sat}(I). \tag{9}$$

PROOF. Using Lemma 48.3 with $\kappa_\alpha=2^\tau$ for all $\alpha<\kappa$, we can see that

$$PT(\kappa,2^\tau,I)>2^\kappa.$$

(6) in Theorem 48.4 implies that

$$PT(\kappa, 2^{\tau}, I) \leq (2^{\tau} \cdot \operatorname{sat}(I))^{+}$$

as we assumed that $2^{\tau} < \kappa^{+\kappa}$. These two inequalities imply (9), completing the proof.

Our next result sharpens Theorem 47.4 for certain specific φ's provided the ideal I has good saturation properties. This result will give us estimates of powers of certain regular limit cardinals. In order to state our result, we define $\lambda^{-1+\alpha}$, where λ is a cardinal and $\alpha \geq 1$ an ordinal, as $\lambda^{+(\alpha-1)}$ if $\alpha < \omega$ and as $\lambda^{+\alpha}$ if $\alpha \geq \omega$. In particular, for an ordinal μ we have $\lambda^{-1+(\mu+1)} = \lambda^{+\mu}$ if $\mu < \omega$ and $\lambda^{-1+(\mu+1)} = \lambda^{+(\mu+1)}$ if $\mu \geq \omega$. We shall simply write $\lambda^{-1+\mu+1}$ for $\lambda^{-1+(\mu+1)}$ We have

THEOREM 48.6. *Let κ be an uncountable regular cardinal, I a normal ideal on κ, and $\langle \sigma_\alpha : \alpha < \kappa \rangle$ a nondecreasing continuous sequence of infinite cardinals* (*i.e.*, $\sigma_\alpha = \bigcup_{\beta < \alpha} \sigma_{\beta+1}$ *for* $\alpha > 0$). *Let $\mu < \kappa^{+}$ be an ordinal, and let $\varphi_\mu \in {}^{\kappa}On$ be the function described by Lemma* 47.3(a). *Define $\psi_\mu \in {}^{\kappa}On$ by putting $\psi_\mu(\alpha) = \sigma_\alpha^{+\varphi_\mu(\alpha)}$ for all $\alpha < \kappa$ and let*

$$\Delta_I = \sup \left\{ \left(\sum_{\alpha < \kappa} \rho_\alpha \right)^{+} : \rho_\alpha < PT(\kappa, \sigma_\alpha, I) \right\}. \tag{10}$$

Then

$$PT(\psi_\mu, I) \leq \Delta_I^{-1+\mu+1}. \tag{11}$$

PROOF. The proof bears some similarity to that of Theorem 47.4. We use induction on μ. Assume first that $\mu = 0$. According to Lemma 47.3(b), $\{\alpha < \kappa : \varphi_0(\alpha) \neq 0\} \in I$, so we may just as well assume that $\varphi_0 \equiv 0$, i.e., that $\psi_0(\alpha) = \sigma_\alpha$ for all $\alpha < \kappa$. Let F be a sad pat for ψ_0 and I; we have to prove that $|F| < \Delta_I$. To this end let $f \in F$ be arbitrary. Putting $A = \{\alpha < \kappa : \alpha$ is a limit ordinal$\}$, $\kappa \setminus A$ is nonstationary, and so $\kappa \setminus A \in I$ since, as is easily seen, a normal ideal contains every nonstationary set. As $\operatorname{dom}(f) \notin I$, we therefore have $\operatorname{dom}(f) \cap A \notin I$ as well. If $\alpha \in \operatorname{dom}(f) \cap A$, then $f(\alpha) < \psi_0(\alpha) = \sigma_\alpha$, and, as α is limit, we have $f(\alpha) < \sigma_{h_f(\alpha)}$ for some $h_f(\alpha) < \alpha$ by the continuity of the sequence $\langle \sigma_\alpha : \alpha < \kappa \rangle$. By virtue of the normality of I, the function h_f assumes a constant value $\beta_f < \kappa$ on a set $A_f \notin I$ ($A_f \subseteq \operatorname{dom}(f) \cap A$). Put $\bar{f} = f \restriction A_f$. If $f, f' \in F$ and $f \neq f'$, then clearly $\bar{f}$ and $\bar{f}'$ are almost disjoint with respect to I, since $\{\alpha : \bar{f}(\alpha) = \bar{f}'(\alpha)\} \subseteq \{\alpha : f(\alpha) = f'(\alpha)\} \in I$; in particular, $\bar{f} \neq \bar{f}'$. Writing

$$F_\beta = \{\bar{f} : f \in F \;\&\; \beta_f = \beta\}$$

for any $\beta < \kappa$, we have $|F_\beta| < PT(\kappa, \sigma_\beta, I)$ and $|F| = \sum_{\beta < \kappa} |F_\beta|$; hence $|F| < \Delta_I$, proving the assertion in case $\mu = 0$.

Assume now that $0<\mu<\kappa^+$, and that the assertion to be proved holds with any $\mu'<\mu$ replacing μ. Let F be a sad pat for φ_μ. For each $f\in F$ define $\varphi_f\in {}^\kappa On$ as follows:

$$\varphi_f(\alpha)=\min\{\beta: |f(\alpha)|\leq\sigma_\alpha^{+\beta}\}$$

provided $\alpha\in\operatorname{dom}(f)$ and $\varphi_f(\alpha)=0$ if $\alpha\notin\operatorname{dom}(f)$. Put $\mu_f=\|\varphi_f\|_I$; then we have

$$\mu_f<\mu \tag{12}$$

by Lemma 47.3(a), as we have either $\varphi_f(\alpha)<\varphi_\mu(\alpha)$ or $\varphi_f(\alpha)=0$ for every $\alpha<\kappa$. Moreover, Lemma 47.3(a) also implies that

$$X_f=\{\alpha<\kappa: \varphi_f(\alpha)>\varphi_{\mu_f}(\alpha)\}\in I. \tag{13}$$

For $\nu<\mu$ put

$$F_\nu=\{f\in F: \mu_f=\nu\};$$

we have $F=\bigcup_{\nu<\mu}F_\nu$ according to (12). To see (11), we have to show that $|F|<$ $<\Delta_I^{-1\dot{+}\mu\dot{+}1}$. As $\mu<\Delta_I^{-1\dot{+}\mu\dot{+}1}$, and the latter here is a regular cardinal, it will suffice to show that $|F_\nu|<\Delta_I^{-1\dot{+}\mu\dot{+}1}$ holds for each $\nu<\mu$. We shall in fact show that

$$|F_\nu|\leq\Delta_I^{-1\dot{+}\nu\dot{+}1} \tag{14}$$

holds for every $\nu<\mu$.

To see (14), let $\nu<\mu$ be fixed. Define a set mapping H on F_ν by putting

$$H(f)=\{g\in F_\nu: \{\alpha\in\operatorname{dom}(f)\cap\operatorname{dom}(g): \quad g(\alpha)<f(\alpha)\}\notin I\}$$

for every $f\in F_\nu$. We want to have a good estimate for the cardinality of $H(f)$. To this end, fix $f\in F_\nu$ and put

$$\bar{g}=g\upharpoonright(\operatorname{dom}(f)\cap\operatorname{dom}(g)\setminus X_f)$$

for any $g\in H(f)$ (see (13) for X_f). Clearly, if $g\neq g'$, then $\bar{g}\neq\bar{g}'$ (cf. (13) and the definition of $H(f)$), and so $|H(f)|=$ the cardinality of the set $\bar{H}(f)=$ $=\{\bar{g}: g\in H(f)\}$. This latter set is a sad pat for $f\upharpoonright(\operatorname{dom}(f)\setminus X_f)$ and I, and so

$$|H(f)|=|\bar{H}(f)|<PT(f\upharpoonright(\operatorname{dom}(f)\setminus X_f), I). \tag{15}$$

To estimate the right-hand side here, define $\psi_\nu\in{}^\kappa On$ by putting $\psi_\nu(\alpha)=\sigma_\alpha^{+\varphi_\nu(\alpha)}$ for all $\alpha<\kappa$, and observe that

$$|f(\alpha)|\leq\psi_\nu(\alpha)$$

for all $\alpha\in\operatorname{dom}(f)\setminus X_f$ by (13), as $\mu_f=\nu$ in the present case. Hence, in virtue of (3) above, the right-hand side of (15) is

$$\leq PT(\psi_\nu\upharpoonright(\operatorname{dom}(f)\setminus X_f, I)\leq PT(\psi_\nu, I)\leq\Delta_I^{-1\dot{+}\nu\dot{+}1},$$

where the last inequality follows from (11) with ν replacing μ, which holds according to the induction hypothesis. Comparing this to (15) we obtain that

$$|H(f)| < \Delta_I^{-1+\nu+1}$$

for every $f \in F_\nu$. Assume now that (14) fails, i.e., that $|F_\nu| > \Delta_I^{-1\dot{+}\nu\dot{+}1}$. Then using the above estimate for $|H(f)|$, it follows by Hajnal's Theorem 44.3 that there is a set $F' \subseteq F_\nu$ of cardinality $\Delta_I^{-1\dot{+}\nu\dot{+}2}$ that is free with respect to H. Then the set

$$\operatorname{dom}(f) \cap \operatorname{dom}(g) = \{\alpha < \kappa : f(\alpha) < g(\alpha)\} \cup \{\alpha < \kappa : f(\alpha) = \\ = g(\alpha)\} \cup \{\alpha < \kappa : g(\alpha) < f(\alpha)\}$$

belongs to I for any $f, g \in F'$; in fact, the first set on the right-hand side belongs to I because $f \notin H(g)$, the second one, because f and g are almost disjoint partial transversals, and the third one, because $g \notin H(f)$. So the elements of the set

$$\{\operatorname{dom}(f) : f \in F'\}$$

are almost disjoint with respect to I; in other words, defining $\tilde{f}$ by stipulating that $\operatorname{dom}(\tilde{f}) = \operatorname{dom}(f)$ and $\tilde{f}(\alpha) = 0$ for any $\alpha \in \operatorname{dom}(f)$, the set $\{\tilde{f} : f \in F'\}$ is a sad pat for the indentically 1 function on κ; i.e.

$$|F'| = |\{\tilde{f} : f \in F'\}| < PT(\kappa, 1, I) \leq \Delta_I,$$

which is a contradiction, since $|F'| = \Delta_I^{-1\dot{+}\nu\dot{+}2}$. The proof is complete.

We are now going to list some corollaries of the above result. The first one is an analogue of Theorem 47.5; in fact, the only change here is that we use Δ_I instead of Δ, but then we have to assume that $\|\varphi\|_I < \kappa^+$.

THEOREM 48.7. *Let κ be an uncountable regular cardinal, I a normal ideal on κ, $\langle \sigma_\alpha : \alpha < \kappa \rangle$ a nondecreasing continuous sequence of infinite cardinals, and define Δ_I according to* (10).

Let $\varphi \in {}^\kappa On$ be a function, write

$$\mu = \sup\{\|\varphi\|_{I,X} : X \notin I\}, \tag{16}$$

and assume that $\mu < \kappa^+$. Let $\langle \kappa_\alpha : \alpha < \kappa \rangle$ be a sequence of cardinals such that

$$\prod_{\beta < \alpha} \kappa_\beta \leq \sigma_\alpha^{+\varphi(\alpha)} \tag{17}$$

holds for every $\alpha < \kappa$. Then

$$\prod_{\alpha < \kappa} \kappa_\alpha \leq \Delta_I^{-1\dot{+}\mu}. \tag{18}$$

PROOF. Define the functions ψ and ψ_μ on κ by putting $\psi(\alpha)=\sigma_\alpha^{+\varphi(\alpha)}$ and $\psi_\mu(\alpha)=\sigma_\alpha^{+\varphi_\mu(\alpha)}$ for every $\alpha<\kappa$, where φ_μ is the function described in Lemma 47.3(a). As $\mu<\kappa^+$, (16) implies via Lemma 47.3(a) that $\{\alpha<\kappa\colon \varphi(\alpha)>\varphi_\mu(\alpha)\}\in I$. Hence we also have $\{\alpha<\kappa\colon \psi(\alpha)>\psi_\mu(\alpha)\}\in I$, and so, as is easily seen,

$$PT(\psi, I)\leq PT(\psi_\mu, I)$$

holds. The right-hand side is here is $\leq \Delta^{-1\dot{+}\mu+1}$ by (11), and the left-hand side is $> \prod_{\alpha<\kappa} \kappa_\alpha$ by Lemma 48.3 in view of (17). Thus (18) is established, completing the proof.

COROLLARY 48.8. *Let $\kappa>\omega$ be a regular cardinal, I a normal ideal on κ, and assume $\lambda>\kappa$ is a cardinal such that I is λ-saturated. Let $\varphi\in{}^\kappa On$ be a function with $\|\varphi\|_I<\kappa^+$ such that*

$$2^{|\alpha|}\leq|\alpha|^{+\varphi(\alpha)} \tag{19}$$

holds for any ordinal $\alpha<\kappa$. Then

$$2^\kappa\leq\lambda^{-1\dot{+}\|\varphi\|_I}. \tag{20}$$

PROOF. We are going to use Theorem 48.7 with κ_α, σ_α $(\alpha<\kappa)$ defined as follows: for each $\alpha<\kappa$ put $\kappa_\alpha=|\alpha|$. The definition of the σ_α's is slightly more complicated, as we have to make sure that the sequence $\langle\sigma_\alpha\colon \alpha<\kappa\rangle$ is continuous. To this end, given $\alpha<\kappa$, put $\sigma_\alpha=\alpha$ if α is a cardinal, put $\sigma_\alpha=|\alpha|^+$ if α is not a cardinal and $|\alpha|^+<\kappa$, and, finally, put $\sigma_\alpha=|\alpha|$ if $|\alpha|^+=\kappa$ (note that we have $\sigma_\alpha=|\alpha|$ for a closed unbounded set of α's because either κ is a successor cardinal, or the cardinals $<\kappa$ form a closed unbounded set).

Observe that (19) implies (17) with the above choices of κ_α and σ_α. Writing $\mu=\|\varphi\|_I$, (16) is, however, not necessarily satisfied, and to make it valid we shall have to replace the ideal I with another one. To this end let φ_μ be the function described in Lemma 47.3(a) (note that $\mu<\kappa^+$). We have

$$A=\{\alpha<\kappa\colon \varphi(\alpha)\leq\varphi_\mu(\alpha)\}\notin I$$

according to that lemma. Put

$$I'=\{Y\subseteq\kappa\colon\ A\cap Y\in I\}.$$

Then I' is a λ-saturated normal ideal on κ; as

$$\{\alpha<\kappa\colon\ \varphi(\alpha)>\varphi_\mu(\alpha)\}=\kappa\setminus A\in I',$$

we have $\|\varphi\|_{I',X}\leq\|\varphi_\mu\|_{I',X}$ for any $X\in\mathscr{P}(\kappa)\setminus I'$. Note here that the function φ_μ

in Lemma 47.3(a) is independent of the ideal I; hence, by that lemma, $\|\varphi_\mu\|_{I',X} = \|\varphi_\mu\|_{I''} = \mu$, where $I'' = \{Y \subseteq \kappa\colon X \cap Y \in I'\}$. That is, (16) is satisfied with I' replacing I. Hence we have

$$2^\kappa = \prod_{\alpha<\kappa} |\alpha| \leq \Delta_{I'}^{-1\dotplus\mu}$$

according to (18) with I' replacing I, as $\kappa_\alpha = |\alpha|$ in the present case. Thus, to see (20) we shall only have to show that $\Delta_{I'} \leq \lambda$. In doing so, we may choose λ the least possible cardinal compatible with the assumptions above, i.e., we may assume that $\lambda = \max\{\kappa^+, \mathrm{sat}\,(I)\}$. Lemma 48.1 then implies that λ is regular. Noting that $\sigma_\alpha < \kappa$ holds for all $\alpha < \kappa$ in the present case, $\Delta_{I'} \leq \lambda$ will therefore follow if we show that

$$PT(\kappa, \tau, I') \leq \lambda \tag{21}$$

holds for each cardinal $\tau < \kappa$ (cf. the definition of Δ_I in (10) above). To this end let F be a sad pat for the constant function having τ as value on κ and for I'; we have to show that $|F| < \lambda$. For each $f \in F$ and $\xi < \tau$ put

$$X_{f,\xi} = \{\alpha \in \mathrm{dom}\,(f)\colon f(\alpha) = \xi\},$$

and write

$$F_\xi = \{f \in F\colon X_{f,\xi} \notin I\}.$$

We have $|F_\xi| < \lambda$ in view of the λ-saturatedness of I', as the sets $X_{f,\xi}$, $f \in F_\xi$, are almost disjoint with respect to I'; and, moreover, we have $F = \bigcup_{\xi<\tau} F_\xi$ by the κ-completeness of I'. So $|F| < \lambda$ by the regularity of λ, which we wanted to snow. The proof is complete.

We mention some corollaries of the above result. The first of these was obtained by J. Ketonen [1974] in case $\mu = 1$.

Corollary 48.9. *Let κ, λ be cardinals, μ an ordinal, and assume $\kappa > \omega$ is regular, $\lambda > \kappa$, and $\mu < \kappa$. Assume, further, that I is a λ-saturated normal ideal on κ. If $\{\alpha < \kappa\colon 2^{|\alpha|} \leq |\alpha|^{+\mu}\} \notin I$, then $2^\kappa \leq \lambda^{-1\dotplus\mu}$.*

Proof. The result follows from the preceding corollary and Lemma 47.3(b).

Corollary 48.10 (D. S. Scott). *If κ is measurable (see* Definition 34.10(i)) *and $2^\rho = \rho^+$ for any $\rho < \kappa$, then $2^\kappa = \kappa^+$. In other words, the* GCH *cannot first fail at a measurable cardinal.*

Proof. This follows from the preceding result with $\lambda = \kappa^+$. (Note that saying that κ is measurable means it carries a 2-saturated κ-complete nontrivial ideal; the existence of a normal 2-saturated ideal on κ follows from Theorem 34.9).

The next result is now only of historical interest, because Theorem 48.5 gives the same conclusion under weaker assumptions.

COROLLARY 48.11 (J. Ketonen). *If* $2^{\aleph_0}=\aleph_1$ *and* $2^{\aleph_1}=\aleph_3$, *then there is no* $\aleph_2$-*saturated* $\aleph_1$-*complete ideal on* $\aleph_1$.

PROOF. Use Corollary 48.9 with $\kappa=\aleph_1$, $\lambda=\aleph_2$, and $\mu=1$; then it follows that there is no $\aleph_2$-saturated normal ideal on $\aleph_1$, and so the result follows via Theorem 34.9.

Theorem 48.5 gives the same conclusion under the assumption $2^{\aleph_0}\leq\aleph_2$ instead of $2^{\aleph_0}=\aleph_1$. In general, it may be observed that Theorem 48.5 is stronger than Corollary 48.9 in case κ is a successor cardinal.

CHAPTER XI

A BRIEF SURVEY OF THE SQUARE BRACKET RELATION

Negative square bracket relations can be used to express stronger combinatorial assertions than the corresponding negative ordinary partition relations. We start with a discussion of the classical Erdős–Hajnal–Rado results on such relations under the assumption of GCH, and will then go on to consider a result of Jensen and Shore about negative relations which follow from the existence of a Suslin tree. The theorems of Galvin and Shelah give important nontrivial negative relations without the assumption of GCH. Then we take a look at what positive relations can be obtained with the aid of the General Canonization Lemma; this will be the only application of the case $r \geq 3$ of this lemma to partition relations in this book. The chapter is concluded with a discussion of infinitary Jónsson algebras.

49. NEGATIVE SQUARE BRACKET RELATIONS AND THE GCH

A theorem of Erdős, Hajnal, and Rado [1965] says that if κ is an infinite cardinal such that $2^\kappa = \kappa^+$, then

$$\kappa^+ \nrightarrow [(\kappa : \kappa^+)]^2_{\kappa^+} \tag{1}$$

holds; this relation is much stronger than the Kuratowski–Sierpiński relation $2^\kappa \nrightarrow (\kappa^+, \kappa^+)$ (cf. (19.14)). A further improvement on (1) is the following theorem, which we shall give with proof. This result cannot be expressed in terms of the partition relations introduced above:

THEOREM 49.1 (Erdős–Hajnal–Milner). *If κ is an infinite cardinal with $2^\kappa = \kappa^+$, then there is a coloring $f: [\kappa^+]^2 \to \kappa^+$ such that for every X, $Y \subseteq \kappa^+$ with $|X| = \kappa$, $|Y| = \kappa^+$ there is a $\xi \in X$ such that $\{f(\{\xi\alpha\}) : \alpha \in Y\} = \kappa^+$ (i.e., such that for every color there is an edge of that color from ξ into Y).*

PROOF. Using the assumption $2^\kappa = \kappa^+$, let $\langle h_\alpha : \alpha < \kappa^+ \rangle$ be an enumeration of all functions of cardinality κ (i.e., $|h_\alpha| = |\mathrm{dom}\,(h_\alpha)| = \kappa$) the domains and ranges of which are subsets of κ^+. Given $\alpha < \kappa^+$, we are going to specify the colors $f(\{\xi\alpha\})$,

$\xi<\alpha$, in such a way that if $\beta<\alpha$ and dom $(h_\beta)\subseteq\alpha$, then

$$f(\{\xi_\beta\alpha\})=h_\beta(\xi_\beta) \tag{2}$$

for some $\xi_\beta\in$ dom (h_β) (depending on α as well). This can easily be done as follows. Let g_α be a one-to-one mapping of a cardinal $\rho_\alpha\leq\kappa$ onto $\{\beta<\alpha$: dom $(h_\beta)\subseteq\alpha\}$; for $\nu<\rho_\alpha$, pick a

$$\xi_{g_\alpha(\nu)}\in \text{dom}\,(h_\alpha(\nu))\backslash\{\xi_{g_\alpha(\mu)}:\mu<\nu\}$$

by transfinite recursion, and specify $f(\{\xi_{g_\alpha(\nu)}\alpha\})$ according to (2); for any $\xi<\alpha$ for which $f(\{\xi\alpha\})$ thus remains undefined, it can be defined arbitrarily. This completes the definition of the coloring f.

To show that f has the desired properties, let X, $Y\subseteq\kappa^+$ be sets with $|X|=\kappa$ and $|Y|=\kappa^+$, and assume that, contrary to the assertion of the theorem, for every $\xi\in X$ there is a color $h(\xi)<\kappa^+$ such that $f(\{\xi\alpha\})\neq h(\xi)$ for every $\alpha\in Y$. h is a function from X into κ^+, so $h=h_\beta$ for some $\beta<\kappa^+$. Pick an $\alpha\in Y$ for which $\beta<\alpha$ and dom $(h_\beta)=X\subseteq\alpha$. Then $f(\{\xi\alpha\})=h_\beta(\xi)=h(\xi)$ for some $\xi\in$ dom $(h_\beta)=X$ according to (2); but this contradicts the definition of h, completing the proof.

In Section 19 we proved that

$$\kappa^{\text{cf}(\kappa)}\nrightarrow(\kappa^+,\text{cf}(\kappa)^+)^2 \tag{3}$$

holds for any singular cardinal κ (cf. (19.12) in Corollary 19.7). Here we are going to show that under the assumption $2^\kappa=\kappa^+$ the following strengthening of this relation holds:

$$\kappa^+\nrightarrow[\text{cf}(\kappa)^+,(\kappa^+:\kappa^+)_{\kappa^+}]^2 \tag{4}$$

(this result holds of course also in case κ is regular but then it says less than (1)). Similarly as in case of (1) and Theorem 49.1, we shall prove a result stronger than (4).

THEOREM 49.2. *Let κ be a singular cardinal with $2^\kappa=\kappa^+$. Then there is a coloring $f:[\kappa^+]^2\to\kappa^+$ such that* (i) *for any $X\subseteq\kappa^+$ of cardinality* cf $(\kappa)^+$ *we have $0\in f``[X]^2$ (i.e., there is an edge of color 0 in X), and* (ii) *for any sets $X_0, X_1\subseteq\kappa^+$ of cardinality κ^+, there is an $i=0$ or 1 and a $\xi\in X_i$ such that $\kappa^+\setminus\{0\}\subseteq\{f(\{\xi\alpha\}):\alpha\in X_{1-i}\}$ (i.e., such that for every nonzero color there is an edge of that color from ξ into X_{1-i}).*

PROOF. We have $\kappa^{\text{cf}(\kappa)}=\kappa^+$ under the assumption $2^\kappa=\kappa^+$. Let $<'$ be a vellordering of type κ^+ of ${}^{\text{cf}(\kappa)}\kappa$, and let $\prec'$ be its lexicographic ordering (see Definition 19.2). Defining the Sierpiński partition $I'=\langle I'_0, I'_1\rangle$ by

$$I'_0=\{\{xy\}:x,y\in{}^{\text{cf}(\kappa)}\kappa\;\&\;x<'y\;\&\;x\prec'y\} \tag{5}$$

and

$$I'_1 = \{\{xy\}: x, y \in {}^{\mathrm{cf}(\kappa)}\kappa \,\&\, x <' y \,\&\, y \prec' x\}, \tag{6}$$

we know from Corollary 19.7 that this partition establishes relation (3). Let $g: \kappa^+ \to {}^{\mathrm{cf}(\kappa)}\kappa$ be the 1-1 onto mapping that takes the ordering 'less than' of ordinals over into the wellordering $<'$ (i.e., $\xi < \eta \Leftrightarrow g(\xi) <' g(\eta)$ for any $\xi, \eta < \kappa^+$), and define the ordering $\prec$ by

$$\xi \prec \eta \Leftrightarrow g(\xi) \prec' g(\eta); \tag{7}$$

then the partition $I = \langle I_0, I_1 \rangle$ defined by

$$I_0 = \{\{\xi\eta\}: \xi < \eta < \kappa^+ \,\&\, \xi \prec \eta\} \tag{8}$$

and

$$I_1 = \{\{\xi\eta\}: \xi < \eta < \kappa^+ \,\&\, \eta \prec \xi\} \tag{9}$$

clearly also establishes relation (3). By further subdividing the class I_1 in this partition we shall be able to establish the stronger relation in (4), and, in fact, the assertion of the theorem to be proved.

In order to simplify our notation, if $X \subseteq \kappa^+$ and $\xi < \kappa^+$ then write $X \prec \xi$ for $\forall \eta \in X[\eta \prec \xi]$; the expressions $\xi \prec X$, $X < \xi$, $\xi < X$, $X \prec Y$ (where $Y \subseteq \kappa^+$) are interpreted similarly (some of these expressions will not be used since e.g. $X < \xi$ can also be written as $X \subseteq \xi$). Analogously as in the proof of the preceding theorem, let $\langle h_\alpha : \alpha < \kappa^+ \rangle$ be an enumeration of all functions of cardinality κ with range $\subseteq \kappa^+$ and domain $\subseteq \kappa^+ \setminus \{0\}$. We are about to define the coloring $f: [\kappa^+]^2 \to \kappa^+$. Given $\alpha < \kappa^+$, we are going to define the colors $f(\{\xi\alpha\})$, $\xi < \alpha$, as follows: if $\xi \prec \alpha$ (i.e., if $\{\xi\alpha\} \in I_0$), then we put $f(\{\xi\alpha\}) = 0$. The case $\alpha \prec \xi$ is more complicated: in this case we make sure that for every $\beta < \alpha$ with $\mathrm{dom}(h_\beta) \subseteq \alpha$ and $\alpha \prec \mathrm{dom}(h_\beta)$ there is a $\xi \in \mathrm{dom}(h_\beta)$ such that

$$f(\{\xi\alpha\}) = h_\beta(\xi); \tag{10}$$

this can easily be done in a way similar to what we did in order to make (2) valid in the proof of the preceding theorem. This completes the definition of the coloring f.

We are now going to show that this f satisfies assertions (i) and (ii) of the theorem to be proved. As for (i) this is obvious from the fact that, as remarked above, the partition I defined in (8) and (9) establishes relation (3).

To see (ii), note first that if $X \subseteq \kappa^+$ and $|X| = \kappa^+$, then there is a $\xi < \kappa^+$ such that

$$|\{\eta \in X: \eta \prec \xi\}| = |\{\eta \in X: \xi \prec \eta\}| = \kappa^+. \tag{11}$$

In fact, putting

$$X' = \{\xi \in X: |\{\eta \in X: \eta \prec \xi\}| \leq \kappa\},$$

there is, by the Hausdorff Cofinality Lemma, a set $Y \subseteq X'$ wellordered in the ordering $\prec$ that is cofinal to X' in this ordering. We may also assume here that the order type of Y in $\prec$ is a cardinal; this assumption immediately implies that $|Y| \leq \kappa$, since $\prec$ is isomorphic to $\prec'$ (cf. (7)), and there is no increasing sequence of type κ^+ in the lexicographic order $\prec'$; we proved this latter fact when in Corollary 19.7 we showed that the partition in (5) and (6) establishes relation (3). As

$$X' = \bigcup_{\xi \in Y} \{\eta \in X : \eta \preceq \xi\}$$

and each of the sets on the right-hand side has cardinality $\leq \kappa$ by the definition of X', and $|Y| \leq \kappa$ as remarked above, it follows that $|X'| \leq \kappa$. Similarly, putting

$$X'' = \{\xi \in X : |\{\eta \in X : \xi \prec \eta\}| \leq \kappa\},$$

we can conclude that $|X''| \leq \kappa$. Any element ξ of the set $X \setminus (X' \cup X'')$, which has now been shown to have cardinality κ^+, will satisfy (11).

We are now ready to verify (ii). To this end, let X_0, $X_1 \subseteq \kappa^+$ be sets of cardinality κ^+, and let $\xi_0 \in X_0$ and $\xi_1 \in X_1$ be elements for which (11) holds with these sets replacing X; assume e.g. that $\xi_0 \preceq \xi_1$. We are going to show that then (ii) holds with $i = 1$. Assume the contrary, i.e., that for every $\xi \in X_1$ there is a color $h(\xi) \in \kappa^+ \setminus \{0\}$ such that $f(\{\xi\alpha\}) \neq h(\xi)$ holds with any $\alpha \in X_0$. Put

$$Y_0 = \{\eta \in X_0 : \eta \prec \xi_0\}$$

and

$$Y_1 = \{\eta \in X_1 : \xi_1 \prec \eta\};$$

then $|Y_0| = |Y_1| = \kappa^+$ by (11), and, moreover $Y_0 \prec Y_1$. Let Y_1' be a subset of cardinality κ of Y_1, and write $h' = h \restriction Y_1'$; then $h' = h_\beta$ for some $\beta < \kappa^+$. Take an $\alpha \in Y_0$ such that $\beta < \alpha$ and $Y_1' = \text{dom}(h_\beta) \subseteq \alpha$. We have $\alpha \prec \text{dom}(h_\beta)$ as $Y_0 \prec Y_1$; so we have $f(\{\xi\alpha\}) = h_\beta(\xi) = h(\xi)$ for some $\xi \in \text{dom}(h_\beta) = Y_1' \subseteq X_1$ in view of (10). This contradicts the definition of h, establishing (ii). The proof is complete.

Some remarks concerning relations similar to (4) above were made at the end of Section 20, when discussing relations (20.7)–(20.9). An interesting problem that remains open is whether or not the relation (see explanation below)

$$\kappa^+ \to \left[\left[\begin{matrix}\kappa^+ \\ \kappa^+\end{matrix}\right], \text{cf}(\kappa)^+\right]^2 \tag{12}$$

holds for a singular cardinal under the assumption of GCH. Here $\left[\begin{matrix}\kappa^+ \\ \kappa^+\end{matrix}\right]$ denotes the set of all graphs

$$\{\{\xi\eta\} : \xi < \eta \;\&\; \xi \in A \;\&\; \eta \in B\}$$

where A, $B \subseteq \kappa^+$ are sets of cardinality κ^+, and the relation in (12) is interpreted according to the convention set down in Subsection 8.8.

In Section 25 we proved that for any infinite cardinal κ the relation

$$2^\kappa \not\rightarrow (\kappa^+, 4)^3 \tag{13}$$

holds (cf. Corollary 25.2). Under the assumption $2^\kappa = \kappa^+$ this can be strengthened to

$$\kappa^+ \not\rightarrow [4, (\kappa^+)_{\kappa^+}]^3 . \tag{14}$$

In fact, similarly as before, we can prove an even stronger result:

THEOREM 49.3. *Let κ be an infinite cardinal with $2^\kappa = \kappa^+$. Then there exists a coloring $f: [\kappa^+]^3 \rightarrow \kappa^+$ such that* (i) *for any four-element set $X \subseteq \kappa^+$ there are at least two triples $u, v \subseteq X$ such that $f(u) = f(v) = 0$, and* (ii) *for any sets $X, Y \subseteq \kappa^+$ with $|X| = \kappa$, $|Y| = \kappa^+$ there is an edge $\{\xi\eta\} \subseteq X$ such that $\kappa^+ \setminus \{0\} \subseteq \{f(\{\xi\eta\alpha\}) : \alpha \in Y\}$.*

PROOF. Let $\langle h_\alpha : \alpha < \kappa^+ \rangle$ be an enumeration of all functions of cardinality κ with domain $\subseteq [\kappa^+]^2$ and range $\subseteq \kappa^+ \setminus \{0\}$. In defining the colors $f(\{\xi\eta\alpha\})$ with $\xi, \eta < \alpha$ for an $\alpha < \kappa^+$, for each $\beta < \alpha$ with dom $(h_\beta) \subseteq [\alpha]^2$ pick ordinals ξ_β, η_β (depending also on α) with $\{\xi_\beta \eta_\beta\} \in$ dom (h_β) in such a way that all these ordinals are pairwise distinct and put

$$f(\{\xi_\beta \eta_\beta \alpha\}) = h_\alpha(\{\xi_\beta \eta_\beta\}) ; \tag{15}$$

otherwise put $f(\{\xi\eta\alpha\}) = 0$ $(\xi < \eta < \alpha)$. This completes the definition of f.

If is easy to see (i). To this end let $\xi_0 < \xi_1 < \xi_2 < \alpha < \kappa^+$ be four ordinals. Out of the three triples $\{\xi_0\, \xi_1\, \alpha\}$, $\{\xi_0\, \xi_2\, \alpha\}$ and $\{\xi_1\, \xi_2\, \alpha\}$ only one can have a nonzero color; in fact, it is clear from the above definition that if the triples $\{\xi\, \eta\, \alpha\}$ and $\{\xi'\, \eta'\, \alpha\}$ $(\xi, \eta, \xi', \eta' < \alpha)$ have nonzero colors, then the edges $\{\xi\, \eta\}$ and $\{\xi'\, \eta'\}$ must be disjoint (i.e., we must have $\{\xi\, \eta\} \cap \{\xi'\, \eta'\} = 0$). So at least two of the above triples must have color zero, which is what we claimed.

To see (ii), let $X, Y \subseteq \kappa^+$ be sets with $|X| = \kappa$ and $|Y| = \kappa^+$. Assume that, contrary to what we claimed, for every distinct ξ, $\eta \in X$ there is a color $h(\{\xi\eta\}) \subseteq \kappa^+ \setminus \{0\}$ such that $f(\{\xi\eta\alpha\}) \neq h(\{\xi\eta\})$ for any $\alpha \in Y$. We have $h = h_\beta$ for some $\beta < \kappa^+$; let $\alpha \in Y$ be such that dom $(h_\beta) \subseteq [\alpha]^2$ (i.e., $X \subseteq \alpha$) and $\beta < \alpha$. Then (15) implies

$$f(\{\xi\, \eta\, \alpha\}) = h_\beta(\{\xi\, \eta\}) = h(\{\xi\, \eta\})$$

for some distinct $\xi, \eta \in X$; this contradicts the choice of h, establishing (ii). The proof is complete.

It is clear from the above proof that (ii) in our last theorem can be strengthened as follows: (ii′) for any set $E \subseteq [\kappa^+]^2$ of cardinality κ of pairwise disjoint edges and any set $Y \subseteq \kappa^+$ of cardinality κ^+, there is an edge $e \in E$ such that $\kappa^+ \setminus \{0\} \subseteq \{f(e \cup \{\alpha\}) : \alpha \in Y\}$.

Using exactly the same method as above, the reader can easily establish the following generalization of (14): if κ is an infinite cardinal with $2^\kappa = \kappa^+$, then

$$\kappa^+ \not\to [(r+1), (\kappa^+)_{\kappa^+}]^r \tag{16}$$

holds for any integer $r \geq 3$. There is also an analogous generalization of Theorem 49.3.

50. THE EFFECT OF A SUSLIN TREE ON NEGATIVE RELATIONS

Trees were defined in Section 13. Given a tree $\langle T, < \rangle$, a set $X \subseteq T$ is called an *antichain* it if consists of elements parirwise incomparable in the partial ordering $<$. For an infinite cardinal κ, the tree $\langle T, < \rangle$ is called a *Suslin κ-tree* if it has length κ and has no chain or antichain of cardinality κ. We prove

THEOREM 50.1 (R.B. Jensen [1972], R.A. Shore [1974]). *If κ is a regular cardinal such that there is a Suslin κ-tree, then $\kappa \not\to [\kappa]^2_\kappa$ holds.*

According to a well-known theorem of Jensen, under the assumption of $V = L$ there is a Suslin κ-tree for regular κ exactly if $\kappa \not\to (\kappa)^2_2$ holds (see e.g. Jech [1978], Theorem 49 on p. 226, and the remarks on p. 336; cf. also our Theorem 29.5). The proof relies on the following lemma:

LEMMA 50.2. *Let κ be a regular cardinal and assume that there is a Suslin κ-tree. Then there is a Suslin κ-tree T such that, for every $\alpha < \kappa$, each element in the αth level of T has at least $|\alpha|$ immediate successors.*

PROOF. Let $\langle S, < \rangle$ be a Suslin κ-tree. We are going to construct T as a subtree of S. To this end, let S_0 be the set of elements of S that have κ successors, i.e., put

$$S_0 = \{x \in S : |\{y \in S : x < y\}| = \kappa\} . \tag{1}$$

We claim that

$$|S \setminus S_0| < \kappa . \tag{2}$$

In fact, for each $x \in S \setminus S_0$, there is a minimal $y \leq x$ with $y \in S \setminus S_0$; that is, putting

$$A = \{y \in S : |\{u \in S : y < u\}| < \kappa \;\&\; \forall z < y \, |\{u \in S : z < u\}| = \kappa\} ,$$

we have

$$S \setminus S_0 = \bigcup_{y \in A} \{x \in S : y \leq x\} .$$

Each summand on the right-hand side here has cardinality $<\kappa$, so by the regularity of κ, (2) will follow if we show that $|A|<\kappa$. But A is obviously an antichain, so this inequality must hold, as S is a Suslin tree.

By (1) and (2) it is clear that $\langle S_0, <\rangle$ is a Suslin tree where each element has κ successors, that is

$$|\{y\in S_0: x<y\}|=\kappa \tag{3}$$

for every $x\in S_0$. To get closer to our goal of constructing T we want to find a 'large' subtree S_1 of S_0 such that each element of S_1 splits, i.e., such that $|\text{ims}\ (x, S_1)|\geq 2$ for each $x\in S_1$, where ims (x, S_1) denotes the set of immediate successors of x in S_1 (cf. Section 13). To this end, define the equivalence relation $\equiv$ on S_0 as follows: for each $x, y\in S_0$, $x\equiv y$ if either (i) $x\leq y$ and no element $z\in S_0$ with $x\leq z<y$ splits, or (ii) $y\leq x$ and no $z\in S_0$ with $y\leq z<x$ splits. Clearly, each equivalence class C is a chain, and so $|C|<\kappa$. Define S_1 as a set of the least elements of the equivalence classes. Since, as was remarked just before, each equivalence class has cardinality $<\kappa$, it is easy to see from (3) that

$$|\{y\in S_1: x<y\}|=\kappa \tag{4}$$

for each $x\in S_1$. Moreover, it is also easy to see that each element of S_1 splits, i.e., that

$$|\text{ims}\ (x, S_1)|\geq 2 \tag{5}$$

for each $x\in S_1$. In fact, let $x\in S_1$ and define α as the least ordinal $<\kappa$ such that the set

$$S_0[\alpha]\cap\{y\in S_0: x<y\} \tag{6}$$

has at least two elements, where $S_0[\alpha]$ denotes the αth level of the tree $\langle S_0, <\rangle$. There must be such an α since otherwise the set on the left-hand side of (3) would be a chain of cardinality κ. Noting that each element of S_1 is the least in its equivalence class, it is easy to see then that the equivalence class under $\equiv$ of the x in question is $\{y\in S_0: x\leq y\ \&\ \text{o}(y, S_0)<\alpha\}$, where $\text{o}(y, S_0)$ is the order of the element y in the tree S_0 (see Section 13). This means that the elements of the set in (6) belong to S_1 and are immediate successors of x. This establishes (5).

(4) says that any element x of S_1 has κ successors in S_1. Our next goal is to show that, for any cardinal $\lambda<\kappa$, there are λ pairwise incomparable ones among them. To this end, we first show that there are λ successors of x in S_1 that form a chain. In fact, for any $\alpha<\kappa$, the set

$$S_1^{x,\alpha}=\{y\in S_1:\ x<y\ \&\ \text{o}(y, S_1)=\alpha\}$$

is an antichain, and so it has cardinality $<\kappa$. So by (4) and the regularity of κ, $S^{x,\alpha}$ is nonempty for every α with $\text{o}(x, S_1)<\alpha<\kappa$. Choose an element u in S_1^{x,α_0}, where

$\alpha_0 = o(x, S_1) \dot{+} 1 \dot{+} \lambda$. Then

$$C_{x,\lambda} = \{y \in S_1 : x < y < u\}$$

is a chain of cardinality λ. Using (5), each $y \in C_{x,\lambda}$ has an (immediate) successor $s_y \in S_1$ that is not comparable with u; then

$$A_{x,\lambda} = \{s_y : y \in C_{x,\lambda}\} \subseteq S_1 \tag{7}$$

is an antichain of cardinality λ consisting of successors of x. Thus we proved that, for any $\lambda < \kappa$, any $x \in S_1$ has a set of λ pairwise incomparable successors in S_1.

It is now easy to define a Suslin κ-tree $T \subseteq S_1$ as required by the lemma to be proved. T will be defined as a union $\bigcup_{\alpha<\kappa} T_\alpha$. Put $T_0 = S_1[0]$; let α be an ordinal with $0 < \alpha < \kappa$, and assume that T_β has already been defined for all $\beta < \alpha$. If α is a successor ordinal, i.e., $\alpha = \gamma \dot{+} 1$ for some γ, then write

$$T_\alpha = \bigcup \{A_{x,|\gamma|} : x \in T_\gamma\}, \tag{8}$$

where the sets $A_{x,\lambda}$ were defined in (7). If α is a limit ordinal, then let

$$\gamma_\alpha = \sup \{o(x, S_1) \dot{+} 1 : \; x \in \bigcup_{\beta<\alpha} T_\beta\},$$

and put

$$T_\alpha = S_1[\gamma_\alpha]. \tag{9}$$

It is easy to see that the subtree $T = \bigcup_{\alpha<\kappa} T_\alpha$ of T so defined satisfies the requirements of the lemma. In fact, (9) makes sure that $\langle T, < \rangle$ has length κ. Moreover, it is clear that if $\beta \leq \alpha < \kappa$ and $x \in T_\beta$, $y \in T_\alpha$, then $y < x$ cannot hold. Hence if $x \in T_\alpha$ $(\alpha < \kappa)$, then $o(x, T) \leq \alpha$; all elements of $A_{x,|\alpha|}$ are immediate successors of x (there may be others, added at a limit step later, cf. (9)), and so x does indeed have at least $|o(x, T)|$ immediate successors. The proof of the lemma is complete.

Proof of Theorem 50.1. Let T be a Suslin κ-tree such as described by the preceding lemma, and for each $x \in T$ let $\langle x_\xi : \xi < \alpha_x \rangle$ be an enumeration of the immediate successors of x, where $\alpha_x \geq o(x, T)$. As $|T| = \kappa$, we may define the coloring f confirming the assertion of the theorem on $[T]^2$ instead of $[\kappa]^2$. So define $f : [T]^2 \to \kappa$ as follows: for any distinct $x, y \in T$ put (i) $f(\{xy\}) = \xi$ if $x_\xi \leq y$ or (ii) $f(\{xy\}) = \xi$ if $y_\xi \leq x$, and (iii) define $f(\{xy\})$ arbitrarily (e.g. put $f(\{xy\}) = 0$) if neither (i) nor (ii) holds for any $\xi < \kappa$. We are going to show that f verifies the relation $\kappa \not\to [\kappa]^2_\kappa$. In fact, assume, on the contrary, that $X \subseteq T$ is a set of cardinality κ such that there is a $\xi < \kappa$ with $\xi \notin f``[X]^2$. Write $X' = X \setminus \{x \in T : o(x, T) \leq \xi\}$. Then X' still has cardinality κ, and any $x \in X'$ has an immediate successor x_ξ. If $x, y \in X'$ and $x \neq y$, then $x_\xi < y_\xi$ would mean that $x_\xi \leq y$, i.e., that $f(\{xy\}) = \xi$, which we assumed was not the case. Hence $\{x_\xi : x \in X'\}$ is an antichain of cardinality κ, which is a contradiction, since T is a Suslin tree. The proof is complete.

51. THE KUREPA HYPOTHESIS AND NEGATIVE STEPPING UP TO SUPERSCRIPT 3

We start with the definition of a Kurepa family:

DEFINITION 51.1. Let κ be an infinite cardinal. A Kurepa family on κ is a set $F \subseteq \mathscr{P}(\kappa)$ such that (i) $|F| = \kappa^+$, and (ii) for every ordinal α with $\omega \leq \alpha < \kappa$, we have $|F \cap \mathscr{P}(\alpha)| \leq |\alpha|$. The *$\kappa$-Kurepa Hypothesis*, abbreviated as KH(κ), says that there is a κ-Kurepa family.

According to a theorem of Solováy, if $V = L$, then the KH(κ) holds with every successor cardinal $\kappa > \omega$. Jensen extended this result by proving that if $V = L$, then KH(κ) holds if and only if $\kappa > \omega$ is not ineffable (κ is called *ineffable* if for every coloring $f : [\kappa]^2 \to 2$ there is a homogeneous set stationary in κ). The following theorem connects the Kurepa Hypothesis with negative square-bracket relations.

THEOREM 51.2. (R. A. Shore [1974]). *Let κ be an infinite cardinal such that the relation*

$$\kappa^+ \nrightarrow [(\kappa : \kappa^+)]^2_{\kappa^+} \tag{1}$$

and the κ^+-Kurepa Hypothesis hold. Then

$$\kappa^{++} \nrightarrow [\kappa^+]^3_{\kappa^+} . \tag{2}$$

The relation in (1) will hold e.g. if $2^\kappa = \kappa^+$ (cf. (49.1)); if $V = L$ then KH(κ^+) also holds, as remarked above, and so (2) always holds if $V = L$.

PROOF. Let F be a κ^+-Kurepa family; for any two distinct $x, y \in F$, write

$$d(\{xy\}) = \min \{\alpha < \kappa^+ : x \cap (\alpha \dot{+} 1) \neq y \cap (\alpha \dot{+} 1)\} .$$

For any three distinct $x, y, z \in F$ put

$$d(\{xyz\}) = \{d(\{xy\}), d(\{xz\}), d(\{yz\})\} .$$

The three elements on the right-hand side here are not all distinct; in fact, it is clear that $d(\{xyz\})$ contains exactly two distinct elements. Let now $g : [\kappa^+]^2 \to \kappa^+$ be a coloring verifying relation (1). Then we claim that the coloring $f : [F]^3 \to \kappa^+$, defined by putting

$$f(u)=g(d(u)) \tag{3}$$

for any $u\in[F]^3$, verifies relation (2) (note that $|F|=\kappa^{++}$).

To see this, let $E\subseteq F$ be an arbitrary set of cardinality κ^+, and consider the set

$$X=\{d(\{xy\}): x, y\in E \mathbin{\&} x\neq y\}.$$

Note that $X\subseteq\kappa^+$; moreover, X has cardinality κ^+. In fact, assume, on the contrary, that $X\subseteq\alpha$ for some $\alpha<\kappa^+$; then the mapping $x\longmapsto x\cap(\alpha\dot{+}1)$ $(x\in E)$ is 1-1; that is

$$\kappa^+=|E|=|E\cap\mathscr{P}(\alpha\dot{+}1)|\leq|F\cap\mathscr{P}(\alpha\dot{+}1)|,$$

which contradicts the assumption that F is a κ^+-Kurepa family.

Next put

$$H=\{\langle x\cap\alpha, \alpha\rangle: x\in E \mathbin{\&} \exists y\in E[x\neq y \mathbin{\&} d(\{xy\})=\alpha]\}.$$

Clearly, $|H|=\kappa^+$, since for each $\alpha\in X$ there is an $x\in E$ such that $\langle x\cap\alpha, \alpha\rangle\in H$. Define the partial ordering $\prec$ on H as follows: if $\langle u, \alpha\rangle$, $\langle v, \beta\rangle\in H$, then put

$$\langle u, \alpha\rangle\prec\langle v, \beta\rangle \text{ iff } \alpha<\beta \text{ and } u=v\cap\alpha.$$

It is easy to see that $\langle H, \prec\rangle$ is a tree of length $\leq\kappa^+$. We claim that for each $\xi<\kappa^+$, its ξth level $H[\xi]$ has cardinality $\leq\kappa$. Assuming the contrary, let $\xi<\kappa^+$ be the least ordinal for which this claim fails. Recalling that $H|\xi$ stands for $\bigcup_{\gamma<\xi} H[\gamma]$, put

$$\eta=\sup\{\alpha: \exists u\,\langle u, \alpha\rangle\in H|\xi\};$$

note that $\eta<\kappa^+$ by our assumption on ξ. Observing that for each $\alpha<\kappa^+$ we have

$$\{u: \langle u, \alpha\rangle\in H[\xi]\}\subseteq\{x\cap\alpha: x\in E\}\subseteq F\cap\mathscr{P}(\alpha),$$

and the set on the right-hand side has cardinality $\leq\kappa$, the assumption $|H[\xi]|=\kappa^+$ implies that the set

$$D=\{\langle u, \alpha\rangle: \langle u, \alpha\rangle\in H[\xi] \mathbin{\&} \alpha>\eta\dot{+}1\}$$

has cardinality κ^+. Note, however, that if $\langle u, \alpha\rangle$, $\langle v, \beta\rangle\in D$ are distinct, then $u\cap(\eta\dot{+}1)\neq v\cap(\eta\dot{+}1)$ since otherwise $\langle u, \alpha\rangle$ or $\langle v, \beta\rangle$ would be above the ξth level in H; this, however, means that the set

$$\{u\cap(\eta\dot{+}1): \langle u, \alpha\rangle\in D\}$$

also has cardinality κ^+. As this set is a subset of $F\cap\mathscr{P}(\eta\dot{+}1)$, this contradicts the assumption that F is a κ^+-Kurepa family. This establishes our claim that $|H[\xi]|\leq\kappa$ for each $\xi\leq\kappa$.

Observing that the set

$$\bigcup_{s\in H[\kappa]} \{t\in H: s\prec t\}$$

contains all but $\leq\kappa$ elements of H, and so it has cardinality κ^+, there is an $s\in H[\kappa]$ for which $\{t\in H: s\prec t\}$ has cardinality κ^+. Pick such an $s=\langle x\cap\zeta, \zeta\rangle$, where $x\in E$ and $\zeta<\kappa^+$. By the definition of H, for each $r\prec s$ there is an $x_r\in E$ such that $r=\langle x\cap d(\{xx_r\}), d(\{xx_r\})\rangle$, and for each $t\succ s$ there are $x_t^0, x_t^1\in E$ such that $t=\langle x_t^0\cap d(\{x_t^0x_t^1\}), d(\{x_t^0x_t^1\})\rangle$; $r\prec s\prec t$ here obviously implies

$$d(\{x_t^ix_r\})=d(\{xx_r\}) \qquad (i=0,1). \tag{4}$$

Put now

$$A=\{d(\{xx_r\}): r\prec s\} \quad (=\{\gamma: \exists r\prec s\exists u[r=\langle u,\gamma\rangle]\})$$

and

$$B=\{d(\{x_t^0x_t^1\}): s\prec t\} \quad (=\{\gamma: \exists t\succ s\exists u[t=\langle u,\gamma\rangle]\}).$$

We have $|A|=\kappa$ and $|B|=\kappa^+$ by the choice of s (and, clearly, $A, B\subseteq\kappa^+$).

We can now easily show that for every $\gamma<\kappa^+$ there is an element u of $[E]^3$ such that $f(u)=\gamma$. In fact, the coloring $g: [\kappa^+]^2\to\kappa^+$ establishes relation (1), and so there are $\alpha\in A$ and $\beta\in B$ such that $g(\{\alpha\beta\})=\gamma$. Pick $r\prec s$ and $t\succ s$ such that $\alpha=d(\{xx_r\})$ and $\beta=d(\{x_t^0x_t^1\})$, and put $u=\{x_r x_t^0x_t^1\}$. Then

$$d(u)=\{d(\{x_rx_t^0\}), d(\{x_rx_t^1\}), d(\{x_t^0x_t^1\})\}=\{\alpha\beta\}$$

(cf. (4)), and so, by (3),

$$f(u)=g(\{\alpha\beta\})=\gamma,$$

which we wanted to show. This establishes our claim that f verifies relation (2). The proof of the theorem is complete.

It is an open problem whether similar relations hold with higher superscripts under certain assumptions. The simplest instance of this problem is the following: assuming $V=L$, does the relation

$$\aleph_3\not\to[\aleph_1]^4_{\aleph_1} \tag{5}$$

hold.

Added in proof: Recently S. B. Todorčević [1982] proved the following negative stepping up implication. Assume $\kappa\geq\omega$ and $\square_\kappa$. Let $v\geq 2$, $2\leq r<\omega$ and $\omega\leq\lambda_\xi=\mathrm{cf}(\lambda_\xi)\leq\kappa$ for all $0<\xi<v$. Then $\kappa\not\to[\lambda_\xi]^r_{\xi<v}$ implies $\kappa^+\not\to[\lambda_\xi+1]^{r+1}_{\xi<v}$.

The simplest instance of Theorem 51.2 is that

$$\aleph_2\not\to[\aleph_1]^3_{\aleph_1} \tag{6}$$

provided $V=L$. The following result throws some light on the strength of this relation:

THEOREM 51.3 (Rowbottom). *If* (6) *holds, then*

$$\aleph_2 \nrightarrow [\aleph_1]^2_{\aleph_1,\aleph_0}. \tag{7}$$

As can be seen from Subsection 8.5, relation (7) means the following: there is an $f: [\aleph_2]^2 \to \aleph_1$ such that for any set $X \subseteq \aleph_2$ of cardinality $\aleph_1$ we have $|f``[X]^2| = \aleph_1$. Relation (7) is well known to be equivalent to the negation of Chang's conjecture, also discussed in Subsection 8.5. According to a theorem of Silver, Chang's conjecture is consistent with ZFC+GCH provided the existence of a cardinal κ with $\kappa \to (\kappa)_2^{<\omega}$ is consistent with ZFC+GCH (as we saw in Section 34, a measurable κ satisfies this partition relation). So the above theorem shows that (6) cannot be expected to be provable in ZFC+GCH (unless, of course, large cardinal hypotheses generally considered 'sound' will prove to be inconsistent).

PROOF OF THEOREM 51.3. Let $g: [\aleph_2]^3 \to \aleph_1$ be a coloring that establishes (6), and for each $\alpha < \aleph_2$ let $f_\alpha: \alpha \to \aleph_1$ be a 1-1 mapping (into $\aleph_1$ if $\alpha < \aleph_1$, of course). For each α, $\beta < \aleph_2$ with $\beta < \alpha$ put

$$f(\{\beta\alpha\}) = \sup \{g(\{\gamma\beta\alpha\}) : \gamma < \alpha \;\&\; f_\alpha(\gamma) < f_\alpha(\beta)\}.$$

First note that the condition $f_\alpha(\gamma) < f_\alpha(\beta)$ allows countably many γ's only, and so $f(\{\beta\alpha\}) < \aleph_1$. We claim that the coloring $f: [\aleph_2]^2 \to \aleph_1$ verifies relation (7). In fact, if $X \subseteq \aleph_2$ has cardinality $\aleph_1$ then $g``[X]^3 = \aleph_1$. On the other hand, every triple in $[X]^3$ can be obtained in the form $\{\gamma\beta\alpha\}$, where β, $\gamma < \alpha$ and $f_\alpha(\gamma) < f_\alpha(\beta)$. Hence f cannot be bounded by an ordinal less than $\aleph_1$ on the set $[X]^2$. This completes the proof.

52. ARONSZAJN TREES AND SPECKER TYPES (PREPARATIONS FOR RESULTS WITHOUT GCH)

In Section 53 we shall prove two relations, due to Galvin and Shelah, derivable in ZFC without any additional assumptions:

$$\aleph_1 \nrightarrow [\aleph_1]^2_4 \quad \text{and} \quad 2^{\aleph_0} \nrightarrow [2^{\aleph_0}]^2_{\aleph_0}.$$

The proof of the first relation relies on results derived in this section; that of the second does not need any of these results.

The definition of Aronszajn κ-trees was given in Definition 29.2; to recapitulate, an Aronszajn κ-tree is a tree of length κ having no branch of length κ such that each of its levels has cardinality $<\kappa$. The main result that we need here is the following

THEOREM 52.1 (N. Aronszajn–E. Specker [1951]; for Aronszajn's proof see Kurepa [1935]). *There is an Aronszajn $\aleph_1$-tree.*

PROOF. The elements of the Aronszajn $\aleph_1$-tree $\langle T, \prec\rangle$ will be countable sets of rationals wellordered by the 'less than' relation $<$ on reals, and an element of T will precede another if it is an initial segment of the latter in this ordering. We shall construct T level by level by transfinite recursion. The bottom level $T[0]$ of T will be defined as $\{0\}$, and $T[1]$ as $\{\{r\}: r \in Q\}$, where Q is the set of rational numbers. The elements in the αth level $(\alpha<\omega_1)$ will have order type α in the ordering 'less than'. Assume that $\alpha<\omega_1$ and $T|\alpha=\bigcup_{\gamma<\alpha} T[\gamma]$ has been defined in such a way that for any $\gamma<\alpha$ and any $x \in T[\gamma]$ the set

$$\{\sup \bigcup p\colon p \text{ is a path of length } \alpha \text{ in } T|\alpha \,\&\, x \in p\} \tag{1}$$

contains all rationals $>\sup x$.

We are now going to construct the level $T[\alpha]$; for this, we distinguish two cases: a) $\alpha=\eta\dot{+}1$ for some η, and b) α is a limit ordinal. In case a) put

$$T[\alpha]=\{x\cup\{r\}: x \in T[\eta] \,\&\, r \in Q \,\&\, \sup x<r\},$$

and in case b), for each $x \in T|\alpha$ and for each rational $r>\sup x$ choose a path $p_{r,x}$ of length α in $T|\alpha$ such that $x \in p_{r,x}$ and $\sup \bigcup p_{r,x}=r$ (this is possible according to what we said immediately after (1)), and put

$$T[\alpha]=\{(\bigcup p_{r,x})\cup\{r\}: x \in T|\alpha \,\&\, r \in Q \,\&\, \sup x<r\}.$$

This finishes the construction of the tree $\langle T, \prec\rangle$, provided we can show that the assertion about the set in (1) containing all rationals $>\sup x$ is valid. Using transfinite induction, suppose that this is the case if we replace α by $\alpha'<\alpha$. If α is a successor ordinal, then the assertion is obviously true for α as well in view of the way $T[\alpha]$ was constructed in case a). Assume now that α is a limit ordinal $<\omega_1$ and fix $x \in T|\alpha$. Let $\langle\alpha_n: n<\omega\rangle$ be an increasing sequence of ordinals tending to α such that $x \in T|\alpha_0$, let $r>\sup x$ be a rational, and let $\langle r_n\colon n<\omega\rangle$ be a strictly increasing sequence of rationals with $\sup x<r_n<\sup\{r_n\colon n<\omega\}=r$. We are going to construct a sequence $\langle x_n\colon n<\omega\rangle$ of elements of T, increasing in $\prec$, such that $x\prec x_n \in T[\alpha_n]$ and $\sup x_n=r_n$. Then $p=\{y \in T\colon \exists n<\omega\; y\prec x_n\}$ is clearly a path of length α with $\sup \bigcup p=\sup \bigcup_{n<\omega} x_n=\sup\{r_n\colon n<\omega\}=r$, which will establish our claim; all we need is to construct the sequence x_n as described. To this end put $x_{-1}=x$, and if x_{n-1} has already been constructed for some $n<\omega$ in such a way that $\sup x_{n-1}<r_n$ then let p_n be a path of length $\alpha_n\dot{+}1$ of T such that $x_{n-1} \in p_n$ and $\sup \bigcup p_n=r_n$; there is such a p_n by the induction hypothesis according to which the claim about the set in (1) is valid for any $\alpha'<\alpha$ replacing α.

Then $x_n = \bigcup p_n \in T[\alpha_n]$ is such that $\sup x_n = r_n < r_{n+1}$, and so we can finish the construction of the sequence $\langle x_n: n < \omega \rangle$. Thus our claim about the set in (1) is established, and so the definition of the tree $\langle T, \prec \rangle$ is complete. It is clear by this claim that, for each $\alpha < \omega_1$, $T[\alpha]$ is not empty; it is also clear from the construction of T that each $T[\alpha]$ is countable. Finally, T has no branch of type ω_1; in fact, if p were such a branch then $\bigcup p$ would be a set of rational numbers that has order type ω_1 in the natural ordering of the rationals, which is absurd. The proof of the theorem is complete.

Next we define Specker types:

DEFINITION 52.2. The order type of a totally ordered set $\langle A, \prec \rangle$ is called a Specker type if (i) $|A| = \aleph_1$, (ii) $\omega_1, \omega_1^* \not\leq \text{tp} \langle A, \prec \rangle$, and (iii) there is no uncountable subset B of A such that $\langle B, \prec \rangle$ is order isomorphic to a set of reals (in other words, there is no uncountable order type ψ with $\psi \leq \text{tp}\langle A, \prec \rangle$ and $\psi \leq$ the order type of the reals).

THEOREM 52.3 (Specker). *There is a Specker type; that is, there is a totally ordered set $\langle A, \prec \rangle$ satisfying* (i)–(iii) *in the above definition.*

PROOF. Let $\langle A, < \rangle$ be an Aronszajn $\aleph_1$-tree. For each $\xi < \aleph_1$, let $\prec_\xi$ be an arbitrary (total) ordering of $A[\xi]$. Furthermore, for each $x \in A$ with $\text{o}(x) \geq \xi$ denote by $x|\xi$ the unique $y \in A[\xi]$ with $y < x$. We are now going to define the ordering $\prec$ on A. For any $x, y \in A$ write $x \prec y$ if either $x < y$ or x and y are incomparable in the ordering $<$ and for the least ξ for which $x|\xi \neq y|\xi$ we have $x|\xi \prec_\xi y|\xi$. This is the same ordering as described in the verification of implication (ii)$\Rightarrow$(iii) in the proof of Theorem 29.6 with $\kappa = \aleph_1$, (except that there in case $x < y$ held we did not commit ourselves to make $x \prec y$ so as to be able to handle the proofs of $\kappa \not\leq \text{tp}\langle A, \prec \rangle$ and $\kappa^* \not\leq \text{tp}\langle A, \prec \rangle$ symmetrically). Note that only the regularity, and not the inaccessibility of κ was used there, and so the proof there gives that $\omega_1, \omega_1^* \not\leq \text{tp}\langle A, \prec \rangle$ holds for the present $\langle A, \prec \rangle$, which proves (ii) ((i) is obvious).

As for (iii), let B be an arbitrary uncountable subset of A, and assume, contrary to what we claim, that $\langle B, \prec \rangle$ is order isomorphic to a set of reals. We may assume that B is densely ordered. Then there is a countable set $X \subseteq B$ such that X is dense in B, i.e., such that for any two elements $b_0, b_1 \in B$ with $b_0 \prec b_1$ there is an $x \in X$ such that $b_0 \prec x \prec b_1$. It is easy to show that this is absurd. In fact, let $\sigma = \{\sup \text{o}(x, \langle A, \prec \rangle): x \in X\}$, and let $B' = B \setminus A|(\alpha \dot{+} 1)$. B' has cardinality $\aleph_1$, as $A|(\alpha \dot{+} 1)$ is countable. The set

$$\{b|(\alpha \dot{+} 1): b \in B'\},$$

being a subset of $A[\alpha \dot{+} 1]$, is countable, so there are two elements b_0, b_1 of B' with

$b_0 \prec b_1$ and $b_0|(\alpha\dot{+}1) = b_1|(\alpha\dot{+}1)$. Clearly, there is no $x \in A|(\alpha\dot{+}1)$ with $b_0 \preceq x \preceq b_1$; hence, in particular, there is no such x in X. This is what we wanted to show. The proof of the theorem is complete.

53. NEGATIVE SQUARE BRACKET RELATIONS WITHOUT GCH

After the preparations made in the preceding section, we can now prove

THEOREM 53.1 (Galvin–Shelah [1973]). *We have*

$$\aleph_1 \nrightarrow [\aleph_1]^2_4. \tag{1}$$

For the proof, we need the following lemma:

LEMMA 53.2. *Let* $\langle A, \prec \rangle$ *be an ordered set with* tp $\langle A, \prec \rangle$ *a Specker type (see Definition 52.2), and let* f: $A \to [0, 1]$ *be a* 1-1 *mapping, where* $[0, 1]$ *denotes the interval* $0 \le t \wedge 1$ *of reals. Then there exists an* $x_0 \in A$ *such that* $\{y \in A: y \succ x_0 \,\&\, f(y) > f(x_0)\}$ *has cardinality* $\aleph_1$.

PROOF. For $x \in A$, let

$$g(x) = \inf \{t \in [0, 1]: \{y \in A: f(y) \ge t \,\&\, y \succ x\} \text{ is countable}\}; \tag{2}$$

as is easily seen, the infimum here is assumed, i.e., $g(x)$ belongs to the set on the right-hand side here. g is a nonincreasing real-valued function on A; therefore g can only have countably many different values. In fact, otherwise there would be an uncountable order type ψ with $\psi \le$ tp $\langle A, \prec \rangle$ and $\psi \le$ the order type of the reals, which is not the case, as tp $\langle A, \prec \rangle$ was assumed to be a Specker type. Hence g is constant on an uncountable set $A_0 \subseteq A$. We can choose an $x \in A_0$ so that $\{y \in A_0: y \succ x\}$ is uncountable; otherwise A_0 would have a subset of order type ω_1^*, which is not the case. Now we can choose an $x_0 \in A_0$ so that $x_0 \succ x$ and $f(x_0) < g(x)$ (here we used the fact, that as remarked above, the infimum in (2) is assumed). Using (2), we can see that, since $g(x_0) = g(x) > f(x_0)$, there are uncountably many $y \succ x_0$ such that $f(y) > f(x_0)$. This completes the proof of the lemma.

The next lemma is just a restatement of the one just proved in a form more convenient to our purposes.

LEMMA 53.3. *If* tp $\langle A, <_1 \rangle$ *is a Specker type and* tp $\langle A, <_2 \rangle \le$ *the order type of the reals, then there is an* $x \in A$ *such that* $\{x \in A: x <_1 y \,\&\, x <_2 y\}$ *is uncountable.*

PROOF OF THEOREM 53.1. Let $|A| = \aleph_1$. Choose orderings $<_0$, $<_1$, $<_2$ so that tp $\langle A, <_0 \rangle = \omega_1$, tp $\langle A, <_1 \rangle$ is a Specker type, and tp $\langle A, <_2 \rangle \le$ the type of the reals. Analogously to proofs of negative ordinary partition relations with the aid of Sierpiński partitions (cf. Section 19), we define a disjoint partition $I=\langle I_i: i<4\rangle$ of $[A]^2$ by putting

$$I_0=\{\{xy\}: x<_0 y, x<_1 y, x<_2 y\},$$

$$I_1=\{\{xy\}: x<_0 y, x<_1 y, y<_2 x\},$$

$$I_2=\{\{xy\}: x<_0 y, y<_1 x, x<_2 y\},$$

$$I_3=\{\{xy\}: x<_0 y, y<_1 x, y<_2 x\},$$

where x and y range over elements of A. Consider any uncountable $A'\subseteq A$. Since tp $\langle A', <_1\rangle$ is still a Specker type and tp $\langle A', <_2\rangle\leq$ the type of reals, by Lemma 53.3 there is an $x\in A'$ such that

$$\{y\in A': x<_1 y \,\&\, x<_2 y\}$$

is uncountable. Since tp $\langle A, <_0\rangle=\omega_1$, we can choose an y from this set such that $x<_0 y$; then $\{xy\}\in I_{0'}$, showing that $[A']^2\cap I_0\neq 0$. Since tp $\langle A, >_1\rangle$ is also a Specker type and tp $\langle A, >_2\rangle\leq$ the type of reals, it follows by symmetry that $[A']^2\cap I_i\neq 0$ for $i=1, 2, 3$ as well. This completes the proof of Theorem 53.1.

The proof of the next theorem does not rely on the material discussed in the preceding section.

THEOREM 53.4 (Galvin–Shelah [1973]). *We have*

$$2^{\aleph_0}\nrightarrow[2^{\aleph_0}]^2_{\aleph_0} \tag{3}$$

and

$$\operatorname{cf}(2^{\aleph_0})\nrightarrow[\operatorname{cf}(2^{\aleph_0})]^2_{\aleph_0}. \tag{4}$$

Actually (3) can easily be derived from (4), but we shall not have to make the effort to do so, as (3) will be obtained directly. The proof of the theorem proceeds through several lemmas, all due to Galvin and Shelah.

LEMMA 53.5. *Let n be an integer, let $f_0, \ldots, f_n: 2^{\aleph_0}\to R$ be* 1-1 *mappings, where R is the set of reals, and suppose that for every $B\subseteq 2^{\aleph_0}$ cofinal in $2^{\aleph_0}$ there exist $\mu, \nu\in B$ such that $f_i(\mu)<f_i(\nu)$ for all $i\leq n$. Then there is a* 1-1 *mapping $g: 2^{\aleph_0}\to R$ such that, for every $B\subseteq 2^{\aleph_0}$ cofinal in $2^{\aleph_0}$ we have:*

(i) *there exist $\mu, \nu\in B$ such that $f_i(\mu)<f_i(\nu)$ for all $i\leq n$, and $g(\mu)>g(\nu)$; and*

(ii) *there exist $\mu, \nu\in B$ such that $f_i(\mu)<f_i(\nu)$ for all $i\leq n$, and $g(\mu)<g(\nu)$.*

PROOF. We write $f(\mu)<f(\nu)$ to mean $f_i(\mu)<f_i(\nu)$ for all $i\leq n$. Since $(2^{\aleph_0})^{\aleph_0}=2^{\aleph_0}$, we can write $[2^{\aleph_0}]^{\aleph_0}=\{A_\mu: \mu<2^{\aleph_0}\}$. Let $\nu<2^{\aleph_0}$, and suppose that $g(\mu)$ has been defined for all $\mu<\nu$. Now choose $g(\nu)\in R$ so that

(iii) $g(\nu) \neq g(\mu)$ for all $\mu < \nu$;

(iv) if $\mu < \nu$ and $A_\mu \subseteq \nu$, then $g(\nu)$ is neither the supremum nor the infimum of the set $\{g(\alpha) : \alpha \in A_\mu \;\&\; f(\alpha) < f(\nu)\}$.

So $g: 2^{\aleph_0} \to R$ is defined in a 1-1 way.

Consider any $B \subseteq 2^{\aleph_0}$ cofinal in $2^{\aleph_0}$, and suppose that (i) fails; this means that, for every $\mu, \nu \in B$, $f(\mu) < f(\nu)$ implies $g(\mu) < g(\nu)$. Define $h: 2^{\aleph_0} \to {}^{n+2}R$ by putting $h(\nu) = \langle f_0(\nu), \ldots, f_n(\nu), g(\nu) \rangle$. Since ${}^{n+2}R$ is a separable metric space (i.e., it includes a countable dense subset), we can choose a $\mu < 2^{\aleph_0}$ so that $A_\mu \subseteq B$ and $h``A_\mu$ is dense in $h``B$. Let such a μ be fixed. Let $B' = \{\nu \in B : \mu < \nu \text{ and } A_\mu \subseteq \nu\}$; then B' is cofinal in $2^{\aleph_0}$ and for every $\nu \in B'$,

$$g(\nu) > \sup \{g(\alpha) : \alpha \in A_\mu \;\&\; f(\alpha) < f(\nu)\}\,.$$

Indeed, $=$ cannot hold here by (iv), and $<$ cannot hold either as we assumed that (i) fails with B (cf. the second part of the first sentence in this paragraph). Recall that $\mathrm{cf}\,(2^{\aleph_0}) > \aleph_0$ according to König's theorem. Consequently, if we partition a cofinal subset of $2^{\aleph_0}$ into countably many parts, then at least one of the parts will be cofinal; we are going to use this fact repeatedly. Choose $\varepsilon > 0$ and $B'' \subset B'$ so that B'' is cofinal in $2^{\aleph_0}$ and

$$g(\nu) - \varepsilon > \sup \{g(\alpha) : \alpha \in A_\mu \;\&\; f(\alpha) < f(\nu)\} \tag{5}$$

holds for any $\nu \in B''$. Choose a set $B''' \subseteq B''$ cofinal in $2^{\aleph_0}$ so that $|g(\nu_0) - g(\nu_1)| < \varepsilon/2$ for all $\nu_0, \nu_1 \in B'''$. Choose $\alpha_0, \nu \in B'''$ so that $f(\alpha_0) < f(\nu)$; there are such α_0 and ν according to the assumptions of the lemma. Since $h``A_\mu$ is dense in $h``B$, we can choose an $\alpha \in A_\mu$ so that $|g(\alpha) - g(\alpha_0)| < \varepsilon/2$ and $|f_i(\alpha) - f_i(\alpha_0)| < f_i(\nu) - f_i(\alpha_0)$ for all $i \leq n$. Then $f(\alpha) < f(\nu)$ and

$$g(\alpha) > g(\nu) - |g(\nu) - g(\alpha_0)| - |g(\alpha_0) - g(\alpha)| > g(\nu) - \varepsilon\,;$$

but this contradicts (5) since $\nu \in B''$. This proves that (i) holds; (ii) follows by symmetry. The proof of the lemma is complete.

Lemma 53.6. *There are* 1-1 *mappings* $f_n: 2^{\aleph_0} \to R$ $(n < \omega)$ *such that, for any* $B \subseteq 2^{\aleph_0}$ *cofinal in* $2^{\aleph_0}$ *and for any* $n < \omega$, *there exist* $\mu, \nu \in B$ *such that* $f_i(\mu) \subset f_i(\nu)$ *for all* $i \leq n$ *but* $f_{n+1}(\mu) > f_{n+1}(\nu)$.

Proof. Start with a 1-1 mapping $f_0: 2^{\aleph_0} \to R$, and apply the preceding lemma repeatedly to obtain $f_1, f_2, \ldots$.

Lemma 53.7. *There are pairwise disjoint sets* $I_n \subseteq [2^{\aleph_0}]^2$ $(n < \omega)$ *such that* $[B]^2 \cap I_n \neq 0$ *for every* $n < \omega$ *and for every* $B \subseteq 2^{\aleph_0}$ *that is cofinal in* $2^{\aleph_0}$.

PROOF. Using the mappings $f_n: 2^{\aleph_0} \to R$ described in the preceding lemma, put

$$I_n = \{\{\mu\nu\}: \mu, \nu < 2^{\aleph_0} \; \& \; \mu \neq \nu \; \& \; \forall i \leq n [f_i(\mu) < f_i(\nu)] \; \&$$
$$\& \; f_{n+1}(\mu) > f_{n+1}(\nu)\}.$$

The sets I_n fulfill all our requirements according to the preceding lemma.

PROOF OF THEOREM 53.4. The result in (3) immediately follows from the preceding lemma. As for (4), take a cofinal subset A of $2^{\aleph_0}$ of cardinality cf $(2^{\aleph_0})$; then the sets $I_n \cap [A]^2$, $n < \omega$, with I_n as in the preceding lemma, form a disjoint partition of a subset of $[A]^2$ that verifies (4). The proof is complete.

The results in this section should be compared with (49.1), and also to a result of Baumgartner according to which $2^{\aleph_0} > \aleph_1$ is anything reasonable and

$$2^{\aleph_0} \not\to [(\aleph_0 : \aleph_1)]^2_{\aleph_0} \tag{6}$$

is consistent with ZFC. Moreover, a result of Solovay says that if $2^{\aleph_0}$ is real-valued measurable, then

$$2^{\aleph_0} \to [2^{\aleph_0}]^n_{\aleph_1} \tag{7}$$

holds for every $n < \omega$; hence (3) above seems to be best possible. As for possible strengthenings of (1) it is an open problem whether any of the relations

$$\aleph_1 \not\to [\aleph_1]^2_\alpha \tag{8}$$

is provable in ZFC for $5 \leq \alpha \leq \aleph_1$. As for higher superscripts, Shelah, using a result of Galvin, proved that

$$\aleph_1 \not\to [\aleph_1]^3_{\aleph_1} \tag{9}$$

holds in ZFC.

54. POSITIVE RELATIONS FOR SINGULAR STRONG LIMIT CARDINALS

The following result is an easy consequence of the General Canonization Lemma:

THEOREM 54.1. *Let κ be a singular strong limit cardinal with*

$$\text{cf}(\kappa) \to (\text{cf}(\kappa))^2_2. \tag{1}$$

Let $r > 1$ be an integer and let τ be a cardinal with $2^{r-1} < \tau < \kappa$. Then

$$\kappa \to [\kappa]^r_{\tau, 2^{r-1}}. \tag{2}$$

The definition of this symbol is given in Section 8.5, but the first sentence of the proof will also clarify its meaning.

PROOF. Let $f:[\kappa]^r\to\tau$ be a coloring; we have to prove that there is an $X\subseteq\kappa$ of cardinality κ such that $|f``[X]^r|\le 2^{r-1}$. Using Corollary 28.2 of the General Canonization Lemma, we can see that there are pairwise disjoint sets $B_\xi\subseteq\kappa$, $\xi<\mathrm{cf}(\kappa)$, such that $B=\bigcup_{\xi<\mathrm{cf}(\kappa)} B_\xi$ has cardinality κ and $\langle B_\xi:\xi<\mathrm{cf}(\kappa)\rangle$ is canonical with respect to f. We may assume here that $|B_\xi|$ increases with ξ. For any $x\in[B]^r$ define the type typ x of x as follows: let $\xi_0<\ldots<\xi_k$ be the elements of the set $\{\xi<\mathrm{cf}(\kappa): x\cap B_\xi\neq 0\}$, and put

$$\mathrm{typ}\,x=\langle \mathrm{typ}_1\,x,\mathrm{typ}_2\,x\rangle=\langle\{\xi_i:i\le k\},\langle|x\cap B_{\xi_i}|:i\le k\rangle\rangle. \tag{3}$$

Here $\mathrm{typ}_1\,x\in[\mathrm{cf}(\kappa)]^{\le r}$ and $\mathrm{typ}_2\,x$ is a finite sequence of positive integers whose sum equals r. It is easy to show by induction on r that the number N_r of possibilities for $\mathrm{typ}_2\,x$ is 2^{r-1}, since we clearly have $N_0=1$ (the empty sequence is the only possibility), $N_1=1$, and $N_r=\sum_{i=1}^{r} N_{r-i}$, where the ith term here represents the case when the first integer in $\mathrm{typ}_2\,x$ is i. The canonicity of $\langle B_\xi:\xi<\mathrm{cf}(\kappa)\rangle$ means that, for $x\in[B]^r$, $f(x)$ depends only on typ x; hence it is sound to define a function g: $\{\mathrm{typ}\,x: x\in[B]^r\}\to\tau$ by putting $g(\mathrm{typ}\,x)=f(x)$ for every $x\in[B]^r$. According to the implication (vi)⇒(iv) in Theorem 29.6 (cf. also Theorem 29.1), we have $\mathrm{cf}(\kappa)\to(\mathrm{cf}(\kappa))^r_\tau$ in view of (1). Using this relation (e.g. repeatedly, once for each value of $\mathrm{typ}_2\,x$), we can see that there is a set $H\subseteq\mathrm{cf}(\kappa)$ of cardinality $\mathrm{cf}(\kappa)$ such that whenever $\mathrm{typ}_1\,x\subseteq H$, then $g(\mathrm{typ}\,x)$ is completely determined by $\mathrm{typ}_2\,x$. Putting $X=\bigcup_{\xi\in H}B_\xi$, this means, in particular, that

$$|f``[X]^r|=|g``\{\mathrm{typ}\,x:\mathrm{typ}_1\,x\subseteq H\}|\le$$
$$\le\text{the number of possibilities for } \mathrm{typ}_2\,x = 2^{r-1},$$

where the last equality here was established above. As X has cardinality κ, this completes the proof of the lemma.

Note that (2) is best possible here in a sense, since we have

$$\kappa\not\to[\kappa]^r_{2^{r-1}} \tag{4}$$

for every singular κ and every integer $r>1$. To see this, represent κ as a sum $\bigcup_{\xi<\mathrm{cf}(\kappa)} B_\xi$ of pairwise disjoint sets of cardinality less than κ, and define the function f on $[\kappa]^r$ by putting $f(x)=\mathrm{typ}_2\,x$ for any $x\in[\kappa]^r$. Then, as remarked above, the cardinality of the range of f is 2^{r-1}. If $X\subseteq\kappa$ has cardinality κ, then there are at least r different ordinals $\xi<\mathrm{cf}(\kappa)$ such that $|X\cap B_\xi|\ge r$; hence all possible values for $\mathrm{typ}_2\,x$ will occur if x is restricted to elements of $[X]^r$. Thus f verifies relation (4).

55. INFINITARY JÓNSSON ALGEBRAS

The following problem is due to B. Jónsson: Given an infinite cardinal κ, is there an algebra of cardinality κ with countably many operations that has no proper subalgebra of cardinality κ? An algebra with this property is called a Jónsson algebra. As far as finitary algebras are concerned, i.e., when the operations are assumed to be finitary, this question has not yet been answered completely. H. J. Keisler and F. Rowbottom [1965] showed that the answer is yes for all κ if $V=L$ is assumed, P. Erdős and A. Hajnal [1966] proved this for successor κ under the assumption of GCH; they also proved this in ZFC without any additional assumption if $\kappa=\aleph_n$, $n<\omega$. The question of the existence of a Jónsson algebra of cardinality $\aleph_\omega$ (with or without the assumption of GCH) seems to be a very difficult open problem.

If ω-ary algebras are considered, i.e., if ω-place operations are also allowed, then the question becomes much simpler, and we shall prove in ZFC that then the answer is yes for all $\kappa\geq\omega$; we shall actually prove the (apparently) stronger result that we have $\kappa\not\to[\kappa]^\omega_\kappa$ for all infinite κ. A more comprehensive treatment of the topics dicussed in this chapter is to be found in Galvin–Prikry [1976]. Our first result confirms the existence of an ω-ary Jónsson algebra for all $\kappa\geq\omega$ with finitely many operations.

Lemma 55.1 (Erdős–Hajnal [1966]). *Let $\kappa\geq\omega$ be a cardinal. Then there are an integer n, a set A of cardinality κ, and functions $f_0,\ldots,f_{n-1}:[A]^{\aleph_0}\to A$ such that whenever $X\subseteq A$ and $|X|=\kappa$ we have* $\bigcup_{i<n} f_i``[X]^{\aleph_0}=A$.

Proof. According to Theorem 45.1, we have $(\kappa,\aleph_0,2)\not\to\aleph_0$, i.e., there is a set mapping $g:[\kappa]^{\aleph_0}\to[\kappa]^1$ such that there is no infinite free set with respect to g. Define $f:[\kappa]^{\aleph_0}\to\kappa$ by writing $g(x)=\{f(x)\}$ for all $x\in[\kappa]^{\aleph_0}$. Assuming that the assertion of the lemma to be proved fails, we are going to define a sequence $\langle\xi_k:k<\omega\rangle$ of pairwise distinct ordinals and a sequence of sets $\kappa=A_0\supseteq A_1\supseteq\ldots\supseteq A_k\supseteq\ldots$ $(k<\omega)$ as follows. Assuming that ξ_i, $i<k$ and A_i, $i\leq k$, have already been defined in such a way that $\{\xi_i:i<k\}\cap A_k=0$, for each $b\subseteq\{\xi_i:i<k\}$ and for each $x\in[\kappa\setminus b]^{\aleph_0}$ put $f_b(x)=f(b\cup x)$. According to our assumption, the assertion of the lemma fails in particular with $A=A_k$ and with the functions f_b, $b\subseteq\{\xi_i:i<k\}$ (restricted to $[A_k]^{\aleph_0}$). Hence there is a $\xi_k\in A_k$ and a set $A_{k+1}\subseteq A_k\setminus\{\xi_k\}$ of cardinality κ such that

$$\xi_k\neq f_b(x)=f(b\cup x) \tag{1}$$

for every $b\subseteq\{\xi_i:i<k\}$ and $x\in[A_{k+1}]^\omega$. This finishes the construction of the sequences of the ξ's and the A's. Noting that

$$\{\xi_i:k+1\leq i<\omega\}\subseteq A_{k+1}\,,$$

(1) means in particular that $\xi_k \neq f(y)$, i.e., that $\xi_k \notin g(y)$, holds for every $k<\omega$ and every infinite set $y \subseteq \{\xi_i : i<\omega\}$ with $\xi_k \notin y$. Thus $\{\xi_i : i<\omega\}$ is a free set with respect to g, which is a contradiction, proving the lemma.

THEOREM 55.2 (Erdős–Hajnal [1966]). *We have*

$$\kappa \not\to [\kappa]^{\aleph_0}_{\kappa} \tag{2}$$

for every infinite cardinal κ.

PROOF. (2) says that there is a function $f: [\kappa]^{\aleph_0} \to \kappa$ such that $f``[X]^{\aleph_0} = \kappa$ holds for any set $X \subseteq \kappa$ of cardinality κ; in other words, it says that the preceding lemma is true already with $n=1$ (we may obviously assume $A=\kappa$ there). I.e., we have to show that the functions $f_0, \ldots, f_{n-1}$ in the preceding lemma can be 'coded' as a single function f. We shall have to do this only in case $\kappa > \aleph_0$ as we have $\aleph_0 \not\to [\aleph_0]^{\aleph_0}_{2^{\aleph_0}}$ according to Theorem 12.1, which implies (2) in case $\kappa=\omega$.

So assume $\kappa > \aleph_0$, and let $f_0, \ldots, f_{n-1} : [\kappa]^{\aleph_0} \to \kappa$ be functions as described in the preceding lemma (i.e., we assume $A=\kappa$ there). Let $g: [\kappa]^{\aleph_0} \to n$ be a function verifying the relation $\kappa \not\to [\aleph_0]^{\aleph_0}_n$, which holds according to Theorem 12.1, in fact, even the stronger relation $\kappa \not\to [\aleph_0]^{\aleph_0}_{2^{\aleph_0}}$ holds according to that theorem. I.e., g is such that, for any infinite set $x \subseteq \kappa$, $g``[x]^{\aleph_0} = n$ holds. If $x \subseteq \kappa$ is a countable set of order type $\geq \omega \dot{+} \omega$, then denote by x' the set of its first ω elements and put $f(x) = f_{g(x')}(x \setminus x')$; if $x \in [\kappa]^{\aleph_0}$ has order type $< \omega \dot{+} \omega$, then define $f(x) < \kappa$ arbitrarily. Let now $x \subseteq \kappa$ be a set of cardinality κ and let Z be the set of its first ω elements. Then

$$f``[X]^{\aleph_0} \supseteq \{f(z \cup x): z \in [Z]^{\aleph_0} \,\&\, x \in [X \setminus Z]^{\aleph_0}\} =$$
$$= \{f_{g(z)}(x): z \in [Z]^{\aleph_0} \,\&\, x \in [X \setminus Z]^{\aleph_0}\}.$$

As $g``[Z]^{\aleph_0} = n$ by the choice of g, the last set here equals

$$\{f_i(x): i<n \,\&\, x \in [X \setminus Z]^{\aleph_0}\} = \kappa,$$

where this latter equality holds since $|X \setminus Z| = \kappa$ and in view of the fact that the functions f_i satisfy the requirements of the preceding lemma. This means that $f``[X]^{\aleph_0} = \kappa$, completing the proof of the theorem.

BIBLIOGRAPHY

A. V. Arhangel'skiĭ

[1969] The power of bicompacta with first axiom of countability. Dokl. Akad. Nauk SSSR **187** (1969), 967–968. (Soviet Math. Dokl. **10** (1969), 951–955.)

J. E. Baumgartner

[1970] Results and independence proofs in combinational set theory. Doctoral Dissertation, University of California, Berkeley, Calif. (1970).

[1976] A new class of order types. Annals of Math. Logic **9** (1976), 187–222.

J. E. Baumgartner–A. Hajnal–A. Máté

[1975] Weak saturation properties of ideals. In: Infinite and finite sets. (Colloq., Keszthely 1973; dedicated to P. Erdős on his 60th birthday), Part I, 137–158. Colloq. Math. Soc. János Bolyai, Vol. **10**, North Holland, Amsterdam, 1975.

N. G. de Bruijn–P. Erdős

[1951] A color problem for infinite graphs and a problem in the theory of relations. Akademia Amsterdam **13** (1951), 371–373.

L. Bukovský

[1965] The continuum problem and powers of alephs. Commentationes Math. Univ. Carolinae, **6** (1965), 181–197.

J. P. Burgess

[1979] On a set-mapping problem of Hajnal and Máté, Acta Sci. Math. **41** (1979), 283–288.

P. J. Cohen

[1963] The independence of the continuum hypothesis. Proc. Nat. Acad. Sci. **50** (1963), 1143–1148.

[1964] The independence of the continuum hypothesis. Proc. Nat. Acad. Sci. **51** (1964), 105–110.

[1966] Set theory and the continuum hypothesis. W. A. Benjamin, New York (1966).

K. J. Devlin

[1973] Aspect of constructibility. Lecture Notes in Math. **354**, Springer-Verlag, Berlin–Heidelberg–New York 1973.

[1973a] Some weak versions of large cardinal axioms. Annals of Math. Logic **5** (1973), 291–325.

K. J. Devlin–R. Jensen

[1975] Marginalia to a theorem of Silver, in: Proceedings of ISILC, Kiel 1974, Lecture Notes Vol. 499 (Springer Verlag, New York–Heidelberg–Berlin, 1975), 115–142.

K. J. Devlin–J. B. Paris

[1973] More on the free subset problem. Annals of Math. Logic **5** (1973), 327–336.

A. Dodd–R. Jensen

[1981] The core model, Annals of Math. Logic **20** (1981), 43–75.

F. R. Drake

[1974] Set theory. North Holland, Amsterdam, 1974.

B. Dushnik–E. W. Miller

[1941] Partially ordered sets. Amer. J. of Math. **63** (1941), 605.

W. B. Easton

[1970] Powers of regular cardinals. Annals of Math. Logic **1** (1970), 139–178.

P. Erdős

[1947] Some remarks on the theory of graphs. Bull. Amer. Math. Soc. **53** (1947), 292–294.

P. Erdős–A. Hajnal

[1958] On the structure of set mappings. Acta Math. Acad. Sci. Hung. **9** (1958), 111–131.

[1966] On a problem of B. Jónsson. Bull. Acad. Polon. Sci. Ser. Sci. Math. Astronom. Phys. **14** (1966), 19–23.

[1970] Unsolved problems in set theory. In: Proc. Sympos. Pure Math. Vol. 13, Part 1, Amer. Math. Soc. Providence, R. I. 1971, pp. 17–48.

[1974] Unsolved and solved problems in set theory. In: Proc. Sympos. Pure Math. 25, Amer. Math. Soc. Providence, R. I. 1974, pp. 269–287.

P. Erdős–A. Hajnal–E. C. Milner

[1968] On almost disjoint subsets of a set. Acta Math. Acad. Sci. Hung. **19** (1968), 209–218.

P. Erdős–A. Hajnal–R. Rado

[1965] Partition relations for cardinal numbers. Acta Math. Acad. Sci. Hung. **16** (1965), 93–196.

P. Erdős–R. Rado

[1952] Combinatorial theorems on classifications of subsets of a given set. Proc. London Math. Soc. **3** (1952), 417–439.

[1956] A partition calculus in set theory. Bull. Amer. Math. Soc. **62** (1956), 427–489.

P. Erdős–G. Szekeres

[1935] A combinatorial problem in geometry. Comp. Math. **2** (1935), 463–470.

P. Erdős–A. Tarski

[1943] On families of mutually exclusive sets. Annals of Math. **44** (1943), 315–329.

[1961] On some problems involving inaccessible cardinals. In: Essays on the foundations of mathematics, Jerusalem, 1961, 50–82.

G. Fodor

[1952] Proof of a conjecture of P. Erdős. Acta Sci. Math. **14** (1952), 219–227.

[1956] Eine Bemerkung zur Theorie der regressiven Funktionen. Acta Sci. Math. **17** (1956), 139–142.

F. Galvin–A. Hajnal

[1975] Inequalities for cardinal powers. Annals of Math. **101** (1957), 491–498.

F. Galvin–K. Prikry

[1976] Infinitary Jónsson algebras and partition relations. Algebra Universalis **6** (1976), 367–376.

F. Galvin–S. Shelah

[1973] Some counterexamples in the partition calculus. J. of Combinational Theory **15** (1973), 167–174.

K. Gödel

[1940] The consistency of the axiom of choice and of the generalized continuum hypothesis. Princeton, 1940.

[1947] What is Cantor's continuum problem. Amer. Math. Monthly **54** (1947), 515–525.

R. E. Greenwood–A. M. Gleason

[1955] Combinatorial relations and chromatic graphs. Canadian J. of Math. **7** (1955), 1–7.

J. de Groot

[1965] Discrete subspaces of Hausdorff spaces. Bull. Acad. Polon. Sci. Ser. Sci. Math. Astronom. Phys. **13** (1965), 537–544.

A. Hajnal

[1960] Some results and problems in set theory. Acta Math. Acad. Sci. Hung. **11** (1960), 277–298.

[1961] Proof of a conjecture of S. Ruziewicz. Fund. Math. **50** (1961), 123–128.

[1964] Remarks on a theorem of W. P. Hanf. Fund. Math. **54** (1964), 109–113.

[1969] Ulam matrices for inaccesible cardinals. Bull. Acad. Polon. Sci. Ser. Sci. Math. Astronom. Phys. **17** (1969), 683–688.

A. Hajnal–I. Juhász

[1967] Discrete subspaces of topological spaces. Indag. Math. **29** (1967), 343–356.

A. Hajnal–A. Máté

[1975] Set mappings, partitions, and chromatic numbers. In: Logic Colloq. '73, Proc. Bristol 1973. North Holland, Amsterdam, 1975. 347–379.

W. P. Hanf

[1964] Incompactness in languages with infinitely long expressions. Fund. Math. **53** (1964), 309–324.

[1964a] On a problem of Erdős and Tarski. Fund. Math. **53** (1964), 325–334.

W. P. Hanf–D. S. Scott

[1965] Classifying inaccessible cardinals. Amer. Math. Soc. Notices **12** (1965), 723.

F. Hausdorff

[1904] Der Potenzbegriff in der Mengenlehre. Jahrber. d. Deutsch. Math. Ver. **13** (1904), 569–571.

S. H. Hechler

[1973] Powers of singular cardinals and a strong form of negation of the generalized continuum hypothesis. Z. Math. Logik Grundlagen Math. **19** (1973), 83–84.

T. J. Jech

[1971] Lectures in set theory with particular emphasis on the method of forcing. Lectur Notes in Math. **217**, Springer–Verlag, Berlin–Heidelberg–New York, 1971.

[1973] Some combinatorial problems concerning uncountable cardinals. Annals of Math. Logic **5** (1973), 165–198.

[1978] Set theory. Academic Press, New York–San Francisco–London, 1978.

T. J. Jech–K. Prikry

[1976] On ideals of sets and the power set operation. Bull. Amer. Math. Soc. **82** (1976), 593–595.

R. B. Jensen

[1972] The fine structure of the constructible hierarchy, Ann. Math. Logic **4** (1972), 229–308. Erratum: Ann. Math. Logic. **4** (1972), 443.

[1974] Marginalia to a theorem of Silver. Dittoed notes.

[1974a] More Marginalia. Dittoed notes.

[1974b] Marginalia III. Dittoed notes.

I. Juhász

[1971] Cardinal functions in topology. Math. Centre Tracts **34** (1971), Math. Centrum, Amsterdam.

H. J. Keisler–F. Rowbottom

[1965] Constructible sets and weakly compact cardinals. Amer. Math. Soc. Notices **12** (1965), 373.

H. J. Keisler–A. Tarski

[1964] From accessible to inaccessible cardinals. Fund. Math. **53** (1964), 225–308.

J. L. Kelley

[1955] General topology. Van Nostrand, New York, 1955.

J. Ketonen

[1974] Some combinatorial principles. Trans. Amer. Math. Soc. **188** (1974), 387–394.

J. König

[1905] Zum Kontinuum-Problem. Math. Annalen **60** (1905), 177–180.

D. König

[1927] Über eine Schlussweise aus dem Endlichen ins Unendliche. Acta Sci. Math. **3** (1927), 121–130.

K. Kunen

[1970] Some applications of iterated ultrapowers in set theory. Annals of Math. Logic **1** (1970), 179–227.

[1971] A partition theorem. Notices of Amer. Math. Soc. **19** (1971), 425.

[1978] Saturated ideals. J. Symbolic Logic **43** (1978), 65–76.

K. Kuratowski

[1922] Une méthode d'élimination des nombres transfinis des raisonnements mathématiques. Fund. Math. **3** (1922), 76–108.

[1951] Sur une caractérisation des alephs. Fund. Math. **38** (1951), 14–17.

D. Kurepa

[1935] Ensembles ordonnés et ramifiés. Publ. Math. Belgrade **4** (1935), 1–138.

R. Laver

[1975] Partition relations for uncountable cardinals $\leq 2^{\aleph_0}$. In: Infinite and finite sets. (Colloq., Keszthely 1973; dedicated to P. Erdős on his 60th birthday), Part II, 1029–1042. Colloq. Math. Soc. János Bolyai Vol. **10**, North Holland, Amsterdam, 1975.

D. Lázár

[1936] On a problem in the theory of aggregates. Compositio Math. **3** (1936), 304.

M. Magidor

[1977] Chang's conjecture and powers of singular cardinals. J. Symbolic Logic **42** (1977), 272–276.

[1977a] On the singular cardinal problem I. Israel J. Math. **28** (1977), 1–31.

[1977b] On the singular cardinal problem. Ann. of Math. **106** (1977), 517–547.

P. Mahlo

[1911] Über lineare transfinite Mengen. Berichte über die Verhandlungen der Königliche Sächischen Gesellschaft der Wissenschaften zu Leipzig, Mathematische-Physische Klasse. **63** (1911), 187–225.

D. A. Martin–R. M. Solovay

[1972] Internal Cohen extensions. Annals of Math. Logic, **4** (1972), 121–130.

A. Mostowski

[1949] An undecidable arithmetical statement. Fund. Math. **36** (1949), 1–164.

W. Neumer

[1951] Verallgemeinerung eines Satzes von Alexandrov und Urysohn. Math. Z., **54** (1951), 254–261.

F. P. Ramsey

[1930] On a problem of formal logic. Proc. London Math. Soc. (2) **30** (1930), 264–286.

W. N. Reinhardt–R. M. Solovay–A. Kanamori

[1978] Strong axioms of infinity and elementary embeddings. Ann. Math. Logic. **13** (1978), 73–116.

F. Rowbottom

[1971] Some strong axioms of infinity incompatible with the axiom of constructibility. Ann. of Math. Logic **13** (1971), 1–44.

S. Ruziewicz

[1936] Une généralisation d'un théorème de M. Sierpiński. Publications Mathématiques de l'Université de Belgrade **5** (1936), 23–27.

S. Shelah

[1975] Notes on partition calculus. In: Infinite and finite sets. (Colloq., Keszthely 1973; dedicated to P. Erdős on his 60th birthday), Part III, 1257–1276. Colloq. Math. Soc. János Bolyai Vol. **10**, North Holland, Amsterdam, 1975.

[1979] Weakly compact cardinals: A combinatorial proof. J. Symbolic Logic **44**, (1979), 559–562

[1980] A note on cardinal exponentiation. J. Symbolic Logic **45** (1980), 56–66.

[1981] Canonization theorems and applications. J. Symbolic Logic **46** (1981), 345–353.

[1982] Proper forcing, Lecture Notes in Mathematics **940** (Springer, New York–Heidelberg–Berlin, 1982).

J. R. Shoenfield

[1971] Unramified forcing. In: D. Scott, ed., Axiomatic Set Theory. Proc. Sympos. Pure Math. XIII, Part 1 (Amer. Math. Soc., Providence, R. I., 1971), 357–382.

R. A. Shore

[1974] Square bracket relations in L. Fund. Math. **84** (1974), 101–106.

W. Sierpiński

[1933] Sur un problème de la théorie des relations. Ann. Scuolo Norm. Sup. Pisa **2** (1933), 285–287.

R. Sikorski

[1960] Boolean algebras. Springer-Verlag, Berlin–Göttingen–Heidelberg, 1960.

J. H. Silver

[1966] Some applications of model theory in set theory. Doctoral dissertation, University of California, Berkeley, 1966.

[1971] Some applications of model theory in set theory. Annals of Math. Logic **3** (1971), 45–110.

[1975] On the singular cardinal problem. Proc. Internat. Congr. Mathematicians (Vancouver 1974) Vol. **1**, 256–268, 1975.

R. M. Solovay

[1967] A nonconstructible Δ_3^1-set of integers. Trans. Amer. Math. Soc. **127** (1967). 50–75.

[1971] Real-valued measurable cardinals. In: Axiomatic set theory. Proc. of Symposia in Pure Math., Vol. **XIII**, Part 1, (Amer. Math. Soc., Providence, R. I., 1971), 397–428.

[1974] Strongly compact cardinals and the GCH. In: Proceedings of the Tarski Symposium (Proc. Sympos. Pur Math. Vol. **XXV**, Univ. of California, Berkeley, Calif. 1971), (Amer. Math. Soc. Providence, R. I., 1974), 365–372.

R. M. Solovay–S. Tennenbaum

[1971] Iterated Cohen extensions and Souslin's problem. Annals of Math., **94** (1971), 201–245.

E. Specker

[1951] Sur un problème de Sikorski, Collouq. Math. **2** (1951), 9–12.

J. Spencer

[1975] Ramsey's theorem — a new lower bound. J. of Combinatorial Theory Ser. A **18** (1975), 108–115.

G. Takeuti–W. M. Zaring

[1971] Introduction to axiomatic set theory. Springer, New York (1971).

A. Tarski

[1925] Quelques théorèmes sur les alephs. Fund. Math. **7** (1925), 1–14.

S. B. Todorčević

[1979] Some results in set theory. II. Abstract 79T–E39, Notices Amer. Math. Soc. **26** (1979). A-440.

[∞] Trees, subtrees and order types, to appear.

[1982] Some results in Set Theory, III. Abtract. Notices Amer. Math. Soc. (1982), 185–186.

S. Ulam

[1930] Zur Masstheorie in der allgemeinen Mengenlehre. Fund. Math. **16** (1930), 140–150.

N. H. Williams

[1977] Combinatorial set theory. North Holland, Amsterdam–New York–Oxford, 1977.

M. Zorn

[1935] A remark on method in transfinite algebra. Bull. Amer. Math. Soc. **41** (1935), 667–670.

AUTHOR INDEX

SUBJECT INDEX